AF773939

Foliations in Cauchy-Riemann Geometry

Mathematical Surveys and Monographs

Volume 140

Foliations in Cauchy-Riemann Geometry

Elisabetta Barletta
Sorin Dragomir
Krishan L. Duggal

American Mathematical Society

EDITORIAL COMMITTEE

Jerry L. Bona   Ralph L. Cohen
Michael G. Eastwood   Michael P. Loss
J. T. Stafford, Chair

2000 *Mathematics Subject Classification.* Primary 53C12, 53C50, 32T15, 32V05, 32V15; Secondary 53D10, 32T27, 32V20, 32V30, 32V35.

For additional information and updates on this book, visit
www.ams.org/bookpages/surv-140

Library of Congress Cataloging-in-Publication Data

Barletta, E. (Elisabetta), 1957-
Foliations in Cauchy-Riemann geometry / E. Barletta, S. Dragomir, K. L. Duggal.
p. cm. — (Mathematical surveys and monographs, ISSN 0076-5376 ; v. 140)
Includes bibliographical references.
ISBN 978-0-8218-4304-8 (alk. paper)
1. Foliations (Mathematics). 2. Cauchy-Riemann equations. I. Dragomir, Sorin, 1955- II. Duggal, Krishan L., 1929- III. Series.

QA613.62.B37 2007
514′.72—dc22   2007060684

Copying and reprinting. Individual readers of this publication, and nonprofit libraries acting for them, are permitted to make fair use of the material, such as to copy a chapter for use in teaching or research. Permission is granted to quote brief passages from this publication in reviews, provided the customary acknowledgment of the source is given.

Republication, systematic copying, or multiple reproduction of any material in this publication is permitted only under license from the American Mathematical Society. Requests for such permission should be addressed to the Acquisitions Department, American Mathematical Society, 201 Charles Street, Providence, Rhode Island 02904-2294, USA. Requests can also be made by e-mail to `reprint-permission@ams.org`.

© 2007 by the American Mathematical Society. All rights reserved.
The American Mathematical Society retains all rights
except those granted to the United States Government.
Printed in the United States of America.

∞ The paper used in this book is acid-free and falls within the guidelines
established to ensure permanence and durability.
Visit the AMS home page at `http://www.ams.org/`

10 9 8 7 6 5 4 3 2 1   12 11 10 09 08 07

Contents

Preface

The present monograph is an attempt to a better understanding of an interdisciplinary question, namely the impact of foliation theory on the geometry and analysis on CR manifolds. To start with, any Levi-flat CR manifold M carries a complex foliation $\mathcal{F}$ (the *Levi foliation*) tangent to the null space of the Levi form of the manifold. At least in the real analytic case, if M is embedded then $\mathcal{F}$ extends to a holomorphic foliation of an open neighborhood of M (*Rea's theorem*, [**203**]). Complex foliations occur in a natural way on certain nondegenerate CR manifolds, as well. To give a simple example, R. Penrose's manifold $\mathbb{P}(\mathbb{T}_0)$ (the boundary of the manifold $\mathbb{P}(\mathbb{T}_+)$ of all right-handed spinning photons, cf. [**199**]) is a nondegenerate CR manifold of hypersurface type foliated by $\mathbb{C}P^1$'s and this situation generalizes to C. Le Brun's twistor CR manifolds (cf. [**165**]-[**166**]). As shown in [**84**], there are also natural CR analogues of complex Monge-Ampére foliations (in the sense of [**30**], for instance) occurring on strictly pseudoconvex CR manifolds. Each leaf of such a tangential Monge-Ampére foliation is a CR manifold and the inclusion in the ambient space is a pseudohermitian immersion. Finally, let us mention that each nondegenerate CR manifold carries a flow defined by its contact vector field. This is evidence enough to the interrelation between foliation theory and CR geometry, and that an overall use of the former is liable to clear up certain questions in complex analysis. For instance, let $\Omega \subset \mathbb{C}^{n+1}$ be a strictly pseudoconvex domain with real analytic boundary $\partial\Omega$. Let $\mathcal{O}(\overline{\Omega})$ be the algebra of functions on $\overline{\Omega}$ which admit a holomorphic extension to some neighborhood of $\overline{\Omega}$. Let $M \subset \partial\Omega$ be a real analytic submanifold which is not $\mathbb{C}$-tangent at any of its points. By a result of L. Boutet de Monvel and A. Iordan (cf. [**56**]) M is locally a maximum modulus set for $\mathcal{O}(\overline{\Omega})$ (in the sense of T. Duchamp and E.L. Stout, [**93**]) if and only if $\mathcal{L}(X, Y)$ is real valued for any sections X, Y in $L = T(M) \cap H(\partial\Omega)$. Here $\mathcal{L}$ is the Levi form of $\partial\Omega$ and $H(\partial\Omega)$ is its maximal complex distribution. If this is the case then L is completely integrable and gives rise to a $\mathbb{C}$-tangent foliation $\mathcal{F}$ of M of codimension one and the paper [**22**] studies the interplay between the properties of $\mathcal{F}$ and the geometry of the second fundamental form of M in $\partial\Omega$. M turns out to be a Levi flat contact CR submanifold of $\partial\Omega$ and $\mathcal{F}$ is its Levi foliation. When M is minimal $\mathcal{F}$ is harmonic.

Let us add that, besides from the very interest in interdisciplinary problems, in a series of papers (cf. [**16**], [**19**] and [**21**]) the first two authors developed an idea of E.M. Chirka, [**67**], regarding foliations with transverse CR structure (which contain the class of CR manifolds as the special case of transversally CR foliations by points) which led to Chapter 6 of this monograph.

Sections 1.1 and 1.2 review the notions of foliation theory needed through the text. We only sketch the essentials, as many monographs on the subject have been available for quite a few years (such as [**179**], or [**243**], which are the most frequently referred to).

The next seven chapters form the main core of this book. The case of foliated CR manifolds is considered in Chapter 2. Sections 2.3 to 2.5 are imitative of P. Tondeur's exposition of the geometry of foliations on Riemannian manifolds, cf. [**243**], p. 47-73, and the similarity comes from the fact that in the nondegenerate case CR manifolds possess a canonical metric (the *Webster metric*) and connection (the *Tanaka-Webster connection*) of which the latter resembles the Chern connection in Hermitian geometry and the Levi-Civita connection in Riemannian geometry.

Chapter 3 is dedicated to Levi foliations and their holomorphic extendibility. We give a proof of a beautiful result referred to as *Rea's theorem*. It is based on a theorem by F. Severi and G. Tomassini (cf. [**219**] and [**242**]) about holomorphic extension of CR functions in the real analytic case. There are many other CR extension theorems available in today's mathematical literature (cf. [**50**] and references therein) yet it seems to the authors that Rea's is the only attempt (cf. [**203**]) to apply a CR extension result in order to get a holomorphic extension of a Levi (or semi-holomorphic) foliation.

Related to Rea's theorem we present the solution (due to D.E. Barrett, [**28**]) to the problem of the existence of a pluriharmonic defining function for a Levi-flat real analytic hypersurface in a complex manifold. Next, we exhibit a characterization of Levi flatness of real analytic hypersurfaces in $\mathbb{C}^n$ in terms of *holomorphic degeneracy*, cf. Theorem 3.22 in Section 3.4 (due to N.K. Stanton, [**227**]).

An active research field in complex analysis (in several complex variables) is that related to the problem of global regularity of the Neumann operators N_q, $1 \leq q \leq n$, and of the Bergman projections P_q, $0 \leq q \leq n$, for a smoothly bounded pseudoconvex domain $\Omega \subset \mathbb{C}^n$. Precisely, the question is whether N_q and P_q are continuous on the space $W^s_{(0,q)}(\Omega)$, $s \geq 0$, of all $(0,q)$-forms with coefficients in the Sobolev space $W^s(\Omega)$, $s \geq 0$ (cf. e.g. [**48**]). The state of the art is represented by Theorem 3.32 (due to H.P. Boas and E.J. Straube, [**49**]) in Section 3.5 of this monograph. The estimates leading to the result in Theorem 3.32 were known (by a result of D. Catlin, [**61**]) at the points of finite type, yet required a new technique, based on the existence of complex vector fields commuting approximately with $\overline{\partial}$ (cf. Definition 3.33) on the set $K \subset \partial\Omega$ of all boundary points of infinite type. Such vector fields were shown to exist when Ω admits a plurisubharmonic defining function (cf. Definition 3.30), a fact which led to Theorem 3.31 (due again to H.P. Boas et al., [**48**]). When the set K of all infinite type points is contained in a real submanifold $M \subset \partial\Omega$ of the boundary which is tangent (i.e. $T(M) \subset \text{Null}(G_\theta)$) to the Levi null distribution (e.g. when $K = \overline{\mathring{K}}$ and the Levi form of $\partial\Omega$ vanishes at each point of $\mathring{K}$) the beautiful (from a differential geometric viewpoint) finding by H.P. Boas et al., [**49**] (and further examined by E.J. Straube and M.K. Sucheston, [**233**]) is the existence of a de Rham cohomology class $a(M) \in H^1(M, \mathbb{R})$ (the *D'Angelo class*, under the terminology adopted in this monograph) which is an obstruction to the existence of the special vector fields mentioned above (cf. Theorem 3.36). Section 3.5 concludes with a discussion of the D'Angelo class within foliation theory (i.e. the relationship among $a(M)$ and the infinitesimal holonomy of the leaf M of the Levi foliation on $\mathring{K}$) and a few open problems.

Chapter 4 reports on the known results about the nonexistence of Levi flat CR submanifolds in a complex projective space, such as Y-T. Siu's result (cf. [**221**]-[**222**]) with the lower differentiability requirements due to J. Cao and M-C. Shaw and L. Wang, [**60**] (cf. Theorem 4.1 in Section 4.1 of this book), the result of L. Ni and J. Wolfson, [**188**] (based on a Lefschetz type result for CR submanifolds of a Kählerian manifold of positive holomorphic bisectional curvature, established by themselves, and the classical theorem of A. Haefliger, [**131**], on the inexistence of real analytic codimension one foliations on compact simply connected manifolds), and the purely differential geometric approach of M. Djorić and M. Okumura, [**81**].

Chapter 5 is about foliations with tangential CR structure i.e. each of whose leaves is a CR manifold. We look at foliations by level hypersurfaces of the defining function of a strictly pseudoconvex domain in $\mathbb{C}^n$ such as occurring in C.R. Graham and J.M. Lee's paper [**124**] (and studied by them in connection with the Dirichlet problem for certain degenerate Laplacians of which the prototype is the Bergman Laplacian on the unit ball in $\mathbb{C}^n$). We give a new axiomatic description of the canonical connection there (the *Graham-Lee connection*) and use it to look at the boundary values of a Yang-Mills field in a Hermitian holomorphic vector bundle $\pi : F \to \Omega$ over a smoothly bounded strictly pseudoconvex domain $\Omega \subset \mathbb{C}^n$ (cf. [**26**]). Precisely we endow Ω with the Bergman metric and consider the Dirichlet problem for the Yang-Mills equations

$$\delta^D R^D = 0 \text{ in } \Omega, \quad D = D^0 \text{ on } \partial\Omega, \tag{0.1}$$

where the boundary data D^0 is a C^∞ Hermitian connection in the Hermitian CR-holomorphic vector bundle $E = \pi^{-1}(\partial\Omega) \to \partial\Omega$. It is then shown that the boundary values D^0 of a solution D to (0.1) must be a pseudo Yang-Mills field on $\partial\Omega$ (cf. our Theorem 5.22). Section 5.6 is based on our own work on tangential Monge-Ampére foliations (cf. [**84**] and fitting into the theory of pseudohermitian immersions, cf. also [**89**]).

Chapters 6 to 8 are based on work on transversally CR foliations by the first two authors (cf. *op. cit.*). Chapter 6 is devoted to the basics while Chapters 7 and 8 present two main applications. The first regards the interrelation between G. Gigante and G. Tomassini's theory of CR Lie algebras (cf. [**116**]) and F. Fedida's $\mathcal{G}$-Lie foliations (cf. [**103**]) and includes a homotopy classification of transverse f-structures. The second is devoted to solving a transverse Beltrami equation, which is a foliated analogue of the Beltrami equation in the work of A. Korányi and H.M. Reimann, [**159**]. The effect is producing foliated quasiconformal mappings (cf. E. Barletta, [**16**]). These results extend A. Korányi and H.M. Reimann's considerations - originally holding only on strictly pseudoconvex CR manifolds - to certain degenerate CR manifolds where the degeneracy may be 'factored out' by an algebraic process leading to a strictly pseudoconvex transversally CR foliation. The authors hope that Section 6.4 may contribute to a better understanding of the features of degenerate CR manifolds.

At least for compact Hausdorff foliations i.e. foliations with all leaves compact and the leaf space Hausdorff, the leaf space has (by a result of J. Girbau and M. Nicolau, [**120**], relying itself on a result by D.B.A. Epstein, [**100**]) a natural structure of an *orbifold* (or *V-manifold* in the terminology of I. Satake, [**213**], to whom the notion is due). If this is the case a given transverse CR structure induces a CR structure (in the sense of Chapter 11) on the leaf space, the latter becoming a *CR orbifold*. Chapter 11 aims to a motivation of the need for a theory of CR orbifolds and states some open problems. On the other hand, there is a growing theory of orbifolds, among whose contributors one finds W.L. Baily, [**8**]-[**10**], J.E. Borzellino, [**51**], J.E. Borzellino and B.G. Lorica, [**53**], J.E. Borzellino and S-H. Zhu, [**52**], J.E. Borzellino and V. Brunsden, [**54**], M. Carlotti, [**62**]-[**63**], J. Girbau and M. Nicolau, [**120**], T.D. Jeffres, [**141**], H. Kitahara, [**154**], L-K. Koh, [**172**], T. Shioya, [**220**], and I. Satake himself, [**213**]-[**215**], but to the knowledge of the authors no monograph is available on this subject except for a portion of [**239**], confined to the 3-dimensional case, and of J.E. Borzellino's Ph.D. thesis, [**51**]. There are many differences in style and notations between the above quoted papers and also some inadequacies (for instance [**62**] *postulates* the existence of the monomorphism η while that may be proved, cf. Section 9.3 of this monograph). We choose to expose carefully the basics of the theory of orbifolds in Chapter 9, following mainly the paper [**120**] and hoping to remedy to the mentioned inadequacies and hinting to a further development of differential geometry and analysis on CR orbifolds. Ending these comments, we would like to mention the work by Y-J. Chiang, [**67**], on harmonic maps from a Riemannian orbifold to an ordinary Riemannian manifold (and showing that in the homotopy class of a map of a Riemannian orbifold into a Riemannian manifold of negative sectional curvature there is a harmonic representative). Y-J. Chiang's result is generalized by the work of A. El Kacimi-Alaoui and E.G. Gomez's Theorem 6 in [**148**], p. 121, as $W/\mathrm{SO}(q)$ (where W is the base of the fibration giving rise to the basic foliation associated with the lifted foliation, cf. our section 1.2) is not an orbifold unless the action of $\mathrm{SO}(q)$ on W is locally free. This means that the open problem (of which only the local part is dealt with in Section 11.5) regarding the existence of a parametrix for the Kohn-Rossi operator on a CR orbifold may find its proper and more general setting in a theory of *transversally subelliptic operators* eventually parallelling A. El Kacimi-Alaoui's work [**145**].

The authors are grateful to all scientists whose works have been used in this Monograph. Special thanks are due to Robert Wolak (Jagiellonian University, Kraków, Poland) for his kind advice on the contents of Chapters 1 and 9. E. Barletta and S. Dragomir were partially supported by INdAM (Italy) within the project *Nonlinear subelliptic equations of variational origin in contact geometry* and by MURST (Italy) within the project PRIN *Riemannian metrics and Differentiable Manifolds.* The final version of the manuscript was completed while S. Dragomir was a guest of the Department of Mathematics and Statistics of the University of Windsor (Ontario, Canada) and he wishes to expresses his gratitude for the excellent scientific atmosphere and working conditions there. K.L. Duggal is grateful to the Natural Sciences and Engineering Research Council (NSERC) of Canada for financial support. The authors are grateful to the anonymous referees whose constructive suggestions led to the improvement of this Monograph.

Elisabetta Barletta
Sorin Dragomir
Krishan L. Duggal

CHAPTER 1

Review of foliation theory

We devote this chapter to recalling the differential geometric notions of foliation theory needed throughout. There are quite a few books on the geometry and topology of foliations, e. g. P. Molino, [**179**], P. Tondeur, [**243**]-[**244**], C. Godbillon, [**121**], H.B. Lawson, [**164**], A. Candel & L. Conlon, [**59**], V.Y. Rovenskii, [**209**], I. Tamura, [**234**], C.C. Moore & C. Schochet, [**185**], G. Hector & U. Hirsch, [**132**], C. Camacho & A. Lins Neto, [**58**], and B.L. Reinhart, [**205**], the first two of which are largely quoted through these notes. Therefore, the material on general foliation theory we include is kept to a minimum, intentionally with the scope to fix notations and conventions, hint to some developements (cf. e. g. the finite dimensionality of the basic chomology of a Riemannian foliation on a compact manifold, [**147**], or the study of harmonic foliated maps, [**148**]) not described in the above quoted texts, and to some unsolved problems.

1.1. Basic notions

This section describes the very basic notions of foliation theory, such as leaves, saturated sets, transverse manifolds, and holonomy. For details, see [**179**].

1.1.1. The model foliation. To start with we think of a foliation of a given manifold M in a quantitative way, as a partition of M whose individual subsets are referred to as *leaves*. There is a remarkable foliation of $M = \mathbb{R}^m$ which turns out to be a local model for foliations of arbitrary manifolds.

DEFINITION 1.1. *Let us consider* $\mathbb{R}^m = \mathbb{R}^p \times \mathbb{R}^q$ *with the Cartesian coordinates* $(x, y) = (x^1, \cdots, x^p, y^1, \cdots, y^q)$. *The* model foliation *of dimension* p *and codimension* q *is the family of all affine subspaces (the* leaves *of the model foliation) of* $\mathbb{R}^m$ *parallel to* $\mathbb{R}^p$. □

The foliation of $\mathbb{R}^m$ in Definition 1.1 may be thought of as the foliation by level sets of the map

$$\pi_q : \mathbb{R}^m \to \mathbb{R}^q, \quad \pi_q(x, y) = y.$$

Foliations by level sets of a given function possess remarkable geometric properties and will be given appropriate room later in this book (cf. Section 5.3). The local automorphisms of the model foliation, i.e. the local diffeomorphisms of $\mathbb{R}^m$ mapping leaves to leaves, have a particularly simple structure.

DEFINITION 1.2. Let $\phi : U \to U'$, $\phi = (f^1, \cdots, f^p, g^1, \cdots, g^q)$, be a local diffeomorphism of $\mathbb{R}^m$. Then ϕ is a *local automorphism* of the model foliation if

$$\frac{\partial g^j}{\partial x^\alpha} = 0, \quad 1 \le j \le q, \quad 1 \le \alpha \le p.$$

□

When analyzing the local structure of a given foliation, one is interested in the connected components of the intersection of a leaf with some distinguished open set, the so called *plaques* of the foliation. For the model foliation these have a simple aspect, yet a leaf may intersect the same open set along several plaques. Let $\pi_q : \mathbb{R}^m \to \mathbb{R}^q$ be the natural projection, as above. Let $U \subseteq \mathbb{R}^m$ be an open subset and $\xi \in U$. Set $y = \pi_q(\xi)$.

DEFINITION 1.3. The *plaque* of ξ in U is the connected component of $U \cap \pi_q^{-1}(y)$ containing ξ. □

If $\phi : U \to U'$, $\phi = (f, g)$, is a local automorphism of the model foliation and $\xi \in U$ then it is an easy exercise that

$$\phi(U \cap \pi_q^{-1}(y)) \subseteq U' \cap \pi_q^{-1}(g(y))$$

i.e. ϕ maps the plaque of ξ in U onto the plaque of $\phi(\xi)$ in U'.

Let $\Gamma_{m,q}$ be the set of all local automorphisms of the model foliation of $\mathbb{R}^m$. This is a pseudogroup of transformations of $\mathbb{R}^m$. For arbitrary manifolds M a foliation will be specified on M by indicating an atlas of M whose transition functions belong to $\Gamma_{m,q}$.

DEFINITION 1.4. Let M be a real m-dimensional C^∞ manifold. A *foliated atlas* of codimension q on M is a C^∞ atlas of M whose transition functions are local automorphisms of the model foliation of codimension q of $\mathbb{R}^m$. A *foliation* of *codimension* q of M is a maximal foliated atlas $\mathcal{F}$ of codimension q of M and a pair $(M, \mathcal{F})$ is a *foliated manifold.* □

DEFINITION 1.5. *Let $(M, \mathcal{F})$ be a foliated manifold. Local coordinate systems $(U, \varphi) \in \mathcal{F}$ are referred to as* foliated charts *and U is a* distinguished *open set. Also $\varphi = (x^1, \cdots, x^p, y^1, \cdots, y^q)$ are* distinguished coordinates *and $(y^1, \cdots, y^q)$ are* transverse coordinates. □

DEFINITION 1.6. *Let (U, φ) be a foliated chart of $(M, \mathcal{F})$. If $x \in U$ let α be the plaque of $\varphi(x)$ in $\varphi(U)$ with respect to the model foliation. Then $\varphi^{-1}(\alpha)$ is the* plaque *of x in U.* □

Clearly given a foliated chart (U, φ) with $\varphi = (x^1, \cdots, x^p, y^1, \cdots, y^q)$ and given a plaque $\varphi^{-1}(\alpha) \subset U$ there are constants $(c^1, \cdots, c^q) \in \mathbb{R}^q$ such that $\varphi^{-1}(\alpha)$ is a connected component of $\{x \in U : y^1(x) = c^1, \cdots, y^q(x) = c^q\}$.

Let $x_0 \in M$ and let (U, φ) be a foliated chart such that $x_0 \in U$. Let P_{x_0} be the span of $\{(\partial/\partial x^\alpha)(x_0) : 1 \leq \alpha \leq p\}$. Clearly the definition of P_{x_0} does not depend upon the choice of distinguished local coordinates at x_0 and the assignment $x \in M \mapsto P_x$ is a C^∞ distribution of rank p on M.

DEFINITION 1.7. P_x is the *tangent space* to $\mathcal{F}$ at x and is also denoted by $T(\mathcal{F})_x$. The distribution $T(\mathcal{F})$ is the *tangent bundle* of $\mathcal{F}$. □

Each $X \in T(\mathcal{F})$ is locally a linear combination of the $\partial/\partial x^\alpha$'s hence $T(\mathcal{F})$ is involutive i.e. $[X, Y] \in T(\mathcal{F})$ for any $X, Y \in T(\mathcal{F})$. The converse is a difficult yet classical result, the Frobenius theorem: if P is an involutive p-dimensional C^∞ distribution on a m-dimensional manifold M then P is *integrable* i.e. there is a codimension $q = m - p$ foliation $\mathcal{F}$ of M such that $P = T(\mathcal{F})$. An integrable subbundle of $T(M)$ is often referred to as a foliation of M, as well.

When his branch is yet tender, and putteth
forth *leaves*, ye know that summer is nigh.
Mathew 24-23

1.1.2. Leaves. Let $(M, \mathcal{F})$ be a foliated manifold, of codimension q. Let $\Omega(x_0, x)$ be the set of all piecewise C^1 differentiable curves in M joining the points x_0 and x. Let P be the tangent bundle of $\mathcal{F}$.

DEFINITION 1.8. The *leaf* through x_0 is the set L_{x_0} consisting of all points $x \in M$ which may be joined to x_0 by a curve $\gamma \in \Omega(x_0, x)$ tangent to P i.e. $(d\gamma/dt)(t) \in P_{\gamma(t)}$ for each value of the parameter t for which $(d\gamma/dt)(t)$ is defined. $\square$

PROPOSITION 1.9. *If $x \in L_{x_0}$ and α_x is the plaque through x in a distinguished open set then α_x is contained in L_{x_0}.*

Proof. Let $\Omega_P(x, z)$ be the set of all piecewise C^1 curves $\gamma : [0,1] \to M$ joining the points $x, z \in M$ (i.e. $\gamma(0) = x$ and $\gamma(1) = z$) and such that $(d\gamma/dt)(t) \in P_{\gamma(t)}$ for all values of the parameter t for which $(d\gamma/dt)(t)$ is defined. Then the leaf L_{x_0} of $\mathcal{F}$ passing through x_0 is given by $L_{x_0} = \{x \in M : \Omega_P(x_0, x) \neq \emptyset\}$. Let $x \in L_{x_0}$ and $(U, \varphi) \in \mathcal{F}$ such that $x \in U$. We set $p := \varphi(x) \in \mathbb{R}^m$ and $p = (\xi, \eta) \in \mathbb{R}^p \times \mathbb{R}^q$. Let C be the connected component of p in $\Omega \cap (\mathbb{R}^p \times \{\eta\})$ where $\Omega = \varphi(U) \subseteq \mathbb{R}^m$. Then the plaque α_x in U through x is the set $\alpha_x = \varphi^{-1}(C)$. Let $y \in \alpha_x$ and consider the curve $a : [0,1] \to \mathbb{R}^m$ given by $a(t) = (1-t)p + tq$ for any $0 \leq t \leq 1$, where $q = \varphi(y)$. We may arrange from the very beginning that Ω is a convex set so that a is a curve in $\Omega \cap (\mathbb{R}^p \times \{\eta\})$. Moreover $a([0,1])$ is a connected set containing the point p hence $a([0,1]) \subseteq C$. Therefore the curve

$$\gamma : [0,1] \to U, \quad \gamma(t) = \varphi^{-1}(a(t)), \quad 0 \leq t \leq 1,$$

is well defined and $\gamma(t) \in \alpha_x$ for any $0 \leq t \leq 1$. If the distinguished local coordinates are $\varphi = (x^1, \cdots, x^p, y^1, \cdots, y^q)$ then

$$\frac{d\gamma}{dt}(t) = (x^\alpha(q) - \xi^\alpha) \left. \frac{\partial}{\partial x^\alpha} \right|_{\gamma(t)} \in P_{\gamma(t)}$$

for all t (where the tangent to γ is defined). Consequently $\gamma \in \Omega_P(x, y)$ and we are done.

The plaques in L_{x_0} form a base of open sets for the *leaf topology* on L_{x_0}. In this topology L_{x_0} is arcwise connected. Also L_{x_0} admits a structure of p-dimensional C^∞ manifold whose underlying topology is the leaf topology. Traces of open sets in M on a leaf L are open in the leaf topology.

DEFINITION 1.10. Let S be a submanifold of M i.e. the inclusion $i : S \hookrightarrow M$ is an immersion. Then S is *weakly embedded* if for any C^∞ manifold N and any C^∞ map $f : N \to M$ with $f(N) \subset S$ the corestriction $f : N \to S$ is a C^∞ map. $\square$

Each leaf of a foliation $\mathcal{F}$ of M is a weakly embedded submanifold of M. An important example of foliation is that of a foliation defined by a surjective C^∞ submersion $f : M \to N$. One checks easily that the *vertical distribution* $P = \mathrm{Ker}(df)$ is involutive and hence integrable. It determines a foliation $\mathcal{F}$ of M whose leaves are the connected components of the fibres of f. The leaves of $\mathcal{F}$ are closed embedded submanifods of M. More room will be dedicated to the argument in section 1.1.5.

1.1.3. Foliated maps. Let $(M, \mathcal{F})$ and $(M', \mathcal{F}')$ be two foliated manifolds and $P = T(\mathcal{F})$, $P' = T(\mathcal{F}')$ the tangent bundles of $\mathcal{F}$ and $\mathcal{F}'$, respectively.

DEFINITION 1.11. A C^∞ map $f : M \to M'$ is a *foliated map* if $(d_x f)P_x \subseteq P'_{f(x)}$ for any $x \in M$. An *automorphism* of $(M, \mathcal{F})$ is a C^∞ diffeomorphism of M and a foliated map. □

Let $\mathrm{Aut}(M, \mathcal{F})$ be the group of all (global) automorphisms of $(M, \mathcal{F})$. Each $\phi \in \mathrm{Aut}(M, \mathcal{F})$ maps locally leaves to leaves.

1.1.4. Saturated sets. Let $(M, \mathcal{F})$ be a foliated manifold.

DEFINITION 1.12. A subset $A \subseteq M$ is a *saturated set* if A is a union of leaves of $\mathcal{F}$. □

DEFINITION 1.13. Two points in M are *equivalent* if they belong to the same leaf of $\mathcal{F}$. □

One checks easily that the relation in Definition 1.13 is an equivalence relation on M and that the corresponding equivalence classes are the leaves of $\mathcal{F}$ themselves. The corresponding quotient space is denoted by $M/\mathcal{F}$ and is thought of as carrying the quotient topology.

DEFINITION 1.14. $M/\mathcal{F}$ is the *leaf space* of $(M, \mathcal{F})$. □

In general $M/\mathcal{F}$ may fail to admit a manifold structure whose underlying topology is the quotient topology. Let $\pi : M \to M/\mathcal{F}$ be the natural projection.

DEFINITION 1.15. The set $\pi^{-1}(\pi(A))$ is the *saturation* of $A \subseteq M$. □

For an arbitrary subset $A \subseteq M$ its saturation $\pi^{-1}(\pi(A))$ is a saturated set containing A. The closure and interior of a saturated set are saturated sets. The saturation of an open set is open.

DEFINITION 1.16. The *saturated topology* of M is the topology whose open sets are the saturated open sets. □

Let us mention that, besides from the saturated topology and the topology underlying the C^∞ manifold structure, a foliated manifold M admits a third topology, the so called *leaf topology* for which the plaques in M are a base of open sets.

1.1.5. Simple foliations. Let $(M, \mathcal{F})$ be a foliated manifold and $P = T(\mathcal{F})$ its tangent bundle. We adopt the following

DEFINITION 1.17. The foliation $\mathcal{F}$ is said to be *simple* if there is a surjective submersion f of M onto some C^∞ manifold N such that each fibre of f is connected and $T(\mathcal{F})_x = \mathrm{Ker}(d_x f)$ for any $x \in M$ i.e. $T(\mathcal{F})$ is the vertical bundle associated to f. □

For a simple foliation $\mathcal{F}$ there is a natural identification of the leaf space $M/\mathcal{F}$ with N. Hence the leaf space of a simple foliation admits a C^∞ manifold structure whose underlying topology is the quotient topology. Let us look at the following

EXAMPLE 1.18. Let $\mathbb{C}^{n+1}$ with the complex coordinates $(z^1, \cdots, z^n, w)$, $w = u + iv$, and $\alpha : \mathbb{R} \to \mathbb{R}$ a C^∞ function such that $\alpha(0) = 0$ and $\alpha'(t) < 0$, for any $t \in \mathbb{R}$. Define $f : \mathbb{C}^{n+1} \to \mathbb{R}$ by

$$f(z^1, \cdots, z^n, w) = \alpha(|z^1|^2 + \cdots + |z^n|^2 - v)e^u.$$

Then f is a submersion so that it defines a simple foliation $\mathcal{F}$ of $\mathbb{C}^{n+1}$ whose leaves are the level sets of f. Note that

$$f^{-1}(c) = \begin{cases} \{(z,w) \in \Omega_{n+1} : u = \log(c/\alpha(\rho))\} & \text{if } c > 0 \\ \partial\Omega_{n+1} \approx \mathbb{H}_n & \text{if } c = 0 \\ \{(z,w) \in \mathbb{C}^{n+1} \setminus \overline{\Omega}_{n+1} : u = \log(c/\alpha(\rho))\} & \text{if } c < 0 \end{cases}$$

where $\rho = \sum_{\alpha=1}^{n} |z^\alpha|^2 - v$ and $\Omega_{n+1} \subset \mathbb{C}^{n+1}$ is the Siegel domain $\Omega_{n+1} = \{(z,w) : \rho < 0\}$. Thus $\mathcal{F}$ is a foliation of $\mathbb{C}^{n+1}$ by real hypersurfaces one of whose leaves is the Heisenberg group $\mathbb{H}_n$ (cf. Chapter 2). □

1.1.6. Transverse submanifolds. Let $(M, \mathcal{F})$ be a foliated manifold. Let $\nu(\mathcal{F}) = T(M)/T(\mathcal{F})$ be the *normal*, or *transverse*, bundle. In general there isn't a natural choice of complement E to $T(\mathcal{F})$ in $T(M)$ such that $E \approx \nu(\mathcal{F})$ (a vector bundle isomorphism). Of course the possibility of choosing an integrable distribution E on M such that $T_x(M) = T_x(\mathcal{F})_x \oplus E_x$ for any $x \in M$ is even scarcer. However it turns out that such a decomposition is feasible along certain submanifolds of M.

DEFINITION 1.19. A submanifold T of a foliated manifold $(M, \mathcal{F})$ is a *transverse submanifold* if $T_x(M) = T(\mathcal{F})_x \oplus T_x(T)$ for any $x \in T$. A *total transversal* is a transverse submanifold which meets all leaves of $\mathcal{F}$. □

Let T be a transverse submanifold of $(M, \mathcal{F})$. If $U \subseteq M$ is an open subset and $\mathcal{F}_U$ the foliation induced by $\mathcal{F}$ on U (i.e. $T(\mathcal{F}_U)$ is the portion of $T(\mathcal{F})$ over U) then $T \cap U$ is a transverse submanifold of $(U, \mathcal{F}_U)$.

DEFINITION 1.20. An open set $U \subseteq M$ is said to be *simple* if i) the leaf space $U/\mathcal{F}_U$ has a C^∞ manifold structure whose underlying topology is the quotient topology, ii) the quotient map $\pi_U : U \to U/\mathcal{F}_U$ is a C^∞ submersion, and iii) $\mathcal{F}_U$ is the simple foliation associated with π_U. □

REMARK 1.21. If $V \subseteq M$ is a simple open set then for any point $x_0 \in V$ there is an open subset $U \subseteq V$ such that $x_0 \in U$ which is both simple and distinguished. Indeed let $(U, \varphi) \in \mathcal{F}$ such that $x_0 \in U \subseteq V$. Let L be a leaf of $\mathcal{F}$ passing through $x \in U$. Let $p := \pi_V(x)$ where $\pi : V \to V/\mathcal{F}_V$ is the C^∞ submersion locally defining $\mathcal{F}$, according to Definition 1.20. Let π_U be the restriction of π_V to U. Then

$$\pi_U^{-1}(p) = U \cap \pi_V^{-1}(p) = U \cap (V \cap L) = U \cap L$$

i.e. U is simple, too. □

The following local reformulation[1] of the Frobenius theorem is also useful in practice

THEOREM 1.22. *Let $U \subseteq \mathbb{R}^n$ be an open subset and $\eta_1, \cdots, \eta_q \in \Omega^1(U)$ real-valued linearly independent differential 1-forms. Let $x_0 \in U$. The following statements are equivalent*

i) *There is an open neighborhood $W \subseteq U$ of x_0 and there exist real-valued 1-forms $\alpha_{ij} \in \Omega^1(W)$ such that*

$$d\eta_i = \sum_{j=1}^{q} \alpha_{ij} \wedge \eta_j \,, \quad 1 \le i \le q. \tag{1.1}$$

---

[1]Resembling superficially to the deeper complex Frobenius-Nirenberg theorem (cf. [**194**]).

ii) *There is an open neighborhood* $W \subseteq U$ *of* x_0 *and there exist smooth functions* $\sigma_i, \beta_{ij} \in C^\infty(W, \mathbb{R})$, $1 \leq i, j \leq q$, *such that* $\det[\beta_{ij}] \neq 0$ *everywhere in* W *and*

$$\eta_i = \sum_{j=1}^{q} \beta_{ij} d\sigma_j \,, \quad 1 \leq i \leq q. \tag{1.2}$$

Let us show that Theorem 1.22 follows from the Frobenius theorem as stated above. Let us assume that (i) in Theorem 1.22 holds and set

$$P_x = \{v \in T_x(\mathbb{R}^n) : \eta_{i,x}(v) = 0, \quad 1 \leq i \leq q\}, \quad x \in U.$$

Then $x \in U \mapsto P_x$ is a $(n-q)$-dimensional smooth distribution on U. As a consequence of (i) the distribution P is involutive. Indeed, for any $X, Y \in P$

$$\eta_i([X,Y]) = -2(d\eta_i)(X,Y) = -2\sum_{j=1}^{q}(\alpha_{ij} \wedge \eta_j)(X,Y) =$$

$$= -\sum_{j=1}^{q}\{\alpha_{ij}(X)\eta_j(Y) - \alpha_{ij}(Y)\eta_j(X)\} = 0$$

i.e. $[X,Y] \in P$. By the Frobenius theorem P must be integrable. Let then $\mathcal{F}$ be a codimension q foliation such that $P = T(\mathcal{F})$. As $\mathcal{F}$ may be described locally by C^∞ submersions (cf. Definition 1.20 and Remark 1.21 above) given $x_0 \in U$ there is an open neighborhood $W \subseteq U$ of x_0 and there exists a C^∞ submersion $\sigma = (\sigma_1, \cdots, \sigma_q) : W \to \mathbb{R}^q$ such that

$$T(\mathcal{F})_x = \bigcap_{i=1}^{q} \operatorname{Ker}(d\sigma_i)_x \,, \quad x \in W.$$

Then it is a matter of linear algebra that $\eta_{i,x} = \sum_{j=1}^{q} \beta_{ij}(x)\,(d\sigma_j)_x$ for some $\beta_{ij}(x) \in \mathbb{R}$ (depending differentiably of x and such that $\det[\beta_{ij}(x)] \neq 0$), which is (ii) in Theorem 1.22. Finally we may take the exterior differential of (1.2) to obtain (1.1) with $\alpha_{ij} = \sum_{k=1}^{q} \gamma_{kj} d\beta_{ik}$ and $[\gamma_{ij}] = [\beta_{ij}]^{-1}$. □

We emphasize that the statements in Theorem 1.22 are really about the differential ideal $\mathcal{I}^\bullet_\eta(W)$ of $\Omega^\bullet(W)$ (the de Rham algebra of all real-valued differential polynomials on W) spanned by $\eta_1, \cdots, \eta_q$ that is

$$\mathcal{I}^\bullet_\eta = \{\sum_{i=1}^{q} \eta_i \wedge \omega_i : \omega_i \in \Omega^\bullet(W), \quad 1 \leq i \leq q\}.$$

Indeed (i) in Theorem 1.22 is equivalent to $d\mathcal{I}^\bullet_\eta(W) \subseteq \mathcal{I}^\bullet_\eta(W)$, while (ii) is equivalent to the statement that $\{d\sigma_1, \cdots, d\sigma_q\}$ spans $\mathcal{I}^\bullet_\eta(W)$.

If $U \subseteq M$ is a simple open set and $T \subset M$ is a transversal such that $T \cap U \neq \emptyset$ then the differential

$$d_x\pi_U : T_x(T \cap U) \to T_{\pi_U(x)}(U/\mathcal{F}_U)$$

is a linear isomorphism for any $x \in T \cap U$. Hence π_U is a local diffeomorphism of $T \cap U$ into $U/\mathcal{F}_U$. Consequently, the saturation $\pi_U^{-1}(\pi_U(T \cap U))$ of $T \cap U$ is an open set in U hence the saturation $\pi^{-1}(\pi(T))$ of T in M is an open set.

For any $x \in M$ there is a transverse submanifold T through x. This may be constructed locally, in a foliated local chart (U, φ) at x by merely pulling back

$(\{\xi\} \times \mathbb{R}^q) \cap \varphi(U)$ via φ, where $\varphi(x) = (\xi, \eta) \in \mathbb{R}^p \times \mathbb{R}^q$. It is noteworthy that the resulting transverse submanifold T meets each plaque of U in but one point i.e. T is a (connected) total transversal of $(U, \mathcal{F}_U)$. In general, this does not imply that T meets each leaf L of $\mathcal{F}$ in a single point (as two plaques of U may lie on the same leaf) unless L is *proper* i.e. the leaf topology and the induced topology of L actually coincide. It should be emphasized that given a proper leaf L and a point $x \in L$ there is a transversal T at x such that $L \cap T = \{x\}$ (in general a proper leaf may meet a transversal at infinitely many points). Every closed leaf is proper.

1.1.7. Holonomy. Let $(M, \mathcal{F})$ be a foliated manifold. Let $L \in M/\mathcal{F}$ be a fixed leaf and T, T' two transversals passing through the points $x, x' \in L$. Let $\gamma : [0,1] \to L$ be a continuous curve joining x and x' i.e. $\gamma(0) = x$ and $\gamma(1) = x'$. Next, let us consider a partition $t_0 = 0 < t_1 < \cdots < t_{k-1} < t_k = 1$ of the interval $[0,1]$ such that each curve segment $\gamma([t_{i-1}, t_i])$ is contained in a simple open set U_i, for $1 \leq i \leq k$. In other words, we cover the curve γ by a finite set $\{U_1, \cdots, U_k\}$ of simple open sets. Of course, this is possible because $\gamma([0,1])$ is compact and $M \supset \gamma([0,1])$ admits a covering consisting of simple open sets. By Remark 1.21 one may assume w.l.o.g. that the open sets U_i are both simple and distinguished. Let $f_i : U_i \to N_i := U_i/\mathcal{F}_{U_i}$ be the local defining submersions and $\varphi_i : U_i \to \Omega_i := \varphi_i(U_i) \subseteq \mathbb{R}^m$ the corresponding distinguished local charts. Let us set $x_i := \gamma(t_i)$ and $p_i = f_i(x_i) \in N_i$ for $1 \leq i \leq k$ so that $L \cap U_i = f_i^{-1}(p_i)$. Let α_i be the plaque in U_i passing through the point x_i i.e. $\alpha_i = \varphi^{-1}(C_i)$ and C_i is the connected component of $\varphi_i(x_i) = (\xi_i, \eta_i) \in \mathbb{R}^p \times \mathbb{R}^q$ in $\Omega_i \cap (\mathbb{R}^p \times \{\eta_i\})$. By Proposition 1.9 one has $x_i \in \alpha_i \subseteq L$. Moreover $K_i := \gamma([t_{i-1}, t_i])$ is a connected set containing the point x_i hence $K_i \subseteq \alpha_i$. Let now T_i be just any transverse submanifold of $(U_i, \mathcal{F}_{U_i})$ passing through x_i, for $1 \leq i \leq k-1$. For instance one may consider

$$T_i := \varphi_i^{-1}(\Omega_i \cap (\{\xi_i\} \times \mathbb{R}^q)).$$

Also we set $T_0 = T$ and $T_k = T'$. Both $d_{x_{i-1}} f_i : T_{x_{i-1}}(T_{i-1}) \to T_{p_i}(N_i)$ and $d_{x_i} f_i : T_{x_i}(T_i) \to T_{p_i}(N_i)$ are $\mathbb{R}$-linear isomorphisms hence there are open neighborhoods of x_{i-1} in T_{i-1}, respectively of x_i in T_i

$$x_{i-1} \in \mu_{i-1} \subseteq T_{i-1}\,, \quad x_i \in \mu_i \subseteq T_i\,,$$

such that $f_i : \mu_{i-1} \to N_i$ and $f_i : \mu_i \to N_i$ are diffeomorphisms on their images. Then $A_i := f_i(\mu_{i-1}) \cap f_i(\mu_i)$ is an open neighborhood of p_i in N_i. Let F_{i-1} and F_i be the restrictions of f_i to μ_{i-1} and μ_i, respectively. We consider the open sets

$$\nu_{i-1} = F_{i-1}^{-1}(A_i) \subseteq T_{i-1}\,, \quad \nu_i = F_i^{-1}(A_i) \subseteq T_i\,,$$

and the diffeomorphism

$$\phi_i := (F_i : A_i \to \nu_i)^{-1} \circ (F_{i-1} : \nu_{i-1} \to A_i) : \nu_{i-1} \to \nu_i\,.$$

DEFINITION 1.23. The diffeomorphism $\phi_i : \nu_{i-1} \approx \nu_i$ is said to be obtained by *sliding along the plaques* in U_i. □

Moreover let us set $\nu = \nu_0$ and $\nu' = \nu_k$. We may consider the diffeomorphism

$$\phi_k \circ \phi_{k-1} \circ \cdots \circ \phi_1 : \nu \to \nu'. \tag{1.3}$$

and denote its germ at x by h_γ. It may be shown that the definition of the diffeomorphism (1.3) doesn't depend upon the choice of the transversal submanifolds $T_1, \cdots, T_{k-1}$.

DEFINITION 1.24. h_γ is called *sliding along the leaves along* γ. $\square$

It may be shown that the definition of h_γ doesn't depend upon the choice of the simple sets U_i covering γ but only on γ itself. Also, by considering a continuous deformation $\gamma_s(t)$ of γ with end points fixed such that the previously chosen chain $\{U_1, \cdots, U_k\}$ covers $t \mapsto \gamma_s(t)$ for each value of the parameter s, one may show that h_γ depends only on the homotopy class of γ.

Let $\gamma : [0,1] \to L$ be a loop at x i.e. $\gamma(0) = \gamma(1) = x$. Let T be a transversal at x. Then h_γ built above (with $x' = x$ and $T' = T$) is a germ at x of a local diffeomorphism of T which maps x to itself. One may check that given another loop τ in L at x one has $h_{\gamma\cdot\tau} = h_\gamma \circ h_\tau$ (where $\gamma \cdot \tau$ is the juxtaposition of the two loops) so that there is a natural group homomorphism

$$h_x : \pi_1(L, x) \to \mathrm{Diff}_x(T), \quad [\gamma] \mapsto h_\gamma\,.$$

Here $\pi_1(L, x)$ is the fundamental group of L with base point x and $[\gamma] \in \pi_1(L, x)$ is the homotopy class of the loop γ at x. Also $\mathrm{Diff}_x(T)$ denotes the group of germs at x of local diffeomorphisms of T in itself.

DEFINITION 1.25. h_x *is the* holonomy representation *of* L *at* x *and its image* $h_x(\pi_1(L, x))$ *is the* holonomy group *of* L *at* x. $\square$

1.2. Transverse geometry

In this section we discuss basic forms and the corresponding basic cohomology of a foliated manifold, cf. [**179**]. Also we recall the facts we need from the geometry of transverse G-structures on foliated manifolds, cf. L. Conlon, [**70**], and P. Molino, [**180**]. See also F. Kamber & P. Tondeur, [**149**]-[**150**]. Moreover, we state the first structure theorem on Riemannian foliations (cf. P. Molino, [**182**]) and review some of its applications.

1.2.1. Basic functions, foliate vector fields, and basic forms. Let $(M, \mathcal{F})$ be a foliated manifold and $P = T(\mathcal{F})$ the tangent bundle of the foliation.

DEFINITION 1.26. A C^∞ function f on M is *basic* if $X(f) = 0$ for any $X \in P$. $\square$

Let $\Omega^0_B(\mathcal{F})$ be the ring of all real valued basic functions, a subring of $\Omega^0(M) = C^\infty(M)$. A function $f \in \Omega^0(M)$ is basic if and only if its local expression with respect to any foliated chart (U, x^α, y^i) is a function of the variables y^i only.

DEFINITION 1.27. A tangent vector field $Y \in \mathcal{X}(M)$ is *foliate* if $[X, Y] \in P$ for any $X \in P$. $\square$

A vector field $Y \in \mathcal{X}(M)$ is foliate if and only if the components Y^{j+p} of Y with respect to an arbitrary foliated chart (U, x^α, y^i)

$$Y = Y^\alpha \frac{\partial}{\partial x^\alpha} + Y^{j+p} \frac{\partial}{\partial y^j}$$

depend on the variables y^i only. Let $V(\mathcal{F})$ be the Lie algebra of all foliate vector fields. $V(\mathcal{F})$ is a Lie subalgebra of $\mathcal{X}(M)$.

Let $Q = \nu(\mathcal{F}) = T(M)/P$ be the normal bundle of $(M, \mathcal{F})$ and let $\Pi : T(M) \to Q$ be the natural bundle map.

DEFINITION 1.28. The elements of $\ell(\mathcal{F}) = \Pi\, V(\mathcal{F}) = \{\Pi(Y) : Y \in V(\mathcal{F})\}$ are the *transverse vector fields* of $(M, \mathcal{F})$. □

Clearly

$$0 \to \Gamma^\infty(P) \hookrightarrow V(\mathcal{F}) \stackrel{\Pi}{\to} \ell(\mathcal{F}) \to 0$$

is a short exact sequence of Lie algebras and Lie algebra homomorphisms.

DEFINITION 1.29. A C^∞ differential k-form η on M is *basic* if $X \rfloor \eta = 0$ and $X \rfloor d\eta = 0$ for any $X \in P$. □

Let $\Omega^k_B(\mathcal{F})$ be the space of all basic k-forms, a $\Omega^0_B(\mathcal{F})$-submodule of $\Omega^k(M) = \Gamma^\infty(\Lambda^k T^* M)$. Note that the differential of a basic form is again a basic form, hence d induces a differential operator

$$d_B : \Omega^k_B(\mathcal{F}) \to \Omega^{k+1}_B(\mathcal{F}).$$

DEFINITION 1.30. The *basic cohomology* of $(M, \mathcal{F})$ is the cohomology of the complex $\{\Omega^\bullet_B(\mathcal{F}), d_B\}$ i.e.

$$H^k_B(\mathcal{F}) = \frac{Ker\{d_B : \Omega^k_B(\mathcal{F}) \to \cdot\}}{d_B \Omega^{k-1}_B(\mathcal{F})}, \quad k \geq 0.$$

Here $\Omega^{-1}_B(\mathcal{F}) = (0)$. □

Roughly speaking $H^\bullet_B(\mathcal{F})$ is the 'de Rham cohomology' of the leaf space $M/\mathcal{F}$. Therefore, it is a natural question whether M compact implies that $H^\bullet_B(\mathcal{F})$ is finite dimensional for every $\mathcal{F}$. For instance, if $\mathcal{F}$ is the simple foliation defined by the surjective submersion $f : M \to N$ then $H^\bullet_B(\mathcal{F}) \approx H^\bullet(N, \mathbb{R})$ (an algebra isomorphism); if M is compact then N is compact and $H^\bullet_B(\mathcal{F})$ is finite dimensional.

Note that in general $H^1_B(\mathcal{F})$ injects into $H^1(M, \mathbb{R})$ (for if $\eta \in \Omega^1_B(\mathcal{F})$ and $\eta = df$ then $f \in \Omega^0_B(\mathcal{F})$). Hence if we assume that M is compact then $H^1_B(\mathcal{F})$ is finite dimensional. G.W. Schwarz, [**217**], has built examples of n-dimensional foliations $\mathcal{F}$ on compact $(n+3)$-dimensional manifolds M so that $\dim H^k_B(\mathcal{F}) = \infty$, $2 \leq k \leq n+2$. See also [**179**] (the appendix authored by V. Sergiescu) for an example (of a foliation with infinite dimensional basic cohomology) due to E. Ghys.

1.2.2. Transverse G-structures. Let $\mathcal{F}$ be a codimension q foliation of a C^∞ manifold M and $Q = T(M)/T(\mathcal{F})$. Let $\Pi : T(M) \to Q$ be the projection.

DEFINITION 1.31. A $\mathbb{R}$-linear isomorphism $z : \mathbb{R}^q \to Q_x$, $x \in M$, is a *transverse frame* on M at x. □

Let $B^1_T(M, \mathcal{F})_x$ be the space of all transverse frames at x. Then $B^1_T(M, \mathcal{F})$ is a principal $\mathrm{GL}(q, \mathbb{R})$-bundle over M.

DEFINITION 1.32. The *canonical* 1-*form* $\theta_T \in \Gamma^\infty(T^*(B^1_T(M, \mathcal{F})) \otimes \mathbb{R}^q)$ is given by the commutative diagram

$$\begin{array}{ccc} T_z(B^1_T(M,\mathcal{F})) & \stackrel{(\theta_T)_z}{\longrightarrow} & \mathbb{R}^q \\ d_z p^1_T \downarrow & & \uparrow z^{-1} \\ T_x(M) & \stackrel{\Pi_x}{\longrightarrow} & Q_x \end{array}$$

for any $z \in B^1_T(M, \mathcal{F})_x$, where $p^1_T : B^1_T(M, \mathcal{F}) \to M$ is the natural projection. □

The p-dimensional distribution P_T^1 on $B_T^1(M,\mathcal{F})$ given by

$$(P_T^1)_z = \{X \in T_z(B_T^1(M,\mathcal{F})) : X \rfloor (\theta_T^1)_z = 0\ ,\ X \rfloor (d\theta_T^1)_z = 0\}$$

is integrable hence gives rise to a foliation $\mathcal{F}_T^1$ of $B_T^1(M,\mathcal{F})$ such that $P_T^1 = T(\mathcal{F}_T^1)$.

DEFINITION 1.33. $\mathcal{F}_T^1$ is called the *lifted foliation.* □

It is invariant by right translations and each leaf of $\mathcal{F}_T^1$ is a Galois covering[2] of a leaf of $\mathcal{F}$.

DEFINITION 1.34. A connection 1-form $\omega \in \Gamma^\infty(T^*(B_T^1(M,\mathcal{F})) \otimes \mathbf{gl}(q,\mathbb{R}))$ is a *transverse connection* if $P_T^1 \subset \text{Ker}(\omega)$. □

By a partition of unity argument, transverse connections always exist.

DEFINITION 1.35. A connection 1-form ω in $B_T^1(M,\mathcal{F})$ is *projectable* if $\omega \in \Omega_B^1(\mathcal{F}_T^1)$. □

Clearly projectable connections are transverse. Let $G \subset \text{GL}(q,\mathbb{R})$ be a Lie subgroup and $B_G(M) \to M$ a principal G-subbundle of $B_T^1(M,\mathcal{F})$.

DEFINITION 1.36. $B_G(M)$ is a *transverse G-structure* on $(M,\mathcal{F})$ if $B_G(M)$ is a saturated subset of the foliated manifold $(B_T^1(M,\mathcal{F}),\mathcal{F}_T^1)$. □

The notion is due to L. Conlon, [**70**]. A principal G-subbundle $B_G(M)$ of $B_T^1(M,\mathcal{F})$ is a transverse G-structure if and only if there is a transverse connection ω on $B_T^1(M,\mathcal{F})$ which is adapted to $B_G(M)$ i.e. if $\mathbf{g}$ is the Lie algebra of G then the pullback of ω to $B_G(M)$ is $\mathbf{g}$-valued.

DEFINITION 1.37. If $G = \text{GL}^+(q,\mathbb{R})$ then a transverse G-structure on $(M,\mathcal{F})$ is a *transverse orientation.* If $G = \{e\}$ a transverse G-structure is a *transverse parallelism.* If $G = \text{O}(q)$ then a transverse G-structure is a *transverse Riemannian structure* and a foliation carrying such a structure is a *Riemannian foliation.* Moreover, if $q = 2r$ then a transverse $\text{GL}(r,\mathbb{C})$-structure is a *transverse holomorphic structure* and a foliation with such a structure is a *transversally holomorphic* foliation. □

Therefore, one may discuss first order geometric structures on $Q = \nu(\mathcal{F})$ in the unifying language of transverse G-structures, very much in the spirit of the theory of ordinary G-structures on C^∞ manifolds, cf. e.g. S. Sternberg, [**230**], M. Crampin, [**74**], P. Molino, [**183**]. However, a general theory of equivalence (structure functions, prolongations, etc.) of transverse G-structures has not been fully developed as yet. The existing results in this sense belong to R. Wolak, [**252**].

1.2.3. $d_\mathcal{F}$-Cohomology, the Atiyah class.

Let $(M,\mathcal{F})$ be a foliated manifold.

DEFINITION 1.38. A principal G-bundle $p : E \to M$ is a *foliated principal bundle* over $(M,\mathcal{F})$ if E carries a foliation $\mathcal{F}_E$ such that 1) $T(\mathcal{F}_E)$ is invariant by right translations, 2) $T(\mathcal{F}_E) \cap \text{Ker}(dp) = (0)$, and 3) $(dp)T(\mathcal{F}_E) = T(\mathcal{F})$. Given an open subset $U \subseteq M$, a section $s : U \to E$ is *foliate* if $(d_x s)T(\mathcal{F})_x \subseteq T(\mathcal{F}_E)_{s(x)}$ for any $x \in U$. □

---

[2] That is if $S \in B_T^1(M,\mathcal{F})/\mathcal{F}_T^1$ then $L := p_T^1(S) \in M/\mathcal{F}$ and $p_T^1 : S \to L$ is an étale mapping and a locally trivial fibration. Cf. [**179**], p. 47.

Given a point $x_0 \in M$ there is a neighborhood U of x_0 such that foliate sections $s : U \to E$ exist. Let (E, p, M) be a foliated principal G-bundle over $(M, \mathcal{F})$ and $\omega \in \Gamma^\infty(T^*(E) \otimes \mathbf{g})$ a connection 1-form on E.

DEFINITION 1.39. ω is *adapted* to $\mathcal{F}_E$ if $T(\mathcal{F}_E) \subset \mathrm{Ker}(\omega)$. □

Connections adapted to $\mathcal{F}_E$ always exist.

DEFINITION 1.40. ω is *projectable* if $\omega \in \Omega^1_B(\mathcal{F}_E)$. □

Clearly projectable connections are adapted. Let $P = T(\mathcal{F})$ and let P' be a direct summand to P in $T(M)$ i.e. $T(M) = P \oplus P'$. Let $\{X_1, \cdots, X_p\}$ be a local frame of P and $\{Y_1, \cdots, Y_q\}$ a local frame of P' defined on the same open set. Moreover let $\{\theta^1, \cdots, \theta^p, \eta^1, \cdots, \eta^q\}$ be dual to $\{X_\alpha, Y_j : 1 \le \alpha \le p,\ 1 \le j \le q\}$.

DEFINITION 1.41. A *pure form* of type (r, s) on $(M, \mathcal{F})$ is a k-form $\alpha \in \Omega^k(M)$ whose restriction to U admits a representation of the form

$$\alpha = \sum_{\substack{1 \le \beta_1 < \cdots < \beta_r \le p \\ 1 \le j_1 < \cdots < j_s \le q}} \alpha_{\beta_1 \cdots \beta_r j_1 \cdots j_s}\, \theta^{\beta_1} \wedge \cdots \wedge \theta^{\beta_r} \wedge \eta^{j_1} \wedge \cdots \wedge \eta^{j_s}$$

for some local C^∞ functions $\alpha_{\beta_1 \cdots \beta_r j_1 \cdots j_s}$ on M, where $r + s = k$. □

Any differential form on M admits a decomposition in terms of pure forms. If $\alpha \in \Omega^k(M)$ then $\alpha_{r,s}$ will denote its pure component of type (r, s).

Let (U, x^α, y^j) be a foliated chart of $(M, \mathcal{F})$ and let Y_j be the P'-component of $\partial/\partial y^j$. Then $\{Y_1, \cdots, Y_q\}$ is a local frame of P' on U and we may take $\eta_\alpha = dy^j$, $1 \le j \le q$. With this choice of local coframes, it follows that the differential $d\alpha$ of any pure form α of type (r, s) is a sum of pure components of type $(r - 1, s + 2)$, $(r, s+1)$, and $(r+1, s)$. Let $d_{\mathcal{F}}\alpha$ be the pure component of type $(r+1, s)$ of $d\alpha$. Let $\Omega^{r,s}(\mathcal{F})$ be the $\Omega^0(M)$-module of pure forms of type (r, s). We built the operators

$$d_{\mathcal{F}} : \Omega^{r,s}(\mathcal{F}) \to \Omega^{r+1,s}(\mathcal{F}) \tag{1.4}$$

such that $d^2_{\mathcal{F}} = 0$ i.e. (1.4) is a cochain complex.

DEFINITION 1.42. The cohomology

$$H^{r,s}_{\mathcal{F}}(M) = \frac{\mathrm{Ker}\{d_{\mathcal{F}} : \Omega^{r,s}(\mathcal{F}) \to \cdot\}}{d_{\mathcal{F}}\Omega^{r-1,s}(\mathcal{F})}$$

of the cochain complex (1.4) is the *$d_{\mathcal{F}}$-cohomology* of $(M, \mathcal{F})$. □

See also C. Roger, [**207**], where the cohomology groups $H^{r,s}_{\mathcal{F}}(M)$ are studied in relation to infinitesimal deformations of foliations, classifying spaces for foliations, and characteristic classes. Let $(E, M, p, G, \mathcal{F}_E)$ be a foliated principal bundle over $(M, \mathcal{F})$ and $\rho : G \to \mathrm{End}_{\mathbb{R}}(V)$ a representation of G in a finite dimensional linear space V. We shall need the following notion

DEFINITION 1.43. A k-form $\alpha \in \Gamma^\infty(\Lambda^k T^*(E) \otimes V)$ is *tensorial of type* $\rho(G)$ if 1) $\mathrm{Ker}(dp) \rfloor \alpha = 0$, and 2) $R^*_g \alpha = \rho(g^{-1})\alpha$ for any $g \in G$. A tensorial form α of type $\rho(G)$ is *pure of type* (r, s) if $s^*\alpha$ is a pure form of type (r, s) for any local section $s : U \to E$. □

A tensorial form α on E is completely determined by the local forms $\alpha_i = s_i^*\alpha$, where $\{s_i\}_{i\in I}$ is a family of local sections $s_i : U_i \to E$ such that $\{U_i\}_{i\in I}$ is an open cover of M. For any $i, j \in I$ with $U_{ij} = U_i \cap U_j \neq \emptyset$ there is $\gamma_{ij} : U_i \cap U_j \to G$ such that $s_j = s_i\gamma_{ij}$ on U_{ij}. Consequently $\alpha_j = \rho(\gamma_{ij}^{-1})\alpha_i$. Viceversa, let $\gamma_i : p^{-1}(U_i) \to U_i \times G$ be a local trivialization atlas of E with the corresponding transition functions $\gamma_{ij} : U_{ij} \to G$. Any family $\{\alpha_i\}_{i\in I}$ of local forms $\alpha_i \in \Gamma^\infty(U_i, \Lambda^k T^*(M) \otimes V)$ satisfying $\alpha_j = \rho(\gamma_{ij}^{-1})\alpha_i$ on U_{ij} determines a tensorial form of type $\rho(G)$ on E.

As E is a foliated principal bundle, one may choose a family $\{s_i\}_{i\in I}$ of foliate local sections such that γ_{ij} are G-valued basic functions. Let α be a tensorial form of type $\rho(G)$, which is also pure of type (r, s). As $d_\mathcal{F} f = 0$ for any $f \in \Omega_B^0(\mathcal{F})$ it follows that

$$d_\mathcal{F}\alpha_j = \rho(\gamma_{ij}^{-1})d_\mathcal{F}\alpha_i$$

hence the local forms $\{d_\mathcal{F}\alpha_i\}_{i\in I}$ determine a tensorial form of type $\rho(G)$ on E which is also a pure form of type $(r+1, s)$. We built an operator

$$d_\mathcal{F} : \Omega_\rho^{r,s}(E, V) \to \Omega_\rho^{r+1,s}(E, V)$$

where $\Omega_\rho^{r,s}(E, V)$ is the space of pure tensorial forms of type (r, s). Once again $d_\mathcal{F}^2 = 0$.

DEFINITION 1.44. The cohomology

$$H^{r,s}_{\rho,\mathcal{F}}(E, V) = \frac{\operatorname{Ker}\{d_\mathcal{F} : \Omega_\rho^{r,s}(E, V) \to \cdot\}}{d_\mathcal{F}\Omega_\rho^{r-1,s}(E, V)}$$

is the *$d_\mathcal{F}$-cohomology* of (E, V). □

Let ω be an adapted connection in the foliated principal bundle $(E, p, M, G, \mathcal{F}_E)$ and Ω the curvature 2-form of ω. Then Ω is a tensorial form on E of type $\operatorname{ad}(G)$, where $\operatorname{ad} : G \to \operatorname{End}_\mathbb{R}(\mathbf{g})$ is the adjoint representation of G in its Lie algebra. An argument based on the second structure equation (cf. e.g. [**155**], vol. I, p. 7) shows that Ω is a sum of a pure form $\Omega_{1,1}$ of type $(1, 1)$ and a pure form $\Omega_{0,2}$ of type $(0, 2)$, i.e. $\Omega = \Omega_{1,1} + \Omega_{0,2}$. In particular $\Omega_{1,1} \in \Omega^{1,1}_{\rm ad}(E, \mathbf{g})$ and $d_\mathcal{F}\Omega_{1,1} = 0$. The corresponding cohomology class $[\Omega_{1,1}] \in H^{1,1}_{{\rm ad},\mathcal{F}}(E, \mathbf{g})$ does not depend upon the choice of an adapted connection ω in E.

DEFINITION 1.45. $[\Omega_{1,1}]$ is the *Atiyah class* of $(E, \mathcal{F}_E^1)$. □

The notion is due to P. Molino, [**181**]. The Atiyah class of $(E, \mathcal{F}_E)$ vanishes if and only if there exists a projectable connection on $(E, \mathcal{F}_E)$. Note that $(B_T^1(M, \mathcal{F}), \mathcal{F}_T^1)$ is a foliated $GL(q, \mathbb{R})$-principal bundle over $(M, \mathcal{F})$.

DEFINITION 1.46. The Atiyah class

$$A(M, \mathcal{F}) \in H^{1,1}_{{\rm ad},\mathcal{F}}(B_T^1(M, \mathcal{F}), \mathbf{gl}(q, \mathbb{R}))$$

of the foliated bundle $(B_T^1(M, \mathcal{F}), \mathcal{F}_T^1)$ is the *Atiyah class* of $(M, \mathcal{F})$. □

1.2.4. Riemannian foliations. Let M be a C^∞ manifold and $\mathcal{F}$ a codimension q foliation of M.

DEFINITION 1.47. A Riemannian bundle metric g_Q in $Q = \nu(\mathcal{F})$ is *holonomy invariant* if $\mathcal{L}_X g_Q = 0$ for any $X \in T(\mathcal{F})$. □

Any holonomy invariant metric g_Q in Q gives rise naturally to a transverse O(q)-structure $O(Q) \to M$ hence $(\mathcal{F}, g_Q)$ is a Riemannian foliation of M.

Let g be a Riemannian metric on M and let $P^\perp$ be the orthogonal complement of $P = T(\mathcal{F})$ in $T(M)$. There is a natural bundle isomorphism $\sigma_g : Q \to P^\perp$. Let us set

$$g_Q(r, s) = g(\sigma_g(r), \sigma_g(s))$$

for any $r, s \in Q$.

DEFINITION 1.48. The Riemannian metric g is said to be *bundle-like* if g_Q is holonomy invariant. □

By a geometric interpretation due to B. Reinhart, [**204**], g is bundle-like if and only if any geodesic of (M, g) orthogonal to some leaf of $\mathcal{F}$ is orthogonal on every other leaf it meets.

Riemannian foliations of compact manifolds are described by the following *first structure theorem*, cf. [**179**], p. 155-156. Let $(\mathcal{F}, g_Q)$ be a Riemannian foliation on a compact connected manifold M. Then i) the closures of the leaves of $\mathcal{F}$ form a partition $\overline{\mathcal{F}}$ of M by compact embedded submanifolds which are integral manifolds of an involutive distribution of variable dimension, ii) the closure of the leaves of $\mathcal{F}$ are the projections on M of the closures of the leaves of the lifted foliation $\mathcal{F}_T^1$ on $O(Q)$; the space $M/\overline{\mathcal{F}}$ of closures of leaves of $\mathcal{F}$ may be identified with the quotient space $W/\mathrm{O}(q)$, where W is the basic manifold of $(O(Q), \mathcal{F}_T^1)$, and iii) when restricted to the closure of a leaf, $\mathcal{F}$ induces a transversally homogeneous foliation (in the sense of R.A. Blumenthal, [**46**]). The key fact in the proof of the first structure theorem is that $\mathcal{F}_T^1$ is transversally parallelizable (hence the *basic foliation* associated with $\mathcal{F}_T^1$, in the sense of [**179**], p. 105, is simple and defined by a locally trivial fibration $O(Q) \to W$). Cf. also [**182**].

The first structure theorem has several important applications. For instance, it plays a key role in the proof of the A. El Kacimi-Alaoui & V. Sergiescu & G. Hector result (cf. [**147**]) that for any codimension q Riemannian foliation $\mathcal{F}$ of a closed (i.e. compact, without boundary) manifold M the basic cohomology groups $H_B^\bullet(\mathcal{F})$ are finite dimensional; moreover $H_B^q(\mathcal{F})$ is either 0 or $\mathbb{R}$. See also A. El Kacimi-Alaoui & G. Hector, [**146**], and V. Sergiescu, [**218**].

As an other application, A. El Kacimi-Alaoui & G. Gomez, [**148**], have established the following result. Let M be a compact manifold and $\mathcal{F}$ a Riemannian foliation of M of codimension $q \geq 2$. Let N be a compact orientable Riemannian manifold and $\varphi : M \to N$ a foliated map (N is thought of as endowed with the trivial foliation by points). If N has negative sectional curvature then there is a harmonic foliated map homotopic to φ. We recall that given a foliated map $\varphi : M \to N$ of a foliated manifold $(M, \mathcal{F})$ into an ordinary manifold N (i.e. $(d_x\varphi)P_x = (0)$, for any $x \in M$) one may define a section

$$d_T\varphi \in \Gamma^\infty(Q^* \otimes \varphi^* T(N))$$

by setting $(d_T\varphi)s = (d\varphi)Y$ for some $Y \in T(M)$ with $\pi Y = s$. Then, if $\mathcal{F}$ and N are Riemannian, as above, we set

$$e_T(\varphi) = \frac{1}{2}\|d_T\varphi\|^2$$

where the norm in $Q^* \otimes \varphi^{-1}TN$ is induced by g_Q and by the Riemannian structure of N. As $e_T(\varphi) \in \Omega_B^0(\mathcal{F})$, i.e. $e_T(\varphi)$ is constant along the leaves of $\mathcal{F}$, it follows that $e_T(\varphi)$ is constant along the closures of the leaves of $\mathcal{F}$. Hence $e_T(\varphi)$ may be lifted to a $\mathrm{O}(q)$-invariant function on $O(Q)$, which in turn induces a $\mathrm{O}(q)$-invariant function $\bar{e}_T(\varphi) : W \to \mathbb{R}$. The *transverse energy* $E_T(\varphi)$ of φ is given by

$$E_T(\varphi) = \int_W \bar{e}_T(\varphi) dw$$

where dw is the volume form of the metric on W induced by the metric on $O(Q)$. Finally $\varphi : M \to N$ is *transversally harmonic* if φ is an extremal of E_T for all variations through foliated maps. Holomorphic maps of generalized Hopf, or Vaisman, manifolds (in the sense of [**88**], p. 33) are transversally harmonic (with respect to the transversally Kählerian foliations determined by the Lee and anti-Lee fields), cf. [**23**].

CHAPTER 2

Foliated CR manifolds

Through this section $(M, T_{1,0}(M))$ denotes a strictly pseudoconvex CR manifold, of CR dimension n. For all the notions of CR and pseudohermitian geometry needed in this text we refer to the monograph [**89**]. Nevertheless, to facilitate reading we recall a few basic concepts (such as CR structures, Levi forms, contact forms, the Tanaka-Webster connection, etc.) as they are requested by our exposition. Let M be a real $(2n+k)$-dimensional C^∞ manifold. A complex subbundle $T_{1,0}(M) \subset T(M) \otimes \mathbb{C}$ of the complexified tangent bundle over M, of complex rank n, is called a *CR structure* on M if

$$T_{1,0}(M) \cap T_{0,1}(M) = (0), \tag{2.1}$$

$$Z, W \in T_{1,0}(M) \Longrightarrow [Z, W] \in T_{1,0}(M). \tag{2.2}$$

The integer n is referred to as the *CR dimension* of the given CR structure $T_{1,0}(M)$ while k is its *CR codimension*. Here $T_{0,1}(M) = \{\overline{Z} : Z \in T_{1,0}(M)\}$ and $\overline{Z}$ is the complex conjugate of Z. If $\{P_1, \cdots, P_n\}$ is a local frame of $T_{1,0}(M)$ defined on some open set $U \subseteq M$ then the meaning of (2.1) is that $\{P_1, \cdots, P_n, \overline{P}_1, \cdots, \overline{P}_n\}$ are linearly independent at each point of U, and hence a local frame of $T_{1,0}(M) \oplus T_{0,1}(M)$. The property (2.2) is commonly referred to as the (*Frobenius*) *formal integrability property*. The notation $Z \in T_{1,0}(M)$ is similar to that used for real distributions on manifolds: Z is actually a complex vector field on M defined on some open subset $U \subseteq M$ such that $Z_x \in T_{1,0}(M)_x$ for any $x \in U$. Hence (2.2) states that $T_{1,0}(M)$ is involutive. However $T_{1,0}(M)$ is a complex distribution so that the classical Frobenius theorem doesn't apply i.e. in general given a point $x \in M$ there might be no submanifold $S \hookrightarrow M$ passing through x and such that $T_{1,0}(M) = T(S) \otimes \mathbb{C}$ along S. By a *complex distribution* $\mathcal{D}$ on M we mean an assignment $x \in M \mapsto \mathcal{D}_x \subseteq T_x(M) \otimes_{\mathbb{R}} \mathbb{C}$ such that i) each $\mathcal{D}_x$ is a complex subspace of $T_x(M) \otimes_{\mathbb{R}} \mathbb{C}$ of complex dimension r, and ii) $\mathcal{D}$ is C^∞ differentiable in the sense that for any point $x \in M$ there is an open neighborhood $U \subseteq M$ and a set $\{Z_1, \cdots, Z_r\}$ of complex vector fields of class C^∞ defined on U such that $\{Z_1(y), \cdots, Z_r(y)\}$ span $\mathcal{D}_y$ for any $y \in U$. The integer r is the *rank* of the complex distribution.

A pair $(M, T_{1,0}(M))$ consisting of a $(2n+k)$-dimensional manifold M and a CR structure $T_{1,0}(M)$ of CR dimension n is called a *CR manifold* of *type* (n, k). When $k = 0$ the condition (2.2) implies that $(M, T_{1,0}(M))$ is a complex manifold, of complex dimension n. The case of CR codimension $k = 1$ is particularly interesting as such CR manifolds appear often as real hypersurfaces of complex manifolds. If for instance $M \subset \mathbb{C}^{n+1}$ is a real hypersurface then

$$T_{1,0}(M) = T^{1,0}(\mathbb{C}^{n+1}) \cap [T(M) \otimes \mathbb{C}]$$

defines a CR structure on M, of CR dimension n. CR manifolds of type $(n,1)$, and therefore of real dimension $2n+1$, are also said to be of *hypersurface type.* Let us briefly discuss the case of real submanifolds $M \subset \mathbb{C}^{n+1}$ of arbitrary real codimension $m \geq 1$. If this is the case then for any $x_0 \in M$ there is an open subset $U \subseteq \mathbb{C}^{n+1}$ such that $x_0 \in U$ and there is a C^∞ map $\rho = (\rho_1, \cdots, \rho_m) : U \to \mathbb{R}^m$ such that

$$M \cap U = \{x \in U : \rho(x) = 0\}$$

and $d_x\rho : T_x(\mathbb{C}^{n+1}) \to T_{\rho(x)}(\mathbb{R}^m)$ has rank m for any $x \in M \cap U$ (equivalently $(d\rho_1 \wedge \cdots \wedge d\rho_m)_x \neq 0$ for any $x \in M \cap U$). We may still set as above

$$(2.3) \qquad (d_x j)T_{1,0}(M)_x = T^{1,0}(\mathbb{C}^{n+1})_x \cap [(d_x j)T_x(M) \otimes_{\mathbb{R}} \mathbb{C}], \quad x \in M,$$

where $j : M \to \mathbb{C}^{n+1}$ is the canonical inclusion, yet in general the dimension of $T_{1,0}(M)_x$ may vary with x. If the dimension is constant then $T_{1,0}(M)$ is a CR structure on M and

$$(2.4) \qquad (d_x j)T_{1,0}(M)_x = T^{1,0}(\mathbb{C}^{n+1})_x \cap \operatorname{Ker}(\partial\rho)_x\,, \quad x \in M \cap U,$$

where $j : M \cap U \subset \mathbb{C}^{n+1}$ and $\partial\rho = (\partial\rho_1, \cdots, \partial\rho_m)$. We set as customary

$$\partial\rho_j = (d\rho_j) \circ \pi^{1,0}, \quad \overline{\partial}\rho_j = (d\rho_j) \circ \pi^{0,1}$$

and $\pi^{1,0} : T(\mathbb{C}^{n+1}) \otimes \mathbb{C} \to T^{1,0}(\mathbb{C}^{n+1})$ and $\pi^{0,1} : T(\mathbb{C}^{n+1}) \otimes \mathbb{C} \to T^{0,1}(\mathbb{C}^{n+1})$ are the canonical projections associated to the direct sum decomposition $T(\mathbb{C}^{n+1}) \otimes \mathbb{C} = T^{1,0}(\mathbb{C}^{n+1}) \oplus T^{0,1}(\mathbb{C}^{n+1})$. To prove (2.4) let $Z \in T_{1,0}(M)_x$ with $x \in M \cap U$ then (by (2.3)) $(d_x j)Z \in T^{1,0}(\mathbb{C}^{n+1})_x$ and

$$(\partial\rho_j)_x(d_x j)Z = (d_x(\rho_j \circ j))Z = 0$$

because $\rho_j \circ j = 0$. Viceversa, if $W \in T^{1,0}(\mathbb{C}^{n+1})_x$ is a complex tangent vector such that $(\partial\rho)_x W = 0$ then

$$W(\rho_j) = (d_x\rho_j)W = (\partial\rho_j)_x W + (\overline{\partial}\rho_j)_x W = 0$$

hence W is tangential i.e. $W = (d_x j)Z$ for some $Z \in T_x(M) \otimes_{\mathbb{R}} \mathbb{C}$. Then (again by (2.3)) $Z \in T_{1,0}(M)_x$. □

It should be observed that, unlike the case of complex manifolds, the involutivity condition (2.2) doesn't imply in general the existence of special local coordinates, such as the local complex coordinates on a complex manifold, a question to which we shall get back later on. The overall conclusion is that the integrability property (2.2) is rather formal and the question arises whether the whole theory might not be developed for an *almost CR structure* $T_{1,0}(M)$ that is one for which but the axiom (2.1) is required to hold. This is indeed the case when dealing with contact Riemannian manifolds (cf. e.g. D.E. Blair, [**43**]) which possess natural almost CR structures which aren't in general integrable. As shown for instance by E. Barletta et al., [**24**], and D.E. Blair et al., [**45**], a theory similar to that developed for studying pseudohermitian geometry (cf. S.M. Webster, [**250**]) may indeed be built.

Let $(M, T_{1,0}(M))$ be a CR manifold of type (n,k). The *Levi distribution* is the subbundle $H(M) \subset T(M)$ defined by

$$H(M) := \operatorname{Re}\{T_{1,0}(M) \oplus T_{0,1}(M)\}.$$

It carries a natural complex structure i.e. a bundle endomorphism $J : H(M) \to H(M)$ such that $J^2 = -I$. It is given by $J(Z + \overline{Z}) = i(Z - \overline{Z})$ for any $Z \in T_{1,0}(M)$

(with $i = \sqrt{-1}$). The *conormal bundle* of $H(M)$ is

$$H(M)_x^\perp = \{\omega \in T_x^*(M) : \mathrm{Ker}(\omega) \supseteq H(M)_x\}, \quad x \in M.$$

When $k = 1$, i.e. when M is a CR manifold of hypersurface type, $H(M)^\perp \to M$ is a real line subbundle of the cotangent bundle. If this is the case and M is assumed to be oriented then $H(M)^\perp$ is oriented hence trivial. Indeed the orientation of M induces an orientation of $T(M)$ as a vector bundle; also $H(M)$ is oriented by its complex structure J so that the quotient bundle $T(M)/H(M)$ is oriented, as well. Finally there is a bundle isomorphism $H(M)^\perp \approx T(M)/H(M)$ (although in general there is no *canonical* choice of such an isomorphism) and oriented line bundles over connected manifolds are known to be trivial i.e. $H(M)^\perp \approx M \times \mathbb{R}$. Consequently there exist globally defined nowhere zero sections $\theta \in \Gamma^\infty(H(M)^\perp)$ each of which is called a *pseudohermitian structure* on M.

Let M be a CR manifold of type (n,k). The *Levi form* is given by

$$L_x(z,w) := \pi_x[Z, \overline{W}]_x\,, \quad z,w \in T_{1,0}(M)_x, \quad x \in M.$$

Here $Z, W \in T_{1,0}(M)$ are complex vector fields extending z, w i.e. $Z_x = z$ and $W_x = w$. Also $\pi : T(M) \otimes \mathbb{C} \to [T(M) \otimes \mathbb{C}]/[H(M) \otimes \mathbb{C}]$ is the natural projection. The definition of $L_x(z,w)$ doesn't depend upon the choice of extensions because of the integrability property (2.2). We say M is *nondegenerate* if L is nondegenerate i.e. $L_x(z,w) = 0$ for any $w \in T_{1,0}(M)_x$ implies $z = 0$. There is yet another approach to the Levi form and nondegeneracy when $k = 1$ and M is orientable. Indeed we may consider a pseudohermitian structure θ on M and set

$$L_\theta(Z, \overline{W}) = -i(d\theta)(Z, \overline{W}), \quad Z, W \in T_{1,0}(M),$$

and call M nondegenerate when L_θ is nondegenerate. It is easily shown that L_θ and L coincide up to a bundle isomorphism. Although L_θ is not uniquely determined (in fact it is determined only up to a conformal factor i.e. if $\hat{\theta} = f\theta$ for some C^∞ function $f : M \to \mathbb{R} \setminus \{0\}$ then $L_{\hat{\theta}} = fL_\theta$) nondegeneracy is non ambiguously defined. We say nondegeneracy is a *CR invariant* notion. More precisely, an object defined in terms of $T_{1,0}(M)$ and θ is a *CR invariant* if it is invariant under a transformation of the pseudohermitian structure of the form $\hat{\theta} = f\theta$. A less naive approach requires the notion of CR automorphism.

Let M and N be CR manifolds, of arbitrary but fixed types. A *CR map* is a C^∞ map $f : M \to N$ such that $(d_x f)T_{1,0}(M)_x \subseteq T_{1,0}(N)_{f(x)}$ for any $x \in M$. A *CR transformation* is a C^∞ diffeomorphism and a CR map. A *CR automorphism* of M is a CR transformation of M in itself. The set $\mathrm{Aut}_{CR}(M)$ of all CR automorphisms of M is a group under composition. Then CR invariants are invariants under the natural action of $\mathrm{Aut}_{CR}(M)$ on M.

As a first example of foliations naturally occurring on a CR manifold let us look at

EXAMPLE 2.1. (Levi foliations)
Let M be a real submanifold of $\mathbb{C}^n$, of real codimension m, so that for any $x_0 \in M$ there is an open neighborhood $U \subseteq \mathbb{C}^n$, $x_0 \in U$, and a C^∞ map $\rho = (\rho_1, \cdots, \rho_m) : U \to \mathbb{R}^m$ such that $M \cap U = \{x \in U : \rho(x) = 0\}$ and the map $d_x\rho : T_x(\mathbb{C}^n) \to T_{\rho(x)}(\mathbb{R}^m)$ has rank m for any $x \in M \cap U$. Let $\mathcal{F}$ be a foliation of M by k-dimensional complex submanifolds of $\mathbb{C}^n$. Let $L \in M/\mathcal{F}$ be a leaf of $\mathcal{F}$ and $x_0 \in L \cap U$. Let $\varphi : V \to \mathbb{C}^k$ be a local chart on L such that $x_0 \in V \subseteq L \cap U$. As L is a complex

submanifold of $\mathbb{C}^n$ the map $f : \Omega \to \mathbb{C}^n$ given by

$$f(\zeta) = (\varphi : V \to \Omega)^{-1}(\zeta), \quad \zeta \in \Omega = \varphi(V),$$

is holomorphic. Obviously $f(\Omega) \subseteq L \cap U \subset M \cap U$ hence $\rho_j \circ f = 0$ on Ω where from

$$(d_z \rho_j) \circ (d_\zeta f) = 0, \quad z = f(\zeta), \quad \zeta \in \Omega.$$

As $d = \partial + \overline{\partial}$ the preceding identity may also be written as

$$(\partial \rho_j)_z \circ (d_\zeta f) + (\overline{\partial} \rho_j)_z \circ (d_\zeta f) = 0.$$

Let us apply this identity at $v \in T^{1,0}(\mathbb{C}^k)_\zeta$. Yet $(d_\zeta f)v \in T^{1,0}(\mathbb{C}^n)_z$ (as f is holomorphic) and $(\overline{\partial}\rho_j)_z$ vanishes on complex vectors of type $(1,0)$ hence

$$(\partial \rho_j)_z (d_\zeta f) v = 0.$$

On the other hand $(d_\zeta f) T^{0,1}(\mathbb{C}^k)_\zeta \subset T^{0,1}(\mathbb{C}^n)_z$ (again because f is holomorphic) and $(\partial \rho_j)_z$ vanishes on complex vectors of type $(0,1)$ hence we also have for free $(\partial \rho_j)_z (d_\zeta f)\overline{v} = 0$. We may conclude that

$$(\partial \rho_j)_{f(\zeta)} \circ (d_\zeta f) = 0, \quad \zeta \in \Omega, \quad 1 \le j \le m. \tag{2.5}$$

The simple relations (2.5) are a necessary first order differential condition on ρ_j for the existence of a complex foliation $\mathcal{F}$ on M as above (discovered by M. Freeman, [**112**], p. 1). Moreover (2.5) may also be written

$$f^* \partial \rho_j = 0$$

and applying the exterior differentiation operator

$$d\,(f^* \partial \rho_j) = 0.$$

Yet the pullback by f and exterior differentiation commute so that (again by $d = \partial + \overline{\partial}$ and by $\partial^2 = 0$, $\overline{\partial}\partial = -\partial\overline{\partial}$)

$$f^*\, \partial \overline{\partial} \rho_j = 0, \quad 1 \le j \le m. \tag{2.6}$$

This is a second set of necessary differential conditions to be satisfied by ρ_j in order that M be foliated by complex k-submanifolds of $\mathbb{C}^n$ (due again to [**112**], p. 2). Let us assume that M is a generic CR submanifold (cf. e.g. Definition 5 in [**50**], p. 102) of $\mathbb{C}^n$. Let $N_x(M)$ be the orthogonal complement of $T_x(M)$ in $T_x(\mathbb{C}^n)$ for any $x \in M$ (the normal space at x) and let $\tilde{\mathcal{L}}_x$ be the extrinsic Levi form at x i.e.

$$\tilde{\mathcal{L}}_x(w) = \frac{1}{2i} \pi_x \left(J[Z, \overline{Z}] \right)_x, \quad w \in T_{1,0}(M)_x,$$

where $\pi_x : T_x(M) \to N_x(M)$ is the natural projection (associated to the direct sum decomposition $T_x(\mathbb{C}^n) = T_x(M) \oplus N_x(M)$) and Z is any local C^∞ section in $T_{1,0}(M)$ extending w i.e. $Z_x = w$. Also J denotes the complex structure on $\mathbb{C}^n$. Given a point $x_0 \in M$ let $U \subseteq \mathbb{C}^n$ be a neighborhood of x_0 and $\rho \in C^\infty(U, \mathbb{R}^m)$ a defining function of M such that $\{\nabla \rho_1(x_0), \cdots, \nabla \rho_m(x_0)\}$ is a orthonormal basis of $N_{x_0}(M)$ (both the gradient $\nabla \rho_j$ of ρ_j and orthonormality are meant with respect to the canonical flat Euclidean metric on $\mathbb{R}^{2n}$). Then (by Theorem 1 in [**50**], p. 160) the extrinsic Levi form at x_0 is given by

$$\tilde{\mathcal{L}}_{x_0}(w) = -\sum_{j=1}^m \left(\frac{\partial^2 \rho_j}{\partial z^A \partial \overline{z}^B}(x_0) w^A \overline{w}^B \right) \nabla \rho_j(x_0),$$

for any $w = w^A(\partial/\partial z^A)_{x_0} \in T_{1,0}(M)_{x_0}$. Here $(z^1, \cdots, z^n)$ are the natural complex coordinates on $\mathbb{C}^n$. Then (2.6) is equivalent to

$$\tilde{\mathcal{L}}_{x_0}(w) = 0, \quad w \in T^{1,0}(L)_{x_0}, \quad x_0 \in L,$$

i.e. the extrinsic Levi form is degenerate and the holomorphic tangent space to a leaf of $\mathcal{F}$ is contained in the null space of the Levi form. For this reason the foliations of M by complex submanifolds of $\mathbb{C}^n$ as above are called *Levi foliations.* Levi foliations are discussed in detail in Chapters 3 and 4 of this monograph. □

Let M be a CR manifold of CR dimension n and of CR codimension $k = 1$. If M is nondegenerate then it may be shown that each pseudohermitian structure θ is a *contact form* on M i.e. $\theta \wedge (d\theta)^n$ is a volume form on M. A real version of the Levi form L_θ is also of common use, namely we set

$$G_\theta(X,Y) = (d\theta)(X, JY), \quad X, Y \in H(M).$$

Clearly L_θ and the extension by $\mathbb{C}$-linearity of G_θ to $H(M) \otimes \mathbb{C}$ coincide on $T_{1,0}(M) \otimes T_{0,1}(M)$. It is easily verified that G_θ is a semi-Riemannian bundle metric in $H(M)$ and that

$$G_\theta(JX, JY) = G_\theta(X,Y), \quad X, Y \in H(M).$$

A very important consequence of nondegeneracy is that once a contact form θ is fixed there is a unique nowhere zero tangent vector field T on M such that

$$\theta(T) = 1, \quad T \rfloor d\theta = 0.$$

T is the *characteristic direction* of $d\theta$. It immediately furnishes a natural complement to $H(M)$ in $T(M)$ i.e. one has the direct sum decomposition $T(M) = H(M) \oplus \mathbb{R}T$ and it allows us to extend G_θ to a natural semi-Riemannian metric on M namely the *Webster metric* g_θ given by

$$g_\theta(X,Y) = G_\theta(X,Y), \quad g_\theta(X,T) = 0, \quad g_\theta(T,T) = 1,$$

for any $X, Y \in H(M)$. The Webster metric is a semi-Riemannian metric on M of index $2s$ where (r,s) is the signature of L_θ $(r+s=n)$. By a fundamental result due to N. Tanaka, [**235**], and S.M. Webster, [**250**], for any nondegenerate CR manifold of hypersuperface type on which a contact form θ has been fixed there is a unique linear connection ∇ on M (the *Tanaka-Webster connection* of (M,θ)) such that i) $H(M)$ is parallel with respect to ∇, ii) $\nabla J = 0$ and $\nabla g_\theta = 0$, and iii) the torsion T_∇ of ∇ is *pure* i.e.

$$T_\nabla(Z,W) = 0, \quad T_\nabla(Z,\overline{W}) = 2iL_\theta(Z,\overline{W})T, \quad Z, W \in T_{1,0}(M), \tag{2.7}$$

$$\tau \circ J + J \circ \tau = 0, \tag{2.8}$$

where $\tau(X) := T_\nabla(T,X)$ for any $X \in T(M)$. τ is the *pseudohermitian torsion* of the Tanaka-Webster connection. It should be underlined that both the Levi form and the torsion of just any linear connection parallelizing $H(M)$ may be looked at as obstructions towards the integrability of $H(M)$. Indeed $L_\theta = 0$ (i.e. M is *Levi flat*) if and only if $H(M)$ is involutive, as it may be easily checked (see also Theorem 3.1 in Chapter 3 of this monograph). Also, by an elementary result in differential geometry, any distribution which is parallel with respect to a torsion-free linear connection is involutive. On the other hand, under the assumptions of the Tanaka-Webster result, $H(M)$ is not involutive as L_θ is nondegenerate and in particular $L_{\theta,x} \neq 0$ for any $x \in M$. Therefore it comes to no surprise the fact that the torsion of the Tanaka-Wester connection encodes the Levi form, cf. our (2.7).

An orientable CR manifold of hypersurface type is *strictly pseudoconvex* if L_θ is positive definite for some θ. An important example of strictly pseudoconvex CR manifold is the Heisenberg group. We recall it in some detail. Let $\mathbb{H}_n = \mathbb{C}^n \times \mathbb{R}$ be endowed with the natural coordinates $(z,t) = (z^1, \cdots, z^n, t)$. $\mathbb{H}_n$ is a group with the group law

$$(z,t)\cdot(w,s) = (z+w, t+s+2\mathrm{Im}\langle z,w\rangle),$$

where $\langle z,w\rangle = \delta_{jk} z^j \overline{w^k}$. This makes $\mathbb{H}_n$ into a Lie group referred to as the *Heisenberg group*. Let us consider the complex vector fields on $\mathbb{H}_n$

$$T_j = \frac{\partial}{\partial z^j} + i\,\overline{z}^j\,\frac{\partial}{\partial t} \tag{2.9}$$

where $\partial/\partial z^j = \frac{1}{2}\left(\partial/\partial x^j - i\,\partial/\partial y^j\right)$ and $z^j = x^j + iy^j$, $1 \leq j \leq n$. Let us define $T_{1,0}(\mathbb{H}_n)_{(z,t)}$ as the space spanned by the $T_{j,(z,t)}$'s i.e.

$$T_{1,0}(\mathbb{H}_n)_{(z,t)} = \sum_{j=1}^{n} \mathbb{C}T_{j,(z,t)}, \quad (z,t) \in \mathbb{H}_n. \tag{2.10}$$

Since $[T_j, T_k] = 0$, $1 \leq j,k \leq n$, it follows that $(\mathbb{H}_n, T_{1,0}(\mathbb{H}_n))$ is a CR manifold of hypersurface type. Next let us consider the real 1-form θ_0 on $\mathbb{H}_n$ defined by

$$\theta_0 = dt + i\sum_{j=1}^{n}\left(z^j d\overline{z}^j - \overline{z}^j dz^j\right). \tag{2.11}$$

Then θ_0 is a pseudohermitian structure on $(\mathbb{H}_n, T_{1,0}(\mathbb{H}_n))$. By differentiating (2.11) we obtain $d\theta_0 = 2i\sum_{j=1}^{n} dz^j \wedge d\overline{z}^j$. Then $L_{\theta_0}(T_j, T_{\overline{k}}) = \delta_{jk}$ where $T_{\overline{j}} = \overline{T_j}$, $1 \leq j \leq n$, i.e. L_{θ_0} is positive definite. It may be easily checked that $T = \partial/\partial t$ is the characteristic direction of $d\theta_0$.

We continue this brief preparation of CR geometry with a few remarks regarding the tangential Cauchy-Riemann equations. Let $M \subset \mathbb{C}^{n+1}$ be a real hypersurface given by the defining function ρ i.e. we are given a function $\rho \in C^\infty(\mathbb{C}^{n+1})$ such that $(\rho_{z_1}(z), \cdots, \rho_{z_{n+1}}(z)) \neq 0$ for any $z \in M$ and $M = \{z \in \mathbb{C}^{n+1} : \rho(z) = 0\}$. Here $(z_1, \cdots, z_{n+1})$ are the natural complex coordinates on $\mathbb{C}^{n+1}$ and $\rho_{z_j} = \partial\rho/\partial z_j$ for $1 \leq j \leq n+1$. Clearly M is a CR manifold of type $(n,1)$ with the *induced CR structure* $T_{1,0}(M)$ given by

$$(d_x j)T_{1,0}(M)_x = T^{1,0}(\mathbb{C}^{n+1})_x \cap [(d_x j)T_x(M) \otimes_{\mathbb{R}} \mathbb{C}], \quad x \in M.$$

Here $j : M \hookrightarrow \mathbb{C}^{n+1}$ is the inclusion and $T^{1,0}(\mathbb{C}^{n+1})_x$ is the span of $\{(\partial/\partial z_j)(x) : 1 \leq j \leq n+1\}$. If $\{P_\alpha : 1 \leq \alpha \leq n\}$ is a local frame of $T_{1,0}(M)$ defined on the open set U then

$$P_\alpha = \sum_{j=1}^{n+1} a_{\alpha j}\,\frac{\partial}{\partial z_j},$$

for some C^∞ functions $a_{\alpha j} : U \to \mathbb{C}$ such that

$$\sum_{j=1}^{n+1} a_{\alpha j}(z)\frac{\partial\rho}{\partial z_j}(z) = 0, \quad z \in U.$$

The first order system of PDEs

$$\overline{P}_\alpha u = 0, \quad 1 \leq \alpha \leq n, \tag{2.12}$$

is the *tangential Cauchy-Riemann system* on U and a C^1 function $u : U \to \mathbb{C}$ satisfying (2.12) is a *CR function* on U. A global approach on a not necessarily immersed CR manifold is as follows. Let M be nondegenerate CR manifold, θ a contact form on M, and T the corresponding characteristic direction. A complex s-form ω on M is a *$(0,s)$-form*, or a *form of type* $(0,s)$, if $T_{1,0}(M) \rfloor \omega = 0$ and $T \rfloor \omega = 0$. Locally, if $\{P_\alpha : 1 \leq \alpha \leq n\}$ is a local frame of $T_{1,0}(M)$ defined on the open set $U \subseteq M$ and $\{\theta^\alpha : 1 \leq \alpha \leq n\}$ are the locally defined complex 1-forms given by

$$\theta^\alpha(P_\beta) = \delta^\alpha_\beta\,, \quad \theta^\alpha(P_{\overline{\beta}}) = 0, \quad \theta^\alpha(T) = 0, \tag{2.13}$$

then a $(0,s)$-form ω is given by

$$\omega = f_{\overline{\alpha}_1,\cdots,\overline{\alpha}_s}\,\theta^{\overline{\alpha}_1} \wedge \cdots \wedge \theta^{\overline{\alpha}_s}$$

on U, for some C^∞ functions $f_{\overline{\alpha}_1,\cdots,\overline{\alpha}_s} : U \to \mathbb{C}$. Here $P_{\overline{\alpha}} = \overline{P}_\alpha$ and $\theta^{\overline{\alpha}} = \overline{\theta^\alpha}$. The local coframe $\{\theta^\alpha : 1 \leq \alpha \leq n\}$ given by (2.13) is commonly referred to as the *adpated coframe* associated to $\{P_\alpha : 1 \leq \alpha \leq n\}$ and θ. Let $\Lambda^{0,s}(M) \to M$ be the bundle of all $(0,s)$-forms on M and let us set $\Omega^{0,s}(M) = \Gamma^\infty(\Lambda^{0,s}(M))$. Let us consider the first order differential operator

$$\overline{\partial}_M : \Omega^{0,s}(M) \to \Omega^{0,s+1}(M), \quad s \geq 0,$$

built as follows. If $\omega \in \Omega^{0,s}(M)$ then $\overline{\partial}_M\omega$ is the unique $(0,s+1)$-form on M coinciding with $d\omega$ on $T_{0,1}(M) \otimes \cdots \otimes T_{0,1}(M)$ ($s+1$ terms). For instance, if $f \in \Omega^{0,0}(M) = C^\infty(M) \otimes \mathbb{C}$ then $(\overline{\partial}_M f)(\overline{Z}) = \overline{Z}(f)$ for any $Z \in T_{1,0}(M)$. $\overline{\partial}_M$ is the *tangential Cauchy-Riemann operator* and

$$\overline{\partial}_M u = 0 \tag{2.14}$$

are the *tangential Cauchy-Riemann equations* on M. A *CR function* is a C^1 solution $u : M \to \mathbb{C}$ to (2.14). Clearly, if $\{P_\alpha : 1 \leq \alpha \leq n\}$ is a local frame of $T_{1,0}(M)$ then (2.14) assume the form (2.12) on U. It may be argued that the so called global approach is less general as one assumed nondegeneracy. Indeed the notion of tangential Cauchy-Riemann operator may be introduced on an arbitrary CR manifold of type (n,k) as the differential operator

$$\overline{\partial}_M : \Gamma^\infty(\Lambda^s T_{0,1}(M)^*) \to \Gamma^\infty(\Lambda^{0,s+1} T_{0,1}(M)^*)$$

given by

$$(\overline{\partial}_M\omega)(\overline{Z}_0, \overline{Z}_1, \cdots, \overline{Z}_s) =$$

$$= \sum_{j=0}^{n} (-1)^j \overline{Z}_j(\omega(\overline{Z}_1, \cdots, \hat{\overline{Z}}_j, \cdots, \overline{Z}_n) +$$

$$+ \sum_{0 \leq j < k \leq n} (-1)^{j+k} \omega([\overline{Z}_j, \overline{Z}_k], \overline{Z}_0, \cdots, \hat{\overline{Z}}_j, \cdots, \hat{\overline{Z}}_k, \cdots, \overline{Z}_n)$$

for any $Z_j \in T_{1,0}(M)$, $0 \leq j \leq n$. A hat indicates as usual the suppression of a term. Here ω is not a differential form but only a field of skew-symmetric multilinear forms

$$\omega_x \in T_{0,1}(M)_x \otimes \cdots \otimes T_{0,1}(M)_x\,, \quad x \in M,$$

and $(\overline{\partial}_M\omega)(\overline{Z}_0, \cdots, \overline{Z}_n)$ is well defined duc to the involutivity property of $T_{0,1}(M)$. It is easy to see that when M is a nondegenerate CR manifold of type $(n,1)$ and a contact form θ is fixed there is a natural[1] bundle monomorphism from $\Lambda^s T_{0,1}(M)^*$ into $\Lambda^{0,s}(M)$ and the two approaches to $\overline{\partial}_M$ actually coincide. Since the point of view of the present monograph is that of pseudohermitian geometry we shall assume in most cases that M is a nondegenerate CR manifold of hypersurface type, a case in which the domain $\Omega^{0,s}(M)$ of the tangential Cauchy-Riemann operator consists of genuine differential forms. The space of CR functions of class C^r on M is denoted by $CR^r(M)$.

Let M be a CR manifold, of arbitrary but fixed type (n, k).

DEFINITION 2.2. A smooth vector field $X \in \mathcal{X}(M)$ is called an *infinitesimal CR automorphism* if the local 1-parameter group $\{\varphi_t\}_{|t|<\epsilon}$ of local transformations got by (locally) integrating X consists of (local) CR automorphisms $\varphi_t : U_t \to \varphi_t(U_t)$ i.e. for each $|t| < \epsilon$ one has $(d_x\varphi_t)T_{1,0}(M)_x = T_{1,0}(M)_{\varphi_t(x)}$ for any $x \in U_t$. □

We denote by $\text{aut}(M, x)$ the space of all infinitesimal CR automorphisms defined on some open neighborhood of the point $x \in M$. If $M \subset \mathbb{C}^{n+1}$ is a CR submanifold passing through the origin $(0 \in M)$ we also set $\text{aut}(M) = \text{aut}(M, 0)$.

DEFINITION 2.3. (M.S. Baouendi & L.P. Rothschild & F. Treves, [**11**])
A real hypersurface $M \subset \mathbb{C}^{n+1}$ is called *rigid* at a point $x_0 \in M$ if there are complex coordinates $(z, w) = (z_1, \cdots, z_n, w)$ in a neighborhood $U \subseteq \mathbb{C}^{n+1}$ of x_0 such that

$$M \cap U = \{(z, w) \in U : \text{Im}(w) = F(z, \overline{z})\}$$

for some smooth function F (not necessarily holomorphic[2]). Moreover M is *rigid* if it is rigid at each point $x_0 \in M$. Among the rigid hypersurfaced the *homogeneous* ones are those locally equivalent to a hypersurface of the form $\text{Im}(w) = p(z, \overline{z})$ where $p \in \mathbb{C}[z, \overline{z}]$ is a homogenous polynomial. □

Rigidity is known (cf. [**11**]) to be equivalent to the existence of a transversal infinitesimal CR automorphism on M. Rigid hypersurfaces are referred to as *regular* by N. Tanaka, [**236**], and as *T-regular* by J.P. D'Angelo, [**77**].

Let $U \subseteq \mathbb{C}^{n+1}$ be an open set and $Z \in \mathcal{X}(U)$ a holomorphic vector field i.e. $Z = f^A \partial/\partial z^A$ and $f^A \in \mathcal{O}(U)$ are holomorphic functions. Then the local 1-parameter group of local transformations associated to the real vector field $X = \text{Re}(Z)$ consists of local biholomorphisms. Therefore if X is tangent to M then $X|_{M\cap U}$ is an infinitsimal CR automorphism on $M \cap U$. Let $\text{hol}(M)$ be the space of all infinitesimal CR automorphisms on M, defined on a neighborhood of the origin, which are of the form $X = \text{Re}(Z)$ for some holomorphic vector field Z as above. Thus $\text{hol}(M) \subset \text{aut}(M)$.

The following questions were posed by N.K. Stanton (in a series of papers [**226**]-[**229**])

Problem. (N.K. Stanton, *op. cit.*)

i) Given a real analytic real hypersurface $M \subset \mathbb{C}^2$ through the origin, compute $\text{aut}(M)$ provided that M is globally given by a rigid equation of the form $\text{Im}(w) = F(z, \overline{z})$.

ii) Characterize the real analytic homogenous hypersurfaces among the rigid ones.

---

[1] One identifies an element $\omega \in \Gamma^\infty(\Lambda^s T_{0,1}(M)^*)$ with the $(0, s)$-form $\tilde{\omega} \in \Omega^{0,s}(M)$ given by $\tilde{\omega} = \omega$ on $T_{0,1}(M) \otimes \cdots \otimes T_{0,1}(M)$ (s terms), $T_{1,0}(M) \rfloor \tilde{\omega} = 0$ and $T \rfloor \tilde{\omega} = 0$.

[2] A fact which is usually emphasized by writing $F(z, \overline{z})$ rather than $F(z)$.

iii) Let $M \subset \mathbb{C}^{n+1}$ be a rigid real analytic real hypersurface with $0 \in M$. When is $\text{aut}(M)$ finite dimensional? or (a weaker question) when is $\text{hol}(M)$ finite dimensional?

iv) Under the hypothesis of (iii) when $\text{hol}(M) = \text{aut}(M)$?

The answers to the above questions (cf. **[226]**-**[229]**) are related to the study of holomorphic degeneracy of a real hypersurfce at a point.

DEFINITION 2.4. (N.K. Stanton, **[227]**)
A real hypersurface $M \subset \mathbb{C}^{n+1}$ such that $0 \in M$ is called *holomorphically nondegenerate* at the origin if there is no nonzero holomorphic vector field tangent to M in a neighborhood of the origin. Otherwise M is *holomorphically degenerate* at 0. □

Holomorphic degeneracy is related to the local existence of Levi foliations. Precisely the following results are known

THEOREM 2.5. (N.K. Stanton, **[227]**) *Let $M \subset \mathbb{C}^{n+1}$ be a real analytic real hypersurfce with $0 \in M$. If M is essentially finite or has a somewhere nondegenerate Levi form then M is holomorphically nondegenerate. In particular if $M \subset \mathbb{C}^2$ then M is holomorphically nondegenerate if and only if M is not flat.*

It should be remarked that in higher dimensions holomorphic nondegeneracy doesn't coincide with nonflat, finite type or essentially finite. As to the terminology used in the preceding remark, we recall the following

DEFINITION 2.6. i) A hypersurface $M \subset \mathbb{C}^{n+1}$ is of *finite type* m at a point x if m is the smallest positive integer such that the tangent space $T_x(M)$ is spanned by commutators of length[3] m of sections in $H(M)$. If $m = \infty$ we say M is of *infinite type*. ii) Let $M \subset \mathbb{C}^{n+1}$ be a real analytic hypersurface through the origin, given by the equation $\rho(z, \overline{z}) = 0$, for some function $\rho(z, \tilde{z})$ of the arguments $z, \tilde{z} \in \mathbb{C}^{n+1}$. Let us set

$$\mathcal{V} = \{\xi \in \mathbb{C}^{n+1} : \rho(\xi, \eta) = 0, \quad \forall \eta \in \mathbb{C}^{n+1} \text{ with } \rho(0, \eta) = 0\}.$$

We say M is *essentially finite* at 0 if $\mathcal{V} = \{0\}$. iii) A real hypersurface $M \subset \mathbb{C}^{n+1}$ is said to be *flat* if M is CR equivalent to a real hyperplane in $\mathbb{C}^{n+1}$. □

The relationship among holomorphic degeneracy and foliation theory is examined in Section 3.4 of this monograph.

2.1. The normal bundle

Let $(M, T_{1,0}(M))$ be a strictly pseudoconvex CR manifold, of CR dimension n. Let θ be a contact form on M such that G_θ is positive definite. Let $\mathcal{F}$ be a codimension q foliation of M. Let us extend G_θ to the whole of $T(M)$ as the symmetric $(0,2)$-tensor field $\tilde{G}_\theta$ built by requesting that T is orthogonal to each $V \in T(M)$ i.e.

$$\tilde{G}_\theta(X, T) = 0, \quad X \in T(M).$$

In particular $\tilde{G}_\theta(T, T) = 0$ i.e. $\tilde{G}_\theta$ is a degenerate metric. We start by studying the geometry of $\mathcal{F}$ in $(M, \tilde{G}_\theta)$. The following concepts and terminology should be kept in mind as to degenerate metrics (cf. **[97]**). Let $E \to M$ be a real vector bundle of rank q ($q \geq 2$) over a C^∞ manifold M.

[3]A commutator of the form $[X, Y]$ has length 2.

DEFINITION 2.7. By a (*bundle*) *metric* in E we intend a C^∞ section $g : x \in M \mapsto g_x \in E_x^* \otimes_{\mathbb{R}} E_x^*$ in $E^* \otimes E^*$ such that g_x is symmetric and has constant index $\text{ind}(g_x) = \sigma$, for any $x \in M$. If each g_x is nondegenerate and $1 \leq \sigma \leq q-1$ (respectively if each g_x is positive definite) then g is a *semi-Riemannian* metric (respectively a *Riemannian* metric) in E. □

The index $\text{ind}(g_x)$ is the dimension $\dim_{\mathbb{R}} W$ of the highest dimensional subspace $W \subseteq E_x$ such that g_x is negative definite on W. An arbitrary metric g in E is therefore allowed to be degenerate i.e. $\text{Rad}\,(E,g)_x \neq (0)$ for some $x \in M$ where

$$\text{Rad}\,(E,g)_x := \{v \in E_x : g_x(v,w) = 0, \quad w \in E_x\}, \quad x \in M.$$

It is customary to work under the assumption that Rad E is a subbundle of E of rank ρ with $1 \leq \rho \leq q$. If this is the case then

DEFINITION 2.8. g is referred to as a *ρ-lightlike* metric while Rad E is called the *radical bundle* of (E,g). □

EXAMPLE 2.9. Let (N,G) be a n-dimensional semi-Riemannian manifold ($n \geq 2$) where G is the semi-Riemannian metric of index $1 \leq s \leq n-1$ and let M be a m-dimensional submanifold of N. Let us consider the bundle $E = T(M)$ and the metric $g = \iota^* G$ where $\iota : M \to N$ is the inclusion. Let us assume the distribution $\text{Rad}\,(E,g) = \text{Rad}\,(T(M),g)$ has constant rank ρ with $1 \leq \rho \leq \min\{m,k\}$ where $k = n - m$ is the codimension of M in N. Then g is a ρ-lightlike metric on M. If $T(M)^\perp = \{V \in T(N) : G(V,X) = 0, \quad X \in T(M)\}$ then $\text{Rad}\,(T(M),g) = T(M) \cap T(M)^\perp$. A differential geometric study of $\iota : M \to N$ similar to the geometry of the second fundamental form of an isometric immersion between two Riemannian manifolds has been performed in [**97**]. □

Going back to $\tilde{G}_\theta$ note that

$$\text{Rad}\,(T(M),\tilde{G}_\theta) = \{X \in T(M) : \tilde{G}_\theta(X,Y) = 0, \quad Y \in T(M)\} = \mathbb{R}T$$

hence $\tilde{G}_\theta$ is a 1-lightlike metric on M. As a continuation of the ideas in [**97**] one wishes to study the geometry of the leaves of $\mathcal{F}$ in $(M,\tilde{G}_\theta)$. Let us consider

$$T(\mathcal{F})_0 := \{Y \in T(M) : \tilde{G}_\theta(X,Y) = 0, \quad \text{for all} \quad X \in T(\mathcal{F})\}.$$

We collect a few elementary facts in the following

PROPOSITION 2.10. (S. Nishikawa et al., [**87**])
The tangent bundle $T(\mathcal{F})$ is nondegenerate in $(T(M),\tilde{G}_\theta)$ if and only if the characteristic direction T of (M,θ) is transverse to $T(\mathcal{F})$. In general, let $T(\mathcal{F})_{H(M)} = \{\pi_H X : X \in T(\mathcal{F})\}$ be the projection of $T(\mathcal{F})$ on $H(M)$. Then

$$T(\mathcal{F})_0 = \left[T(\mathcal{F})_{H(M)}\right]^\perp \oplus \mathbb{R}T. \tag{2.15}$$

If T is tangent to $\mathcal{F}$ then
i) $T(\mathcal{F})_{H(M)} = H(M) \cap T(\mathcal{F})$,
ii) the natural bundle map $\sigma : \nu(\mathcal{F}) \to T(\mathcal{F})_0$ is a bundle monomorphism and corestricts to a bundle isomorphism

$$\nu(\mathcal{F}) \approx \left[T(\mathcal{F})_{H(M)}\right]^\perp,$$

and
iii) $H_\theta(r,s) := \tilde{G}_\theta(\sigma(r),\sigma(s))$, $r,s \in \nu(\mathcal{F})$, *is a Riemannian metric in* $\nu(\mathcal{F}) \to M$.

Here $\pi_H : T(M) \to H(M)$ is the natural projection associated to the direct sum decomposition $T(M) = H(M) \oplus \mathbb{R}T$. Also the orthogonal complement $[T(\mathcal{F})_{H(M)}]^{\perp}$ of $T(\mathcal{F})_{H(M)}$ is taken in $(H(M), G_\theta)$. The analysis of the case where $T(\mathcal{F})$ is degenerate is left as an open problem.

Proof of Proposition 2.10. Let us prove the first statement in Proposition 2.10. Assume that T is transverse to $T(\mathcal{F})$. Let $X \in T(\mathcal{F})$ such that $\tilde{G}_\theta(X, Y) = 0$ for any $Y \in T(\mathcal{F})$. Then

$$0 = \tilde{G}_\theta(X, X) = G_\theta(\pi_H X, \pi_H X) = \|\pi_H X\|^2,$$

i.e. $\pi_H X = 0$. Thus $T(\mathcal{F}) \ni X = \theta(X)T$, which yields $\theta(X) = 0$, i.e. $X = 0$.

Viceversa, assume that $T(\mathcal{F})$ is nondegenerate in $(T(M), \tilde{G}_\theta)$. The proof is by contradiction. If $T_x \in T(\mathcal{F})_x$, for some $x \in M$, then $\tilde{G}_{\theta,x}(v, T_x) = 0$, for any $v \in T_x(M) \supset T(\mathcal{F})_x$, yields [by the nondegeneracy of $T(\mathcal{F})_x$ in $(T_x(M), \tilde{G}_{\theta,x})$] $T_x = 0$, a contradiction.

To prove the second statement in Proposition 2.10, let $T(\mathcal{F})_{H(M)}$ be the projection of $T(\mathcal{F})$ on $H(M)$ i.e.

$$T(\mathcal{F})_{H(M)} = \{X - \theta(X)T : X \in T(\mathcal{F})\}.$$

Note that

$$\left[T(\mathcal{F})_{H(M)}\right]^{\perp} \cap \mathbb{R}T \subseteq H(M) \cap \mathbb{R}T = (0),$$

hence the sum in (2.15) is direct. To check (2.15) note first that $T \in T(\mathcal{F})_0$. Next if $Z \in \left[T(\mathcal{F})_{H(M)}\right]^{\perp}$ then $\tilde{G}_\theta(Z, Y) = 0$ for any $Y \in T(\mathcal{F})_{H(M)}$, in particular for $Y := X - \theta(X)T$, with $X \in T(\mathcal{F})$. Thus

$$0 = \tilde{G}_\theta(Z, Y) = \tilde{G}_\theta(Z, X),$$

i.e. $Z \in T(\mathcal{F})_0$. To check the opposite inclusion, let $Z \in T(\mathcal{F})_0 \subset H(M) \oplus \mathbb{R}T$. Then $Z = Y + fT$, for some $Y \in H(M)$ and $f \in C^\infty(M)$. As $\tilde{G}_\theta(Z, X) = 0$ for any $X \in T(\mathcal{F})$, it follows that

$$G_\theta(Y, X - \theta(X)T) = \tilde{G}_\theta(Z, X) = 0,$$

i.e. $Y \in \left[T(\mathcal{F})_{H(M)}\right]^{\perp}$.

Consider the bundle map

$$\sigma_0 : \nu(\mathcal{F}) \to T(\mathcal{F})_0, \quad \sigma_0(\Pi Y) := (Y - \theta(Y)T)^{\perp}, \quad Y \in T(M).$$

Here $(Y - \theta(Y)T)^{\perp}$ is the $\left[T(\mathcal{F})_{H(M)}\right]^{\perp}$-component of $Y - \theta(Y)T$ in $H(M)$. To see that $\sigma_0(\Pi Y)$ is well defined, assume that $\Pi Y = \Pi Z$. Then $Y - Z \in T(\mathcal{F})$ hence $Y - Z - \theta(Y - Z)T \in T(\mathcal{F})_{H(M)}$, i.e. $(Y - Z - \theta(Y - Z)T)^{\perp} = 0$.

Assume now that $T \in T(\mathcal{F})$. The proof of (i) is immediate. To check that σ_0 is a bundle monomorphism, let $\sigma_0(\Pi Y) = 0$, i.e.

$$Y - \theta(Y)T \in T(\mathcal{F})_{H(M)} = H(M) \cap T(\mathcal{F}) \subseteq T(\mathcal{F}).$$

Thus [by $T \in T(\mathcal{F})$]

$$0 = \Pi(Y - \theta(Y)T) = \Pi Y.$$

The isomorphism claimed in (ii) of Theorem 2.10 follows by a dimension argument. Indeed, $\dim_{\mathbb{R}} \nu(\mathcal{F})_x = q$. Also $H(M) + T(\mathcal{F}) \supseteq H(M) + \mathbb{R}T = T(M)$ hence

$$2n + 1 = \dim_{\mathbb{R}} H(M)_x + \dim_{\mathbb{R}} T(\mathcal{F})_x - \dim_{\mathbb{R}} \left[H(M)_x \cap T(\mathcal{F})_x\right],$$

i.e. $\dim_{\mathbb{R}} \left[T(\mathcal{F})_{H(M)}\right]_x = 2n - q$, for any $x \in M$.

Let us prove (iii) in Proposition 2.10. As the image of σ_0 lies in $H(M)$, $H_\theta(r,r) = \|\sigma_0(r)\|^2 \geq 0$ and $= 0$ if and only if $Y - \theta(Y)T \in T(\mathcal{F})_{H(M)}$, for each $Y \in T(M)$ such that $\Pi(Y) = r$. Therefore $H_\theta(r,r) = 0$ if and only if $r \in \mathbb{R}\,\Pi T$. In particular, if $T \in T(\mathcal{F})$ then H_θ is a Riemannian metric in $\nu(\mathcal{F})$. Proposition 2.10 is proved.

REMARK 2.11. As the Webster metric g_θ is a Riemannian metric on M one may consider as well the normal bundle

$$T(\mathcal{F})^\perp := \{Y \in T(M) : g_\theta(Y,X) = 0, \text{ for all } X \in T(\mathcal{F})\}.$$

With the notations in Example 2.9 the portion of $T(\mathcal{F})^\perp$ over a leaf $L \in M/\mathcal{F}$ is the normal bundle $T(L)^\perp \to L$ of $\iota : L \hookrightarrow (M, g_\theta)$. The corresponding bundle isomorphism is

$$\sigma : \nu(\mathcal{F}) \to T(\mathcal{F})^\perp, \quad \sigma(\Pi\, Y) := Y^\perp,$$

where $Y^\perp$ is the $T(\mathcal{F})^\perp$-component of $Y \in T(M) = T(\mathcal{F}) \oplus T(\mathcal{F})^\perp$. Then the purpose of the remark is the following

PROPOSITION 2.12. *If $T \in T(\mathcal{F})$ then $T(\mathcal{F})^\perp = \left[T(\mathcal{F})_{H(M)}\right]^\perp$ and $\sigma = \sigma_0$. Also the metric induced by g_θ in $\nu(\mathcal{F})$ is precisely H_θ.*

Indeed let $Y \in T(\mathcal{F}^\perp$ i.e. $g_\theta(Y,X) = 0$ for any $X \in T(\mathcal{F})$. In particular $g_\theta(Y,T) = 0$, i.e. $Y \in H(M)$. Therefore

$$\tilde{G}_\theta(Y, X - \theta(X)T) = \tilde{G}_\theta(Y,X) = g_\theta(Y,X) - \underbrace{\theta(Y)}_{=0}\,\theta(X) = 0,$$

for any $X \in T(\mathcal{F})$, hence $T(\mathcal{F})^\perp \subseteq \left[T(\mathcal{F})_{H(M)}\right]^\perp$. The opposite inclusion may be proved in a similar manner. Next

$$\begin{aligned} \sigma(\Pi\, Y) = & \qquad\qquad\qquad\qquad (\text{as } T \in T(\mathcal{F})) \\ = \sigma(\Pi(Y - \theta(Y)T)) = (Y - \theta(Y)T)^\perp = & \ \left(\text{as } T(\mathcal{F})^\perp = [T(\mathcal{F})_{H(M)}]^\perp\right) \\ = (Y - \theta(Y)T)^\perp = \sigma_0(\Pi\, Y), & \end{aligned}$$

for any $Y \in T(M)$. □

2.2. Foliations of CR manifolds and the Fefferman metric

The first statement in Proposition 2.10 shows that, under the natural assumption that T is tangent to the leaves of $\mathcal{F}$, $T(\mathcal{F})$ is degenerate in $(T(M), \tilde{G}_\theta)$. However, the pullback of $\mathcal{F}$ to the (total space of the) principal S^1-bundle

$$C(M) := [K(M) \setminus \{\textit{zero section}\}]/\mathbb{R}_+$$

turns out to be nondegenerate in $(C(M), F_\theta)$, where F_θ is the Fefferman metric of (M, θ). One may see C. Fefferman, [**104**], and J.M. Lee, [**168**], or the monograph [**89**] for a detailed description of the Fefferman metric. Nevertheless, to facilitate reading we collect a few notions and results below.

DEFINITION 2.13. Let M be a $(2n+1)$-dimensional CR manifold of CR dimension n. A complex valued differential s-form ω on M is a $(r,0)$-form on M if $T_{0,1}(M) \rfloor \omega = 0$. Let $\Lambda^{r,0}(M) \to M$ be the bundle of all $(r,0)$-forms on M. Then $K(M) := \Lambda^{n+1,0}(M)$ is the *canonical line bundle* over M. □

A remark is in order. The concept of a $(r,0)$-form is clearly similar to that of a form of type $(r,0)$ on a complex n-dimensional complex manifold X. If $(U, z^1, \cdots, z^n)$ is a local system of complex coordinates on X then a form ω of type $(r,0)$ is locally represented as $\omega = f_{j_1 \cdots j_r}\, dz^{j_1} \wedge \cdots \wedge dz^{j_r}$ (for some C^∞ functions $f_{j_1 \cdots j_r} : U \to \mathbb{C}$) hence the top degree forms of type $(r,0)$ on a complex n-dimensional manifold are the forms of type $(n,0)$. The case of a CR manifold M of dimension $\dim_{\mathbb{R}} M = 2n+1$ and CR dimension n is rather different in this respect. Indeed, let us assume that M is nondegenerate and fix as customary a contact form θ and the corresponding characteristic direction T. If $\{\theta^\alpha : 1 \leq \alpha \leq n\}$ is the adapted local coframe associated to the local frame $\{T_\alpha : 1 \leq \alpha \leq n\}$ of $T_{1,0}(M)$ and to θ then a $(s,0)$-form on ω is a sum of monomials of the form

$$\theta^{\alpha_1} \wedge \cdots \wedge \theta^{\alpha_s}, \quad \theta \wedge \theta^{\alpha_1} \wedge \cdots \wedge \theta^{\alpha_{s-1}},$$

with locally defined $C^\infty(U) \otimes \mathbb{C}$-coefficients. It turns out that the top degree $(s,0)$-forms are the $(n+1,0)$-forms.

Let $\sigma : M \to K(M)$ be the zero section i.e. $\sigma(x) = 0_x \in K(M)_x$, $x \in M$. Let $(K(M) \setminus \sigma(M)) \times \mathbb{R}_+ \to K(M) \setminus \sigma(M)$ be the natural action of the multiplicative positive reals $\mathbb{R}_+ = (0, +\infty)$ on $K(M) \setminus \sigma(M)$ and let $C(M) = (K(M) \setminus \sigma(M))/\mathbb{R}_+$ be the quotient space. There is a natural action of the circle S^1 on $C(M)$ such that $C(M)$ becomes the total space of a principal S^1-bundle whose projection is denoted by $\pi : C(M) \to M$.

DEFINITION 2.14. The S^1-bundle $S^1 \to C(M) \xrightarrow{\pi} M$ is called the *canonical circle bundle* over M. □

The Fefferman metric is a remarkable Lorentz metric on $C(M)$.

DEFINITION 2.15. Let M be a strictly pseudoconvex CR manifold of CR dimension n and θ a contact form with G_θ positive definite. The *Fefferman metric* of (M, θ) is given by

$$F_\theta = \pi^* \tilde{G}_\theta + 2(\pi^*\theta) \odot \eta$$

where $\eta \in \Omega^1(C(M))$ is the globally defined real 1-form given by

$$\eta = \frac{1}{n+2} \left\{ d\gamma + \pi^* \left(i\, \omega_\alpha{}^\alpha - \frac{i}{2}\, h^{\alpha\overline{\beta}} dh_{\alpha\overline{\beta}} - \frac{1}{4(n+1)} \rho\, \theta \right) \right\}.$$

Here $\gamma : \pi^{-1}(U) \to \mathbb{R}$ is a local fibre coordinate. Moreover $h_{\alpha\overline{\beta}} = L_\theta(T_\alpha, T_{\overline{\beta}})$ and $\omega_\beta{}^\alpha$ are the connection 1-forms of the Tanaka-Webster connection i.e. $\nabla T_\beta = \omega_\beta{}^\alpha \otimes T_\alpha$, with respect to a local frame $\{T_\alpha : 1 \leq \alpha \leq n\}$ in $T_{1,0}(M)$ defined on the open set $U \subseteq M$. Finally $\rho = h^{\alpha\overline{\beta}} R_{\alpha\overline{\beta}}$ is the pseudohermitian scalar curvature of (M, θ). □

For the sake of completeness we recall that the pseudohermitian Ricci tensor $R_{\alpha\overline{\beta}}$ is given by

$$R_{\alpha\overline{\beta}} = \mathrm{Ric}(T_\alpha, T_{\overline{\beta}}),$$

$$\mathrm{Ric}(X, Y) = \mathrm{trace}\{Z \in T(M) \mapsto R(Z, X)Y \in T(M)\},$$

where R is the curvature tensor field of the Tanaka-Webster connection ∇ of (M, θ). It may be shown that

THEOREM 2.16. (J.M. Lee, [**168**])
The Fefferman metric F_θ is a Lorentz metric on $C(M)$ and a change of contact form $\hat{\theta} = e^u \theta$ ($u \in C^\infty(M)$) has the effect $F_{\hat{\theta}} = e^{u \circ \pi} F_\theta$ i.e. F_θ changes conformally. In particular the set of Lorentz metrics $\{e^{u \circ \pi} F_\theta : u \in C^\infty(M)\}$ is a CR invariant.

We also recall that

PROPOSITION 2.17. (R. Graham, [**123**], p. 855)
η is a connection 1-form in the principal S^1-bundle $\pi : C(M) \to M$.

Let then

$$\beta_z := (d_z \pi : \mathrm{Ker}(\eta_z) \to T_x(M))^{-1} \ , z \in C(M)_x, \ x \in M,$$

be the horizontal lift isomorphism with respect to the connection 1-form η in $S^1 \to C(M) \to M$. For a tangent vector field X on M we denote by $X^\uparrow := \beta X$ the horizontal of X (with respect to η).

Let $S = \partial / \partial \gamma$ be the tangent to the S^1-action. Precisely if $z \in C(M)$ and $a_z : \mathbb{R} \to C(M)$ is the curve $a_z(s) = e^{is} z$, $s \in \mathbb{R}$, then

$$S_z = \frac{da_z}{ds}(0) \in T_z(C(M)), \quad z \in C(M).$$

Since $(C(M), F_\theta)$ is a Lorentz manifold we may adopt the ordinary terminology that is

DEFINITION 2.18. A nonzero tangent vector $v \in T_z(C(M))$ is called *timelike* (respectively *nonspacelike, null* or *spacelike*) if $F_{\theta,z}(v,v) < 0$ (respectively if $F_{\theta,z}(v,v) \leq 0$, $F_{\theta,z}(v,v) = 0$ or $F_{\theta,z}(v,v) > 0$). $\square$

DEFINITION 2.19. A continuous tangent vector field V on $C(M)$ is called a *time orientation* of $(C(M), F_\theta)$ if $F_\theta(V,V)_z < 0$ i.e. V_z is a timelike vector for any $z \in C(M)$. $\square$

In general a given Lorentz manifold doesn't necessarily admit a time orientation. If that does happen the Lorentz manifold in question is said to be time oriented (a time oriented Lorentz manifold is commonly referred to as a space-time, cf. e.g. [**32**], p. 17). It turns out that the total space $C(M)$ of the canonical circle bundle over a strictly pseudoconvex CR manifold M is time oriented. Precisely

PROPOSITION 2.20. *The tangent vector field $T^\uparrow - S$ is timelike hence $(C(M), F_\theta)$ is time oriented by $T^\uparrow - S$ i.e. $(C(M), F_\theta)$ is a space-time.*

The time orientation $T^\uparrow - S$ divides the set of all nonspacelike tangent vectors into two subsets. Precisely

DEFINITION 2.21. A nonspacelike vector $v \in T_z(C(M))$ is called *future directed* (respectively *past directed*) if $F_{\theta,z}(T_z^\uparrow - S_z \, , \, v) < 0$ (respectively if $F_{\theta,z}(T_z^\uparrow - S_z \, , \, v) > 0$). $\square$

It is immediate that

PROPOSITION 2.22. *A nonspacelike vector $v \in T_z(C(M))$ is future directed (respectively past directed) if $(n+2)\eta_z(v) < \theta_x((d_z\pi)v)$ (respectively if $(n+2)\eta_z(v) > \theta_x((d_z\pi)v)$) where $x = \pi(z) \in M$.*

Let us also recall the following general notions of causality theory on a space-time.

DEFINITION 2.23. Let (N, g) be a space-time. A smooth curve $C : (a, b) \to N$ is *timelike* (respectively *nonspacelike, null* or *spacelike*) if its tangent vector $\dot{C}(t) \in T_{C(t)}(N)$ is timelike (nonspacelike, null or spacelike) for all values of the parameter $a < t < b$. If $z, w \in N$ and there is a smooth future directed timelike curve joining z and w then one writes $z << w$. Also one writes $z \leq w$ if either $z = w$ or there is a smooth future directed nonspacelike curve from z to w. The set $I^+(z) = \{w \in N : z << w\}$ is the *chronological future* of z whilst $I^-(z) = \{w \in N : w << z\}$ is its *chronological past.* A space-time (N, g) is said to be *chronological* if $z \notin I^+(z)$ for all $z \in N$. □

By Prop. 2.6 in [**32**], p. 23, any compact space-time (N, g) contains some closed timelike curve hence such (N, g) cannot be chronological. On the other hand if M is a compact strictly pseudoconvex CR manifold then $C(M)$ is compact as well. Therefore

PROPOSITION 2.24. *If M is compact then $(C(M), F_\theta)$ is not chronological.*

A systematic application of causality theory to the Fefferman metric, i.e. to the space-time $(C(M), F_\theta)$ oriented by $T^\uparrow - S$, is still missing from the present day mathematical literature. We leave this as an open problem.

Let $\mathcal{F}$ be a foliation of M and $\pi^*\mathcal{F}$ the pullback of $\mathcal{F}$ to $C(M)$ i.e.

$$T(\pi^*\mathcal{F})_z = (d_z\pi)^{-1}T(\mathcal{F})_{\pi(z)}, \quad z \in C(M).$$

The leaves of $\pi^*\mathcal{F}$ are connected components of the inverse images via π of the leaves of $\mathcal{F}$. We may state the following

PROPOSITION 2.25. (S. Nishikawa et al., [**87**])
Let $\mathcal{F}$ be a foliation of the strictly pseudoconvex CR manifold M carrying the contact form θ with G_θ positive definite. Let $T(\mathcal{F})^\uparrow$ be the horizontal lift with respect to η of $T(\mathcal{F})$ i.e. $T(\mathcal{F})^\uparrow_z = \beta_z T(\mathcal{F})_{\pi(z)}$, $z \in C(M)$. Then

$$T(\pi^*\mathcal{F}) = T(\mathcal{F})^\uparrow \oplus \mathrm{Ker}(d\pi). \tag{2.16}$$

Let us assume from now on that T is tangent to $\mathcal{F}$. Then 1) $T(\pi^*\mathcal{F})$ is nondegenerate in $(T(C(M)), F_\theta)$ and each leaf $\tilde{L}$ of $\pi^*\mathcal{F}$ is a Lorentz manifold (with the induced metric ι^*F_θ, where $\iota : \tilde{L} \subset C(M)$). In particular 2) the metric h_θ induced by F_θ in $\nu(\pi^*\mathcal{F}) = T(C(M))/T(\pi^*\mathcal{F})$ is positive definite. Finally 3) the Fefferman metric F_θ is bundle-like for $(C(M), \pi^*\mathcal{F})$ if and only if the Webster metric g_θ is bundle-like for $(M, \mathcal{F})$.

By slightly generalizing Definition 1.47 (or the definition in [**179**], p. 79) given a semi-Riemannian manifold (N, g) and a foliation $\mathcal{F}$ of N

DEFINITION 2.26. g is said to be *bundle-like* for $(N, \mathcal{F})$ if 1) $T(\mathcal{F})$ is nondegenerate in (N, g) and 2) the metric h induced by g in $\nu(\mathcal{F})$ is invariant by holonomy i.e. $\mathcal{L}_X h = 0$ for any $X \in T(\mathcal{F})$. □

Let us prove (2.16) in Proposition 2.25. As η is a connection 1-form

$$T(\mathcal{F})^\uparrow \cap \mathrm{Ker}(d\pi) \subseteq \mathrm{Ker}(\eta) \cap \mathrm{Ker}(d\pi) = (0),$$

hence the sum in (2.16) is direct. The inclusion "$\supseteq$" holds by the very definition of $\pi^*\mathcal{F}$. Viceversa, if

$$V \in T(\pi^*\mathcal{F}) \subseteq T(C(M)) = \mathrm{Ker}(\sigma) \oplus \mathrm{Ker}(d\pi)$$

then $V = X^\uparrow + fS$ for some $X \in T(M)$ and $f \in C^\infty(C(M))$, where $X^\uparrow := \beta X$. Also $V \in T(\pi^*\mathcal{F})$ yields $X = (d\pi)V \in T(\mathcal{F})$ hence $X^\uparrow \in T(\mathcal{F})^\uparrow$ i.e. $V \in T(\mathcal{F})^\uparrow + \mathrm{Ker}(d\pi)$. The identity (2.16) is proved.

Assume now that $T \in T(\mathcal{F})$. Consider $V \in T(\pi^*\mathcal{F})$ such that $F_\theta(V, W) = 0$ for any $W \in T(\pi^*\mathcal{F})$ i.e.

$$(\pi^*\tilde{G}_\theta)(V, W) + (\pi^*\theta)(V)\eta(W) + (\pi^*\theta)(W)\eta(V) = 0,$$

or, by taking into account the decomposition $V = V_H + V_V \in \mathrm{Ker}(\eta) \oplus \mathrm{Ker}(d\pi)$

$$\tilde{G}_\theta((d\pi)V_H, (d\pi)W_H) + \tag{2.17}$$
$$+\theta((d\pi)V_H)\eta(W_V) + \theta((d\pi)W_H)\eta(V_V) = 0,$$

for any $W \in T(\pi^*\mathcal{F})$. Let us set $W = S \in \mathrm{Ker}(d\pi) \subset T(\pi^*\mathcal{F})$. Then $W_H = 0$. Since

$$\eta = \frac{1}{n+2}\{d\gamma + \pi^*\eta_0\}$$

(for some 1-form η_0 on M, determined in terms of θ) and $(d\gamma)S = 1$, it follows that $\eta(W_V) = 1/(n+2)$. Then (2.17) gives $\theta((d\pi)V_H) = 0$ that is

$$(d\pi)V_H \in H(M), \tag{2.18}$$

with the corresponding simpler form of (2.17) i.e.

$$\tilde{G}_\theta((d\pi)V_H, (d\pi)W_H) + \theta((d\pi)W_H)\eta(V_V) = 0. \tag{2.19}$$

Let us set $W = V$. Then [by (2.18)] $\|(d\pi)V_H\|^2 = 0$, i.e. $(d\pi)V_H = 0$ hence $V_H \in \mathrm{Ker}(d\pi) \cap \mathrm{Ker}(\eta) = (0)$. With $V_H = 0$ the identity (2.19) becomes

$$\theta((d\pi)W_H)\eta(V_V) = 0, \tag{2.20}$$

for any $W \in T(\pi^*\mathcal{F})$. Let us set $W = T^\uparrow \in T(\mathcal{F})^\uparrow \subset T(\pi^*\mathcal{F})$. Then (2.20) gives $\eta(V_V) = 0$ i.e. $V_V = 0$. We conclude that $V = 0$ i.e. $T(\pi^*\mathcal{F})$ is nondegenerate in $(T(C(M)), g_\theta)$.

Note that $F_\theta(S, S) = 0$ hence F_θ is indefinite on $T(\pi^*\mathcal{F})$. Yet F_θ is nondegenerate on $T(\pi^*\mathcal{F})$ hence there F_θ has signature $(2n + 1 - q, 1)$. Yet F_θ is a Lorentz metric hence F_θ is positive definite on $T(\pi^*\mathcal{F})^\perp$. Therefore $h_\theta(r, s) := F_\theta(\rho\, r\,, \rho\, s)$, $r, s \in \nu(\pi^*\mathcal{F})$, is positive-definite. Here $\rho : \nu(\pi^*\mathcal{F}) \to T(\pi^*\mathcal{F})^\perp$ is the natural isomorphism.

To prove the last statement in Proposition 2.25 note first that $\mathcal{L}_{\tilde{X}} h_\theta = 0$ if and only if

$$\tilde{X}(g_\theta(V, W)) = g_\theta([\tilde{X}, V], W) + g_\theta(V, [\tilde{X}, W]), \tag{2.21}$$

for any $\tilde{X} \in T(\pi^*\mathcal{F})$ and $V, W \in T(\pi^*\mathcal{F})$. We need the following

LEMMA 2.27. *$T(\pi^*\mathcal{F})^\perp \subseteq \mathrm{Ker}(\eta)$ and consequently*

$$\mathrm{Ker}(\eta) = T(\mathcal{F})^\uparrow \oplus T(\pi^*\mathcal{F})^\perp. \tag{2.22}$$

Moreover $(d\pi)T(\pi^*\mathcal{F})^\perp \subseteq H(M)$.

Proof. For any

$$V \in T(\pi^*\mathcal{F})^\perp \subset T(C(M)) = \mathrm{Ker}(\eta) \oplus \mathrm{Ker}(d\pi)$$

one has the decomposition $V = V_H + fS$, $V_H \in \mathrm{Ker}(\eta)$. On the other hand

$$F_\theta(S, T^\uparrow) = (\pi^*\theta)(T^\uparrow)\eta(S) = \frac{1}{n+2}\,\theta((d\pi)T^\uparrow) = \frac{1}{n+2},$$

hence

$$F_\theta(V, T^\uparrow) = \frac{1}{n+2} f + F_\theta(V_H, T^\uparrow).$$

As $T \in T(\mathcal{F})$ it follows that

$$T^\uparrow \in T(\mathcal{F})^\uparrow \subset T(\pi^*\mathcal{F}),$$

hence $T^\uparrow$ is orthogonal to V. It follows that

$$f = -(n+2)F_\theta(V_H, T^\uparrow) = -(n+2)(\pi^*\tilde{G}_\theta)(V_H, T^\uparrow) =$$

$$= -(n+2)\tilde{G}_\theta((d\pi)V_H, T) = 0$$

i.e. $V \in \operatorname{Ker}(\eta)$. The identity (2.22) follows [by (2.16)] from

$$T(C(M)) = [T(\mathcal{F})^\uparrow \oplus \operatorname{Ker}(d\pi)] \oplus T(\pi^*\mathcal{F})^\perp,$$

$$T(\mathcal{F})^\uparrow \oplus T(\pi^*\mathcal{F})^\perp \subseteq \operatorname{Ker}(\eta).$$

To prove the last statement in Lemma 2.27 let

$$V \in T(\pi^*\mathcal{F})^\perp \subset T(C(M)) = H(M)^\uparrow \oplus \mathbb{R}T^\uparrow$$

i.e. $V = Y^\uparrow + fT^\uparrow$ for some $Y \in H(M)$. As

$$S \in \operatorname{Ker}(d\pi) \subset T(\pi^*\mathcal{F}),$$

S and V are orthogonal. Thus

$$0 = F_\theta(V, S) = F_\theta(Y^\uparrow, S) + \frac{f}{n+2} =$$

$$= (\pi^*\theta)(Y^\uparrow)\eta(S) + \frac{f}{n+2},$$

hence $\theta(Y) = 0$ yields $f = 0$. Q.e.d. Lemma 2.27 is proved.

By Lemma 2.27, (2.21) holds if and only if it is satisfied for vector fields V, W of the form $V = Y^\uparrow$, $W = Z^\uparrow$, for some $Y, Z \in H(M)$. Also, (2.21) is identically satisfied when $\tilde{X} \in \operatorname{Ker}(d\pi)$. Indeed if this is the case then (by a result in [**155**], vol. I, p. 78) one has $[\tilde{X}, Y^\uparrow] = [\tilde{X}, Z^\uparrow] = 0$ hence

$$\tilde{X}(F_\theta(Y^\uparrow, Z^\uparrow)) = X(G_\theta(Y, Z) \circ \pi) = 0,$$

as $(d\pi)X = 0$. Assume from now on that $\tilde{X} \in T(\mathcal{F})^\uparrow$ i.e. $\tilde{X} = X^\uparrow$ for some $X \in T(\mathcal{F})$. By Prop. 1.3 in [**155**], Vol. I, p. 65, $[X, Y]^\uparrow$ is the $\operatorname{Ker}(\eta)$-component of $[X^\uparrow, Y^\uparrow]$. Then [by $\theta(Y) = \theta(Z) = 0$] the identity (2.21) is equivalent to

$$X(F_\theta(Y, Z)) = \tilde{G}_\theta([X, Y], Z) + \tilde{G}_\theta(Y, [X, Z]), \tag{2.23}$$

for any $X \in T(\mathcal{F})$ and any $Y, Z \in H(M)$ such that $Y^\uparrow$, $Z^\uparrow \in T(\pi^*\mathcal{F})^\perp$. Finally note that for any $V = Y^\uparrow \in T(\pi^*\mathcal{F})^\perp$ with $Y \in H(M)$ one has

$$0 = F_\theta(X^\uparrow, V) = \tilde{G}_\theta(X, Y) \circ \pi$$

for any $X \in T(\mathcal{F})$ hence

$$Y \in H(M) \cap T(\mathcal{F})^\perp = \left[T(\mathcal{F})_{H(M)}\right]^\perp.$$

Therefore $\mathcal{L}_{\tilde{X}} h_\theta = 0$ if and only if (2.23) holds for any $X \in T(\mathcal{F})$ and any $Y, Z \in \left[T(\mathcal{F})_{H(M)}\right]^\perp$ i.e. if and only if $\mathcal{L}_X H_\theta = 0$. Q.e.d.

2.3. Foliated Lorentz manifolds

Let N be a C^∞ manifold. Let $\mathcal{F}$ be a codimension q foliation of N. Let $\nu(\mathcal{F}) = T(N)/T(\mathcal{F})$ and let $\Pi : T(M) \to \nu(\mathcal{F})$ be the projection. We shall need the basic complex (cf [**243**], p. 119) $\Omega^0_B(\mathcal{F}) \stackrel{d_B}{\to} \Omega^1_B(\mathcal{F}) \stackrel{d_B}{\to} \cdots \stackrel{d_B}{\to} \Omega^q_B(\mathcal{F}) \stackrel{d_B}{\to} 0$ and the corresponding basic cohomology $H^j_B(\mathcal{F}) = H^j(\Omega^\bullet_B(N), d_B)$, $0 \le j \le q$. Also we consider the following multiplicative filtration of the de Rham complex $\Omega^\bullet(N)$ (a decreasing filtration by differential ideals, cf. [**243**], p. 120)

$$F^r\Omega^m(N) = \{\omega \in \Omega^m(N) : i(X_1)\cdots i(X_{m-r+1})\omega = 0,$$

$$\text{for all } X_1, \cdots, X_{m-r+1} \in T(\mathcal{F})\}.$$

Let $\mathcal{F}$ be a codimension q foliation of a n-dimensional connected Lorentz manifold (N, g) such that $T(\mathcal{F})$ is nondegenerate in $(T(N), g)$. We set $\text{ind}(\mathcal{F}) := -1$ if each leaf L of $\mathcal{F}$ is Lorentz, respectively $\text{ind}(\mathcal{F}) := 1$ if each leaf L of $\mathcal{F}$ is Riemannian with the induced metric $g_L := \iota^* g$, where $\iota : L \to N$ is the inclusion. As N is connected and g is absolutely parallel an argument based on parallel translation along arcs in N shows that no other possibility occurs.

DEFINITION 2.28. The *second fundamental form* of $\mathcal{F}$ in (N, g) is

$$\alpha : T(\mathcal{F}) \otimes T(\mathcal{F}) \to \nu(\mathcal{F}), \quad \alpha(X, Y) := \Pi(\nabla^N_X Y), \quad X, Y \in T(N),$$

where ∇^N is the Levi-Civita connection of (N, g). □

As in the Riemannian case ∇^N is torsion-free hence

PROPOSITION 2.29. *α is symmetric.*

As $T(\mathcal{F})$ is nodegenerate in $(T(N), g)$ by mere linear algebra (cf. e.g. [**193**], p. 49) $T(N) = T(\mathcal{F}) \oplus T(\mathcal{F})^\perp$ and the morphism

$$\sigma : Q := \nu(\mathcal{F}) \to T(\mathcal{F})^\perp, \quad \sigma(s) := Y_s, \quad s \in \nu(\mathcal{F}),$$

for some $Y_s \in T(N)$ such that $\Pi(Y_s) = s$, is a bundle isomorphism. Let us consider

$$g_Q(r, s) := g(\sigma(r), \sigma(s)), \quad r, s \in \nu(\mathcal{F}).$$

Then (Q, g_Q) is a semi-Riemannian bundle.

DEFINITION 2.30. For any $Z \in T(\mathcal{F})^\perp$ the *Weingarten operator* is given by

$$W(Z) : T(\mathcal{F}) \to T(\mathcal{F}),$$

$$g_P(W(Z)X, X') = g_Q(\alpha(X, X'), \sigma^{-1}(Z)), \quad X, X' \in T(\mathcal{F}),$$

where g_P is the semi-Riemannian bundle metric induced by g in $P := T(\mathcal{F})$. □

Then

PROPOSITION 2.31. *$W(Z)$ is self-adjoint.*

DEFINITION 2.32. The *mean curvature* of $\mathcal{F}$ in (N, g) is the 1-form $\kappa \in \Omega^1(N)$ defined by

$$\kappa(Z) := \text{trace}\, W(Z), \quad Z \in \Gamma^\infty(P^\perp),$$

$$X \rfloor \kappa = 0, \quad X \in \Gamma^\infty(P).$$

□

Assume from now on that $\mathcal{F}$ s tangentially oriented i.e. $\mathcal{F}$ is equipped with a principal $\mathrm{GL}^+(p,\mathbb{R})$-subbundle $(p := n-q)$ $B \to N$ of the principal $\mathrm{GL}(p,\mathbb{R})$-bundle $L(P) \to N$ of all frames in the fibres of P i.e. an element $z \in L(P)_x$ is a $\mathbb{R}$-linear isomorphism $z : \mathbb{R}^p \to P_x$, $x \in N$. Let $\{E_1, \cdots, E_p\}$ be a local g_P-orthonormal frame in P adapted to B and defined on an open set $U \subseteq N$ i.e. $g_P(E_i, E_j) = \epsilon_i \delta_{ij}$ with $\epsilon_i^2 = 1$ (thus $\mathrm{ind}(\mathcal{F}) = \epsilon_1 \cdots \epsilon_p$).

DEFINITION 2.33. The *characteristic form* $\chi_{\mathcal{F}} \in \Omega^p(N)$ is given by

$$\chi_{\mathcal{F}}(Y_1, \cdots, Y_p) = \det[g(Y_i, E_j)].$$

□

Note that $P^\perp \rfloor \chi_{\mathcal{F}} = 0$. The Lorentz analogue of Rummler's formula (cf. e.g. [**243**], p. 68) still holds i.e.

$$Z \rfloor d\chi_{\mathcal{F}} + \kappa(Z)\chi_{\mathcal{F}} = 0 \quad \text{along } P, \tag{2.24}$$

for any $Z \in \Gamma^\infty(P^\perp)$. Indeed, as $\chi_{\mathcal{F}}(E_1, \cdots, E_p) = \mathrm{ind}(\mathcal{F})$

$$(\mathcal{L}_Z \chi_{\mathcal{F}})(E_1, \cdots, E_p) = -\sum_{i=1}^p \chi_{\mathcal{F}}(E_1, \cdots, \pi^\perp[Z, E_i], \cdots, E_p) =$$

$$= -\sum_{i=1}^p \epsilon_i \,\mathrm{ind}(\mathcal{F})\, g([Z, E_i], E_i),$$

where $\mathcal{L}_Z$ denotes the Lie derivative and $\Pi^\perp : T(N) \to P$ is the natural projection. On the other hand

$$\kappa(Z) = \sum_{i=1}^p \epsilon_i \, g_P(W(Z)E_i, E_i) = \sum_{i=1}^p \epsilon_i \, g([Z, E_i], E_i),$$

and (2.24) is proved.

Assume that $\mathcal{F}$ is transversally oriented, i.e. $P^\perp$ is oriented, and let ν be the characteristic form of $P^\perp$. Let $\mu = d\,\mathrm{vol}(g)$ be the Lorentz volume form on N. Assume N oriented and let $\{E_A : 1 \le A \le n\}$ be an oriented local g-orthonormal frame $(g(E_A, E_b) = \epsilon_A \delta_{AB})$ of $T(N)$ such that $\{E_i : 1 \le i \le p\}$ and $\{E_\alpha : p+1 \le \alpha \le n\}$ are frames in P and $P^\perp$ respectively. Let $\{\omega_A : 1 \le A \le n\}$ be the dual coframe. Then for any $\alpha \in \Omega^r(N)$

$$(*\alpha)(E_{A_1}, \cdots, E_{A_{n-r}})\,\mu = \epsilon_{A_1} \cdots \epsilon_{A_{n-r}}\, \alpha \wedge \omega_{A_1} \wedge \cdots \wedge \omega_{A_{n-r}}\,,$$

where $* : \Omega^r(N) \to \Omega^{n-r}(N)$ is the Hodge operator. In particular

$$(*\nu)(E_1, \cdots, E_p)\,\mu = \mathrm{ind}(\mathcal{F})\, \nu \wedge \omega_1 \wedge \cdots \wedge \omega_p\,. \tag{2.25}$$

Also a calculation based on

$$\nu(Y_{p+1}, \cdots, Y_n) = \det[g(Y_\alpha, E_\beta)]$$

leads to

$$\nu = (n-p)!\epsilon_{p+1} \cdots \epsilon_n\, \omega_{p+1} \wedge \cdots \wedge \cdots \omega_n$$

hence (2.25) may be written

$$(*\nu)(E_1, \cdots, E_p)\,\mu = (-1)^{n-p+1}\,(n-p)!\,\omega_1 \wedge \cdots \wedge \omega_n\,.$$

As $*\nu$, $\chi_{\mathcal{F}} \in \Omega^p(N)$ and $P^\perp \rfloor *\nu = 0$, $P^\perp \rfloor \chi_{\mathcal{F}} = 0$, there is $f \in C^\infty(N)$ such that $*\nu = f\chi_{\mathcal{F}}$. A calculation shows that $f = q!\,(-1)^{q+1}\,\mathrm{ind}(\mathcal{F})$ hence

$$*\nu = q!\,(-1)^{q+1}\,\mathrm{ind}(\mathcal{F})\,\chi_{\mathcal{F}}\,. \tag{2.26}$$

As a corollary of (2.26)

$$\nu \wedge \chi_{\mathcal{F}} = \frac{1}{q!}\,(-1)^{q+1}\,\mathrm{ind}(\mathcal{F})\,\mu\,. \tag{2.27}$$

At this point we may prove the following

PROPOSITION 2.34. (S. Nishikawa et al., [**87**])
Let $\mathcal{F}$ be a transversally oriented foliation of a compact orientable Lorentz manifold (N, g). Assume that the transverse volume element ν is holonomy invariant, hence $\nu \in \Omega^q_B(\mathcal{F})$ and $d\nu = 0$. If $\mathcal{F}$ is harmonic (i.e. $\kappa = 0$) then $[\nu] \neq 0$ in $H^q_B(\mathcal{F})$.

The proof is a *verbatim* repetition of the proof of Theorem 9.21 in [**243**], p. 124 (and Proposition 2.34 is the Lorentz analogue of a result by F. Kamber & P. Tondeur, [**151**]). Indeed, Rummler's formula (2.24) gives (when $\kappa = 0$)

$$d\chi_{\mathcal{F}} \in F^2\Omega^{p+1}(N)$$

and the assumption that $\nu = d_B\alpha$, for some $\alpha \in \Omega^{q-1}_B(\mathcal{F})$, leads (by (2.27)) to

$$d(\alpha \wedge \chi_{\mathcal{F}}) = \frac{1}{q!}\,(-1)^{q+1}\,\mathrm{ind}(\mathcal{F})\,\mu\,,$$

and then, by Green's lemma, to a contradiction. We may also establish the following

PROPOSITION 2.35. (S. Nishikawa et al., [**87**])
Let $\mathcal{F}$ be a foliation of a strictly pseudoconvex CR manifold M, and assume $\mathcal{F}$ tangent to the characteristic direction T of (M, θ), for some contact form θ on M. Then 1) $\mathcal{F}$ is transversally oriented if and only if $\pi^*\mathcal{F}$ is transversally oriented and, if this is the case, the transverse volume element ν of $\mathcal{F}$ in (M, g_θ) is holonomy invariant if and only if the transverse volume element $\tilde{\nu}$ of $\pi^*\mathcal{F}$ in $(C(M), F_\theta)$ is holonomy invariant. 2) $\mathcal{F}$ is harmonic in (M, g_θ) if and only if $\pi^*\mathcal{F}$ is harmonic in $(C(M), F_\theta)$.

Combining Propositions 2.34 and 2.35 shows that

COROLLARY 2.36. *For any transversally oriented codimension q foliation $\mathcal{F}$ of a compact strictly pseudoconvex CR manifold M, if 1) $\mathcal{F}$ is tangent to the characteristic direction of (M, θ), 2) the transverse volume element ν of $\mathcal{F}$ in (M, g_θ) is holonomy invariant, and 3) $\mathcal{F}$ is harmonic in (M, g_θ), then $[\nu] \neq 0$ in $H^q_B(\mathcal{F})$.*

Indeed, if M is compact then so is $C(M)$ and, given a local coordinate system (U, x^A) on M, $(\pi^{-1}(U),\ u^A := x^A \circ \pi,\ u^{2n+2} := \gamma)$ are local coordinates on $C(M)$, hence an orientation of M induces an orientation of M. While this is only illustrative of our ideas as to the use of the Fefferman metric (and the preceding statement also follows by directly applying the aforementioned result of F. Kamber & P. Tondeur, i.e. Theorem 9.21 in [**151**], p. 124, to $\mathcal{F}$ on (M, g_θ)) we may exploit the relationship between pseudohermitian geometry and conformal Lorentzian geometry to prove the following

COROLLARY 2.37. (S. Nishikawa et al., [**87**])
Let $\mathcal{F}$ be a transversally oriented codimension q foliation of a compact strictly pseudoconvex CR manifold M, tangent to the characteristic direction of (M, θ), for some contact form θ. Assume that the transverse volume element of $\mathcal{F}$ in (M, g_θ) is holonomy invariant and that the mean curvature $\kappa \in \Omega^1(M)$ of $\mathcal{F}$ in (M, g_θ) is closed (i.e. $d\kappa = 0$). If $[\kappa] = 0$ in $H^1_B(\mathcal{F})$ then $H^q_B(\mathcal{F}) \neq 0$.

Proof. We shall need the following

LEMMA 2.38. *Let $(N, g, \mathcal{F})$ be a n-dimensional foliated Lorentz manifold, with $\mathcal{F}$ transversally orientable and $T(\mathcal{F})$ nondegenerate in $(T(N), g)$. Then $\mathcal{F}$ is harmonic in $(N, e^{2u}g)$, with $u \in C^\infty(N)$, if and only if*

$$Z(u) = \frac{1}{n - \mathrm{cod}(\mathcal{F})}\,\kappa(Z), \quad Z \in T(\mathcal{F})^\perp , \tag{2.28}$$

where $\kappa \in \Omega^1(N)$ is the mean curvature of $\mathcal{F}$ in (N, g). Also, the following statements are equivalent 1) u is a basic function, i.e. $u \in \Omega^0_B(\mathcal{F})$, and 2) the tranverse volume element $\hat{\nu}$ of $\mathcal{F}$ in $(N, \hat{g} := e^{2u}g)$ is holonomy invariant if and only if the transverse volume element ν of $\mathcal{F}$ in (N, g) is holonomy invariant.

See also [**184**]. *Proof of Lemma* 2.38. The relationship between the Levi-Civita connections of the Lorentz metrics $\hat{g} := e^{2u}g$ and g

$$\nabla^{\hat{g}} = \nabla^g + (du) \otimes I + I \otimes (du) - g \otimes \nabla^g u$$

shows that the second fundamental forms of $\mathcal{F}$ in $(N, \hat{g})$ and (N, g), respectively, are related by

$$\hat{\alpha} = \alpha - g \otimes \pi \nabla^g u$$

hence

$$\hat{W}(Z) = W(Z) - Z(u)\, I\,, \quad Z \in \Gamma^\infty(P^\perp),$$

for the corresponding Weingarten operators, where I is the identical transformation. Consequently the mean curvature forms satisfy

$$\hat{\kappa}(Z) = \kappa(Z) - p\, Z(u),$$

where $p := n - q$ and $q = \mathrm{cod}(\mathcal{F})$. Therefore $\hat{\kappa} = 0$ if and only if u is a solution to (2.28). The second statement in Lemma 2.38 follows from the formulae

$$\mathcal{L}_X \hat{\nu} = e^{qu}\{q\, X(u)\, \nu + \mathcal{L}_X \nu\}\,, \quad X \in \Gamma^\infty(P).$$

Q.e.d. Let us go back to the proof of Corollary 2.37. As $[\kappa] = 0$ in $H^1_B(\mathcal{F})$ there is a basic function $v \in \Omega^0_B(\mathcal{F})$ such that $\kappa = dv$. Let us set $u := (p+1)v$ where $p := 2n + 1 - q$ and $q := \mathrm{cod}(\mathcal{F})$. Hence (by the proof of Proposition 2.35, i.e. cf. (2.31))

$$Z^\uparrow(u \circ \pi) = \frac{1}{p+1}\,\tilde{\kappa}(Z^\uparrow), \quad Z \in T(\mathcal{F})^\perp .$$

Then Lemma 2.38 implies that $\pi^*\mathcal{F}$ is harmonic in $(C(M), e^{2\,u\circ\pi} F_\theta)$, hence (by the previously mentioned result of J.M. Lee, [**168**], according to which the Fefferman metric changes conformally $F_{\hat{\theta}} = e^{2\,u\circ\pi} F_\theta$, under a transformation $\hat{\theta} = e^{2u}\theta$) harmonic in $(C(M), F_{\hat{\theta}})$. By Proposition 2.35, $\tilde{\nu}$ is holonomy invariant. Also as $u \in \Omega^0_B(\mathcal{F})$ it follows that $u \circ \pi \in \Omega^0_B(\pi^*\mathcal{F})$ so that (again by Lemma 2.38) the transverse volume element of $\pi^*\mathcal{F}$ in $(C(M), F_{\hat{\theta}})$ is holonomy invariant. Finally (by Proposition 2.34) we may conclude that $0 \neq H^q_B(\pi^*\mathcal{F}) \approx H^q_B(\mathcal{F})$. Q.e.d.

Proof of Proposition 2.35. Note that

$$\nu(\mathcal{F}) \approx T(\mathcal{F})^\perp \xrightarrow{\beta} \left[T(\mathcal{F})^\perp\right]^\uparrow = T(\pi^*\mathcal{F})^\perp \approx \nu(\pi^*\mathcal{F}),$$

hence $\mathcal{F}$ is transversally oriented if and only if $\pi^*\mathcal{F}$ is transversally oriented. We only need to justify the last equality. If $\tilde{X} = X^\perp + V$ for some $X \in T(\mathcal{F})$ and $V \in \mathcal{V} := \mathrm{Ker}(d\pi)$, then for any $Y \in P^\perp$

$$F_\theta(\tilde{X}; Y^\uparrow) = (\pi^*\tilde{G}_\theta)(\tilde{X}, Y^\uparrow) + (\pi^*\theta)(Y^\uparrow)\eta(\tilde{X}) =$$
$$= \tilde{G}_\theta(X, Y) + \theta(Y)\eta(V) = \tilde{G}_\theta(\pi_H X + \theta(X)T, Y),$$

as $P^\perp \subset H(M)$. Next

$$F_\theta(\tilde{X}; Y^\uparrow) = g_\theta(\pi_H X, Y) + \theta(X)\tilde{G}_\theta(T, Y) =$$
$$= g_\theta(\pi_H X + \theta(X)T, Y) = g_\theta(X, Y) = 0,$$

where from $\left[P^\perp\right]^\uparrow \subseteq T(\pi^*\mathcal{F})^\perp$ hence one has equality, as both bundles have rank q.

At this point we may relate the Weingarten operators of $\mathcal{F}$ and $\pi^*\mathcal{F}$. Let $\nabla^{C(M)}$ be the Levi-Civita connection of $(C(M), F_\theta)$. Given $Z \in P^\perp$ the Weingarten operator

$$\tilde{W}(Z^\uparrow) : T(\pi^*\mathcal{F}) \to T(\pi^*\mathcal{F})$$

is given by

$$F_\theta(\tilde{W}(Z^\uparrow)\tilde{X}, \tilde{X}') = F_\theta(\nabla^{C(M)}_{\tilde{X}} \tilde{X}', Z^\uparrow),$$

for any $\tilde{X} = X^\uparrow + V$ and $\tilde{X}' = X'^\uparrow + V'$, where $X, X' \in P$ and $V, V' \in \mathcal{V}$. As $\pi : C(M) \to M$ is a principal bundle the map π is a submersion. However $F_\theta(S, S) = 0$ hence S is null, or lightlike, so that π is not a semi-Riemannian submersion (according to the terminology adopted in [**193**], p. 212). Nevertheless, we may relate $\nabla^{C(M)}$ to ∇^M, in the spirit of [**192**]. Another difficulty is that $\mathrm{Ker}(\eta)$ and $\mathcal{V}$ are not orthogonal with respect to the Fefferman metric F_θ, yet $H(M)^\uparrow \perp \mathcal{V}$ does hold. Using also $[Y^\uparrow, V] = 0$ for any $Y \in T(M)$, a calculation leads to

$$2F_\theta(\tilde{W}(Z^\uparrow)\tilde{X}, \tilde{X}') = -Z(\tilde{G}_\theta(X, X'))+$$
$$+\tilde{G}_\theta([Z, X], X') - \tilde{G}_\theta([X', Z], X)+$$
$$+2\theta(X)\Omega(X'^\uparrow, Z^\uparrow) + 2\theta(X')\Omega(X^\uparrow, Z^\uparrow)+$$
$$+2(d\theta)(X, Z)\eta(V') + 2(d\theta)(X', Z)\eta(V),$$

where $\Omega = D\eta$ is the curvature 2-form of η. Let $H(\mathcal{F})$ be the g_θ-orthogonal complement of $\mathbb{R}T$ in P. In particular, for any $X, X' \in H(\mathcal{F})$

$$2F_\theta(\tilde{W}(Z^\uparrow)\tilde{X}, \tilde{X}') =$$
$$= -Z(g_\theta(X, X')) + g_\theta([Z, X], X') - g_\theta([X', Z], X)+$$
$$+2(d\theta)(X, Z)\eta(V') + (d\theta)(X', Z)\eta(V),$$

or (by exploiting the explicit expression of $\nabla^{C(M)}$, cf. e.g. [**155**], p. 160)

$$F_\theta(\tilde{W}(Z^\uparrow)\tilde{X}, \tilde{X}') = g_\theta(W(Z)X, X')+$$
$$+(d\theta)(X, Z)\eta(V') + (d\theta)(X', Z)\eta(V). \tag{2.29}$$

Similarly

$$F_\theta(\tilde{W}(Z^\uparrow)T^\uparrow, T^\uparrow) = 2\Omega(T^\uparrow, Z^\uparrow). \tag{2.30}$$

Next, we may calculate $\tilde{\kappa}(Z^\uparrow) := \mathrm{trace}\tilde{W}(Z^\uparrow)$. Let $\{E_1, \cdots, E_{p-1}\}$ be a g_θ-orthonormal frame. Then

$$\{E_1^\uparrow, \cdots, E_{p-1}^\uparrow,\ T^\uparrow + \frac{n+2}{2}S,\ T^\uparrow - \frac{n+2}{2}S\}$$

is a F_θ-orthonormal frame of $\pi^*\mathcal{F}$ (in particular $\operatorname{ind}(\pi^*\mathcal{F}) = -1$). As $\theta(W(Z)T) = 0$, a calculation based on (2.29)-(2.30) leads to

$$\tilde{\kappa}(Z^\uparrow) = \kappa(Z) \circ \pi \,, \quad Z \in P^\perp \,. \tag{2.31}$$

In particular $\mathcal{F}$ is harmonic in (M, g_θ) if and only if $\pi^*\mathcal{F}$ is harmonic in $(C(M), F_\theta)$. Finally if $\tilde{\nu} \in \Omega^q(C(M))$ is given by

$$\tilde{\nu}(\tilde{Y}_1, \cdots, \tilde{Y}_q) = \det[F_\theta(\tilde{Y}_\alpha, E_\alpha^\uparrow)] \,,$$

for some oriented g_θ-orthonormal frame of $P^\perp$, then $\tilde{\nu} = \pi^*\nu$ and a calculation shows that

$$\mathcal{L}_{\tilde{X}}\tilde{\nu} = \pi^*(\mathcal{L}_X \nu),$$

for any $\tilde{X} = X^\uparrow + V$, with $X \in P$ and $V \in \mathcal{V}$. Proposition 2.35 is proved.

2.4. The second fundamental form

We adopt the following terminology. Let $(M, T_{1,0}(M))$ be a CR manifold of hypersurface type and $\mathcal{F}$ a foliation of M. Let $\nu(\mathcal{F} = T(M)/T(\mathcal{F})$ and let $\Pi : T(M) \to \nu(\mathcal{F})$ be the projection.

DEFINITION 2.39. $\mathcal{F}$ is a *semi-Levi foliation* if $T(\mathcal{F}) \subseteq H(M)$. A semi-Levi foliation $\mathcal{F}$ of M is a *Levi foliation* if $JT(\mathcal{F}) = T(\mathcal{F})$. □

It is immediate that

PROPOSITION 2.40. *If $\mathcal{F}$ is a semi-Levi foliation of codimension one then $\mathcal{F}$ is a Levi foliation and $(M, T_{1,0}(M))$ is Levi flat.*

Let $(M, \mathcal{F})$ be a foliated CR manifold and $\theta \in \Omega^1(M)$ a pseudohermitian structure on M.

PROPOSITION 2.41. *If $\theta \in \Omega^1_B(\mathcal{F})$ then $\mathcal{F}$ is a semi-Levi foliation and M is degenerate.*

DEFINITION 2.42. Let $(N, \mathcal{F})$ be a foliated manifold. Then $\mathcal{F}$ is a *semi-Riemannian foliation* if there is a holonomy invariant semi-Riemannian bundle metric g_Q in the transverse bundle $Q = \nu(\mathcal{F})$. □

For any semi-Riemannian foliation $(\mathcal{F}, g_Q)$ on N there is a bundle-like semi-Riemannian metric h on N inducing g_Q.

We wish to study foliations of nondegenerate CR manifolds, on which a contact form θ has been fixed. For any foliation $\mathcal{F}$ of a Riemannian manifold M there is a natural connection in the normal bundle $\nu(\mathcal{F})$, induced by the Bott (partial) connection of $\mathcal{F}$ and the Levi-Civita conection of the given Riemannian metric on M (cf. e.g. (5.3) in [**243**], p. 48). In the spirit of pseudohermitian geometry, when M is a CR manifold, we replace the Riemannian connection by the Tanaka-Webster connection of θ and investigate the resulting theory of the "second fundamental form" of $\mathcal{F}$ in M (cf. also [**84**] and [**20**], where similar ideas led to a study of the geometry of the second fundamental form of a CR immersion).

Let $(M, T_{1,0}(M))$ be a nondegenerate CR manifold and θ a contact form on M. Let $\mathcal{F}$ be a foliation of M such that $P = T(\mathcal{F})$ is nondegenerate in $(T(M), g_\theta)$.

REMARK 2.43. The case where Rad $T\mathcal{F} := \{X \in T(\mathcal{F}) : g_\theta(X, Y) = 0, \ \ Y \in T(\mathcal{F})\} \neq 0$ is not studied here. Nevertheless we shall devote the end of this section to illustrating the lines along which a theory of foliations $\mathcal{F}$ of a semi-Riemannian manifold may be developed when Rad $T\mathcal{F} \neq 0$. There we shall adopt certain ideas from [**97**] regarding the use of screen distributions. □

DEFINITION 2.44. Let $\mathring{\nabla}$ be the *Bott connection* of $(M, \mathcal{F})$ i.e.

$$\mathring{\nabla}_X s = \Pi[X, Y_s], \quad X \in \Gamma^\infty(P), \quad s \in \Gamma^\infty(Q),$$

for some $Y_s \in \mathcal{X}(M)$ such that $\Pi(Y_s) = s$. A connection D in the normal bundle $Q \to M$ is *adapted* if $D_X = \mathring{\nabla}_X$ for any $X \in P$. □

Let $P^\perp$ be the orthogonal complement of P in $T(M)$ with respect to g_θ. Let ∇ be the Tanaka-Webster connection of (M, θ). Let g_Q be the bundle metric induced by g_θ in $Q = \nu(\mathcal{F})$. Let D be the connection in Q defined by

$$D_X s = \begin{cases} \mathring{\nabla}_X s & \text{if } X \in \Gamma^\infty(P), \\ \Pi \nabla_X \sigma(s) & \text{if } X \in \Gamma^\infty(P^\perp). \end{cases}$$

We may state the following

PROPOSITION 2.45. (S. Nishikawa et al., [**87**])
Let M be a nondegenerate CR manifold and θ a contact form on M. Let $\mathcal{F}$ be a foliation of M so that $T(\mathcal{F})$ is nondegenerate in $(T(M), g_\theta)$. Then D is an adapted connection in Q and its torsion T_D satisfies

$$P \rfloor T_D = 0, \quad T_D(Z, Z') = \Pi T_\nabla(Z, Z'),$$

for any $Z, Z' \in \Gamma^\infty(P^\perp)$. Moreover $\mathcal{F}$ is semi-Riemannian and g_θ bundle-like if and only if g_Q is parallel with respect to D.

A pseudohermitian analogue of the second fundamental form (of a foliation of a Riemannian manifold) is given by

DEFINITION 2.46. The *pseudohermitian second fundamental form* of $\mathcal{F}$ in (M, θ) is the $C^\infty(M)$-bilinear form $\alpha : P \otimes P \to Q$ defined by

$$\alpha(X, X') = \Pi \nabla_X X'$$

for any $X, X' \in \Gamma^\infty(P)$. □

Also

DEFINITION 2.47. If $Z \in \Gamma^\infty(P^\perp)$ the *pseudohermitian Weingarten operator* $W(Z) : P \to P$ is given by

$$g_\theta(W(Z)X, X') = g_Q(\alpha(X, X'), \Pi Z).$$

□

Note that in general α is not symmetric and $W(Z)$ is not self-adjoint with respect to g_θ (as ∇ has nontrivial torsion). Next let us define $\alpha^\perp : P^\perp \otimes P^\perp \to P$ by setting

$$\alpha^\perp(Z, Z') = \frac{1}{2}\Pi^\perp(\nabla_Z Z' + \nabla_{Z'} Z)$$

for any $Z, Z' \in \Gamma^\infty(P^\perp)$. Also

DEFINITION 2.48. The *pseudohermitian mean curvature form* $\kappa \in \Omega^1(M)$ of $\mathcal{F}$ in (M, θ) is defined by

$$P \rfloor \kappa = 0, \quad \kappa(Z) = \operatorname{trace} W(Z), \quad Z \in \Gamma^\infty(P^\perp).$$

□

Finally a pseudohermitian analogue of the mean curvature vector (of a foliation of a Riemannian manifold) is given by

DEFINITION 2.49. Let $t \in \Gamma^\infty(P^\perp)$ be defined by

$$g_\theta(t, Z) = \kappa(Z), \quad Z \in \Gamma^\infty(P^\perp).$$

t is called the *pseudohermitian mean curvature vector* of $\mathcal{F}$ in (M, θ). □

Let g_P be the bundle metric induced by g_θ on P. As $\nabla g_\theta = 0$

$$(\mathcal{L}_Z g_P)(X, X') = -2 g_Q(\alpha(X, X'), \Pi Z) +$$

$$+ g_\theta(T_\nabla(X, X'), Z) + g_\theta(T_\nabla(Z, X), X') + g_\theta(T_\nabla(Z, X'), X), \tag{2.32}$$

for any $X, X' \in \Gamma^\infty(P)$ and $Z \in \Gamma^\infty(P^\perp)$. Also (by a similar calculation)

$$(\mathcal{L}_X g_Q)(\pi Z, \pi Z') = -2 g_\theta(\alpha^\perp(Z, Z'), X) +$$

$$+ g_\theta(T_\nabla(X, Z), Z') + g_\theta(T_\nabla(X, Z'), Z), \tag{2.33}$$

for any $X \in \Gamma^\infty(P)$ and $Z, Z' \Gamma^\infty(P^\perp)$. To illustrate our ideas we give an application of these notions to foliations all of whose leaves are tangent to the characteristic direction T of (M, θ). If this is the case, i.e. $T \in \Gamma^\infty(P)$, let $P(\theta)$ be the orthogonal complement with respect to g_θ of $\mathbb{R}T$ in P. Then

$$H(M) = P(\theta) \oplus P^\perp .$$

We may state the following

THEOREM 2.50. (S. Nishikawa et al., [**87**])
Let M be a nondegenerate CR manifold and θ a contact form on M of vanishing pseudohermitian torsion ($\tau = 0$). Let $\mathcal{F}$ be a foliation of M such that $T(\mathcal{F})$ is nondegenerate in $(T(M), g_\theta)$. Let us assume that T is tangent to the leaves of $\mathcal{F}$. Then

i) D is torsion-free.

ii) α is symmetric and $W(Z)$ is self-adjoint, for any $Z \in \Gamma^\infty(P^\perp)$.

iii) The induced metric g_P along the leaves is invariant under flows of vector fields orthogonal to the foliation if and only if $\alpha = 0$ and $P(\theta)$ is J-invariant.

iv) $\mathcal{F}$ is semi-Riemannian and g_θ bundle-like if and only if $\alpha^\perp = 0$.

Proof. The fact that $T_D = 0$ follows from $\tau = 0$ together with Proposition 2.45 (as $\Pi T = 0$). For the moment let us drop the assumption $\tau = 0$. As $\nabla T = 0$ one has

$$\alpha(X, T) = 0, \tag{2.34}$$

for any $X \in \Gamma^\infty(P)$. Moreover

$$\alpha(T, X) = \Pi \, \tau(X) \tag{2.35}$$

as a consequence of (2.32) and of

$$\alpha(X, X') = \alpha(X', X) + \Pi \, T_\nabla(X, X')$$

for any $X, X' \in \Gamma^\infty(P)$. Then $W(Z)$ is $P(\theta)$-valued (by (2.34)) and

$$W(Z)T = \Pi^\perp \tau(Z), \tag{2.36}$$

for any $Z \in \gamma^\infty(P^\perp)$ (by (2.35)). Finally note that

$$\alpha(X, X') = \alpha(X', X), \quad g_\theta(W(Z)X, X') = g_\theta(X, W(Z)X'),$$

for any $X, X' \in \Gamma^\infty(P(\theta))$. Then ii) follows from the more general statements

PROPOSITION 2.51. *1) α is symmetric if and only if $\tau(P) \subseteq P$, and 2) $W(Z)$ is self-adjoint with respect to g_θ if and only if $\tau(P^\perp) \subseteq P^\perp$, provided that $T \in \Gamma^\infty(P)$.*

One may compute the torsion terms in (2.32) so that to obtain

$$(\mathcal{L}_Z g_P)(X, T) = -2g_\theta((J + \tau)X, Z)$$

for any $X \in \Gamma^\infty(P)$, and

$$(\mathcal{L}_Z g_P)(X, X') = -2g_Q(\alpha(X, X'), \pi Z)$$

for any $X, X' \in \Gamma^\infty(P(\theta))$. Then iii) is a corollary of the following more general statement

PROPOSITION 2.52. *$\mathcal{L}_Z g_P = 0$ if and only if $\alpha = 0$ on $P(\theta) \otimes P(\theta)$ and $J + \tau$ is a bundle endomorphism of $P(\theta)$.*

Finally note that (2.33) furnishes

$$(\mathcal{L}_T g_Q)(\pi Z, \pi Z') = -2\theta(\alpha^\perp(Z, Z')) + 2g_\theta(\tau Z, Z'),$$

$$(\mathcal{L}_X g_Q)(\pi Z, \pi Z') = -2g_\theta(\alpha^\perp(Z, Z'), X),$$

for any $X \in P(\theta)$. Then $\mathcal{L}_X g_Q = 0$ for any $X \in P$ if and only if $\alpha^\perp = A \otimes T$, hence iv) holds when $\tau = 0$.

As announced earlier in this section we outline the use of screen distributions in producing a theory of foliations $\mathcal{F}$ with Rad $T\mathcal{F} \neq 0$ in a semi-Riemannian manifold. Let $\mathcal{F}$ be a codimension q foliation of a real n-dimensional manifold M. Let $\nu(\mathcal{F}) = T(M)/T(\mathcal{F})$ be the transverse bundle and $\Pi : T(M) \to \nu(\mathcal{F})$ the projection. Let g be a r-lightlike metric in $T(\mathcal{F})$ where $1 \leq r \leq \min\{m, q\}$ and $m = n - q$. Then

DEFINITION 2.53. $(\mathcal{F}, g)$ is a *tangentially lightlike foliation* of M and Rad $T\mathcal{F}$ is its *tangential radical distribution.* □

It is this sort of foliations that the end of this section is devoted to. If M is a semi-Riemannian manifold of index $1 \leq s \leq n-1$ and g is induced in $T(\mathcal{F})$ by the ambient metric then each leaf of $\mathcal{F}$ is a degenerate or lightlike submanifold of M (cf. [**97**], p. 140). We also adopt the terminology in Table 1. Similarly

DEFINITION 2.54. If g_Q is a ρ-lightlike $(1 \leq \rho \leq q)$ metric in $Q = \nu(\mathcal{F})$ such that $\mathcal{L}_X g_Q = 0$ for any $X \in T(\mathcal{F})$ then $(\mathcal{F}, g_Q)$ is a *transversally lightlike foliation.* □

Let $\mathcal{F}$ be a tangentially lightlike foliation of the semi-Riemannian manifold (M, g). We set

$$T(\mathcal{F})^\perp = \{V \in T(M) : g(V, X) = 0, \quad X \in T(\mathcal{F})\}.$$

Let $S(T\mathcal{F})$ and $S(T\mathcal{F}^\perp)$ be complements to the tangential radical distribution in $T(\mathcal{F})$ and $T(\mathcal{F})^\perp$, respectively. Then

$$T(\mathcal{F}) = S(T\mathcal{F}) \oplus \operatorname{Rad} T\mathcal{F}, \tag{2.37}$$

$$T(\mathcal{F})^\perp = S(T\mathcal{F}^\perp) \oplus \operatorname{Rad} T\mathcal{F}, \tag{2.38}$$

and (by Proposition 2.1 in [**97**], p. 5) both $S(T\mathcal{F})$ and $S(T\mathcal{F}^\perp)$ are nondegenerate. Consequently

$$T(M) = S(T\mathcal{F}) \oplus S(T\mathcal{F})^\perp. \tag{2.39}$$

If $T(\mathcal{F})$ were nondegenerate then the ambient semi-Riemannian metric g would induce a bundle metric in $\nu(\mathcal{F})$ by the natural isomorphism $\nu(\mathcal{F}) \approx T(\mathcal{F})^\perp$. As to the study of tangentially lightlike foliations we circumvent the difficulties (arising from the failure to decompose $T(M) = T(\mathcal{F}) \oplus T(\mathcal{F})^\perp$) by using lightlike transversal bundles (as in the theory of lightlike submanifolds, cf. [**97**], p. 139-148).

$\mathcal{F}$	r
(I) r − lightlike	$1 \le r < \min\{m, q\}$
(II) co − isotropic	$1 \le r = q < m$
(III) isotropic	$1 \le r = m < q$
(IV) totally lightlike	$1 \le r = m = q$

Table 1. Classification of tangentially lightlike foliations of semi-Riemannian manifolds according to the rank of their tangential radical distribution.

Let us start by noticing that

$$S(T\mathcal{F})^\perp \supseteq S(T\mathcal{F}^\perp). \tag{2.40}$$

Indeed if $X \in S(T\mathcal{F}^\perp) \subseteq T(\mathcal{F})^\perp$ then X is orthogonal to $T(\mathcal{F}) \supseteq S(T\mathcal{F})$ hence $X \in S(T\mathcal{F})^\perp$. Next, since $S(T\mathcal{F}^\perp)$ is nondegenerate

$$S(T\mathcal{F})^\perp = S(T\mathcal{F}^\perp) \oplus S(T\mathcal{F}^\perp)^\perp. \tag{2.41}$$

We shall need the following

LEMMA 2.55. *Let $\{\xi_1, \cdots, \xi_r\}$ be a local frame of* $\operatorname{Rad} T\mathcal{F}$ *defined on the open set $U \subseteq M$. There exist $N_i \in \Gamma^\infty(U, S(T\mathcal{F}^\perp)^\perp)$, $1 \le i \le r$, such that $g(N_i, \xi_j) = \delta_{ij}$ and $g(N_i, N_j) = 0$.*

Proof. Note first that

$$\operatorname{Rad} T\mathcal{F} \subseteq S(T\mathcal{F}^\perp)^\perp. \tag{2.42}$$

Indeed if $X \in \operatorname{Rad} T\mathcal{F}$ then X is orthogonal on $T(\mathcal{F})^\perp \supseteq S(T\mathcal{F}^\perp)$ hence $X \in S(T\mathcal{F}^\perp)^\perp$. Next we choose a complement E to $\operatorname{Rad} T\mathcal{F}$ so that

$$S(T\mathcal{F}^\perp)^\perp = (\operatorname{Rad} T\mathcal{F}) \oplus E. \tag{2.43}$$

Consequently $\dim_{\mathbb{R}} E_x = r$ for any $x \in M$. Let then $\{V_1, \cdots, V_r\}$ be a local frame of E on U. One may look for the N_i's in the form

$$N_i = A_i^k \xi_k + B_i^k V_k$$

for some C^∞ functions $A_i^k, B_i^k : U \to \mathbb{R}$ with the requirement

$$\delta_{ij} = g(N_i, \xi_j) = B_i^k g_{jk}$$

where $g_{jk} = g(\xi_j, V_k)$. Let us set $G = \det[g_{jk}]$. We claim that $G(x) \neq 0$ for any $x \in U$. The proof is by contradiction. If $G(x_0) = 0$ for some $x_0 \in U$ then there is $v = (v^1, \cdots, v^r) \in \mathbb{R}^r \setminus \{0\}$ such that

$$g_{jk}(x_0)v^j = 0, \quad 1 \leq k \leq r. \tag{2.44}$$

Let us set $w = v^j \xi_{j,x_0} \in (\text{Rad}\, T\mathcal{F})_{x_0} \subset S(T\mathcal{F}^\perp)^\perp$. Then (by (2.44)) $g_{x_0}(w, V_{k,x_0}) = 0$. Also $g_{x_0}(w, \xi_{k,x_0}) = 0$ by the very definition of w. Then w sits in $S(T\mathcal{F}^\perp)^\perp_{x_0} \setminus \{0\}$ and (by (2.43)) it is perpendicular on $S(T\mathcal{F}^\perp)^\perp_{x_0}$ i.e. $S(T\mathcal{F}^\perp)^\perp_{x_0}$ is degenerate, a contradiction. Therefore it is legitimate to consider $[g^{jk}] := [g_{jk}]^{-1}$. Then $B_i^k = g^{ki}$ and the requirement $g(N_i, N_j) = 0$ yields

$$A_i^j + A_j^i + g^{ki} g^{\ell j} g(V_k, V_\ell) = 0$$

and we may choose $A_j^i := -\frac{1}{2} g^{ki} g^{\ell j} g(V_k, V_\ell)$. Lemma 2.55 is proved. In particular (with the notations of Lemma 2.55) $\{\xi_1, \cdots, \xi_r, N_1, \cdots, N_r\}$ is a local frame of $S(T\mathcal{F}^\perp)^\perp$ on U. Let us set

$$ltr(T\mathcal{F})_x = \sum_{i=1}^r \mathbb{R} N_{i,x}, \quad x \in U.$$

The next lemma is imitative of Theorem 1.4 in [**97**], p. 147.

LEMMA 2.56. *$ltr(T\mathcal{F})_x$ is well defined i.e. its definition doesn't depend upon the choice of local frames $\{\xi_j\}$ of* Rad $T\mathcal{F}$ *and $\{V_k\}$ of E at x. Moreover $ltr(T\mathcal{F}) = \cup_{x \in M} ltr(T\mathcal{F})_x$ is a vector bundle over M and*

$$S(T\mathcal{F}^\perp)^\perp = (\text{Rad}\, T\mathcal{F}) \oplus ltr(T\mathcal{F}). \tag{2.45}$$

We call $ltr(T\mathcal{F}) \to M$ a *lightlike transversal* vector bundle with respect to the pair $(S(T\mathcal{F}),\ S(T\mathcal{F}^\perp))$. Also

$$tr(T\mathcal{F}) := ltr(T\mathcal{F}) \oplus S(T\mathcal{F}^\perp) \tag{2.46}$$

is a *transversal* vector bundle. Then (by (2.39), (2.41) and (2.45))

$$T(M) = S(T\mathcal{F}) \oplus S(T\mathcal{F}^\perp) \oplus (\text{Rad}\, T\mathcal{F}) \oplus ltr(T\mathcal{F})$$

hence

$$T(M) = T(\mathcal{F}) \oplus tr(T\mathcal{F}). \tag{2.47}$$

Let $\sigma : \nu(\mathcal{F}) \to tr(T\mathcal{F})$ be the bundle isomorphism given by

$$\sigma(s) = \text{tra}(Y), \quad \Pi(Y) = s, \quad Y \in T(M),$$

where tra : $T(M) \to tr(T\mathcal{F})$ is the natural projection associated to the decomposition (2.47). Let us set

$$g_{\text{tra}}(s, r) = g(\sigma(s)\,,\, \sigma(r)), \quad s, r \in \nu(\mathcal{F}).$$

DEFINITION 2.57. If g_{tra} is holonomy invariant, i.e. $\mathcal{L}_X g_{\text{tra}} = 0$ for any $X \in T(\mathcal{F})$, then g is said to be *bundle-like.* $\square$

Let $Q = \nu(\mathcal{F})$ for simplicity. One expects g_{tra} to be degenerate, as well. Indeed, if we set

$$\text{Rad}\, Q = \{s \in Q : g_{\text{tra}}(s, r) = 0, \ \ r \in Q\}$$

then we have

PROPOSITION 2.58. *Let $\mathcal{F}$ be a lightlike foliation of the semi-Riemannian manifold (M, g) and $ltr(T\mathcal{F}) \to M$ a lightlike transversal vector bundle associated with the screen distributions $S(T\mathcal{F})$ and $S(T\mathcal{F}^\perp)$. Then*

$$\sigma(\text{Rad } Q) = ltr(T\mathcal{F}). \tag{2.48}$$

Proof. Let $N \in ltr(T\mathcal{F})$ and $r \in Q$. As N is orthogonal to $tr(T\mathcal{F})$

$$g_{\text{tra}}(\sigma^{-1}(N), r) = g(N, \sigma(r)) = 0$$

it follows that $\sigma^{-1}(N) \in \text{Rad } Q$. For the opposite inclusion let $s \in \text{Rad } Q$ and $Z \in tr(T\mathcal{F})$. If we set $r = \sigma^{-1}(Z) \in Q$ then

$$0 = g_{\text{tra}}(s, r) = g(\sigma(s), Z). \tag{2.49}$$

We have $s = \Pi(Y)$ for some $Y \in T(M)$. As a consequence of (2.46)-(2.47) $Y = X+N+V$ for some $X \in T(\mathcal{F})$, $N \in ltr(T\mathcal{F})$ and $V \in S(T\mathcal{F}^\perp)$. Then $\sigma(s) = N+V$. Let $W \in S(T\mathcal{F}^\perp)$. Applying (2.49) for $Z = W$ gives $g(V, W) = 0$ and then $V = 0$ since $S(T\mathcal{F}^\perp)$ is nondegenerate. It remains that $\sigma(s) = N \in \text{Rad } Q$ and Proposition 2.58 is proved.

By (2.46) and Proposition 2.58 a canonical choice of screen distribution in Q is $S(Q) := \sigma^{-1} S(T\mathcal{F}^\perp)$ so that

$$Q = S(Q) \oplus \text{Rad } Q.$$

Let $\mathring{\nabla}$ be the Bott connection of $\mathcal{F}$ and ∇^g the Levi-Civita connection of (M, g). We also consider

$$\nabla_X s = \begin{cases} \mathring{\nabla}_X s, & X \in T(\mathcal{F}), \\ \Pi \nabla^g_X \sigma(s), & X \in ltr(T\mathcal{F}). \end{cases}$$

One checks easily that

PROPOSITION 2.59. *∇ is a connection in Q and $T_\nabla = 0$. Moreover g is bundle-like if and only if $\nabla g_{\text{tra}} = 0$.*

Here $T_\nabla(Y, Z) := \nabla_Y \Pi Z - \nabla_Z \Pi Y - \Pi[Y, Z]$ for any $Y, Z \in T(M)$. See also A. Bejancu & H.R. Farran, [**37**], where a few basic tools (necessary for the investigation of degenerate foliations) are developed [the main application there is to show that any degenerate (i.e. $\text{Rad}\, T(\mathcal{F}) \neq 0$) foliation $\mathcal{F}$ with $\text{cod}(\mathcal{F}) = 1$ on a 3-dimensional semi-Riemanian manifold is either totally umbilical or totally geodesic]. A more developed theory of foliations (perhaps with nontrivial radical bundle) in semi-Riemannian manifolds may contribute to a better understanding of the geometry of the foliations on nondegenerate CR manifolds M (where in general the Webster metric g_θ is only a semi-Riemannian metric) or the geometry of their pullbacks to $C(M)$ (which carries a natural Lorentz metric, the Fefferman metric of (M, θ)). See also [**91**].

2.5. The characteristic form

If $E \to M$ is a vector bundle of standard fibre $\mathbb{R}^k$ let $L(E) \to M$ denote the principal $\text{GL}(k, \mathbb{R})$-bundle of frames in the fibres of E. Let $(\nu, p-\nu)$ be the signature of g_P, where $p = \dim_{\mathbb{R}} P_x, x \in M$. Let $O(P) \to M$ be the principal $\text{O}(\nu, p-\nu)$-subbundle of $L(P) \to M$ determined by g_P. From now on let us assume $\mathcal{F}$ to be tangentially oriented i.e. $\text{O}(\nu, p-\nu)$ reduces to $\text{SO}(\nu, p-\nu)$.

DEFINITION 2.60. Let $\chi_{\mathcal{F}} \in \Omega^p(M)$ be the *characteristic form* of $(M, \mathcal{F})$ i.e.

$$\chi_{\mathcal{F},x}(v_1, \cdots, v_p) = \det[g_{\theta,x}(v_i, u(e_j))]$$

for any $v_i \in T_x(M)$ and some frame $u : \mathbb{R}^p \to T_x(M)$ adapted to the tangential $\mathrm{SO}(\nu, p-\nu)$-structure. □

Clearly, the definition of $\chi_{\mathcal{F},x}(v_1, \cdots, v_p)$ doesn't depend upon the choice of adapted frame at x. Also

$$P^\perp \rfloor \chi_{\mathcal{F}} = 0.$$

We shall need the following

LEMMA 2.61. *Let M be a nondegenerate CR manifold, θ a choice of contact form on M, and $\mathcal{F}$ a tangentially oriented foliation of M whose tangent bundle P is nondegenerate in $(T(M), g_\theta)$. Then*

$$\mathcal{L}_Z \chi_{\mathcal{F}}|_P = \{-\kappa(Z) + \theta(\mathrm{trace}\,(\tau_P)Z + \Pi^\perp(2J - \tau)Z\}\ \chi_{\mathcal{F}}|_P \tag{2.50}$$

for any $Z \in \Gamma^\infty(P^\perp)$.

Here $\tau_P : P \to P$ is given by $\tau_P X = \Pi^\perp \tau X$, for any $X \in P$. The identity (2.50) is the pseudohermitian analogue of a formula in [**211**] (cf. also (6.17) in [**243**], p. 66).

DEFINITION 2.62. The identity (2.50) is called *Rummler's formula.* □

To prove Lemma 2.61, let $\{E_1, \cdots E_p\}$ be an oriented local orthonormal frame of P, i.e. $g_P(E_i, E_j) = \epsilon_i \delta_{ij}$, where $\epsilon_1 = \cdots = \epsilon_\nu = -1$ and $\epsilon_{\nu+1} = \cdots = \epsilon_p = 1$. Then

$$(\mathcal{L}_Z \chi_{\mathcal{F}})(E_1, \cdots, E_p) = -\sum_{i=1}^p \chi_{\mathcal{F}}(E_1, \cdots, \Pi^\perp[Z, E_i], \cdots, E_p)$$

and

$$\Pi^\perp[Z, E_i] = \sum_{j=1}^p \epsilon_j g_\theta([Z, E_i], E_j)E_j$$

yield

$$\mathcal{L}_Z \chi_{\mathcal{F}}|_P = -\sum_{i=1}^p \epsilon_i g_\theta([Z, E_i], E_j)\ \chi_{\mathcal{F}}|_P\ . \tag{2.51}$$

On the other hand

$$\kappa(Z) = \sum_{i=1}^p \epsilon_i g_\theta(W(Z)E_i, E_i) =$$

$$= \sum_{i=1}^p \epsilon_i g_Q(\Pi\nabla_{E_i} E_i, \Pi Z) = -\sum_{i=1}^p \epsilon_i g_\theta(E_i, \nabla_{E-i} Z) =$$

$$= \sum_{i=1}^p \epsilon_i g_\theta([Z, E_i] + T_\nabla(Z, E_i), E_i)$$

as $2g_\theta(E_i, \nabla_Z E_i) = Z(\epsilon_i) = 0$. Finally

$$\kappa(Z) = \sum_{i=1}^p \epsilon_i g_\theta([Z, E_i], E_i) + \theta\left(\mathrm{trace}\,(\tau_P)Z + \Pi^\perp(2J\tau - \tau Z)\right)$$

hence (2.51) yields (2.50).

By Rummler's formula the p-form

$$\eta = i_Z d\chi_{\mathcal{F}} + \{\kappa(Z) - \theta(\mathrm{trace}\,(\tau_P)Z + \Pi^\perp(2J - \tau)Z)\}\chi_{\mathcal{F}}$$

vanishes along the leaves of $\mathcal{F}$. As an immediate application we may look at the case of a foliation tangent to the characteristic direction of (M, θ) and orthogonal to a semi-Levi foliation.

PROPOSITION 2.63. *Let $\mathcal{F}$ be tangent to T. If $P^\perp$ is involutive then the following statements are equivalent*
i) $\kappa = 0$
ii) $\mathcal{L}_Z\chi_{\mathcal{F}} = 0$ *for any* $Z \in \Gamma^\infty(P^\perp)$
iii) $d\chi_{\mathcal{F}} = 0$.

The proof mimics closely that of Theorem 6.23 in [**243**], p. 69, and is therefore omitted.

A foliation $\mathcal{F}$ with $\kappa = 0$ is not necessarily harmonic. For a geometric interpretation of this condition let β be the second fundamental form of (each leaf of) $\mathcal{F}$ in (M, g_θ) i.e.

$$\beta(X, X') = \Pi\nabla^\theta_X X', \quad X, X' \in \Gamma^\infty(P),$$

cf. e.g. (6.1) in [**243**], p. 62. Here ∇^θ is the Levi-Civita connection of (M, g_θ). ∇^θ is related to ∇ hence

$$\beta(X, X') = \alpha(X, X') + \{\Omega_\theta(X, X') - A(X, X')\}\Pi T + \\ + \theta(X')\Pi\tau X + \theta(X)\Pi J X' + \theta(X')\Pi J X \tag{2.52}$$

for any $X, X' \in P$. For $Z \in P^\perp$ let $a(Z) : P \to P$ be the Weingarten map i.e.

$$g_\theta(a(Z)X, X') = g_Q(\beta(X, X'), \Pi Z), \quad X, X' \in P,$$

cf. e.g. (6.3) in [**243**], p. 62. Then (by (2.52))

$$a(Z)X = W(Z)X - \theta(Z)\Pi^\perp(J + \tau)X + \\ + g_\theta(JX, Z)\Pi^\perp T - \theta(X)\Pi^\perp JZ + A(X, Z)\Pi^\perp T. \tag{2.53}$$

Let $\ell \in \Omega^1(M)$ be the mean curvature of $\mathcal{F}$ in (M, g_θ) i.e.

$$P \rfloor \ell = 0, \quad \ell(Z) = \mathrm{trace}\, a(Z), \quad Z \in P^\perp,$$

cf. e.g. (6.13) in [**243**], p. 65. Taking traces in (2.53) we obtain

$$\ell(Z) = \kappa(Z) - \theta\left(\mathrm{trace}\,(\tau_P)Z + \Pi^\perp(2J - \tau)Z\right) \tag{2.54}$$

for any $Z \in P^\perp$.

As an application we may state

PROPOSITION 2.64. *Let $\mathcal{F}$ be tangent to T. Then $\mathcal{F}$ is harmonic if and only if $\kappa = 0$. Or, let us assume that (M, θ) has vanishing pseudohermitian torsion (e.g. M is an odd dimensional sphere, or the Heisenberg group, or the pseudoconvex locus of a pseudo-Siegel domain, cf.* [**19**], p. 84-85). *Then $\mathcal{F}$ is harmonic if and only if $\kappa = \theta \circ (\Pi^\perp J)$.*

CHAPTER 3

Levi foliations

3.1. Existence of Levi foliations

The following elementary result is due to F. Sommer, [**223**].

THEOREM 3.1. *On any Levi flat CR manifold M there is a unique foliation $\mathcal{F}$ by complex manifolds such that $T(\mathcal{F})$ is the Levi distribution of M.*

Here we prove Theorem 3.1 for CR manifolds of hypersurface type (essentially the same proof works in the higher CR codimension case, cf. [**223**]).

Proof of Theorem 3.1. Let θ be a pseudohermitian structure on M and G_θ the corresponding (real) Levi form. As M is Levi flat the null space of the Levi form

$$\mathrm{Null}(G_\theta) = \{X \in H(M) : G_\theta(X, Y) = 0, \quad \forall Y \in H(M)\}$$

actually coincides with the Levi distribution $H(M)$. Then for any $X, Y \in H(M)$

$$\theta([X, Y]) = -2(d\theta)(X, Y) = 2G_\theta(X, JY) = 0$$

so that $[X, Y] \in H(M)$ i.e. $H(M)$ is involutive. By the classical Frobenius theorem $H(M)$ must be completely integrable hence there is a folition $\mathcal{F}$ of M (of codimension 1) such that $T(\mathcal{F}) = H(M)$. To see that any leaf $L \in T(M)/\mathcal{F}$ is a complex manifold let J_L be the pointwise restriction of $J : H(M) \to H(M)$ to L. Then for any $X, Y \in T(L)$ the Nijenhuis tensor of J_L is

$$N_{J_L}(X, Y) = [J_L X, J_L Y] - [X, Y] - J_L\{[J_L X, Y] + [X, J_L Y]\} =$$

$$= [JX, JY] - [X, Y] - J\{[JX, Y] + [X, JY]\} = 0$$

by the formal integrability property of the CR structure $T_{1,0}(M)$ (cf. e.g. (1.9) in [**89**], p. 4). Hence J_L is integrable so that, by the classical Newlander-Nirenberg theorem, (L, J_L) is a complex manifold (of complex dimension n, where n is the CR dimension of M). Q.e.d.

Theorem 3.1 has been generalized by M. Freeman, [**112**], to the case of CR submanifolds M such that $\mathrm{Null}(G_\theta)$ is a smooth distribution of rank $2k$ with $0 < k < n$. Foliations of this sort are encountered again in Section 6.4 (and called Levi foliations as well, cf. Definition 6.25 in Chapter 6) and in Chapter 8 (in connection with the foliated analog of the Beltrami equations in CR geometry, cf. [**16**]). It should be noticed that when M is embedded in some $\mathbb{C}^n$ (hence the complex structure J in $H(M)$ is induced by the complex structure of $\mathbb{C}^n$) the tangent space to each leaf of a given Levi foliation is a complex subspace of the tangent space at $\mathbb{C}^n$. Therefore in the second part of the proof of Theorem 3.1 (that each leaf of $\mathcal{F}$ is a complex manifold) one may avoid the use of the Newlander-Nirenberg theorem and use instead a result of T. Levi-Civita, [**171**]

THEOREM 3.2. *Let $L \subset \mathbb{C}^n$ be a real submanifold. If for each point $x \in L$ the tangent space $T_x(L)$ is a complex subspace of $T_x(\mathbb{C}^n) \approx \mathbb{C}^n$ (that is $J_x T_x(L) = T_x(L)$, where J is the complex structure of $\mathbb{C}^n$) then L is a complex submanifold of $\mathbb{C}^n$.*

Cf. [**111**], p. 5-6 for a simple proof of Levi-Civita's result.

We end this section by briefly discussing M. Freeman's results (cf. [**112**]) as to the existence of local Levi foliations. Let $M \subset \mathbb{C}^n$ be a smooth submanifold of real codimension m and $U \subseteq \mathbb{C}^n$ an open subset such that $M \cap U = \{x \in U : \rho(x) = 0\}$ for some $\rho \in C^\infty(U, \mathbb{R}^m)$ such that $d_x\rho$ has rank m for any $x \in M \cap U$. Let $O(M, U)$ be the ideal of the de Rham algebra $\Omega^\bullet(U)$ defined by

$$O(M, U) = \{\omega \in \Omega^\bullet(U) : \omega|_{M \cap U} = 0\}.$$

Given a set of differential polynomials $\{\eta_1, \cdots, \eta_q\} \subset \Omega^\bullet(U)$ let us denote by $(\eta_1, \cdots, \eta_q)_U$ the smallest ideal of $\Omega^\bullet(U)$ containing both $\{\eta_1, \cdots, \eta_q\}$ and $O(M, U)$ i.e. $(\eta_1, \cdots, \eta_q)_U$ is the ideal spanned by $\{\eta_1, \cdots, \eta_q, \rho_1, \cdots, \rho_m\}$. Moreover, given a C^∞ distribution $P : x \in M \mapsto P_x \subset T_x(M)$ on M, of codimension q, for each $x_0 \in M$ there is an open neighborhood U of x_0 in $\mathbb{C}^n$, a defining function $\rho \in C^\infty(U, \mathbb{R}^m)$ for $M \cap U$, and a system of 1-forms $\{\eta_1, \cdots, \eta_q\} \subset \Omega^1(U)$ of rank q at each point of $M \cap U$ such that

$$P_x = \bigcap_{a=1}^{q} \operatorname{Ker}(\eta_a)_x \,, \quad x \in M \cap U.$$

Then the ideal $(\eta_1, \cdots, \eta_q)_U$ doesn't depend upon the choice of the η_a's and is denoted by $\mathcal{I}_{P,U}$.

DEFINITION 3.3. $\mathcal{I}_{P,U}$ is called the *ideal of the distribution* P on U. If P is the tangent bundle to a foliation $\mathcal{F}$ we adopt the notation $\mathcal{I}_{\mathcal{F},U} := \mathcal{I}_{T(\mathcal{F}),U}$ and call $\mathcal{I}_{\mathcal{F},U}$ the *ideal of the foliation* $\mathcal{F}$ on U. □

A smooth foliation of codimension q of $M \cap U$ may also be thought of as a collection $\mathcal{F}$ of submanifolds of U such that there exists a C^∞ map $\rho : U \to \mathbb{R}^q$ such that $d_x\rho : T_x(\mathbb{C}^n) \to T_{\rho(x)}(\mathbb{R}^q)$ has rank $q > m$ at each $x \in M \cap U$ and $M \cap U = \{x \in U : \rho_j(x) = 0, \quad 1 \le j \le m\}$ and

$$\mathcal{F} = \{M_c : c = (c_{m+1}, \cdots, c_q) \in \mathbb{R}^{q-m}\},$$

$$M_c = \{x \in M \cap U : \rho_j(x) = c_j \,, \quad m < j \le q\}, \quad c \in \mathbb{R}^{q-m}.$$

Then the ideal of the foliation $\mathcal{F}$ is $\mathcal{I}_{\mathcal{F},U} = (d\rho_1, \cdots, \cdots, d\rho_q)_U$ i.e. the ideal spanned by $\{d\rho_1, \cdots, d\rho_q, \rho_1, \cdots, \rho_m\}$.

Since the correspondence $\mathcal{F} \mapsto \mathcal{I}_{\mathcal{F},U}$ is injective, the ideal of a foliation uniquely determines the foliation. M. Freeman's results (cf. *op. cit.*) are that both $\mathcal{I}_{H(M),U}$ on a Levi flat CR submanifold of $\mathbb{C}^n$ and $\mathcal{I}_{\mathrm{Null}(G_\theta),U}$ (provided that $\mathrm{Null}(G_\theta)$ has constant rank) are the ideals of some (Levi) foliation of $M \cap U$ (cf. Theorems 4.2 and 6.1 in [**112**], p. 17 and p. 21).

3.2. Holomorphic extension of Levi foliations

Let X be a complex manifold of complex dimension $n + k$, $k \ge 1$.

DEFINITION 3.4. A *holomorphic foliation* of (complex) *codimension* k of X is an atlas $\mathcal{F}$ of X as a complex manifold such that for any local charts (U, φ), $(U', \varphi') \in \mathcal{F}$ with $\varphi = (z^1, \cdots, z^n, w^1, \cdots, w^k)$ and $\varphi' = (z'^1, \cdots, z'^n, w'^1, \cdots, w'^k)$ if $U \cap U' \neq \emptyset$ then the transition map

$$\varphi' \circ \varphi^{-1} = (F^1, \cdots, F^n, G^1, \cdots, G^k) : \varphi(U \cap U') \to \varphi'(U \cap U')$$

obeys to

$$\frac{\partial G^j}{\partial \zeta^\alpha} = 0, \quad 1 \leq j \leq k, \quad 1 \leq \alpha \leq n,$$

on $\varphi(U \cap U')$. Here $(\zeta^1, \cdots, \zeta^n, \xi^1, \cdots, \xi^k)$ are the natural complex coordinates in $\mathbb{C}^{n+k}$. □

Clearly, as components of a transition function on a complex manifold $F^\alpha(\zeta, \xi)$ and $G^j(\xi)$ are holomorphic functions.

Let $M \subset X$ be a real hypersurface and $\mathcal{F}$ a holomorphic foliation of X of (complex) codimension one ($k = 1$).

PROPOSITION 3.5. *If M is a saturated subset of X then M is Levi flat.*

To prove this statement let $x_0 \in M$ and let L be the leaf of $\mathcal{F}$ through x_0. As M is saturated $L \subseteq M$. Hence $T_x(L) \subseteq T_x(M)$ for any $x \in L$. Next as L is a complex submanifold of X one has

$$T_x(L) = J_x T_x(L) \subseteq J_x T_x(M)$$

where J is the complex structure of X. Finally

$$T_x(L) \subseteq T_x(M) \cap J_x(M) = H(M)_x$$

hence $T_x(L) = H(M)_x$ by a dimension argument i.e. $H(M)$ is integrable Q.e.d.

DEFINITION 3.6. Due to Proposition 3.5 given a holomorphic foliation $\mathcal{F}$ of a complex manifold X and a real hypersurface $M \subset X$ which is a saturated set for $\mathcal{F}$ we say that $\mathcal{F}$ *extends holomorphically* the Levi foliation of M. □

C. Rea has asked, cf. [**203**], whether the converse holds, i.e. given a Levi flat real hypersurface M of a complex manifold X is there a holomorphic foliation of (a neighborhood of M in) X extending the Levi foliation of M? This turns out to be true at least in the real analytic case. Precisely we may state the following

THEOREM 3.7. (C. Rea, [**203**])
Let X be a (open subset of some) complex manifold and $M = \{z \in X : \rho(z) = 0\}$ a Levi flat real hypersurface in X, where $\rho \in C^\infty(X, \mathbb{R})$ and $d\rho \neq 0$ at each point of M. If M is real analytic then there is an open neighborhood Ω of M in X and a holomorphic foliation $\mathcal{F}$ of Ω, of (complex) codimension one, such that M is a saturated subset of Ω.

Rea's result actually holds in the arbitrary CR codimension case (we present the CR codimension one case only, for simplicity). As a crucial property of Levi flat CR manifolds it turns out that they admit nonconstant real valued CR functions (whereas, in the nondegenerate case the only real valued CR functions are the constants, cf. e.g. [**83**], p. 44).

3.2.1. Levi flatness. To prove Theorem 3.7 we need some preparation. We may state

LEMMA 3.8. *Let X be a complex manifold and $M = \{\rho = 0\}$ a real hypersurface in X of class C^r, $r \in \{\infty, \omega\}$. The following assertions are equivalent*

i) M is Levi flat.

ii) For any point $x \in M$ there is a neighborhood V of x in M and a function $f \in C^r(X, \mathbb{R})$ such that $\partial f \wedge \partial \rho = 0$ and $df \wedge d\rho \neq 0$ everywhere in V.

Proof. Let us prove i) $\Longrightarrow$ ii) for first. As M is Levi flat its Levi distribution is integrable. Thus for any $x \in M$ there is an open neighborhood V of x in M and a function $\varphi \in C^r(V, \mathbb{R})$ such that $d\varphi \neq 0$ everywhere in V and

$$H(M)\big|_U = \operatorname{Ker}(d\varphi).$$

Let f be just any C^r extension of φ to X. We wish to show that $df \wedge d\rho \neq 0$ at all points of V. The proof is by contradiction. If $(df \wedge d\rho)_x = 0$ at some $x \in V$ then, as we know that $(d\rho)_x \neq 0$, there is $w \in T_x(X)$ so that $w(\rho) \neq 0$. Let $v \in T_x(M)$ be an arbitrary tangent vector. Then

$$0 = 2(df \wedge d\rho)_{j(x)}((d_x j)v, w) =$$

$$= d_x(f \circ j)(v)\, w(\rho) - w(f)\, d_x(\rho \circ j)(v) = w(\rho)(d_x\varphi)v$$

where $j : M \subset X$ is the inclusion (and $f \circ j = \varphi$), hence $(d\varphi)_x = 0$, a contradiction. It remains to be shown that $\partial f \wedge \partial \rho = 0$ at all points of V. It is an easy fact of linear algebra that given a complex vector space Y and a hyperplane $Z \subset Y$ (i.e. a subspace Z with $\dim_{\mathbb{C}} Z = \dim_{\mathbb{C}} Y - 1$) and two linear forms $A, B : Y \to \mathbb{C}$, if $A\big|_Z = 0$ and $B\big|_Z = 0$ then $B = \lambda A$ for some $\lambda \in \mathbb{C}$. In the situation at hand, given $x \in V$, both linear forms

$$(\partial f)_x, (\partial \rho)_x : (T^{1,0}X)_x \to \mathbb{C}$$

vanish on $T_{1,0}(M)_x$, hence they are proportional. In particular $(\partial f \wedge \partial \rho)_x = 0$. Q.e.d.

Viceversa, to prove ii) $\Longrightarrow$ i) let $Y \in H(M)$ and $x \in M$. Then $Y = Z + \overline{Z}$ for some $Z \in T_{1,0}(M)$. Also by hypothesis there is an open subset $V \subseteq M$ such that $x \in V$ and there is $f \in C^r(X, \mathbb{R})$ such that $\partial f \wedge \partial \rho = 0$ and $df \wedge d\rho \neq 0$ everywhere in V. Let $(z^1, \cdots, z^{n+1})$ be local complex coordinates at $j(x)$ defined on (a eventually smaller neighborhood of $j(x)$, say $V' \subseteq$) V. There is $k \in \{1, \cdots, n+1\}$ such that $\rho_{z_k}(j(x)) \neq 0$ (because of $d\rho \neq 0$ everywhere on M). Then

$$0 = 2(\partial f \wedge \partial \rho)_{j(x)}\Big((d_x j)Y_x, \frac{\partial}{\partial z_k}\Big|_{j(x)}\Big) =$$

$$= [(\partial f)_{j(x)}(d_x j)Z_x]\rho_{z_k}(j(x)) = Z_x(f \circ j) = Z_x(\varphi)$$

because of $\rho \circ j = 0$. Thus $Z_x(\varphi) = 0$. Also as φ is $\mathbb{R}$-valued, $\overline{Z}_x(\varphi) = 0$. Summing up we have $Y_x(\varphi) = 0$ hence $H(M)_x \subseteq \operatorname{Ker}(d\varphi)_x$. Finally $df \wedge d\rho \neq 0$ and $d\rho \neq 0$ at $j(x)$ yield $d\varphi \neq 0$ at x so that $H(M)_x = \operatorname{Ker}(d\varphi)_x$ by a dimension argument. Clearly $\operatorname{Ker}(d\varphi)$ is involutive hence M is Levi flat. Q.e.d.

3.2.2. Holomorphic extension of Levi foliations: the local problem. We may state the following

LEMMA 3.9. *Let X be a complex manifold and $M = \{\rho = 0\} \subset X$ a real hypersurface. Let $(U, z^1, \cdots, z^n, w)$ be a local system of complex coordinates on X such that $U \cap M \neq \emptyset$. The following assertions are equivalent*

i) *There is a holomorphic foliation $\mathcal{F}$ on U, of codimension one, such that a) $(U, z^1, \cdots, z^n, w)$ is a foliated chart of $\mathcal{F}$, and b) $M \cap U$ is a saturated subset of U.*

ii) $\partial\rho/\partial z^\alpha = 0$ *on $M \cap U$, for $1 \leq \alpha \leq n$.*

Proof. We look first at the implication i) $\Longrightarrow$ ii). Let $x_0 \in U \cap M$ and let L be the leaf of $\mathcal{F}$ through x_0. Let α be the plaque of $\varphi(x_0)$ in $\varphi(U)$, i.e. α is the connected component of $\varphi(x_0)$ in $\varphi(U) \cap \pi^{-1}(w_0)$, where $\varphi(x_0) = (z_0, w_0) \in \mathbb{C}^n \times \mathbb{C}$ and $\pi : \mathbb{C}^n \times \mathbb{C} \to \mathbb{C}$ is the natural projection. Here $\varphi = (z^1, \cdots, z^n, w)$. There is $R > 0$ such that $(z_0 + z, w_0) \in \varphi(U)$ for any $|z| < R$. Then

$$V = \varphi^{-1}(\{(z_0 + z, w_0) : |z| < R\})$$

is an open submanifold of the plaque $\varphi^{-1}(\alpha)$ of x_0 in U. One has

$$V \subseteq \varphi^{-1}(\alpha) \subseteq L \subseteq M$$

(the last inclusion holds because $x_0 \in M \cap U$ and $M \cap U$ is saturated). Hence

$$\rho(\varphi^{-1}(z_0 + z, w_0)) = 0$$

for any $|z| < R$, which yields ii).

Viceversa, to prove ii) $\Longrightarrow$ i) let us assume that $\partial\rho/\partial z^\alpha = 0$ on $M \cap U$. Let us consider the model foliation on $\mathbb{C}^n \times \mathbb{C}$, whose leaves are the affine hyperplanes $\xi =$ constant, where $(\zeta^1, \cdots, \zeta^n, \xi)$ are the natural complex coordinates on $\mathbb{C}^{n+1}$. Next let $\mathcal{F}_{\varphi(U)}$ be the foliation induced on $\varphi(U)$ by the model foliation of $\mathbb{C}^n \times \mathbb{C}$ (the leaves of $\mathcal{F}_{\varphi(U)}$ are the connected components of the sets $\varphi(U) \cap \{\xi = \text{constant}\}$). Let $\mathcal{F}_U$ be the pullback of $\mathcal{F}_{\varphi(U)}$ via φ. Then $\mathcal{F}_U$ is a holomorphic foliation of codimension one on U. Let $x \in U \cap M$ and L the leaf of $\mathcal{F}_U$ through x. As L is locally described by the equations $w =$ constant, it follows that $Y \in T(\mathcal{F}_U)$ if and only if $Y = Z + \overline{Z}$ for some Z of the form $Z = Z^\alpha \partial/\partial z^\alpha$. Then (by the hypothesis ii))

$$Y(\rho) = 0 \quad \text{on} \quad M \cap U. \tag{3.1}$$

At this point we need to recall that L consists of all points $y \in U$ which can be joined to x by a piecewise C^1 curve $\gamma : [0,1] \to U$ such that $(d\gamma/dt)(t) \in T(\mathcal{F}_U)_{\gamma(t)}$ for all $0 \leq t \leq 1$ where $(d\gamma/dt)(t)$ is defined. Precisely $\gamma : [0,1] \to U$ is C^0 and there is a partition $t_0 = 0 < t_1 < \cdots < t_{k-1} < t_k = 1$ such that γ is C^1 on each (t_i, t_{i+1}) and $d\gamma_i/dt$ admits a C^0 extension to $[t_i, t_{i+1}]$, $0 \leq i \leq k-1$. Here $\gamma_i = \gamma|_{[t_i, t_{i+1}]}$. Also we set $x_i = \gamma(x_i)$. Then

$$\frac{d\gamma_0}{dt}(t) \in T(\mathcal{F}_U)_{\gamma_0(t)}, \quad 0 < t < t_1,$$

and (by (3.1))

$$\frac{d}{dt}\rho(\gamma_0(t)) = \frac{d\gamma_0}{dt}(t)(\rho) = 0$$

hence $\rho(\gamma_0(t)) = c = \text{constant}$ on $(0, t_1)$. Yet

$$c = \lim_{t \to 0^+} \rho(\gamma_0(t)) = \rho(\gamma(0)) = \rho(x) = 0$$

as $x \in M$. We proved that $\gamma_0(t) \in M$, $0 \le t \le t_1$, hence $x_1 \in M$. A similar argument applied to the C^1 curve γ_1 shows that $x_2 \in M$ and so on. In the end $y = x_k \in M$ hence L lies in $M \cap U$ i.e. $M \cap U$ is a saturated subset of U. Q.e.d.

3.2.3. The global extension problem. We have

PROPOSITION 3.10. *Let X be a complex manifold and $M = \{\rho = 0\} \subset X$ a real hypersurface. The following statements are equivalent*

i) *M is Levi flat and there is an open neighborhood Ω of M in X and a holomorphic foliation $\mathcal{F}$ of Ω of (complex) codimension one extending the Levi foliation of M.*

ii) *For any $x \in M$ there is a neighborhood U of x in X and a holomorphic function $h \in \mathcal{O}(U)$ such that* a) *$\partial h \wedge \partial \rho = 0$ and* b) *$\partial h \neq 0$ everywhere on $M \cap U$.*

To prove Proposition 3.10 we need to establish

LEMMA 3.11. *Let X be a complex manifold, $M = \{\rho = 0\} \subset X$ a real hypersurface, and $(U, z^1, \cdots, z^n, w)$, $(U', z'^1, \cdots, z'^n, w')$ two local coordinate systems on X with the property that $M \cap U \cap U' \neq \emptyset$. If $\partial\rho/\partial z^\alpha = 0$ on $M \cap U$ and $\partial\rho/\partial z'^\alpha = 0$ on $M \cap U'$ then the transition functions $\varphi' \circ \varphi^{-1} = (F^1, \cdots, F^n, G) : \varphi(U \cap U') \to \varphi'(U \cap U')$ obey to*

$$\frac{\partial G}{\partial \zeta^\alpha} = 0 \text{ on } \varphi(U \cap U')$$

where $\varphi = (z^1, \cdots, z^n, w)$ and $\varphi' = (z'^1, \cdots, z'^n, w')$.

Proof. One has

$$0 = \frac{\partial \rho}{\partial z^\alpha}(x) = \frac{\partial(\rho \circ \varphi^{-1})}{\partial \zeta^\alpha}(\varphi(x)) = \frac{\partial(\rho \circ \varphi'^{-1} \circ \varphi' \circ \varphi^{-1})}{\partial \zeta^\alpha}(\varphi(x)) =$$

$$= \frac{\partial(\rho \circ \varphi'^{-1})}{\partial \zeta^\beta}(\varphi'(x)) \frac{\partial F^\beta}{\partial \zeta^\alpha}(\varphi(x)) + \frac{\partial(\rho \circ \varphi'^{-1})}{\partial \xi}(\varphi'(x)) \frac{\partial G}{\partial \zeta^\alpha}(\varphi(x)) =$$

$$= \frac{\partial \rho}{\partial z'^\beta}(x) \frac{\partial F^\beta}{\partial \zeta^\alpha}(\varphi(x)) + \frac{\partial \rho}{\partial w'}(x) \frac{\partial G}{\partial \zeta^\alpha}(G(x))$$

i.e.

$$\frac{\partial \rho}{\partial w'}(x) \frac{\partial G}{\partial \zeta^\alpha}(\varphi(x)) = 0 \tag{3.2}$$

for any $x \in M \cap U \cap U'$. Yet $(\partial\rho)_x \neq 0$ (because of $(d\rho)_x \neq 0$) hence $\rho_{w'}(x) \neq 0$ (as $\rho_{z'^\alpha}(x) = 0$) and (3.2) yields $\partial G/\partial \zeta^\alpha = 0$ on $\varphi(M \cap U \cap U')$. Summing up $h_\alpha = (\partial G/\partial \zeta^\alpha) \circ \varphi$ is a holomorphic function on $U \cap U'$ which vanishes on the real hypersurface $M \cap U \cap U'$. Thus[1] one may conclude that $h_\alpha = 0$ on $U \cap U'$, and Lemma 3.11 is proved.

Before proceeding with the proof of Proposition 3.10 note that the proof of Lemma 3.11 relied on the following well known result

LEMMA 3.12. *Let X be a complex manifold and $M \subset X$ a real hypersurface. Let $U \subseteq X$ be an open subset such that $M \cap U \neq \emptyset$ and $h \in \mathcal{O}(U)$ a holomorphic function on U. If $h = 0$ on $M \cap U$ then h vanishes identically.*

[1] By Lemma 3.12 below.

The proof is well known hence omitted (cf. Lemma 2 in [**50**], p. 142). Let us prove now Proposition 3.10. We deal first with the implication i) $\Longrightarrow$ ii). Let $x \in M$ and let (U, z^α, w) be a foliated chart of $(\Omega, \mathcal{F})$ at x. Let h be the last coordinate function i.e. $h = w$. Then $h \in \mathcal{O}(U)$ and $\partial h \neq 0$ everywhere. Moreover (by Lemma 3.9)

$$\partial \rho = \frac{\partial \rho}{\partial z^\alpha} dz^\alpha + \frac{\partial \rho}{\partial w} dw = \rho_w dw$$

hence $\partial h \wedge \partial \rho = 0$. Viceversa, let us establish the implication ii) $\Longrightarrow$ i). As $\partial h \neq 0$ on $U \cap M$ there exists a system of complex coordinates on (a eventually smaller neighborhood of x say $U' \subseteq) U$, say $(z^1, \cdots, z^n, w)$, such that $h = w$. Hence (by $\partial h \wedge \partial \rho = 0$) it follows that $\partial \rho = \lambda dw$ for some function λ i.e. $\partial \rho / \partial z^\alpha = 0$. As $x \in M$ was arbitrarily chosen it follows that M admits an open cover $\{U_i\}_{i \in I}$ with geometric zones of local coordinate systems (U_i, z_i^α, w_i) of X such that $U_i \cap M \neq \emptyset$ and $\partial \rho / \partial z_i^\alpha = 0$ on $U_i \cap M$ for any $i \in I$. Let us set $\Omega = \cup_{i \in I} U_i$. Then Ω is an open neighborhood of M in X. By Lemmae 3.9 and 3.11 there is a holomorphic foliation $\mathcal{F}$ of Ω of complex codimension one such that M is a saturated set of Ω, hence M is Levi flat and $\mathcal{F}$ extends holomorphically the Levi foliation of M.

3.2.4. Proof of Rea's theorem. As M is Levi flat for any $x \in M$ there is (by Lemma 3.8) an open neighborhood V of x in M and a function $f \in C^\omega(X, \mathbb{R})$ such that $\partial f \wedge \partial \rho = 0$ and $df \wedge d\rho \neq 0$ at each point of V. Let us set $\varphi = f|_M$.

Claim 1. As $\partial f \wedge \partial \rho = 0$ it follows that $\varphi \in CR^\omega(V)$ i.e. φ is a real analytic CR function on V.

To prove Claim 1 let $y \in V$ and $z \in T_{1,0}(M)_y$. As f and ρ are real valued the identity $\partial f \wedge \partial \rho = 0$ may also be written $\overline{\partial} f \wedge \overline{\partial} \rho = 0$. Since

$$(\overline{\partial} \rho)_y (d_y j) \overline{z} = (d\rho)_y (d_y j) \overline{z} = d_y (\rho \circ j) \overline{z} = 0$$

(as $\rho \circ j = 0$) it follows that

$$0 = 2(\overline{\partial} f \wedge \overline{\partial} \rho)_y ((d_y j) \overline{z}, w) = [(\overline{\partial} f)_y (d_y j) \overline{z}](\overline{\partial} \rho)_y w = \overline{z}(\varphi)(\overline{\partial} \rho)_y w$$

for any $w \in T_y(X) \otimes \mathbb{C}$. On the other hand there surely is $w \in T_y(X) \otimes \mathbb{C}$ such that $(\overline{\partial} \rho)_y w \neq 0$ (otherwise $(\overline{\partial} \rho)_y = 0$ and then $(\partial \rho)_y = 0$ by complex conjugation so that $(d\rho)_y = 0$ and in particular $(df \wedge d\rho)_y = 0$, a contradiction) hence $\overline{z}(\varphi) = 0$ and Claim 1 is proved.

By Severi's theorem (cf. [**219**]) there is an open set $U \subseteq X$ such that $U \cap M = V$ and a holomorphic function $h \in \mathcal{O}(U)$ such that $h|_V = \varphi$.

Claim 2. $\partial f \wedge \partial \rho = 0$ yields $\partial h \wedge \partial \rho = 0$ on $M \cap U$ that is property a) in Proposition 3.10 is satisfied.

The proof of Claim 2 is by contradiction. Let us assume that $(\partial h \wedge \partial \rho)_y \neq 0$ for some $y \in V$. Therefore there are $v, w \in T_y(X) \otimes \mathbb{C}$ such that

$$(\partial h \wedge \partial \rho)_y (v, w) \neq 0. \tag{3.3}$$

As ∂h and $\partial \rho$ vanish on complex vectors of type $(0,1)$ it follows that at least one of the vectors v, w belongs to the holomorphic tangent space $T^{1,0}(X)_y$, say $v \in T^{1,0}(X)_y$. Let g be just any Hermitian metric on X and $\nabla \rho$ the gradient of ρ with respect to g i.e. $g(V, \nabla\rho) = V(\rho)$ for any $V \in T(X)$. The direct sum decomposition

$$T^{1,0}(X)_y = [(d_y j)T_{1,0}(M)_y] \oplus \mathbb{C}(\nabla\rho)^{1,0}_y$$

yields $v = (d_y j)z + \lambda(\nabla\rho)^{1,0}_y$ for some $z \in T_{1,0}(M)_y$ and some $\lambda \in \mathbb{C}$. Here $(\nabla\rho)^{1,0}$ is the $T^{1,0}(X)$-component of $\nabla\rho$ with respect to the decomposition $T(X) \otimes \mathbb{C} = T^{1,0}(X) \oplus T^{0,1}(X)$. Then (by (3.3))

$$0 \neq 2(\partial h \wedge \partial\rho)_y(v, w) =$$
$$= 2(\partial h \wedge \partial\rho)_y((d_y j)z, w) + 2\lambda(\partial h \wedge \partial\rho)_y((\nabla\rho)^{1,0}_y, w) =$$
$$= [d_y(h \circ j)z](\partial\rho)_y w + 2\lambda(\partial h \wedge \partial\rho)_y((\nabla\rho)^{1,0}, w^{1,0})$$

where $w^{1,0}$ is the $T^{1,0}(X)_y$-component of w. We may also decompose $w^{1,0}$ as $w^{1,0} = (d_y j)w_{1,0} + \mu(\nabla\rho)^{1,0}_y$ for some $w_{1,0} \in T_{1,0}(M)_y$ and some $\mu \in \mathbb{C}$. Therefore

$$0 \neq [(\partial\varphi)_y z](\partial\rho)_y w + 2\lambda(\partial h \wedge \partial\rho)_y((\nabla\rho)^{1,0}_y, (d_y j)w_{1,0}) =$$
$$= [(df)_y(d_y j)z](\partial\rho)_y w - \lambda[(\partial h)_y(d_y j)w_{1,0}](\partial\rho)_y(\nabla\rho)^{1,0}_y =$$
$$= 2\underbrace{(\partial f \wedge \partial\rho)}_{=0}{}_y((d_y j)z, w) - \frac{\lambda}{2}\left[(d\varphi)_y w_{1,0}\right] \|d\rho\|^2_y = 0$$

because $\overline{w_{1,0}}(\varphi) = 0$ and φ is real so that $w_{1,0}(\varphi) = 0$ as well. This is a contradiction, so that Claim 2 is proved.

Let us set $h = u + iv$ with u and v real valued. Then $v = 0$ on $M \cap U$. Also (as $\overline{\partial} h = 0$) we have $\partial h \wedge d\rho = dh \wedge d\rho$ on $M \cap U$.

Claim 3. $\partial h \wedge d\rho = du \wedge d\rho$ on $M \cap U$.

To prove Claim 3 it suffices to show that $(dv \wedge d\rho)_y = 0$ for any $y \in M \cap U$. Let $\tilde{z}, \tilde{w} \in T_y(X) \otimes \mathbb{C}$. Taking into account the direct sum decomposition

$$T_y(X) \otimes \mathbb{C} = [(d_y j)T(M) \otimes \mathbb{C}] \oplus \mathbb{C}(\nabla\rho)_y$$

we have

$$\tilde{z} = (d_y j)z + \lambda(\nabla\rho)_y, \quad \tilde{w} = (d_y j)w + \mu(\nabla\rho)_y,$$

for some $z, w \in T_y(M) \otimes \mathbb{C}$ and $\lambda, \mu \in \mathbb{C}$. Then

$$(dv \wedge d\rho)_y(\tilde{z}, \tilde{w}) = (dv \wedge d\rho)_y((d_y j)z, (d_y j)w) +$$
$$+ \mu(dv \wedge d\rho)((d_y j)z, (\nabla\rho)_y) + \lambda(dv \wedge d\rho)_y((\nabla\rho)_y, (d_y j)w) = 0$$

because of $\rho \circ j = 0$ and $v \circ j = 0$. Claim 3 is proved.

Claim 4. $du \wedge d\rho \neq 0$ everywhere in $M \cap U$ and in particular $\partial h \neq 0$ everywhere on $M \cap U$ that is property b) in Proposition 3.10 is satisfied.

First, let us observe (by Claim 3) that $\partial h \wedge d\rho = du \wedge d\rho \neq 0$ yields $\partial h \neq 0$ as claimed. The proof of the first statement in Claim 4 is by contradiction. Precisely, we assume that $(du \wedge d\rho)_y = 0$ for some $y \in M \cap U$. Then for any $z \in T_y(M) \otimes \mathbb{C}$ and any $w \in T_y(X) \otimes \mathbb{C}$

$$0 = 2(du \wedge d\rho)_y((d_y j)z, w) =$$
$$= [(df)_y(d_y j)z)(d\rho)_y w = 2(df \wedge d\rho)_y((d_y j)z, w)$$

and (because M has codimension one) this suffices to conclude that $(df \wedge d\rho)_y = 0$, a contradiction. Claim 4 is completely proved.

Finally (by Claim 4) the implication ii) $\Longrightarrow$ i) in Proposition 3.10 yields the statement in Theorem 3.7. Q.e.d.

EXAMPLE 3.13. (*A Levi flat real hypersurface in* $\mathbb{C}^2$ *which is not real analytic at any of its points, yet its Levi foliation extends holomorphically*)

Let $f : \mathbb{R} \to \mathbb{R}$ be a C^∞ function which is not C^ω at any point and let us set

$$M = \{(z, w) \in \mathbb{C}^2 : \mathrm{Im}(w) = f(\mathrm{Re}(w))\}.$$

Then M is Levi flat and the leaves of its Levi foliation are the complex manifolds

$$L_u = \{(z, w) \in \mathbb{C}^2 : w = u + if(u)\}, \quad u \in \mathbb{R}.$$

The Levi foliation of M extends holomorphically to the foliation of $\mathbb{C}^2$ whose leaves are given by $w =$ constant. $\square$

EXAMPLE 3.14. (*A Levi flat real hypersurface in* $\mathbb{C}^2$ *whose Levi foliation does not extend holomorphically*)

By Proposition 3.10 it is enough to find ρ with $d\rho \neq 0$ on $M = \{\rho = 0\}$ and some point $x \in M$ such that any holomorphic function h in a neighborhood U of x and satisfying $\partial h \wedge \partial\rho = 0$ on $M \cap U$, follows to be a constant.

Let $D = \{z \in \mathbb{C} : |z| < 1\}$ and $I = (-1, 1) \subset \mathbb{R}$. Consider the function

$$\Psi : D \times I \to \mathbb{C}^2 \quad \Psi(z, t) = (z, a(t)z + t),$$

where $a \in C^\infty(\mathbb{R})$ is a given $\mathbb{R}$-valued function such that

$$|a'(t)| < \frac{1}{2}, \quad t \in I,$$

$$a'(t)x + 1 \neq 0, \quad (z, t) \in D \times I,$$

(where $x = \mathrm{Re}(z)$). Note that the Jacobian of Ψ at (z, t) is given by

$$\begin{pmatrix} 1 & 0 & a(t) & 0 \\ 0 & 1 & 0 & a(t) \\ 0 & 0 & a'(t)x + t & a'(t)y \end{pmatrix}$$

hence $d\Psi$ has maximal rank at any $(z, t) \in D \times I$. Also Ψ is injective. Indeed if $\Psi(z, t) = \Psi(z', t')$ and $(z, t) \neq (z', t')$, then $z = z'$ and $t \neq t'$, say $t < t'$. Then (by Rolle's theorem) $a'(\tau)z + 1 = 0$ for some $\tau \in (t, t')$. Hence $1 = |a'(\tau)|\,|z| < 1/2$, a contradiction. Therefore there is a function $\rho : \mathbb{C}^2 \to \mathbb{R}$ and an open set $\Omega \subset \mathbb{C}^2$ such that

$$\Psi(D \times I) = \{\rho = 0\} \cap \Omega$$

and $d\rho \neq 0$ on $M = \{\rho = 0\} \cap \Omega$. Then M is Levi flat because for any $x \in M, x = \Psi(z, t)$ the complex line

$$\gamma : \Delta \to \mathbb{C}^2, \quad \gamma(\zeta) = (z + \zeta, a(t)(z + \zeta) + t), \quad \zeta \in \Delta,$$

$$\Delta = \{\zeta \in \mathbb{C} : |\zeta + z| < 1\},$$

(passing through x, i.e. $\gamma(0) = x$) lies on M. Next we choose $a(t)$ as follows

$$\alpha(t) = \begin{cases} \exp(-t^{-2})\sin(\pi/t) & \text{if } t \in \mathbb{R} \setminus \{0\}, \\ 0 & \text{if } t = 0, \end{cases}$$

$$K = 2 \sup_{|t| \leq 1} |\alpha'(t)|, \quad a(t) = K^{-1}\alpha(t), \quad t \in \mathbb{R}.$$

Then the zero set $Z(a)$ of $a(t)$ is precisely $Z(a) = \{t_n : n \geq 0\}$ where $t_n = 1/n$, $n \geq 1$, and $t_0 = 0$. We have

$$\rho(z, a(t)z + t) = 0$$

for any $(z,t) \in D \times I$. Then, by differentiation with respect to z

$$\rho_z(z, a(t)z + t) + a(t)\rho_w(z, a(t)z + t) = 0. \tag{3.4}$$

Let us consider now a holomorphic function $h \in \mathcal{O}(U)$ defined in some neighborhood of the origin, such that $\partial h \wedge \partial \rho = 0$ on $M \cap U$. Consequently

$$h_z \rho_w - h_w \rho_z = 0$$

on $M \cap U$ and substitution from (3.4) leads to

$$\rho_w(z, a(t)z + t)[a(t)h_w(z, a(t)z + t) - h_z(z, a(t)z + t)] = 0. \tag{3.5}$$

By (3.4) we have $\rho_z(z, t_n) = 0$ hence

$$\rho_w(z, t_n) \neq 0 \tag{3.6}$$

for any $z \in D$ and $n \geq 0$ (because of $(d\rho)_{z,t_n)} \neq 0$). Next (by (3.5))

$$\rho_w(z, t_n) h_z(z, t_n) = 0$$

hence (by (3.6))

$$h_z(z, t_n) = 0 \tag{3.7}$$

for any $z \in D$ and any $n \geq 0$. For $z \in D$ fixed let us consider the function $\gamma(\zeta) = h_z(z, \zeta)$. Then γ is holomorphic in a neighborhood of the origin and $\gamma(0) = 0$, $\gamma(1/n) = 0$ (by 3.7)) and $1/n \to 0$ as $n \to \infty$, i.e. 0 is not an isolated zero of γ, so we must have $\gamma \equiv 0$, hence $h_z \equiv 0$. Let us set

$$A = \{(z, a(t)z + t) \in \mathbb{C}^2 : z \in D, t \in I \setminus \{t_n : n \geq 0\}\}.$$

Then A is an open dense subset of M and $\rho_w|_{M \setminus A} \neq 0$ by (3.6). Therefore there is a point of A where $\rho_w \neq 0$ (otherwise, as A is dense in M, one would get $\rho_w = 0$ on M, a contradiction) and hence a whole open subset $B \subset A$ such that $\rho_w \neq 0$ on B. As B lies in M, B itself is a real hypersurface in $\mathbb{C}^2$. By (3.5) we have

$$\rho_w(x) a(t) h_w(x) = 0$$

for any $x \in B$, $x = \Psi(z,t)$. Yet $Z(a) = \{t_n : n \geq 0\}$ hence $a(t) \neq 0$. It follows that $h_w = 0$ on B and then (by Lemma 3.12) $h_w = 0$ on U. Summing up $h_z = 0$ and $h_w = 0$ on U i.e. $h =$ constant. □

3.3. Pluriharmonic defining functions

Let V be a complex manifold and $M \subset X$ a Levi flat real hypersuface. Let $\mathcal{F}$ be the Levi foliation of M. D.E. Barret, [**28**], studies the interplay between the geometric properties of $\mathcal{F}$, the pseudoconvexity properties of $X \setminus M$, and the behavior of the tangential Cauchy-Riemann equations on M. Also [**28**] is devoted to the study of certain classes of compact Levi flat hypersurfaces generalizing an example due to T. Ohsawa, [**197**] (Ohsawa's example is a Levi flat hypersurface occurring as the product of an annulus with the punctured plane $\mathbb{C} \setminus \{0\}$). The following facts are elementary

PROPOSITION 3.15. *Let M be a Levi flat CR manifold and let $\mathcal{F}$ be its Levi foliation. Then* $\mathrm{CR}^\infty(M) = \{f \in C^\infty(M) : f|_L \in \mathcal{O}(L),\ \ L \in M/\mathcal{F}\}$ *i.e. the tangential Cauchy-Riemann equations on M are the ordinary Cauchy-Riemann equations along the leaves of $\mathcal{F}$. Moreover* $\Omega^0_B(\mathcal{F}) = \{f \in \mathrm{CR}^\infty(M) : \overline{f} \in \mathrm{CR}^\infty(M)\}$ *i.e. the basic functions of the Levi foliations are those CR functions of class C^∞ on M whose complex conjugates are CR functions as well.*

Proof. Let $f \in \mathrm{CR}^\infty(M)$. Then $\overline{Z}(f) = 0$ for any $Z \in T_{1,0}(M)$. Let $L \in M/\mathcal{F}$ be a leaf of $\mathcal{F}$ and let $\iota : L \to M$ be the inclusion so that $(d_x\iota)T^{1,0}(L)_x = T_{1,0}(M)_x$ for any $x \in L$. Let $Z \in T^{1,0}(L)_x$. Then

$$\overline{Z}(f|_L) = \overline{Z}(f \circ \iota) = ((d_x\iota)\overline{Z})(f) = 0$$

hence $f|_L \in \mathcal{O}(L)$. The converse is similar. As to the proof of the second statement in Proposition 3.15 let $f \in \Omega^0_B(\mathcal{F})$. Then f is constant on each leaf $L \in M/\mathcal{F}$ and in particular $f|_L$ is holomorphic on L so that, by the first statement in Proposition 3.15, f is a CR function on M. Of course $\overline{f}$ is constant on each leaf as well, hence $\overline{f}$ is CR, too. Conversely, let $f \in \mathrm{CR}^\infty(M)$ such that $\overline{f} \in \mathrm{CR}^\infty(M)$. Let $Y \in T(\mathcal{F}) = H(M)$ so that $Y = Z + \overline{Z}$ for some $Z \in T_{1,0}(M)$. Then

$$Y(f) = \overline{Z}(f) + \overline{\overline{Z}(\overline{f})} = 0$$

so that $f \in \Omega^0_B(\mathcal{F})$. Q.e.d.

THEOREM 3.16. (D.E. Barrett, [**28**])
Let $M \subset X$ be a Levi flat real analytic hypersurface of the complex manifold X and let $\mathcal{F}$ be the Levi foliation of M. The following statements are equivalent

i) *There is a real valued function $r \in C^\infty(\Omega)$ defined on an open neighborhood Ω of M in X such that r is a defining function for M, i.e. $M = \{z \in \Omega : r(z) = 0\}$ and $(dr)(z) \neq 0$ for any $z \in M$, and r is pluriharmonic i.e. $\partial\overline{\partial}r = 0$ in Ω.*

ii) *The Levi foliation $\mathcal{F}$ of M is given by a closed nowhere zero real analytic 1-form i.e. there is $\eta \in \Gamma^\omega(T^*(M))$ such that $d\eta = 0$, $\eta_z \neq 0$ for any $z \in M$, and $T(\mathcal{F}) = \mathrm{Ker}(\eta)$.*

We need the following

LEMMA 3.17. *Let X be a complex manifold and $\Omega \subseteq X$ an open subset. Let $r : \Omega \to \mathbb{R}$ be a C^2 function. Then r is pluriharmonic if and only if the real 1-form $d^c r$ is closed, where $d^c = i(\overline{\partial} - \partial)$.*

The proof of Lemma 3.17 follows from $d(d^c r) = 2i\partial\overline{\partial}r$.

Proof of i) $\Longrightarrow$ ii). Let us assume that M is given by a pluriharmonic defining function i.e. there is an open subset $\Omega \subseteq X$ and a real valued function $r \in C^\omega(\Omega)$ such that $M = \{z \in \Omega : r(z) = 0\}$ and $(dr)(z) \neq 0$ for any $z \in M$, and $\partial\overline{\partial}r = 0$ in Ω. Let $V \in T(\mathcal{F}) = H(M) \subset T(M)$. Then $V(r) = 0$ hence $r \in \Omega^0_B(\mathcal{F})$. Let L be a leaf of $\mathcal{F}$ and $Z \in T^{1,0}(L)_x$, $x \in L$. Then

$$(\overline{\partial}r)_x(d_x\iota)\overline{Z} = (dr)_x(d_x\iota)\overline{Z} = d_x(r \circ \iota)\overline{Z} = 0$$

as $r \circ \iota =$ constant. Therefore $\overline{\partial}r$, and in a similar way ∂r, vanish along the leaves of $\mathcal{F}$. Consequently $d^c r$ vanishes on $T(\mathcal{F})$. Let us consider the real 1-form $\eta := j^* d^c r \in \Gamma^\omega(\Omega, T^*(X))$ where $j : M \to X$ is the inclusion. Then η is closed (by Lemma 3.17) and

$$T(\mathcal{F}) = \mathrm{Ker}(\eta). \tag{3.8}$$

Indeed the inclusion $\subseteq$ in (3.8) has already been proved. On the other hand $(dr)(z) \neq 0$ yields $\eta_z \neq 0$ for any $z \in M$, hence (3.8) follows by comparing dimensions.

Proof of ii) $\Longrightarrow$ i). As the Levi foliation $\mathcal{F}$ has codimension one, for any point $z \in M$ there is an open subset $U \subseteq M$ and a real analytic function $f : U \to \mathbb{R}$ such that $z \in U$ and $T(\mathcal{F})|_U = \mathrm{Ker}(df)$. Therefore, for any $X \in T(\mathcal{F})$ one has $0 = (df)X = X(f)$ hence $f : U \to \mathbb{R}$ is a basic function. Then $f \in \mathrm{CR}^\omega(U)$, as a consequence of Proposition 3.15. Then (by Tomassini's theorem, [**242**]) f extends holomorphically to a neighborhood of z in X i.e. there is an open set $\tilde{U} \subseteq X$ and a holomorphic function $F \in \mathcal{O}(\tilde{U})$ such that $U \subset \tilde{U}$ and $F|_U = f$. Let us consider the function $r : \tilde{U} \to \mathbb{R}$ given by $r = \mathrm{Im}(F)$. Then r is pluriharmonic in $\tilde{U}$ and for any $w \in U$

$$r(w) = \frac{1}{2i}(F(w) - \overline{F(w)}) = \frac{1}{2i}(f(w) - \overline{f(w)}) = 0$$

as f is real i.e. $r = 0$ everywhere in U. Also $(dr)(w) \neq 0$ for any $w \in U$ hence r is a local defining function for M, in a neighborhood of z.

REMARK 3.18. What we just showed is that the mere real analyticity of $\mathcal{F}$ and M already imply the existence of *local* pluriharmonic defining functions for M. This led D.E. Barrett, [**28**], to ask whether given a real analytic Levi flat hypersurface $M \subset X$ there is a *global* pluriharmonic defining function for M and of course Theorem 3.16 is the answer to the question. If we think of $\mathcal{F}$ as a Γ-foliation, where Γ is the pseudogroup of all local C^ω-transformations of $\mathbb{R}$ in itself, and $f' \in C^\omega(U')$ is another choice of local defining submersion for $\mathcal{F}$ such that $z \in U'$ then $f' = \phi \circ f$ for some $\phi \in \Gamma$. Let Φ be the complex analytic extension of ϕ and let us set $F' := \Phi \circ F$. Then F' is a holomorphic function extending f' (and actually the only element of $\mathcal{O}(\Phi(\tilde{U}))$ extending f', by the uniqueness statement in Tomassini's theorem, cf. *op. cit.*). Therefore F and F' have the same level sets near z and as shown in the previous sections (cf. Rea's theorem and its proof) the construction globalizes to show that $\mathcal{F}$ extends to a codimension one holomorphic foliation of a neighborhood of M in X. $\square$

Let us go back to the proof of Theorem 3.16. By hypothesis $T(\mathcal{F}) = \mathrm{Ker}(\eta)$ for some closed real analytic 1-form η. Then (by eventually multiplying f by a nonzero real constant) $\eta|_U = df$. Therefore, if $f' \in C^\omega(U')$ is another local defining submersion as above then $f' = f + c$ in $U \cap U'$ for some $c \in \mathbb{R}$. Hence $\mathrm{Im}(f)$ and $\mathrm{Im}(f')$ coincide on $U \cap U'$. Consequently (by Lemma 3.12) $\mathrm{Im}(F)$ and $\mathrm{Im}(F')$ coincide on $\tilde{U} \cap \tilde{U}'$ so that the local pluriharmonic functions $\mathrm{Im}(F) : \tilde{U} \to \mathbb{R}$ glue up to a pluriharmonic defining function $r : \Omega \to \mathbb{R}$ of M, where Ω is the union of the open neighborhoods $\tilde{U}$ as z varies in M.

3.4. Holomorphic degeneracy

This section is devoted to a characterization of flatness of a real hypersurface in $\mathbb{C}^2$ due to N.K. Stanton, [**227**].

DEFINITION 3.19. For $t > 0$ and $m \in \mathbb{Z}$, $m \geq 0$, we consider the (*nonisotropic*) *dilations* $\delta_t : \mathbb{C}^{n+1} \to \mathbb{C}^{n+1}$ given by $\delta_t(z, w) = (tz\, , t^m w)$ for any $z \in \mathbb{C}^n$ and $w \in \mathbb{C}$. $\square$

If $p \in \mathbb{C}[z,\overline{z}]$ is a homogeneous polynomial of degree m and M_0 is the homogenous real hypersurface in $\mathbb{C}^{n+1}$ given by the equation $\mathrm{Im}(w) = p(z,\overline{z})$ then $\delta_t \in \mathrm{Aut}_{CR}(M_0)$ i.e. $\delta_t(M_0) = M_0$ and $\delta_t : M_0 \to M_0$ is a CR automorphism. Indeed $T_{1,0}(M_0)$ is spanned by

$$Z_j = \frac{\partial}{\partial z^j} + 2ip_{z^j}\frac{\partial}{\partial w}, \quad 1 \leq j \leq n,$$

(where $p_{z^j} = \partial p/\partial z^j$) and an elementary calculation shows that

$$(d_{(z,w)}\delta_t)Z_{j,(z,w)} = t\, Z_{j,\delta_t(z,w)}, \quad (z,w) \in M_0, \;\; t > 0,$$

as $p_{z^j}(tz, t\overline{z}) = t^{m-1}p_{z^j}(z,\overline{z})$. Let us set $\rho_0(z,w) = \frac{1}{2i}(w - \overline{w}) - p(z,\overline{z})$ and consider the pseudohermitian structure $\theta_0 = \frac{i}{2}(\overline{\partial} - \partial)\rho_0$ on M_0. Then

$$\theta_0 = \frac{i}{2}\left(p_{z^j}dz^j - p_{\overline{z}^j}d\overline{z}^j\right) - \frac{1}{4}\left(dw + d\overline{w}\right)$$

and a calculation shows that $\delta_t^*\theta_0 = t^m\theta_0$, that is $\delta_t \in \mathrm{Psh}(M_0, \theta_0)$ for any $t > 0$ i.e. δ_t is a pseudohermitian transformation of (M_0, θ_0).

DEFINITION 3.20. i) A function $f : \mathbb{C}^{n+1} \to \mathbb{C}$ is said to be *homogeneous of weight* k if $f \circ \delta_t = t^k f$. Also if $f \in C^\omega$ then we write $f = O(k)$ if every term in the Taylor expansion of f about 0 is a homogenous function of weight $\geq k$. ii) A vector field $X \in \mathcal{X}(\mathbb{C}^{n+1})$ is said to be *homogeneous of weight* k if $X(f \circ \delta_t) = t^{-k}(Xf) \circ \delta_t$ for any $f \in C^\infty(\mathbb{C}^{n+1})$. Also if X is real analytic, i.e. the coefficients f^j in $X = f^j\partial/\partial z^j + \overline{f}^j\partial/\partial\overline{z}^j$ are of class C^ω, then we write $X = O(k)$ if when we expand the coefficients f^j in Taylor series about 0 each term of X is a homogeneous vector field of weight $\geq k$. If X is a real analytic vector field we say X *has weight* k (and we write $w(X) = k$) if $X = O(k)$ and $X \neq O(k+1)$. □

Let us look at an example of infinitesimal CR automorphism of the homogeneous real hypersurface $M_0 = \{(z,w) \in \mathbb{C}^{n+1} : \mathrm{Im}(w) = p(z,\overline{z})\}$. Precisely, let

$$X_0 = 2\mathrm{Re}\left(z^j\frac{\partial}{\partial z^j} + m\,w\frac{\partial}{\partial w}\right).$$

An integral curve of X_0 is a solution $a(t) = (z(t), w(t))$ to $X_{0,a(t)} = \dot{a}(t)$ i.e.

$$\frac{dz^j}{dt} = z^j(t), \quad \frac{dw}{dt} = mw(t),$$

hence $a(t) = \delta_{e^t}(\lambda, \mu)$, $t \in \mathbb{R}$, for some $\lambda \in \mathbb{C}^n$ and $\mu \in \mathbb{C}$ such that $(\lambda, \mu) \in M_0$. Then $X_0 \in \mathrm{psh}(M_0, \theta_0)$ i.e. X_0 is an infinitesimal pseudohermitian transformation of (M_0, θ_0) and in particular X_0 is an infinitesimal CR automorphism. An elementary argument (based on the expression of the Lie bracket $[X_0, X]$ in terms of the 1-parameter group of transformations of X_0) shows that $[X_0, X] = kX$ for any homogeneous vector field X of weight k.

Let

$$M = \{(z,w) \in \mathbb{C}^{n+1} : \mathrm{Im}(w) = F(z,\overline{z})\} \tag{3.9}$$

be a rigid real analytic real hypersurface in $\mathbb{C}^{n+1}$, passing through the origin ($0 \in M$). Let us assume that M is of finite type m. It is known (cf. e.g. [**12**]) that, due to the finite type condition, one may write $F(z,\overline{z})$ as

$$F(z,\overline{z}) = p(z,\overline{z}) + O(m+1)$$

for some homogeneous polynomial $p \in \mathbb{C}[z,\overline{z}]$ of degree m.

DEFINITION 3.21. The homogenous hypersurface $M_0 = \{(z,w) \in \mathbb{C}^{n+1} : \mathrm{Im}(w) = p(z,\overline{z})\}$ is called the *homogenous part* of the rigid hypersurface M given by (3.9). □

Let us consider the pseudohermitian structure θ on M given by $\theta = \frac{i}{2}\left(\overline{\partial} - \partial\right)\rho$ where

$$\rho = \frac{1}{2i}(w - \overline{w}) - p - h, \quad h = O(m+1).$$

If $h = \sum_{\alpha,\beta\in\mathbb{Z}_+^n} a_{\alpha\overline{\beta}} z^\alpha \overline{z}^\beta$ is the Taylor development of h about 0 then the information $h = O(m+1)$ amounts to saying that $|\alpha| + |\beta| \geq m+1$ for all $\alpha, \beta \in \mathbb{Z}_+^n$ such that $a_{\alpha\overline{\beta}} \neq 0$. Consequently

$$\theta = \theta_0 + O(m).$$

Here a real 1-form on M is $O(k)$ if its coefficients with respect to $\{dz^j, d\overline{z}^j, dw, d\overline{w} : 1 \leq j \leq n\}$ are $O(k)$. The main purpose of this section is to sketch the proof of the following result

THEOREM 3.22. (N.K. Stanton, [**227**])
Let $M \subset \mathbb{C}^{n+1}$ be a C^ω real hypersurface passing through the origin. If M is essentially finite at 0 or the Levi form of M is somewhere nondegenerate then M is holomorphically nondegenerate at 0. If $n = 1$ (i.e. $M \subset \mathbb{C}^2$) then the following statements are equivalent

i) $\dim \mathrm{hol}(M) < \infty$.
ii) *M is not flat.*
iii) *The Levi form of M is somewhere nondegenerate.*
iv) *M is holomorphically nondegenerate at the origin.*

Proof. The proof of the first statement in Theorem 3.22 is by contradiction. Let us assume that M is holomorphically degenerate at 0 i.e. there is an open set $U \subseteq \mathbb{C}^{n+1}$ with $0 \in U$ and there is a holomorphic tangent vector field $W \in \mathcal{O}(U, T^{1,0}(\mathbb{C}^{n+1}))$ such that $W|_{U\cap M} \neq 0$, i.e. $W|_{U\cap M}$ is nontrivial, and $W|_{U\cap M} \in \Gamma^\infty(U \cap M, T_{1,0}(M))$, i.e. $W|_{U\cap M}$ is tangent to M. Thus there is $p \in U \cap M$ such that $W_p \neq 0$ and M is essentially finite at p because it is known (cf. [**13**]) that essential finiteness is an open property. Let us bring W locally to the canonical form. In other words, there is a biholomorphism $\varphi : U \to V$ such that $\varphi(p) = 0$ and $(d\varphi)W = \partial/\partial\zeta^1$. Here we have set $\zeta = \varphi(Z)$ with $Z = (z,w)$. Then $\varphi(M \cap U)$ is given by the equation $\tilde{\rho}(\zeta) = 0$ where the new defining function $\tilde{\rho}$ is given by $\tilde{\rho}(\zeta) = \rho(\varphi^{-1}(\zeta))$. As W is tangent to M

$$0 = W(\rho) = W(\tilde{\rho} \circ \varphi) = [(d\varphi)W]\tilde{\rho} = \frac{\partial\tilde{\rho}}{\partial\zeta^1}\,.$$

Similarly $\partial\tilde{\rho}/\partial\overline{\zeta}^1 = 0$, that is to say $\tilde{\rho}$ doesn't depend upon the arguments ζ^1 and $\overline{\zeta}^1$. To exploit essential finiteness we set as customary

$$\mathcal{V} = \{\xi \in \mathbb{C}^{n+1} : \tilde{\rho}(\xi,\eta) = 0 \text{ if } \eta \in \mathbb{C}^{n+1} \text{ and } \tilde{\rho}(0,\eta) = 0\}.$$

Note that

$$\tilde{\rho}((\zeta^1, 0), \eta) = \tilde{\rho}(0,\eta) = 0$$

hence $(\zeta^1, 0) \in \mathcal{V}$ so that $\mathcal{V} \neq \{0\}$, a contradiction. To prove the second part of Theorem 3.22 (with $n = 1$) we need the following

LEMMA 3.23. (N.K. Stanton, [**227**])
Let $M \subset \mathbb{C}^{n+1}$ be a C^∞ real hypersurface with $0 \in M$. If every neighborhood U of 0 contains a point of $p \in M$ such that the Levi form of M is nondegenerate at p then $\dim \mathrm{hol}(M) \leq (n+2)^2 - 1$.

Proof. Let $X_1, \cdots, X_{(n+2)^2} \in \mathrm{hol}(M)$. Let $U \subseteq \mathbb{C}^{n+1}$ be an open set such that $0 \in U$ and the vector fields X_j are defined on U, for $1 \leq j \leq (n+2)^2$. By hypothesis there is $p \in U \cap M$ such that $L_{\theta,p}$ is nondegenerate, where L_θ is the Levi form of M, associated to some pseudohermitian structure θ. By continuity there is an open set $\tilde{M} \subseteq U \cap M$ such that $p \in \tilde{M}$ and L_θ is nondegenerate everywhere in $\tilde{M}$. Consequently (cf. [**65**]) we may assume (by making a biholomorphic change of coordinates $(z, w) \mapsto (\tilde{z}, \tilde{w})$ if necessary) that $\tilde{M}$ is given by an equation of the form

$$\mathrm{Im}(\tilde{w}) = \sum_{j,k=1}^{n} a_{j\bar{k}} \tilde{z}^j \overline{\tilde{z}}^k + O(3).$$

Let us set

$$\mathcal{W}_j = \{X \in \mathrm{hol}(\tilde{M}) : X = O(j)\}, \quad \mathcal{W} = \bigoplus_{j=-2}^{2} \frac{\mathcal{W}_j}{\mathcal{W}_{j+1}}.$$

Any $X \in \mathcal{W}_j$ may be written uniquely as $X = X_0 + O(j+1)$ where X_0 is homogenous of weight j. Let $\tilde{M}_0$ be the homogeneous part of $\tilde{M}$ i.e. the hypersurface given by the equation

$$\mathrm{Im}(\tilde{w}) = \sum_{j,k=1}^{n} a_{j\bar{k}} \tilde{z}^j \overline{\tilde{z}}^k .$$

The linear map

$$T_j : \frac{\mathcal{W}_j}{\mathcal{W}_{j+1}} \to \mathrm{aut}(\tilde{M}_0), \quad T_j(X + \mathcal{W}_j) := X_0 ,$$

is clearly a monomorphism. Let (p, q) be the signature of the Levi form L_θ restricted to $\tilde{M}$. Then $\mathrm{aut}(\tilde{M}_0) \approx \mathbf{su}(p+1, q+1)$ (cf. e.g. [**140**]) hence on one hand $\dim \mathrm{aut}(\tilde{M}_0) = (n+2)^2 - 1$ an on the other $|w(X_0)| \leq 2$. Hence

$$\mathrm{hol}(\tilde{M}) \approx \mathcal{W}$$

and then $\dim \mathrm{hol}(\tilde{M}) \leq \dim \mathrm{aut}(\tilde{M}_0)$, as $T = \oplus_{j=-2}^{2} T_j$ is injective as well. Therefore $X_1, \cdots, X_{(n+2)^2}$ are linearly dependent at each point of $\tilde{M}$. Yet each X_j is of class C^ω hence $X_1, \cdots, X_{(n+2)^2}$ are linearly dependent at each point of $U \cap M$ i.e. $\dim \mathrm{hol}(M) \leq (n+2)^2 - 1$. Lemma 3.23 is proved.

Let us sketch the proof of the second statement in Theorem 3.22.

(iii) $\Longrightarrow$ (iv). Assume M to be somewhere Levi nondegenerate. Then (by Lemma 3.23 above) $\dim \mathrm{hol}(M) < (n+2)^2$ hence M is holomorphically nondegenerate. Indeed, should $W \neq 0$ be a holomorphic tangent to M at 0 one would have $\mathrm{Re}\,(z^\alpha W) \in \mathrm{hol}(M)$ for any $\alpha \in \mathbb{Z}_+^{n+1}$ so that $\dim \mathrm{hol}(M) = \infty$. This argument also shows that the implication (i) $\Longrightarrow$ (iv) holds true.

(iv) $\Longrightarrow$ (ii). The proof is by contradiction. Let us assume that M is flat. Then M is CR equivalent to a hyperplane i.e. there are complex coordinates $Z = (z, w)$ such that M is given by the equation $\mathrm{Im}(w) = 0$, so that $\partial/\partial z$ is a holomorphic tangent to M, a contradiction.

(ii) $\Longrightarrow$ (iv). If $M \subset \mathbb{C}^2$ is not flat then (cf. e.g. [**65**]) there are complex coordinates $Z = (z, w)$, $w = u + iv$ such that M is given by an equation of the form $v = F(z, \overline{z}, u)$ with $F \neq 0$ and $F(z, 0, u) = F(0, \overline{z}, u) = 0$. Let us show that any holomorphic tangent to M

$$Z = f(z, w)\frac{\partial}{\partial z} + g(z, w)\frac{\partial}{\partial w}$$

vanishes on M, hence M is holomorphically nondegenerate. Indeed $Z(v - F(z, \overline{z}, u)) = 0$ may be written as

$$\frac{1}{2i} g(z, u + iF(z, \overline{z}, u)) = f(z, u + iF(z, \overline{z}, u))F_z(z, \overline{z}, u) + \tag{3.10}$$

$$+ \frac{1}{2} g(z, u + iF(z, \overline{z}, u))F_u(z, \overline{z}, u).$$

Then $F(z, 0, u) = 0$ implies that $F_z(z, 0, u) = 0$ and $F_u(z, 0, u) = 0$ hence (by (3.10)) $g(z, u) = 0$ and then (as g is holomorphic with respect to w) $g(z, w) = 0$. Then (again by (3.10))

$$f(z, u + iF(z, \overline{z}, u))F_z(z, \overline{z}, u) = 0$$

where from $f(z, w) = 0$ i.e. $Z = 0$ as desired.

The implication (ii) $\Longrightarrow$ (iii) holds true because the Levi form of a non flat real hypersurface is always nonzero (cf. e.g. [**14**]).

(iii) $\Longrightarrow$ (i). If L_θ is somewhere nondegenerate then (by Lemma 3.23 with $n = 1$) $\dim \mathrm{hol}(M) \leq 8$. Q.e.d.

It is an open problem whether the proofs above may be refined to give pseudohermitian analogs of Lemma 3.23 and Theorem 3.22 (involving the space $\mathrm{psh}(M, \theta)$ of all infinitesimal pseudohermitian transformations of (M, θ) defined in some neighborhood of 0, rather than $\mathrm{aut}(M)$).

3.5. Twistor CR manifolds

Let M be a n-dimensional manifold and $\mathcal{X} := T^*(M) \otimes \mathbb{C}$. Let $\pi : \mathcal{X} \to M$ be the projection. Let $(U, \tilde{x}^j)$ be a local coordinate system on M and define local coordinates $x^j : \pi^{-1}(U) \to \mathbb{R}$ and $\zeta_j : \pi^{-1}(U) \to \mathbb{C}$ on $\mathcal{X}$ by setting $x^j(\chi) := \tilde{x}^j(\pi(\chi))$ and $\zeta_j(\chi) := \langle \chi, \partial / \partial \tilde{x}^j \rangle$, for any $\chi \in \mathcal{X}$. Let $A \subset T(\mathcal{X}) \otimes \mathbb{C}$ be the span of $\{\partial / \partial \overline{\zeta}^j : 1 \leq j \leq n\}$ i.e. the portion of A over a fibre $\pi^{-1}(x)$ (clearly a complex n-dimensional manifold) is the antiholomorphic tangent bundle over $\pi^{-1}(x)$, for any $x \in M$.

DEFINITION 3.24. The *canonical* 1-*form* θ of $\mathcal{X}$ is given by

$$\theta_\chi(w) := \langle \chi, (d_\chi \pi) w \rangle , \quad w \in T_\chi(\mathcal{X}), \quad \chi \in \mathcal{X}.$$

□

In local coordinates $\theta = \zeta_j dx^j$.

DEFINITION 3.25. The *Hamiltonian form* of $\mathcal{X}$ is $\omega := d\theta = d\zeta_j \wedge dx^j$. □

DEFINITION 3.26. A codimension 2 submanifold $\mathcal{Y} \subset \mathcal{X}$ is an *energy surface* if $\pi^{-1}(x) \cap \mathcal{Y}$ is a complex hypersurface in $\pi^{-1}(x)$, for any $x \in \mathcal{X}$. □

Let

$$N_x^* := \{f \in T_\chi^*(\pi^{-1}(x)) : \mathrm{Ker}(f) \supseteq T_\chi(\pi^{-1}(x) \cap \mathcal{Y})\}, \quad \chi \in \pi^{-1}(x) \cap \mathcal{Y},$$

be the conormal bundle of $\pi^{-1}(x) \cap \mathcal{Y}$ in $\pi^{-1}(x)$. Let us consider the map

$$f \in T_\chi^*(\pi^{-1}(x)) \approx \pi^{-1}(x) = T_x^*(M) \otimes_{\mathbb{R}} \mathbb{C} \to T_x(M) \otimes_{\mathbb{R}} \mathbb{C} \ni z_f ,$$

where the last arrow is the double-dual identification i.e.

$$f = f^j d\zeta_j\big|_\chi + \overline{f}^j d\overline{\zeta}_j\Big|_\chi \longmapsto z_f := f^j \left.\frac{\partial}{\partial \tilde{x}^j}\right|_x . \tag{3.11}$$

DEFINITION 3.27. An energy surface $\mathcal{Y} \subset \mathcal{X}$ is *generic* if

$$\{f \in N_x^* : z_f = \overline{z_f}\} = (0),$$

that is N_x^* contains no nonzero vector corresponding to a real vector in $T_x(M) \otimes_{\mathbb{R}} \mathbb{C}$ under the linear map (3.11). □

If $\iota : \mathcal{Y} \to \mathcal{X}$ is the canonical inclusion we set $\tilde{\omega} := \iota^* \omega$ and consider the bundle morphism

$$\rfloor_{\tilde{\omega}} : T(\mathcal{Y}) \otimes \mathbb{C} \to T^*(\mathcal{Y}) \otimes \mathbb{C} , \quad W \mapsto \tilde{\omega}(W, \cdot), \quad W \in T(\mathcal{Y}) \otimes \mathbb{C} .$$

According to C. LeBrun, [**165**], the following result holds

THEOREM 3.28. (C. LeBrun, [**165**])
Let $\mathcal{Y} \subset \mathcal{X}$ be an energy surface. Then $D := \mathrm{Ker}(\rfloor_{\tilde{\omega}}) \subset T(\mathcal{Y}) \otimes \mathbb{C}$ is involutive. If additionally $\mathcal{Y}$ is generic then $D \cap \overline{D} = (0)$. In particular, if M is a 3-dimensional manifold then D is a CR structure on $\mathcal{Y}$, of CR codimension 1.

C. LeBrun applied (cf. *op. cit.*) his finding (that any generic energy surface associated to a 3-dimensional manifold is a 7-dimensional CR manifold (of hypersurface type)) to the following situation. Let M be a 3-dimensional CR manifold endowed with a conformal structure $G = \{e^u g : u \in C^\infty(M)\}$, where g is a fixed Riemannian metric on M. Next, let us consider the 7-dimensional manifold $\hat{N} := \{\chi \in \mathcal{X} : g^*(\chi, \chi) = 0, \ \chi \neq 0\}$. Then $\mathbb{C}^* := \mathbb{C} \setminus \{0\}$ acts freely on $\hat{N}$ and we may consider the quotient space $N := \hat{N}/\mathbb{C}^*$ (a 5-dimensional manifold). Note that $\hat{N}$, and then N, depends only on the conformal structure G (rather than on the fixed metric g). According to [**165**], $\hat{N} \subset \mathcal{X}$ is a generic energy surface, hence (by Theorem 3.28) carries a CR structure D. Its projection $T_{1,0}(N) := \mathbb{P}_* D$ is a CR structure on N. Here $\mathbb{P} : \hat{N} \to N = \hat{N}/\mathbb{C}^*$ is the natural projection.

DEFINITION 3.29. N is referred to as the *twistor CR manifold* of M. □

There is a natural projection $\pi : N \to M$ all of whose fibers are complex lines. Again together with [**165**], one concludes that $(N, T_{1,0}(N))$ is a nondegenerate CR manifold carrying a smooth foliation by $\mathbb{C}P^1$'s, and $(N, T_{1,0}(N))$ is embeddable if and only if M admits a real analytic atlas with respect to which the conformal structure G contains a real analytic representative. A generalization of the construction and embeddability theorem to n dimensions is due to H. Rossi, [**208**]. When M is a totally umbilical real hypersurface of a real analytic 4-dimensional Lorentzian manifold, the twistor CR manifold N coincides with R. Penrose's CR manifold, cf. [**199**]. A detailed account of the original R. Penrose construction (cf. *op. cit.*) is given in [**89**].

3.6. Analysis on pseudoconvex domains

H. Boas & E.J. Straube, [**48**], have discovered a technique for proving Sobolev estimates for the Bergman projection and the $\overline{\partial}$-Neumann operator on any smoothly bounded weakly pseudoconvex domain $\Omega \subset \mathbb{C}^n$ admitting a defining function that is plurisubharmonic on the boundary. Essential to the proof is the existence of vector fields commuting approximately with $\overline{\partial}$ (in a sense to be made clear in the sequel). The existence of such vector fields turns out to be related to the study of the geometry of the Levi foliation of the boundary $\partial\Omega$, going back to the older discovery, due to A. Noell, that the unit normal to the boundary of a convex domain is constant along Levi-null curves, cf. [**195**]. The reader should consult the monographs [**106**] and [**161**] for a presentation of the $\overline{\partial}$-Neumann problem and the Bergman projection. Roughly speaking the $\overline{\partial}$-Neumann operator N_q is the inverse of the complex Laplacian $\Box_q = \overline{\partial}^* \overline{\partial} + \overline{\partial}\,\overline{\partial}^*$ on $(0,q)$-forms, $1 \leq q \leq n$. The Bergman projection P_q, $0 \leq q \leq n$, is the orthogonal projection from the space of square-integrable $(0,q)$-forms onto the space of $\overline{\partial}$-closed $(0,q)$-forms. P_0 is the usual Bergman projection on functions, cf. [**133**], p. 364-371.

DEFINITION 3.30. A smoothly bounded domain $\Omega \subset \mathbb{C}^n$ is called *pseudoconvex* if the complex Hessian of any smooth defining function ρ for Ω is positive semi-definite on the complex tangent space at each boundary point i.e. $(\partial\overline{\partial}\rho)(Z,\overline{Z}) \geq 0$ for any $Z \in T_{1,0}(\partial\Omega)$. A smooth defining function ρ is said to be *plurisubharmonic on* $\partial\Omega$ if $(\partial\overline{\partial}\rho)(V,\overline{V}) \geq 0$ along $\partial\Omega$ for any $V \in \Gamma^\infty(\partial\Omega, T^{1,0}(\mathbb{C}^n))$. □

Clearly, if Ω admits a defining function which is plurisubharmonic on $\partial\Omega$ then Ω is pseudoconvex (while the converse is false, in general, cf. e.g. the final comments in [**49**]). The finding by H.P. Boas & E.J. Straube, [**48**], is that

THEOREM 3.31. (H.P. Boas & E.J. Straube, *op. cit.*)
Let $\Omega \subset \mathbb{C}^n$ be a smoothly bounded domain admitting a defining function which is plurisubharmonic on the boundary $\partial\Omega$. Then

1) *For any $1 \leq q \leq n$ the $\overline{\partial}$-Neumann operator N_q is continuous on the space of all $(0,q)$-forms with coefficients in $W^s(\Omega)$ for every $s \in [0,+\infty)$.*

2) *For any $0 \leq q \leq n$ the Bergman projection P_q is continuous on the space of all $(0,q)$-forms with coefficients in $W^s(\Omega)$ for every $s \in [0,+\infty)$.*

The two statements (1)-(2) in Theorem 3.31 are actually equivalent (cf. [**47**]). Here $W^s(\Omega)$ is the Sobolev space of order s. Combining Theorem 3.31 and the Sobolev lemma one may show that, under the hypothesis of Theorem 3.31, the operators N_q and P_q are continuous on the space of all $(0,q)$-forms with coefficients in $C^\infty(\overline{\Omega})$.

A key ingredient in the proof of the estimates leading to the conclusion in Theorem 3.31 is the existence, for any $\epsilon > 0$, of a smooth vector field $Z_\epsilon \in \Gamma^\infty(\overline{\Omega}, T^{1,0}(\mathbb{C}^n))$ such that $Z_\epsilon(\rho) = 1$ and

$$\left| g_0([Z_\epsilon, \frac{\partial}{\partial \overline{z}^j}], \overline{L}) \right| < \epsilon, \quad 1 \leq j \leq n, \tag{3.12}$$

along $\partial\Omega$. Here g_0 is the canonical Kähler flat metric on $\mathbb{C}^n$ and

$$L = \frac{2}{|\nabla\rho|^2} \sum_{j=1}^n \frac{\partial\rho}{\partial\overline{z}^j} \frac{\partial}{\partial z^j}.$$

Note that $L(\rho) = 1$, $g_0(L, L) = 1$ and L is transverse to $T_{1,0}(\partial\Omega)$ at each point of $\partial\Omega$. It may be shown (cf. [**48**], p. 83) that whenever Ω possesses a smooth defining function ρ which is plurisubharmonic on $\partial\Omega$, vector fields Z_ϵ as above, in particular satisfying (3.12), do exist. Successively (cf. [**49**]) it has been observed that the normalization condition $Z_\epsilon(\rho) = 1$ isn't really necessary in the proof of Theorem 3.31 and that it actually suffices that $Z_\epsilon(\rho)$ be bounded, and bounded away from zero, independently of ϵ. Also the hypothesis that Ω admits a defining function which is plurisubharmonic on $\partial\Omega$ is used (cf. [**48**]) only in the construction of the vector fields Z_ϵ. One may therefore reformulate Theorem 3.31 in the following more general manner

THEOREM 3.32. (H.P. Boas & E.J. Straube, [**49**])
Let $\Omega \subset \mathbb{C}^n$ be a smoothly bounded pseudoconvex domain and ρ a smooth defining function for Ω. Let $K \subset \partial\Omega$ be the set of all boundary points which are of infinite type. Let us assume that there is a constant $C > 0$ such that for any $\epsilon > 0$ there is a vector field $Z_\epsilon \in \Gamma^\infty(U_\epsilon, T^{1,0}(\mathbb{C}^n))$ defined on an open neighborhood $U_\epsilon \subseteq \mathbb{C}^n$ of K and satisfying the conditions

i) *$Z_\epsilon(\rho)$ is real on K and*

$$\frac{1}{C} < Z_\epsilon(\rho) < C. \tag{3.13}$$

ii) *For any $1 \leq j \leq n$*

$$\left|(\partial\rho)[Z_\epsilon\,,\,\partial/\partial\overline{z}^j]\right| < \epsilon \tag{3.14}$$

everywhere on K. Then the $\overline{\partial}$-Neumann operators N_q, $1 \leq q \leq n$, and the Bergman projections P_q, $0 \leq q \leq n$, are continuous on the Sobolev space $W^s_{(0,q)}(\Omega)$ for any $s \geq 0$.

We recall (cf. Definition 2.6 in this monograph) that a point $x \in \partial\Omega$ has type $m(x)$ if the tangent space $T_x(\partial\Omega)$ is spanned by $m(x)$ sections in $H(\partial\Omega)$ (evaluated at x) and $m(x)$ is the smallest integer with this property. Again using Definition 2.6 we set

$$K = \{x \in \partial\Omega : m(x) = \infty\}.$$

By a result of J.P. D'Angelo, [**76**], K is a compact set. An inspection of (3.14) shows that (3.14) is nothing but an equivalent reformulation of the inequality (3.12), except that it is only required to hold on K, rather than on the whole of $\partial\Omega$. The explanation for the weaker requirement is that local estimates (leading to the conclusion in Theorems 3.31 or 3.32) hold (by subellipticity, cf. D. Catlin, [**61**]) at all points of finite type, so that the new estimation technique, based on the use of the vector fields Z_ϵ, requires the existence of these vector fields only in a neighborhood of the points of infinite type. Also, an analysis of the proof of Theorem 3.32 shows (cf. [**49**]) that $Z_\epsilon(\rho)$ doesn't actually need to be real, but rather $|\arg Z_\epsilon(\rho)|$ should be sufficiently small. One is led to the following

DEFINITION 3.33. Let $U \subseteq \mathbb{C}^n$ be an open set. A smooth vector field $Z \in \Gamma^\infty(U, T^{1,0}(\partial\Omega))$ is said to *ϵ-commute with $\overline{\partial}$* if

$$|\arg Z(\rho)| < \epsilon, \quad C^{-1} < |Z(\rho)| < C \quad \text{on } U,$$

$$\left|(\partial\rho)[Z,\,\partial/\partial\overline{z}^j]\right| < \epsilon \quad \text{on } U, \quad 1 \leq j \leq n,$$

for some $C > 0$ independent of ϵ. □

The choice of terminology is motivated by the last requirement in Definition 3.33 (cf. also (3.14) above). Indeed the inequality says that the modulus of the $(1,0)$-component of the commutator $[Z,\, \partial/\partial \overline{z}^j]$ should be smaller than ϵ, that is Z commutes 'approximately' with $\partial/\partial\overline{z}^j$ for any $1 \leq j \leq n$. If this is the case, one also says loosely that Z (or Z_ϵ) commutes approximately with $\overline{\partial}$. So Theorem 3.32 holds under the assumption that there is $C > 0$ such that for any $\epsilon > 0$ there is an open neighborhood $U_\epsilon \subseteq \mathbb{C}^n$ of the set $K \subset \partial\Omega$ of points of infinite type, and there is a complex vector field Z_ϵ defined on U_ϵ such that Z_ϵ is of type $(1,0)$ and ϵ-commutes with $\overline{\partial}$. In the spirit of the present monograph, we are interested in the differential geometric problem of the existence of such vector fields, and relegate the reader to [**48**]-[**49**] for the estimates proved by employing the vector fields Z_ϵ. The rest of this section is devoted to the obstructions to the existence problem for the vector fields Z_ϵ (the *D'Angelo class*) and a foliation theoretic discussion of these obstructions.

Let $\Omega \subset \mathbb{C}^n$ be a smoothly bounded pseudoconvex domain and θ a pseudohermitian structure on $\partial\Omega$. Let T be a globally defined real vector field on $\partial\Omega$ such that T is transverse to $H(\partial\Omega)$ and $\theta(T) = 1$. Let us consider the 1-form $\alpha = -\mathcal{L}_T\theta$ (cf. J.P. D'Angelo, [**75**]-[**76**]) where $\mathcal{L}_T$ denotes the Lie derivative in the direction T. We need the following

PROPOSITION 3.34. *Let $\Omega \subset \mathbb{C}^n$ be a smoothly bounded pseudoconvex domain. Let $x_0 \in \partial\Omega$ be a boundary point and*

$$\mathcal{N}_{x_0} = \{v \in T_{1,0}(\partial\Omega)_{x_0} : (d\theta)_{x_0}(v, \overline{w}) = 0, \quad \forall\, w \in T_{1,0}(\partial\Omega)_{x_0}\}$$

the null space of the Levi form at x_0. Then

$$(d\alpha)_{x_0}(v, w) = 0$$

for any $v, w \in \mathcal{N}_{x_0} \oplus \overline{\mathcal{N}}_{x_0}$. In particular if $j : M \subset \partial\Omega$ is a real submanifold such that $T_x(M) \subset \mathrm{Null}(G_\theta)_x$ for any $x \in M$ then $A = j^*\alpha$ is a closed 1-form on M, thus giving rise to a de Rham cohomology class $a(M) = [A] \in H^1(M;\mathbb{R})$. Moreover if ρ is a smooth defining function for Ω and

$$\theta = \frac{i}{2}(\overline{\partial} - \partial)\rho, \quad T = i(L - \overline{L}),$$

with $L = 2|\nabla\rho|^{-2}\sum_{j=1}^n \rho_{\overline{z}^j}\, \partial/\partial z^j$, then the definition of $a(M)$ doesn't depend upon the choice of ρ.

Here $\mathrm{Null}(G_\theta) = \{X \in H(\partial\Omega) : G_\theta(X, \cdot) = 0 \quad \text{on} \quad H(\partial\Omega)\}$ so that $\mathrm{Null}(G_\theta) \otimes \mathbb{C} = \mathcal{N} \oplus \overline{\mathcal{N}}$. To prove Proposition 3.34, let $v, w \in \mathcal{N}_{x_0} \oplus \overline{\mathcal{N}}_{x_0}$ and let $X, Y \in \Gamma^\infty(T_{1,0}(\partial\Omega) \oplus T_{0,1}(\partial\Omega)$ such that $X_{x_0} = v$ and $Y_{x_0} = w$. Of course such smooth extensions of v and w may be chosen as $\{x_0\}$ is a closed subset of the boundary. As exterior differentiation and the Lie derivative commute

$$-(d\alpha)(X, Y) = (\mathcal{L}_T d\theta)(X, Y) =$$

$$= T((d\theta)(X, Y)) - (d\theta)([T, X], Y) - (d\theta)(X, [T, Y]).$$

Let us compute the last two terms at the point x_0. As $Y_{x_0} = w \in \mathcal{N}_{x_0} \oplus \overline{\mathcal{N}}_{x_0}$ (and $[T, X] - \theta(X)T$ is the $H(\partial\Omega)$-component of $[T, X]$)

$$-(d\theta)_{x_0}([T, X]_{x_0}, Y_{x_0}) = -(d\theta)_{x_0}(\theta([T, X])_{x_0} T_{x_0}\,,\, w) =$$

$$= 2\theta([T, X])_{x_0}\theta([T, Y])_{x_0}$$

so that

$$-(d\alpha)(X,Y)_{x_0} = T_{x_0}((d\theta)(X,Y)). \tag{3.15}$$

As $(d\alpha)_{x_0}$ is $\mathbb{C}$-linear it suffices to show that $(d\alpha)_{x_0}(v,w) = 0$ in each of the following three cases I) $v, w \in \mathcal{N}_{x_0}$, II) $v, w \in \overline{\mathcal{N}}_{x_0}$ and III) $v \in \mathcal{N}_{x_0}$ and $w \in \overline{\mathcal{N}}_{x_0}$. In the cases I-II, the vectors v and w may be extended to smooth sections X and Y of $T_{1,0}(\partial\Omega)$ (respectively of $T_{0,1}(\partial\Omega)$) so that in both cases (3.15) implies that $(d\alpha)_{x_0}(X,Y) = 0$ due to the formal integrability property of the CR structure on $\partial\Omega$. It remains that we deal with the case III. As $(v,w) \mapsto (d\alpha)_{x_0}(v,\overline{w})$ is a skew Hermitian form on $\mathcal{N}_{x_0}$ it suffices to show that $(d\alpha)_{x_0}(v,\overline{v}) = 0$ for any $v \in \mathcal{N}_{x_0}$. Using a smooth extension $X \in \Gamma^\infty(T_{1,0}(\partial\Omega)$ of v we have (by (3.15)) that $(d\alpha)_{x_0}(v,\overline{v}) = -iT_{x_0}(f)$ where $f \in C^\infty(\partial\Omega)$ is the smooth function defined by $f = -i(d\theta)(X,\overline{X})$. Note that $f(x) \geq 0$ for any $x \in \partial\Omega$ (by the pseudoconvexity of Ω) and $f(x_0) = 0$ (because of $X_{x_0} = v \in \mathcal{N}_{x_0}$) so that f has a minimum point at x_0. Then $T_{x_0}(f) = 0$, as T is tangent to the boundary $\partial\Omega$. The proof of the first statement in Proposition 3.34 is complete. The second statement in Proposition 3.34 follows easily from the first. To end the proof let $\hat{\rho} = \lambda\rho$ be a new defining function with $\lambda \in C^\infty(\mathbb{C}^n)$ so that λ is $\mathbb{R}$-valued and $\lambda(x) \neq 0$ for any $x \in \partial\Omega$. Then

$$|\nabla\hat{\rho}|^2 = \lambda^2|\nabla\rho|^2 + O(\rho), \quad \hat{L} = \lambda\frac{|\nabla\rho|^2}{|\nabla\hat{\rho}|^2}\, L + O(\rho),$$

$$\hat{\theta} = \lambda\theta + O(\rho), \quad \hat{T} = \lambda\frac{|\nabla\rho|^2}{|\nabla\hat{\rho}|^2}\, T + O(\rho),$$

hence along the boundary

$$\hat{\theta} = ce^u\theta, \quad \hat{T} = ce^{-u}T,$$

where we have set $\lambda = ce^u$ on $\partial\Omega$, with $c \in \{\pm 1\}$ and $u \in C^\infty(\partial\Omega)$. Consequently

$$\hat{\alpha} = -\mathcal{L}_{\hat{T}}\hat{\theta} = \alpha + du - T(u)\theta.$$

As $(dj)T(M) \subseteq \text{Null}(G_\theta) \subseteq H(\partial\Omega)$ it follows that $j^*\theta = 0$ hence

$$\hat{A} = j^*\hat{\alpha} = A + dv$$

where $v = u \circ j \in C^\infty(M)$, so that $[\hat{A}] = [A] \in H^1(M;\mathbb{R})$. The proof of Proposition 3.34 is complete. It is often said that a good theorem leads to a valuable new notion. Indeed, using Proposition 3.34 we may give the following

DEFINITION 3.35. Let $\Omega \subset \mathbb{C}^n$ be a smoothly bounded pseudoconvex domain and $M \subset \partial\Omega$ a real submanifold of the boundary, tangent to the null distribution of the Levi form of $\partial\Omega$. The de Rham cohomology class $a(M) \in H^1(M;\mathbb{R})$ is called the *D'Angelo class* of M. $\square$

In the case of interest the D'Angelo class is the obstruction towards the existence of vector fields commuting approximately with $\overline{\partial}$. Precisely

THEOREM 3.36. (H.P. Boas & E.J. Straube, [**49**])
Let $\Omega \subset \mathbb{C}^n$ be a smoothly bounded pseudoconvex domain. Assume that there is a real submanifold $M \subset \partial\Omega$ of the boundary such that i) *M contains all the points of infinite type i.e. $K \subset M$, and* ii) *the tangent space at each point of M lies in the null space of the Levi form of $\partial\Omega$. Then $a(M) = 0$ if and only if there is $C > 0$ such that for any $\epsilon > 0$ there is an open neighborhood $U_\epsilon \subseteq \mathbb{C}^n$ of K and a complex vector field $Z_\epsilon \in \Gamma^\infty(U_\epsilon, T^{1,0}(\mathbb{C}^n)$ such that* a) *$Z_\epsilon(\rho)$ is real on K and $C^{-1} < Z_\epsilon(\rho) < C$*

on K, *and* b) *when* $1 \leq j \leq n$ *the form* $\partial\rho$ *applied to the commutator* $[Z_\epsilon, \partial/\partial\overline{z}^j]$ *has modulus less than* ϵ *on* K.

Combining Theorems 3.32 and 3.36 leads to the following

COROLLARY 3.37. *Let* $\Omega \subset \mathbb{C}^n$ *be a smoothly bounded pseudoconvex domain such that the set of all points of infinite type of the boundary is contained in a real submanifold* $M \subset \partial\Omega$, *tangent to the Levi distribution of the boundary. If all connected components of* M *are simply connected then the Neumann operators* N_q, $1 \leq q \leq n$, *and the Bergman projections* P_q, $0 \leq q \leq n$, *are continuous on* $W^s_{(0,q)}(\Omega)$ *for any* $s \geq 0$.

Indeed under the assumptions in Corollary 3.37 the first de Rham cohomology group of M vanishes ($H^1(M;\mathbb{R}) = 0$). It is an open problem to compute ΔA (where Δ is the Laplace-Beltrami operator on M associated to the Riemannian metric g induced on M by the canonical flat metric g_0 on $\mathbb{C}^n$) so that to use the Hodge-de Rham theorem (to decide whether $a(M) \neq 0$). Another open problem is to compute the sectional curvature $k(M)$ of (M, g) (by using the Gauss equation of M in $\partial\Omega$) and decide when $k(M) > 0$ so that to conclude [by the classical Synge theorem (cf. e.g. [**155**], Vol. II, p. 444)] that $\pi_1(M) = 0$.

Let us now show that the vanishing of the D'Angelo class implies the existence of the vector fields Z_ϵ in Theorem 3.36. If $a(M) = 0$ then $A \in \Omega^1(M)$ is exact i.e. there is $f \in C^\infty(M)$ such that $A = df$. Let $x_0 \in K \subset M$ be a fixed point of infinite type. We set

$$T_{1,0}(M)_{x_0} = T^{1,0}(\mathbb{C}^n)_{x_0} \cap [T_{x_0}(M) \otimes_{\mathbb{R}} \mathbb{C}], \quad m = \dim_{\mathbb{C}} T_{1,0}(M)_{x_0} \geq 0.$$

Let us consider a local frame $\{L_1, \cdots, L_{n-1}\}$ of $T_{1,0}(\partial\Omega)$, defined on an open neighborhood $V \subseteq \partial\Omega$ of x_0, such that the complex vectors $\{L_1(x_0), \cdots, L_m(x_0)\}$ span $T_{1,0}(M)_{x_0}$. Let $F \in C^\infty(U)$ be just any extension of f to an open neighborhood $U \subseteq \mathbb{C}^n$ of x_0 i.e. $F \circ \iota = f$ where $\iota : M \cap U \hookrightarrow U$ is the inclusion. We have

$$\overline{L}_j(x_0) = (d_{x_0}\iota)\overline{w}_j\,, \quad 1 \leq j \leq m,$$

for some $w_j \in T_{1,0}(M)_{x_0}$. Then

$$\begin{aligned} \overline{L}_j(F)_{x_0} &= (d_{x_0}F)\overline{L}_{j,x_0} = (d_{x_0}F)(d_{x_0}\iota)\overline{w}_j = (d_{x_0}f)\overline{w}_j = \\ &= A_{x_0}\overline{w}_j = (j^*\alpha)_{x_0}\overline{w}_j = \alpha_{x_0}(d_{x_0}j)\overline{w}_j \end{aligned}$$

hence

$$(\overline{L}_j F)(x_0) = (\alpha \overline{L}_j)_{x_0}\,, \quad 1 \leq j \leq m. \tag{3.16}$$

On the other hand, for any $1 \leq j \leq n-1$ (as $T_{0,1}(\partial\Omega) \rfloor \theta = 0$)

$$\begin{aligned} \alpha(\overline{L}_j) &= -(\mathcal{L}_T\theta)\overline{L}_j = -T(\theta(\overline{L}_j)) + \theta([T, \overline{L}_j]) = \\ &= \frac{i}{2}\{(\overline{\partial}\rho)[T, \overline{L}_j] - (\partial\rho)[T, \overline{L}_j]\} = \end{aligned}$$

(as $[T, \overline{L}_j]$ are tangent to $\partial\Omega$, for $1 \leq j \leq n-1$)

$$\begin{aligned} &= \frac{i}{2}\{(\overline{\partial}\rho)[T, \overline{L}_j] - (\partial\rho)[T, \overline{L}_j] - (d\rho)[T, \overline{L}_j]\} = \\ &= -i(\partial\rho)[T, \overline{L}_j] = (\partial\rho)[L, \overline{L}_j] \end{aligned}$$

hence (by (3.16))

$$(\overline{L}_j F)(x_0) = (\partial\rho)_{x_0}[L, \overline{L}_j]_{x_0}\,, \quad 1 \leq j \leq m.$$

When $m+1 \leq j \leq n$ then we set $\overline{L}_j(x_0) = u_j + iv_j$ where $u_j, v_j \in T_{x_0}(\mathbb{C}^n)$ are real vectors and $L_n = L$. Let $\sigma_j \subset T_{x_0}(\mathbb{C}^n)$ be the 2-plane spanned by $\{u_j, v_j\}$. Then

$$\dim_{\mathbb{R}}[\sigma_j \cap T_{x_0}(M)] \in \{0, 1\}$$

hence we may extend f off M with prescribed directional derivatives at the point x_0, in the directions $\overline{L}_j$, $m+1 \leq j \leq n$. Summing up, there is an open neighborhood $U \subseteq \mathbb{C}^n$ and there is a $\mathbb{C}$-valued function $\mathfrak{F} \in C^\infty(U)$ such that $\mathfrak{F}|_{U\cap M} = f|_{U\cap M}$ and

$$\overline{L}_j(\mathfrak{F})_{x_0} = (\partial\rho)_{x_0}[L, \overline{L}_j]_{x_0}, \quad 1 \leq j \leq n.$$

Moreover (as $(\partial\rho)L = 1$)

$$e^{-\mathfrak{F}}(\partial\rho)([e^{\mathfrak{F}}L, \overline{L}_j])_{x_0} = (\partial\rho)([L, \overline{L}_j])_{x_0} - \overline{L}_j(\mathfrak{F})(\partial\rho)(L)_{x_0} = 0$$

hence

$$(\partial\rho)\left([e^{\mathfrak{F}}L, \overline{Z}]\right)_{x_0} = 0$$

for any complex vector field Z of type $(1,0)$ on $\mathbb{C}^n$. Hence for any $\epsilon > 0$ there is a sufficiently small open neighborhood $U_\epsilon \subseteq \mathbb{C}^n$ of x_0 such that

$$\left|(\partial\rho)\left[e^{\mathfrak{F}}L\,,\,\partial/\partial\overline{z}^j\right]\right| < \epsilon \quad \text{on } U_\epsilon.$$

Now we piece together the local data by means of a partition of unity. Precisely, as K is compact (cf. [**76**]) there are finitely many points $x_1, \cdots, x_s \in K$ and there are open sets $U_{\epsilon,\alpha} \subseteq \mathbb{C}^n$ and smooth $\mathbb{C}$-valued functions $\mathfrak{F}_\alpha \in C^\infty(U_{\epsilon,\alpha})$ such that

$$x_\alpha \in U_{\epsilon,\alpha}\,, \quad \mathfrak{F}_\alpha|_{M\cap U_{\epsilon,\alpha}} = f|_{M\cap U_{\epsilon,\alpha}}\,,$$

$$\left|(\partial\rho)[e^{\mathfrak{F}_\alpha}L\,,\,\partial/\partial\overline{z}^j]\right| < \epsilon \quad \text{on } U_{\epsilon,\alpha}\,, \tag{3.17}$$

for any $1 \leq \alpha \leq s$. Let $U_\epsilon = \bigcup_{\alpha=1}^s U_{\epsilon,\alpha}$ (so that $K \subset U_\epsilon$) and let $\varphi_\alpha \in C^\infty(U_\epsilon)$ be a partition of unity subordinated to the covering $\{U_{\epsilon,\alpha} : 1 \leq \alpha \leq s\}$ i.e. $0 \leq \varphi_\alpha \leq 1$ and $\text{supp}(\varphi_\alpha) \subset U_{\epsilon,\alpha}$ and $\sum_{\alpha=1}^s \varphi_\alpha = 1$ on U_ϵ. Next let us set

$$Z_\epsilon = \frac{1}{s}\sum_{\alpha=1}^s \varphi_\alpha Z_\alpha\,, \quad Z_\alpha = \begin{cases} e^{\mathfrak{F}_\alpha}L & \text{on } U_{\epsilon,\alpha}\,, \\ 0 & \text{on } U_\epsilon \setminus U_{\epsilon,\alpha}\,, \end{cases}, \quad 1 \leq \alpha \leq s.$$

Then at each point of K

$$\left|(\partial\rho)[Z_\epsilon\,,\,\partial/\partial\overline{z}^j]\right| =$$

$$= \frac{1}{s}\left|\sum_\alpha \varphi_\alpha(\partial\rho)[e^{\mathfrak{F}_\alpha}L\,,\,\partial/\partial\overline{z}^j] - \sum_\alpha \frac{\partial\varphi_\alpha}{\partial\overline{z}^j}e^{\mathfrak{F}_\alpha}(\partial\rho)(L)\right| \leq$$

$$\leq \frac{1}{s}\sum_\alpha \left|(\partial\rho)[e^{\mathfrak{F}_\alpha}L\,,\,\frac{\partial}{\partial\overline{z}^j}]\right| + \frac{1}{s}\left|\sum_\alpha e^f\frac{\partial\varphi_\alpha}{\partial\overline{z}^j}\right| < \epsilon$$

by (3.17) and $\partial\left(\sum_\alpha \varphi_\alpha\right)/\partial\overline{z}^j = 0$. Also $Z_\epsilon(\rho) = (1/s)e^f$ on K hence $Z_\epsilon(\rho)$ is real on K and $C^{-1} < Z_\epsilon(\rho) < C$ for some $C > 0$ independent of ϵ. Q.e.d.

The situation where there is a Levi foliation in the boundary leads to a nice foliation theoretic interpretation of D'Angelo class. Precisely, let $\Omega \subset \mathbb{C}^n$ be a smoothly bounded pseudocnvex domain and $K \subset \partial\Omega$ the set of all boundary points of infinite type. E.J. Straube & M.K. Sucheston study (cf. [**233**]) the case where the compact set K is the closure in $\partial\Omega$ of its interior $\mathring{K}$ in $\partial\Omega$ (one assumes that $\mathring{K} \neq \emptyset$). If additionally, as one also assumes in [**233**], the Levi form of $\partial\Omega$ vanishes at each point of $\mathring{K}$ then (the open set) $\mathring{K}$ is a Levi flat CR manifold of type $(n-1, 1)$ i.e. it carries a foliation $\mathcal{F}$ by complex $(n-1)$-dimensional manifolds tangent to the Levi distribution $H(\partial\Omega)$. Let $M \in \overset{\circ}{K}/\mathcal{F}$ be a leaf of $\mathcal{F}$. Let $x_0 \in M$. By Proposition 3.34 the 1-form $A = j^*\alpha \in \Omega^1(M)$ is closed hence the integral $\int_\gamma A$ depends only on the homotopy class of the loop γ at x_0. The following map (the *period map*) is therefore well defined

$$\text{per}_A : \pi_1(M, x_0) \to \mathbb{R}, \quad \text{per}_A([\gamma]) = \int_\gamma A, \quad [\gamma] \in \pi_1(M, x_0).$$

Moreover, the period map per_A corresponds to the D'Angelo class $a(M) = [A] \in H^1(M, \mathbb{R})$ under the isomorphism

$$\text{Hom}(\pi_1(M, x_0), \mathbb{R}) \approx \text{Hom}(H_1(M, \mathbb{R}), \mathbb{R}) \approx H^1(M, \mathbb{R}). \tag{3.18}$$

Cf. also [**244**], p. 31. On the other hand let $x_0 \in \overset{\circ}{K}$ and let M be the leaf of $\mathcal{F}$ passing through x_0. Let $h_{x_0} : \pi_1(M, x_0) \to \text{Diff}_{x_0}(T)$ be the holonomy representation of the leaf M at x_0 (cf. Section 1.1.7 of this monograph) where T is a transversal at x_0. Let $\Gamma_{x_0} = h_{x_0}(\pi_1(M, x_0))$ be the holonomy group of M at x_0. Any germ $\phi \in \Gamma_{x_0}$ of a diffeomorphism of T in itself (preserving x_0) induces a linear isomorphism $d_{x_0}\phi : T_{x_0}(T) \to T_{x_0}(T)$ and then a linear isomorphism of $\nu(\mathcal{F})_{x_0}$ in itself, via the natural isomorphism $\sigma : \nu(\mathcal{F})_{x_0} \to T_{x_0}(T)$. Here for any $Y \in T_{x_0}(\overset{\circ}{K})$ the vector $\sigma(\pi_{x_0} Y)$ is the $T_{x_0}(T)$-component of Y with respect to the direct sum decomposition $T_{x_0}(\overset{\circ}{K}) = T_{x_0}(M) \oplus T_{x_0}(T)$. Also $\nu(\mathcal{F}) = T(\overset{\circ}{K})/T(\mathcal{F})$ and $\pi : T(\overset{\circ}{K}) \to \nu(\mathcal{F})$ is the projection. Then

$$\Gamma^1_{x_0} = \{\sigma^{-1} \circ \phi \circ \sigma : \phi \in \Gamma_{x_0}\}$$

is a subgroup (the so called *infinitesimal holonomy group* of M at x_0) of $\text{GL}(\nu(\mathcal{F})_{x_0})$. The definition of $\Gamma^1_{x_0}$ doesn't depend upon the choice of transversal T at x_0. Note that $\dim_{\mathbb{R}} \nu(\mathcal{F})_{x_0} = 1$ and $\text{GL}(\mathbb{R}) = \mathbb{R}^*$ where $\mathbb{R}^* = \mathbb{R} \setminus \{0\}$ are the multiplicative reals. Finally the map

$$\pi_1(M, x_0) \to \Gamma_{x_0} \approx \Gamma^1_{x_0} \subset \mathbb{R}^*$$

corresponds to the D'Angelo class $a(M)$ under the isomorphism (3.18), cf. Example 2.3.15 in [**59**], or Remark 3 in [**233**], p. 146.

Let $q : \tilde{M} \to M$ be the universal covering of M and $\tilde{x}_0 \in \tilde{M}$ such that $q(\tilde{x}_0) = x_0$. We set

$$p_A : \tilde{M} \to \mathbb{R}, \quad p_A(\tilde{x}) = \int_C q^* A, \quad \tilde{x} \in \tilde{M},$$

where C is a path in $\tilde{M}$ joining $\tilde{x}_0$ and $\tilde{x}$. As A is closed the definition of $p_A(\tilde{x})$ doesn't depend upon the choice of path C from $\tilde{x}_0$ to $\tilde{x}$. When M is compact and A is nonsingular one may use (by a result of D. Tischler, [**241**]) the fact that p_A is equivariant with respect to the period map

$$p_A([\gamma] \cdot \tilde{x}) = p_A(\tilde{x}) + \text{per}_A([\gamma]), \quad \tilde{x} \in \tilde{M}, \quad [\gamma] \in \pi_1(M, x_0),$$

to build a fibration $p : M \to S^1$ of the leaf M over the circle S^1 such that A (the representative of the D'Angelo class) be arbitrarily close to $p^* dt$ where t is a local coordinate on S^1. Precisely for any $\epsilon > 0$ there is a fibration $p_\epsilon : M \to S^1$ such that $\|A - p_\epsilon^* dt\| < \epsilon$ (cf. also [**244**], p. 30). Although $a(M) = 0$ is here excluded[2] by the compactness assumption on M (and then we may not use the method in the proof of Theorem 3.36 to build the vector fields Z_ϵ) Tischler's theorem (cf. [**241**])

[2]The existence of a function f such that $A = df$ would give rise to singularities of A at the critical points of f.

may prove useful in the understanding of the failure of global regularity for the Bergman projection for specific domains, such as the so called *worm domains* (cf. K. Diederich & J.E. Fornaess, [**79**])

$$\Omega = \{(z_1, z_2) \in \mathbb{C}^2 : \left| z_1 + e^{i \log |z_2|^2} \right|^2 < 1 - \phi(\log |z_2|^2)\}$$

where $\phi(t)$ is a C^∞ function which is chosen such that $\phi(t) \geq 0$, $\phi(-t) = \phi(t)$, $\phi^{-1}(0) = I_{\beta - \pi/2}$ (here $I_a = (-a, a) \subset \mathbb{R}$, $a > 0$) and Ω is smooth, bounded, connected and pseudoconvex. The fact that global regularity for P_0 fails on a worm domain was shown by D.E. Barrett, [**29**].

It is an open problem to study the properties of the lifted foliation $\mathcal{F}_T^1$ on $B_T^1(\overset{\circ}{K}, \mathcal{F})$ (cf. Section 1.2.2 of this book) associated to the Levi foliation $\mathcal{F}$ of $\overset{\circ}{K}$. Note that given a leaf $M_T \in B_T^1(\overset{\circ}{K}, \mathcal{F})/\mathcal{F}_T^1$ projecting on M it follows that $M_T \to M$ is the holonomy bundle (in the sense of [**155**], Vol. I, p. 85) of a flat connection in the portion of $B_T^1(\overset{\circ}{K}, \mathcal{F}) \to \overset{\circ}{K}$ over M. Also the infinitesimal holonomy group $\Gamma_{x_0}^1$ may be identified to the structure group of the principal bundle $M_T \to M$.

Another open problem is whether one may apply the structure theorem for Riemannian foliations (cf. Theorem 5.1 in [**179**], p. 155-156) to the pair $(K, \mathcal{F})$ (under the assumptions on K in [**233**]). Moreover, when $\Omega \subset \mathbb{C}^2$ and $\overset{\circ}{K}$ is a smoothly bounded domain in $\partial\Omega$, so that $\Gamma = \partial K$ is a compact orientable smooth real surface in $\mathbb{C}^2$, the Levi foliation of $\overset{\circ}{K}$ induces a foliation by real curves $\mathcal{F}_\Gamma$ on Γ. The foliation $\mathcal{F}_\Gamma$ has singularities (occurring as the complex tangencies of Γ, cf. E. Bishop, [**40**]) and one ought to clarify both the relationship that the situation at hand bears with the general theory of singular (Riemannian) foliations (cf. e.g. R. Wolak, [**254**]) and the theory of analytic discs, cf. [**135**] or [**50**], p. 206 (when the leaf M is an analytic disc the condition that $a(M) = 0$ is trivially satisfied).

A study of function theory on Levi flat CR manifolds (analogous to the work by J. Noguchi & T. Ochiai, [**196**], and with similar expectations) was started by G. Gigante & G. Tomassini, [**117**]-[**118**]. Cf. also E. Barletta, [**15**].

CHAPTER 4

Levi foliations of CR submanifolds in $\mathbb{C}P^N$

Let $M \subset \mathbb{C}P^N$ be a real submanifold of the complex N-dimensional projective space. Let us assume that M is a CR manifold with the CR structure induced by the complex structure of $\mathbb{C}P^N$ i.e.

$$T_{1,0}(M) = T^{1,0}(\mathbb{C}P^N) \cap [T(M) \otimes \mathbb{C}]$$

has constant rank. It is a natural question whether the Levi form of M may have a nonzero null space of constant rank, equivalently whether M admits Levi foliations $\mathcal{F}$. If this is the case then the next step might be to study the geometry of the second fundamental form of the immersion $M \hookrightarrow \mathbb{C}P^N$ (cf. M. Djorić & M. Okumura, [**81**]) and the topology of $(M, \mathcal{F})$ (cf. L. Ni & J. Wolfson, [**188**]).

4.1. Nonexistence of smooth Levi flat hypersurfaces

Deep results in complex analysis in several complex variables show that smooth Levi flat CR submanifolds in $\mathbb{C}P^N$ are far less abundant then in the complex space $\mathbb{C}^N$. Indeed we may state the following

THEOREM 4.1. (J. Cao & M-C. Shaw & L. Wang, [**60**])
There exists no C^2 Levi flat real hypersurface $M \subset \mathbb{C}P^N$, $N \geq 3$.

Theorem 4.1 improves a result of Y-T. Siu, [**222**], which requires M to be at least C^8 smooth. See also [**221**]. The needed smoothness was subsequently reduced to C^4 by A. Iordan, [**139**]. The real analytic case (therefore establishing the nonexistence of C^ω Levi flat hypersurfaces $M \subset \mathbb{C}P^N$, $N \geq 3$) had been handled earlier by A. Lins Neto, [**174**]. The proof of Theorem 4.1 is too involved to be reproduced in the present monograph. The idea is to assume that a connected C^2 smooth Levi flat real hypersurface $M \subset \mathbb{C}P^N$ exists, so that one may consider a domain $\Omega \subset \mathbb{C}P^N$ ($N \geq 3$) with C^2 smooth Levi flat boundary $\partial\Omega$. As $\partial\Omega$ is Levi flat the domain Ω is both pseudoconvex and pseudoconcave hence one may apply the following Liouville type result

LEMMA 4.2. (J. Cao & M-C. Shaw & L. Wang, [**60**])
Let $\Omega \subset\subset \mathbb{C}P^N$ be a pseudoconcave domain with C^2 smooth boundary $\partial\Omega$, $N \geq 3$. Then $L^2_{(p,0)}(\Omega) \cap \mathrm{Ker}(\overline{\partial}) = \{0\}$ for any $1 \leq p \leq N$ and $L^2_{(0,0)}(\Omega) \cap \mathrm{Ker}(\overline{\partial}) = \mathbb{C}$.

Cf. Proposition 4.5 in [**60**], p. 200. We recall that the domain $\Omega \subset \mathbb{C}P^N$ is *pseudoconcave* if its complement $\mathbb{C}P^N \setminus \overline{\Omega}$ is pseudoconvex. On the other hand one may reach a contradiction by showing that

$$L^2_{(N,0)}(\Omega) \cap \mathrm{Ker}(\overline{\partial}) \neq \{0\}$$

by adapting arguments due to J-P. Demailly, [**78**], and S. Kobayashi & H. Wu, [**156**]. It should be mentioned that J. Cao et al., [**60**], claim that Theorem 4.1

holds for $N = 2$, as well. However the proof of the case $N = 2$ is in doubt[1] and therefore the problem of the non existence of Levi flat real hypersurfaces $M \subset \mathbb{C}P^2$ should be considered open.

4.2. Lefschetz type results for CR submanifolds in $\mathbb{C}P^N$

This section aims to discuss the following result (generalizing the result by A. Lins Neto, [**174**], quoted in the preceding section)

THEOREM 4.3. (L. Ni & J. Wolfson, [**188**])
Let V be a complex ν-dimensional irreducible compact Kähler manifold of nonnegative holomorphic bisectional curvature and complex positivity ℓ. Then V admits no real analytic Levi flat CR submanifolds of dimension $m \geq 2(\nu + 1) - \ell$.

We show how Lefschetz type results may be combined with a well known result of A. Haefliger, [**131**] (about the nonexistence of real analytic foliations of codimension one, cf. also [**164**], p. 377) in order to derive Theorem 4.3. The known results on Kähler manifolds of nonnegative holomorphic bisectional curvaure, cf. [**122**], are recalled in Appendix A. We shall need the following notion

DEFINITION 4.4. Let $M \subset \mathbb{C}P^\nu$ be a CR submanifold and $L : T_{1,0}(M) \otimes T_{1,0}(M) \to [T(M) \otimes \mathbb{C}]/[H(M) \otimes \mathbb{C}]$ its Levi form. The *nullity* of M at $x \in M$ is the dimension of the null space $\mathrm{Null}(L_x)$. $\square$

A method similar to that in the proof of Theorem A.1 (and going back to T. Frankel's original approach, cf. [**110**]) i.e. a study of the relative geometry of a pair of compact submanifolds M and N of $\mathbb{C}P^\nu$, where M is a CR submanifold and N a complex submanifold, leads to the following Lefschetz type result

THEOREM 4.5. (L. Ni & J. Wolfson, [**188**])
Let $M \subset \mathbb{C}P^\nu$ be a compact CR submanifold of real dimension $2p + 1$ and CR dimension p. Assume that the nullity of M is everywhere $\geq r$ with $0 < r \leq p$. Then

$$\pi_j(M, M \cap \mathbb{C}P^{\nu-1}) = 0, \quad j \leq r - 1.$$

Therefore the map induced by the inclusion

$$\pi_j(M \cap \mathbb{C}P^{\nu-1}) \to \pi_j(M)$$

is an isomorphism if $j < r - 1$ and an epimorphism if $j = r - 1$.

In particular

COROLLARY 4.6. *Let $M \subset \mathbb{C}P^\nu$ be a real $(2p + 1)$-dimensional compact Levi flat CR submanifold of CR dimension p. Then*

$$\pi_j(M, M \cap \mathbb{C}P^{\nu-1}) = 0, \quad j \leq p - 1,$$

and the short sequences

$$0 \to \pi_j(M \cap \mathbb{C}P^{\nu-1}) \to \pi_j(M) \to 0, \quad j < p - 1,$$

$$\pi_{p-1}(M \cap \mathbb{C}P^{\nu-1}) \to \pi_{p-1}(M) \to 0,$$

are exact.

[1]According to T. Ohsawa (private communication to M. Eastwood). The papers [**139**] and [**222**] are also in doubt (as far as the case $N = 2$ is concerned).

Similar methods, i.e. a study of the relative geometry of a pair of compact CR submanifolds of a compact Hermitian symmetric space V, lead to necessary topological restrictions that a compact CR submanifold M of V should satisfy, in order that the Levi form of M be degenerate.

THEOREM 4.7. (L. Ni & J. Wolfson, [**188**])
Let V be a complex ν-dimensional irreducible compact Kähler manifold of nonnegative holomorphic bisectional curvature and complex positivity ℓ. Let $M \subset V$ be a real $(2p+1)$-dimensional CR submanifold of CR dimension p and nullity everywhere $\geq r$ with $0 < r \leq p$. Then

$$\pi_j(V, M) = 0, \quad j \leq \ell + 1 - 2(\nu - r).$$

In particular if M is Levi flat of dimension $m \geq 2(\nu + 1) - \ell$ then Theorem 4.7 implies that M is simply connected. At this point Theorem 4.3 follows from

THEOREM 4.8. (A. Haefliger, [**131**])
There are no real analytic codimension one foliations on a compact simply connected manifold.

See also Theorem 4 in [**164**], p. 377, and Lemma 4 in [**164**], p. 388.

We remark that the topology of CR manifolds, as related to the Lefschetz theorem (cf. A. Andreotti & T. Frankel, [**6**]) was also studied by C.D. Hill & M. Nacinovich, [**136**]. Their main finding is the following

THEOREM 4.9. (C.D. Hill et al., [**136**])
Let M be a weakly q-concave Stein CR manifold of type (n, k). Then M has the homotopy type of a CW*-complex of dimension $\leq 2n + k - q$. In particular*

$$H_j(M; \mathbb{Z}) = 0, \quad j > 2n + k - q,$$

and $H_{2n+k-q}(M; \mathbb{Z})$ has no torsion.

We recall that a *Stein CR manifold* is a CR manifold which admits a CR embedding as a closed CR submanifold of some Stein manifold. As the CR codimension k is arbitrary (perhaps $k \geq 2$) we should make clear what is meant by *concavity*. Let $H(M)^\perp \subset T^*(M)$ be the conormal bundle associated to the Levi distribution $H(M)$ (which has codimension k). If $x \in M$ and $\omega \in H(M)^\perp_x$ then for any $w \in T_{0,1}(M)_x$ we set

$$L(\omega, w) := i\,(d\tilde{\omega})_x(w, \overline{w}) = -\frac{i}{2}\omega([W, \overline{W}]_x),$$

where $\tilde{\omega}$ is a local 1-form on M, defined in a neighborhood U of x, such that $\tilde{\omega}_x = \omega$, and $W \in \Gamma^\infty(U, T_{0,1}(M))$ extends w i.e. $W_x = w$. A CR manifold M of type (n, k) is said to be *weakly q-concave* $(0 \leq q \leq n)$ if for any $x \in M$ and any $\omega \in H(M)^\perp_x \setminus \{0\}$ the Levi form $L(\omega, \cdot)$ has at least q nonpositive eigenvalues. A result in [**137**] implies that a weakly 1-concave Stein CR manifold cannot be compact. Any Levi flat CR manifold of type (n, k) is also weakly n-concave. The relationship among Theorem 4.9 and the theory of Levi foliations was not investigated. Cf. also M. Barros & F. Urbano, [**27**], who used Morse theory to show that

THEOREM 4.10. *Let M be a real n-dimensional manifold isometrically imbedded as a closed subset of $\mathbb{C}^m$. If M admits a $2h$-dimensional complex foliation then M has the homotopy type of a CW-complex of dimension $\leq n - h$.*

4.3. CR submanifolds of maximal CR dimension

Let M be a real $(n+2)$-dimensional submanifold of codimension $p \geq 1$ of a Hermitian manifold $(V, \tilde{J}, \tilde{g})$ where $\tilde{J}$ denotes the complex structure and $\tilde{g}$ the Hermitian metric. Let $g = \iota^* \tilde{g}$ be the first fundamental form of $\iota : M \hookrightarrow V$ (the canonical inclusion). Let $T(M)^\perp \to M$ be the normal bundle of ι so that

$$\iota^* T(V) = T(M) \oplus T(M)^\perp. \tag{4.1}$$

Let $\tan : \iota^* T(V) \to T(M)$ and $\mathrm{nor} : \iota^* T(V) \to T(M)^\perp$ be the natural projections associated to the decomposition (4.1). Let $P : T(M) \to T(M)$ and $F : T(M) \to T(M)^\perp$ be the field of endomorphisms and (vector valued) 1-form on M respectively given by

$$P(X) = \tan(\tilde{J}X), \quad F(X) = \mathrm{nor}(\tilde{J}X), \quad X \in T(M).$$

Also we consider the bundle morphisms $t : T(M)^\perp \to T(M)$ and $f : T(M)^\perp \to T(M)^\perp$ given by

$$t(\eta) = \tan(\tilde{J}\eta), \quad f(\eta) = \mathrm{nor}(\tilde{J}\eta), \quad \eta \in T(M)^\perp.$$

Also we shall need the Gauss and Weingarten formulae

$$\tilde{\nabla}_X Y = \nabla_X Y + h(X,Y), \quad \tilde{\nabla}_X \eta = -a_\eta X + \nabla^\perp_X \eta,$$

for any $X, Y \in T(M)$ and any $\eta \in T(M)^\perp$, where $\tilde{\nabla}$ and ∇ are the Levi-Civita connections of $(V, \tilde{g})$ and (M, g) respectively, h is the second fundamental form of ι, a_η is the Weingarten operator associated to the normal section η, and $\nabla^\perp$ is the normal connection (a connection in $T(M)^\perp \to M$).

Let us assume from now on that M is a proper CR submanifold of $\tilde{M}$ in the sense of A. Bejancu (cf. [**33**]-[**34**]) that is M carries a rank $2k$ distribution $H(M)$ such that i) $H(M)$ is holomorphic i.e. $JH(M) = H(M)$, ii) its orthogonal complement $H(M)^\perp$ in $T(M)$ is anti-invariant i.e. $JH(M)^\perp \subseteq T(M)^\perp$, and iii) $k \neq 0$ and $\ell := n + 2 - 2k \neq 0$.

Let $J := P|_{H(M)}$, a complex structure on $H(M)$, and let us set

$$T_{1,0}(M) = \{X - iJX \in T(M) \otimes \mathbb{C} : X \in H(M)\}.$$

By a result in [**44**] the requirements (i)-(iii) imply that $(M, T_{1,0}(M))$ is a CR manifold of type (k, ℓ), of CR dimension $k \leq (n+1)/2$. A series of papers (cf. M. Djorić & M. Okumura, [**80**]-[**81**], J-H. Kwon & J-S. Pak, [**163**], E. Barletta, [**17**]) investigate the geometry of the second fundamental form of a CR submanifold (of a Hermitian manifold) of *maximal* CR dimension i.e. $k = (n+1)/2$. If this is the case then $H(M)^\perp$ is a real line bundle i.e. M is a CR manifold of CR codimension $\ell = 1$ (of hypersurface type) immersed in V with a perhaps larger codimension p.

Let us assume from now on that M is a real $(n+2)$-dimensional orientable CR submanifold, of a Kählerian manifold $(V, \tilde{J}, \tilde{g})$, of codimension p and of maximal CR dimension $(n+1)/2$. As M is orientable and $H(M)$ is oriented by its complex structure there is a unit vector field U on M which spans $H(M)^\perp$ i.e. $H(M)^\perp = \mathbb{R}U$. Let us set $\xi := \tilde{J}U \in T(M)^\perp$. Then

$$T(M) = H(M) \oplus \mathbb{R}U, \quad T(M)^\perp = E(M) \oplus \mathbb{R}\xi,$$

where $E(M) := (\mathbb{R}\xi)^\perp \subset T(M)^\perp$. Note that $E(M)$ is holomorphic i.e. $\tilde{J}E(M) = E(M)$. Also

$$P(U) = 0, \quad f(\xi) = 0.$$

Let θ be the real 1-form on M defined by $\theta(X) = g(X, U)$. Then

$$P^2 = -I + \theta \otimes U, \quad \theta \circ P = 0, \quad F = \theta \otimes \xi.$$

Moreover $\mathrm{Ker}(\theta) = H(M)$ i.e. θ is a pseudohermitian structure on the CR manifold $(M, T_{1,0}(M))$. Let L_θ be the corresponding Levi form

$$L_\theta(Z, \overline{W}) = -i(d\theta)(Z, \overline{W}), \quad Z, W \in T_{1,0}(M).$$

The following results on the nonexistence of Levi flat CR submanifolds of a Kählerian manifold hold true

THEOREM 4.11. (M. Djorić & M. Okumura, [**81**])
Let M be a $(n+2)$-dimensional $(n \geq 1)$ CR submanifold of CR dimension $(n+1)/2$ and codimension p of a complex space form $V^\nu(c)$ of holomorphic sectional curvature c. Let us assume that i) *ξ is parallel in the normal bundle ($\nabla^\perp \xi = 0$),* ii) *U is an eigenvector of the Weingarten operator a_ξ corresponding to some eigenvalue $\alpha \in C^\infty(M)$, and that* iii) *there is an eigenvalue $\lambda \in \mathrm{Spec}(a_\xi)$ such that $\alpha \neq \lambda$ and $\alpha \neq 2\lambda$ everywhere on M.*

i) If $c > 0$, i.e. $V^\nu(c)$ is an elliptic complex space form, then $L_\theta \neq 0$. In particular, if $p = 1$ then L_θ is positive definite, i.e. M is strictly pseudoconvex.

ii) If $L_\theta = 0$ then $c < 0$ i.e. $V^\nu(c)$ is a hyperbolic complex space form.

Proof. Let $\{\xi_1, \cdots, \xi_p\}$ be a local orthonormal frame of the normal bundle $T(M)^\perp$ (i.e. $\tilde{g}(\xi_a, \xi_b) = \delta_{ab}$) such that $\xi_1 = \xi$. Then $E(M)$ is locally the span of $\{\xi_a : 2 \leq a \leq p\}$. Moreover

$$h(X, Y) = \sum_{a=1}^{p} g(a_{\xi_a} X, Y)\xi_a \,. \tag{4.2}$$

Using $F = \theta \otimes \xi$ together with the Gauss and Weingarten formulae, we may conduct the following calculation

$$\nabla_X PY = \tilde{\nabla}_X PY - h(X, PY) =$$
$$= \tilde{\nabla}_X \tilde{J}Y - \tilde{\nabla}_X FY - h(X, PY) =$$

(as $\tilde{g}$ is a Kähler metric, i.e. $\tilde{\nabla}\tilde{J} = 0$)

$$= \tilde{J}\tilde{\nabla}_X Y - X(\theta(Y))\xi - \theta(Y)\tilde{\nabla}_X \xi - h(X, PY) =$$
$$= \tilde{J}\nabla_X Y + \tilde{J}h(X, Y) - X(\theta(Y))\xi + \theta(Y)a_\xi X - \theta(Y)\nabla^\perp_X \xi - h(X, PY) =$$
$$= P\nabla_X Y + \theta(\nabla_X Y)\xi + t\,h(X, Y) + f\,h(X, Y) -$$
$$-X(\theta(Y))\xi + \theta(Y)a_\xi X - \theta(Y)\nabla^\perp_X \xi - h(X, PY).$$

Comparing the tangential components one obtains

$$(\nabla_X P)Y = t\,h(X, Y) + \theta(Y)a_\xi X$$

for any $X, Y \in T(M)$. Note that

$$t\xi_a = \tan(\tilde{J}\xi_a) = \begin{cases} 0, & 2 \leq a \leq p, \\ -U, & a = 1, \end{cases}$$

hence (by (4.2))

$$t\,h(X, Y) = -g(a_\xi X, Y)U.$$

We may conclude that

$$(\nabla_X P)Y = \theta(Y)a_\xi X - g(a_\xi X, Y)U. \tag{4.3}$$

Moreover, on one hand (by the Gauss formula)

$$\nabla_X U = \tilde{\nabla}_X U - h(X, U)$$

and on the other

$$\tilde{\nabla}_X U = -\tilde{\nabla}_X J\xi = -\tilde{J}\tilde{\nabla}_X \xi = \tilde{J} a_\xi X - \tilde{J}\nabla^\perp_X \xi$$

hence

$$\nabla_X U = \tan\left(\tilde{\nabla}_X U\right) = P a_\xi X - t\,\nabla^\perp_X \xi$$

that is

$$\nabla_X U = P a_\xi X \tag{4.4}$$

for any $X \in T(M)$, provided that we show the following

LEMMA 4.12. *On any orientable proper CR submanifold of maximal CR dimension $\iota : M \hookrightarrow V$ in a Kählerian manifold V one has $t\,\nabla^\perp_X \xi = 0$ for any $X \in T(M)$.*

Proof of Lemma 4.12. Let $Y \in H(M)$. Then

$$g(t\,\nabla^\perp_X \xi, Y) = \tilde{g}(\tilde{J}\nabla^\perp_X \xi, Y) = -\tilde{g}(\nabla^\perp_X \xi, \tilde{J}Y) = 0$$

because $\tilde{J}Y \in H(M) \subset T(M)$. So $t\,\nabla^\perp_X \xi$ is orthogonal to $H(M)$. On the other hand

$$g(t\,\nabla^\perp_X \xi, U) = -\tilde{g}(\nabla^\perp_X \xi, \tilde{J}U) = -\tilde{g}(\nabla^\perp_X \xi, \xi) = -\frac{1}{2} X(\|\xi\|^2) = 0$$

so that $t\,\nabla^\perp_X \xi$ is orthogonal to U, as well. Yet $T(M) = H(M) \oplus \mathbb{R}U$ hence the statement in Lemma 4.12. Q.e.d.

We also wish to establish the following

LEMMA 4.13. *If ξ is parallel in the normal bundle (i.e. $\nabla^\perp \xi = 0$) then $a_{\xi_a} U = 0$ for any $2 \leq a \leq p$.*

Proof of Lemma 4.13. By the Weingarten formula

$$g(a_{\xi_a} U, X) = g(U, a_{\xi_a} X) = -\tilde{g}(U, \tilde{\nabla}_X \xi_a) =$$

$$= \tilde{g}(\tilde{J}\xi, \tilde{\nabla}_X \xi_a) = -\tilde{g}(\xi, \tilde{\nabla}_X \tilde{J}\xi_a) = \tilde{g}(\tilde{\nabla}_X \xi, \tilde{J}\xi_a) = \tilde{g}(\nabla^\perp_x \xi, \tilde{J}\xi_a) = 0,$$

as $\tilde{J}\xi_a \in E(M) \subset T(M)^\perp$ for any $2 \leq a \leq p$. Q.e.d.

To complete the proof of Theorem 4.11 let $V = V^\nu(4k)$ be a complex space form of (constant) holomorphic sectional curvature $c = 4k$. The curvature tensor field of $(V, \tilde{g})$ is then given by (cf. e.g. (2.2) in [**257**], p. 9)

$$\tilde{R}(X,Y)Z = k\{\tilde{g}(Y,Z)X - \tilde{g}(X,Z)Y + \tag{4.5}$$

$$+\tilde{g}(\tilde{J}Y,Z)\tilde{J}X - \tilde{g}(\tilde{J}X,Z)\tilde{J}Y - 2\tilde{g}(\tilde{J}X,Y)\tilde{J}Z\}$$

for any $X, Y, Z \in T(V^\nu(4k))$. The covariant derivative of a_ξ is related to the so called Bortolotti-Van der Waerden derivative of h by

$$g((\nabla_X a_\xi)Y, Z) = \tilde{g}((\nabla_X h)(Y,Z), \xi) + \tilde{g}(h(Y,Z), \nabla^\perp_X \xi)$$

for any $X, Y, Z \in T(M)$. Taking into account the last identity the Codazzi equation

$$\operatorname{nor}\left(\tilde{R}(X,Y)Z\right) = (\nabla_X h)(Y,Z) - (\nabla_Y h)(X,Z)$$

yields

$$\tilde{g}(\tilde{R}(X,Y)Z,\xi) = g((\nabla_X a_\xi)Y, Z) - g((\nabla_Y a_\xi)X, Z) + \\ + \tilde{g}(h(X,Z), \nabla^\perp_Y \xi) - \tilde{g}(h(Y,Z), \nabla^\perp_X \xi).$$

Now a calculation based on (4.5) and on Lemma 4.13 leads to

PROPOSITION 4.14. *Let M be an orientable proper CR submanifold of maximal CR dimension of a complex space form $V^\nu(4k)$. If ξ is parallel in the normal bundle then the Codazzi equation of the given immersion $\iota : M \hookrightarrow V^\nu(4k)$ is*

$$(\nabla_X a_\xi)Y - (\nabla_Y a_\xi)X = k\{\theta(X)PY - \theta(Y)PX - 2g(PX,Y)U\}$$

for any $X, Y \in T(M)$.

The calculations are left as an exercise to the reader. Moreover we shall need

LEMMA 4.15. *If ξ is parallel in the normal bundle and U is an eigenvector of the Weingarten operator a_ξ i.e. there is $\alpha \in C^\infty(M)$ such that $U \in$ Eigen$(a_\xi;\alpha)$ then*

$$(2\lambda - \alpha)a_\xi PX = (2k + \alpha\lambda)PX + X(\alpha)U$$

for any $X \in$ Eigen$(a_\xi;\lambda)$ with $\lambda \in C^\infty(M)$ and $\lambda \neq \alpha$ everywhere on M.

Proof of Lemma 4.15. By hypothesis $a_\xi U = \alpha U$. Taking the covariant derivative of this relation we obtain

$$(\nabla_X a_\xi)U + a_\xi P a_\xi X = X(\alpha)U + \alpha\, P a_\xi X \tag{4.6}$$

for any $X \in T(M)$. Let us write the Codazzi equation (cf. Proposition 4.14 above) with $Y = U$. As

$$PU = \tan(\tilde{J}U) = \tan(\xi) = 0,$$

$$g(PX,U) = \tilde{g}(\tilde{J}X,U) = -\tilde{g}(X,\tilde{J}U) = -\tilde{g}(X,\xi) = 0,$$

we obtain the identity

$$(\nabla_X a_\xi)U = (\nabla_U a_\xi)X - k\,PX. \tag{4.7}$$

Let us replace from (4.7) into (4.6) so that to get

$$(\nabla_U a_\xi)X - k\,PX + a_\xi P a_\xi X = X(\alpha)U + \alpha P a_\xi X.$$

Taking the inner product with $Y \in T(M)$ we obtain

$$g((\nabla_U a_\xi)X,Y) - kg(PX,Y) + g(a_\xi P a_\xi X, Y) = \\ = X(\alpha)\theta(Y) + \alpha g(P a_\xi X, Y). \tag{4.8}$$

Let us interchange X and Y in (4.8) to obtain another equation of the sort

$$g((\nabla_U a_\xi)Y,X) - kg(PY,X) + g(a_\xi P a_\xi Y, X) = \\ = Y(\alpha)\theta(X) + \alpha g(P a_\xi Y, X).$$

Subtracting the last equation from (4.8) and using the fact that the operator $\nabla_U a_\xi$ is self-adjoint we obtain

$$2kg(X,PY) + 2g(a_\xi P a_\xi X, Y) = \\ = X(\alpha)\theta(Y) - Y(\alpha)\theta(X) + \alpha g(P a_\xi X + a_\xi PX, Y)$$

for any $X, Y \in T(M)$. The last identity may also be written as

$$-2kPX + 2a_\xi P a_\xi X = X(\alpha)U - \theta(X)\nabla\alpha + \alpha\,(P a_\xi + a_\xi P)\,X.$$

In particular for an eigenvector X of a_ξ corresponding to the eigenvalue $\lambda \in C^\infty(M)$ with $\lambda(x) \neq \alpha(x)$ for any $x \in M$ one has $\theta(X) = 0$ (as the eigenspaces $\text{Eigen}(a_\xi;\alpha)$ and $\text{Eigen}(a_\xi;\lambda)$ are mutually orthogonal) hence

$$(2\lambda - \alpha)a_\xi PX = (2k + \alpha\lambda)PX + X(\alpha)U$$

which is the identity in Lemma 4.15. The last part of the proof consists in computing the Levi form

$$G_\theta(X,Y) = (d\theta)(X,JY) = -\frac{1}{2}\theta([X,JY]) =$$
$$= -\frac{1}{2}g(U,[X,JY]) = -\frac{1}{2}\{\tilde{g}(\xi,\tilde{J}\tilde{\nabla}_X JY) - \tilde{g}(\tilde{J}\tilde{\nabla}_{JX}Y)\} =$$
$$= \frac{1}{2}\{\tilde{g}(\xi,\tilde{\nabla}_X Y) + \tilde{g}(\xi,\tilde{\nabla}_{JY}JX)$$

or (by the Weingarten formula)

$$G_\theta(X,Y) = \frac{1}{2}\tilde{g}(h(X,Y) + h(JX,JY),\xi)$$

for any $X, Y \in H(M)$, cf. also R. Hermann, [**134**]. In terms of the Weingarten operator

$$G_\theta(X,Y) = \frac{1}{2}\{g(a_\xi X,Y) + g(a_\xi JX,JY)\}.$$

Given $X, Y \in H(M)$ there are $V, W \in T(M)$ such that $X = PV$ and $Y = PW$ hence

$$G_\theta(PV,PW) = \frac{1}{2}\{g(a_\xi PV,PW) + g(a_\xi P^2V,P^2W)\} =$$
$$= \frac{1}{2}\{-g(Pa_\xi PV,W) + g(a_\xi V,W) - \theta(V)g(a_\xi U,W) -$$
$$-g(a_\xi V,U)\theta(W) + \theta(V)\theta(W)g(a_\xi U,U)\} =$$
$$= \frac{1}{2}g(-Pa_\xi PV - \theta(V)a_\xi U - \theta(a_\xi V)U + \theta(V)g(a_\xi U,U)U\,,\,W).$$

Next, assuming that $a_\xi U = \alpha U$, we obtain

$$G_\theta(PV,PW) = \frac{1}{2}g(a_\xi V - Pa_\xi PV - \alpha\theta(V)U\,,\,W)$$

for any $V, W \in T(M)$. Therefore, if M is Levi flat ($G_\theta = 0$) then

$$a_\xi V - Pa_\xi PV - \alpha\theta(V)U = 0.$$

Now let $V \in \text{Eigen}(a_\xi;\lambda)$ with $\lambda \in C^\infty(M)$, $\lambda(x) \neq \alpha(x)$ and $\alpha(x) \neq 2\lambda(x)$ for any $x \in M$. As $\alpha \neq \lambda$ everywhere it follows that the eigenspaces $\text{Eigen}(a_\xi;\alpha)$ and $\text{Eigen}(a_\xi;\lambda)$ are orthogonal, hence $\theta(V) = 0$. The last identity becomes (by Lemma 4.15)

$$(2\lambda - \alpha)\lambda V - P\{(2k + \alpha\lambda)PV + V(\alpha)U\} = 0$$

hence (as $PU = 0$ and $P^2V = -V$) $\lambda^2 + k = 0$. Theorem 4.11 is proved. It implies the following weaker (the proof is however simpler and of pure differential geometric nature) version of the result by J. Cao et al., [**60**] (cf. Theorem 4.1 above)

COROLLARY 4.16. *There is no Levi flat C^3 hypersurface $M \subset \mathbb{C}P^N$ admitting a unit normal field ξ such that the shape operator a_ξ satisfies the conditions* ii)-iii) *in Theorem* 4.11.

CHAPTER 5

Tangentially CR foliations

Let $(M, T_{1,0}(M))$ be a CR manifold and $\mathcal{F}$ a foliation of M.

DEFINITION 5.1. We say $\mathcal{F}$ is a (*tangentially*) *CR foliation* if each leaf L of $\mathcal{F}$ is a CR submanifold of M, i.e. L is a CR manifold and the inclusion $\iota : L \subset M$ is a CR map i.e. $(d_x \iota) T_{1,0}(L)_x \subseteq T_{1,0}(M)_x$, $x \in L$. □

The foliation in Example 1.18 is a tangentially CR foliation of $\mathbb{C}^{n+1}$.

Let $\mathcal{F}$ be a CR foliation. Let $H(\mathcal{F}) \to M$ be the subbundle of $T(\mathcal{F})$ whose portion over a leaf L of $\mathcal{F}$ is $H(L)$ i.e. the Levi distribution of L. Similarly, let $T_{1,0}(\mathcal{F}) \to M$ be the complex subbundle of $T(\mathcal{F}) \otimes \mathbb{C}$ whose portion over a leaf L of $\mathcal{F}$ is $T_{1,0}(L)$, the CR structure of L.

5.1. The basic Cauchy-Riemann complex

Let $\mathcal{F}$ be a tangentially CR foliation of a nondegenerate CR manifold M.

DEFINITION 5.2. We say $\omega \in \Omega^{0,s}(M)$ is a *basic* $(0,s)$*-form* if

$$\overline{Z} \rfloor \omega = 0, \quad \overline{Z} \rfloor \overline{\partial}_M \omega = 0,$$

for any $Z \in T_{1,0}(\mathcal{F})$. □

Let $\Omega_B^{0,s}(\mathcal{F})$ denote the space of all basic $(0,s)$-forms on $(M, \mathcal{F})$. As $\overline{\partial}_M^2 = 0$ it follows that

$$\overline{\partial}_M \Omega_B^{0,s}(\mathcal{F}) \subseteq \Omega_B^{0,s+1}(\mathcal{F}).$$

DEFINITION 5.3. A function $f : M \to \mathbb{C}$ is said to be *CR-holomorphic on* $\mathcal{F}$ if its restriction $f\,|_L$ to each leaf L of $\mathcal{F}$ is a CR function on L. □

Let $CR^r(\mathcal{F})$ be the space of all CR-holomorphic functions on $\mathcal{F}$ which are of class C^r so that $f\,|_L \in CR^r(L)$ for any $L \in M/\mathcal{F}$. Note that

$$CR^r(M) \subseteq CR^r(\mathcal{F}). \tag{5.1}$$

Indeed this follows from the identity

$$j^* \, \overline{\partial}_M u = \overline{\partial}_L (u \circ j)$$

for any $u \in C^\infty(M)$. Here $L \in M/\mathcal{F}$ and $j : L \hookrightarrow M$ the inclusion. To prove of the last identity let $Z \in T_{1,0}(L)$. For any $x \in L$ one has

$$\left(\overline{\partial}_L (u \circ j) \overline{Z}\right)_x = \overline{Z}_x (u \circ j) = \left((d_x j) \overline{Z}_x\right)(u).$$

Yet j is a CR immersion hence $(d_x j)\overline{Z}_x \in T_{0,1}(M)_x$ so we may use the definition of the tangential Cauchy-Riemann operator $\overline{\partial}_M$ on functions to conclude that

$$\left(\overline{\partial}_L (u \circ j) \overline{Z}\right)_x = (\overline{\partial}_M u)_x (d_x j) \overline{Z}_x = (j^* \, \overline{\partial}_M u)_x \overline{Z}_x \, .$$

Note that the proof doesn't actually require the nondegeneracy assumption. The statement (5.1) is proved. Viceversa, one may formulate the following open problem. Given a tangentially CR foliation of a CR submanifold $M \subset \mathbb{C}^N$ and given $f \in CR^1(\mathcal{F})$ is there an open subset $U \subseteq \mathbb{C}^N$ and a holomorphic function $h \in \mathcal{O}(U)$ such that $U \supset M$ and $h|_M = f$? Of course, the problem has a local version[1] as well. See [**50**], p. 198-228, for the CR extension theory from a CR submanifold $M \subset \mathbb{C}^N$. However the inclusion (5.1) is strict hence the problem is new.

Next, let us observe that $\Omega_B^{0,0}(\mathcal{F}) = CR^\infty(\mathcal{F})$. Let $\overline{\partial}_B$ be the restriction of $\overline{\partial}_M$ to $\Omega_B^{0,s}(\mathcal{F})$. Then

$$\Omega_B^{0,0}(\mathcal{F}) \xrightarrow{\overline{\partial}_B} \Omega_B^{0,1}(\mathcal{F}) \xrightarrow{\overline{\partial}_B} \cdots \xrightarrow{\overline{\partial}_B} \Omega_B^{0,k}(\mathcal{F}) \xrightarrow{\overline{\partial}_B} 0 \tag{5.2}$$

is a complex. Here $\dim(M) = 2N+1$ and $\mathcal{F}$ has codimension $q := 2k$. For the remainder of this section, we set $n := N - k$ and assume that $n \geq 1$.

DEFINITION 5.4. (5.2) is called the *basic Cauchy-Riemann complex* of $(M, \mathcal{F})$ and its cohomology

$$H_B^{0,s}(\mathcal{F}) := H^s(\Omega_B^{0,\bullet}(\mathcal{F}), \overline{\partial}_B) =$$

$$= \frac{\mathrm{Ker}\{\overline{\partial}_B : \Omega_B^{0,s}(\mathcal{F}) \to \bullet\}}{\overline{\partial}_B \, \Omega_B^{0,s-1}(\mathcal{F})}, \quad 0 \leq q \leq k,$$

is the *basic Kohn-Rossi cohomology* of $(M, \mathcal{F})$. □

Then

$$H_B^{0,0}(\mathcal{F}) = Ker\{\overline{\partial}_B : \Omega_B^{0,0}(\mathcal{F}) \to \bullet\} =$$

$$= \{f \in CR^\infty(\mathcal{F}) : \overline{\partial}_B f = 0\} =$$

$$= CR^\infty(\mathcal{F}) \cap CR^\infty(M) = CR^\infty(M).$$

Let

$$H^{0,s}(M) := H^s(\Omega^{0,\bullet}(M), \overline{\partial}_M)$$

be the ordinary Kohn-Rossi cohomology of the CR manifold M. For any CR foliation $\mathcal{F}$ of a nondegenerate CR manifold there is a natural injection of $H_B^{0,1}(\mathcal{F})$ in the Kohn-Rossi cohomology group $H^{0,1}(M)$ i.e. the map

$$H_B^{0,1}(\mathcal{F}) \hookrightarrow H^{0,1}(M), \quad [\omega] \mapsto [\omega]_{H^{0,1}(M)}, \tag{5.3}$$

is a monomorphism. Here $\omega \in \Omega_B^{0,1}(\mathcal{F})$ with $\overline{\partial}_B \omega = 0$. Indeed if $\omega, \omega' \in \mathrm{Ker}\{\overline{\partial}_B : \Omega_B^{0,1}(\mathcal{F}) \to \cdot\}$ lie in the same Kohn-Rossi cohomology class then $\omega' - \omega = \overline{\partial}_M f$ for some C^∞ function $f : M \to \mathbb{C}$. Then

$$0 = \overline{Z} \rfloor \omega' = \underbrace{\overline{Z} \rfloor \omega}_{=0} + \overline{Z} \rfloor \overline{\partial}_M f,$$

for any $Z \in T_{1,0}(\mathcal{F})$, hence $f \in \Omega_B^{0,0}(\mathcal{F})$. Thus $\omega' - \omega = \overline{\partial}_B f$ i.e. $[\omega] = [\omega']$. Q.e.d.

[1]Given a C^1 function $f : \omega \to \mathbb{C}$ (where $\omega \subseteq M$ is an open subset) which is CR on $L \cap \omega$ for any $L \in M/\mathcal{F}$ and given $x \in \omega$ is there an open set $U \subseteq \mathbb{C}^N$ and a holomorphic function $h \in \mathcal{O}(U)$ such that $x \in U$ and $h|_{U \cap \omega} = f|_{U \cap \omega}$?

REMARK 5.5. When M has CR codimension $k = 0$ (i.e. M is a complex manifold) $\Omega^{0,s}(M)$ is the space of all $(0,s)$-forms (that is the space of all complex valued differential s-forms which are locally spanned by monomials containing s anti-holomorphic differentials $d\overline{z}^\alpha$ with respect to a local system of complex coordinates z^α on M) $H^{0,s}(M)$ is the corresponding Dolbeaut cohomology and given a foliation $\mathcal{F}$ of M by CR submanifolds (5.3) still holds. □

EXAMPLE 5.6. (Example 1.18 continued)
Set $\beta(t) := \alpha'(t)/[\alpha(t) - i\alpha'(t)]$. Then $T_{0,1}(\mathcal{F})$ is the span of

$$\overline{Z}_\alpha := \frac{\partial}{\partial \overline{z}^\alpha} - 2\,\beta(\rho)\,z_\alpha\,\frac{\partial}{\partial \overline{w}}\ , \quad 1 \le \alpha \le n, \tag{5.4}$$

where $z_\alpha = z^\alpha$. Note that $\beta(0) = i$ hence along the leaf $\partial\Omega_{n+1}$ of $\mathcal{F}$ the vector fields (5.4) correspond (under the natural CR isomorphism $\partial\Omega_{n+1} \approx \mathbb{H}_n$) to the Lewy operators. Consider the complex 1-form

$$\Theta := d\overline{w} + 2\,\beta(\rho)\,z_\alpha\,d\overline{z}^\alpha\,.$$

Then $H_B^{0,s}(\mathcal{F}) = 0$, for $s \in \{1,2\}$. First (by (5.3))

$$H_B^{0,1}(\mathcal{F}) \hookrightarrow H^{0,1}(\mathbb{C}^{n+1}) = 0.$$

On the other hand

$$\Omega_B^{0,1}(\mathcal{F}) = \{\lambda\Theta : \lambda \in C^\infty(\mathbb{C}^{n+1}),\ \ \overline{Z}_\alpha(\lambda) = -i\,\beta'(\rho)\,z_\alpha\,\lambda\}\ ,$$

$$\overline{\partial}\,\Omega_B^{0,0}(\mathcal{F}) = \{\frac{\partial f}{\partial \overline{w}}\,\Theta : f \in CR^\infty(\mathcal{F})\}\ ,$$

and $\overline{\partial}\omega = 0$ for any $\omega \in \Omega_B^{0,1}(\mathcal{F})$. Thus the meaning of $H_B^{0,1}(\mathcal{F}) = 0$ is that the system

$$\frac{\partial f}{\partial \overline{z}_\alpha} = 2\beta(\rho) z_\alpha \lambda\,, \quad \frac{\partial f}{\partial \overline{w}} = \lambda\,,$$

admits a solution $f \in C^\infty(\mathbb{C}^{n+1})$ provided λ satisfies the compatibility relations

$$\overline{Z}_\alpha(\lambda) + i\,\beta'(\rho)\,z_\alpha\,\lambda = 0.$$

To compute $H_B^{0,2}(\mathcal{F})$ let $\omega = \omega_{\alpha\beta}\,d\overline{z}^\alpha \wedge d\overline{z}^\beta + \omega_\alpha\,d\overline{z}^\alpha \wedge d\overline{w}$ be a basic $(0,2)$-form. Then

$$0 = \overline{Z}_\alpha \rfloor \omega = (\omega_{\alpha\beta} + \beta(\rho)\,z_\alpha\,\omega_\beta) d\overline{z}^\beta + \frac{1}{2}\,\omega_\alpha\,d\overline{w}$$

yields $\Omega_B^{0,2}(\mathcal{F}) = 0$. □

Another concept of CR geometry which we need is the differential operator ∂_M acting on $(p,0)$-forms, $p \ge 0$. Let M be a nondegenerate CR manifold endowed with a fixed contact form θ. Let T be the characteristic direction of $d\theta$. Let $\Omega^{p,0}(M)$ be the space of $(p,0)$-forms ω such that $T \rfloor \omega = 0$ and let us consider the first order differential operator

$$\partial_M : \Omega^{p,0}(M) \to \Omega^{p+1,0}(M)$$

defined as follows. If $\omega \in \Omega^{p,0}(M)$ then $\partial_M\omega$ is the unique element of $\Omega^{p+1,0}(M)$ coinciding with $d\omega$ on $T_{1,0}(M) \otimes \cdots \otimes T_{1,0}(M)$ ($p+1$ terms). Then $\partial_M^2 = 0$ in all degrees and one may consider the cohomology groups

$$H^{p,0}(M) = H^p(\Omega^{\bullet,0}(M), \partial_M)$$

of the complex at hand. Moreover if $\mathcal{F}$ is a CR foliation of M then one may define the space of basic $(p,0)$-forms $\Omega_B^{p,0}(\mathcal{F})$ consisting of all elements $\omega \in \Omega^{p,0}(M)$ such that $T_{1,0}(\mathcal{F}) \rfloor \omega = 0$ and $T_{1,0}(\mathcal{F}) \rfloor \partial_M \omega = 0$, and the corresponding cohomology

$$H_B^{p,0}(\mathcal{F}) = H^p(\Omega_B^{\bullet,0}(\mathcal{F}), \partial_B),$$

where ∂_B is the restriction of ∂_M to $\Omega_B^{\bullet,0}(\mathcal{F})$. Finally complex cojugation gives isomorphisms $H^{p,0}(M) \approx H^{0,p}(M)$ and $H_B^{p,0}(\mathcal{F}) \approx H_B^{0,p}(\mathcal{F})$.

EXAMPLE 5.7. (*The contact flow*)
Let M be a nondegenerate CR manifold of hypersurface type and θ a contact form on M. Let T be the characteristic direction of (M,θ). Let $\mathcal{F}$ be the flow defined by T (cf. e.g. [**243**], p. 132). Together with [**115**], p. 160, let us consider the space U_h^r of all horizontal r-forms on M, where a r-form ω on M is called *horizontal* if $T \rfloor \omega = 0$ and $\mathcal{L}_T \omega = 0$. Then U_h^r is nothing but $\Omega_B^r(\mathcal{F})$ (and the *horizontal cohomology* groups H_h^r in [**115**] are $H_B^r(\mathcal{F})$). Using J.J. Kohn's approach (cf. [**157**]) to Hodge theory for the $\overline{\partial}_M$ operator on a compact strictly pseudoconvex CR manifold, G. Gigante, [**115**], established the following

THEOREM 5.8. (G. Gigante, [**115**])
Let M be a compact strictly pseudoconvex CR manifold and θ a contact form on M. Let T be the characteristic direction of (M,θ) and $\mathcal{F}$ the flow defined by T. If the Tanaka-Webster connection of (M,θ) has vanishing pseudohermitian torsion ($\tau = 0$) and strictly positive definite pseudohermitian Ricci curvature then $H_B^1(\mathcal{F}) = 0$.

We may give a short proof of Theorem 5.8 based on a result of J.M. Lee, [**169**], and on our previous considerations. Indeed $H_B^1(\mathcal{F}) = H^{0,1}(M) \oplus H^{1,0}(M)$. Yet (by a result in [**169**]) if (M,θ) has strictly positive definite pseudohermitian Ricci tensor then $H^{0,1}(M) = 0$ (note that the assumption $\tau = 0$ was removed). Q.e.d. □

5.2. The filtration $\{F^r\Omega^{0,\bullet} : r \geq 0\}$

We define a multiplicative filtration of the Cauchy-Riemann complex by setting

$$F^r\Omega^{0,m} := \{\omega \in \Omega^{0,m}(M) : i_{\overline{Z}_1} \cdots i_{\overline{Z}_{m-r+1}} \omega = 0,$$
$$\text{for any} \quad Z_j \in T_{1,0}(\mathcal{F}),\ 1 \leq j \leq m-r+1\}.$$

Note that

$$\Omega^{0,m}(M) = F^0\Omega^{0,m} \supseteq F^1\Omega^{0,m} \supseteq \cdots \supseteq F^m\Omega^{0,m} \supseteq F^{m+1}\Omega^{0,m} = 0,$$

for any $0 \leq m \leq N$. Also the following diagrams are commutative

$$\begin{array}{ccccccc}
\Omega^{0,m}(M) & \supseteq \cdots \supseteq & F^r\Omega^{0,m} & \supseteq & F^{r+1}\Omega^{0,m} & \supseteq \cdots \\
\downarrow \overline{\partial}_M & & \downarrow \overline{\partial}_M & & \downarrow \overline{\partial}_M & \\
\Omega^{0,m+1}(M) & \supseteq \cdots \supseteq & F^r\Omega^{0,m+1} & \supseteq & F^{r+1}\Omega^{0,m+1} & \supseteq \cdots
\end{array}$$

Indeed if

$$\omega \in F^r\Omega^{0,m} \subset \Omega^{0,m}(M) \xrightarrow{\overline{\partial}_M} \Omega^{0,m+1}(M)$$

then

$$i_{\overline{Z}_1} \cdots i_{\overline{Z}_{m-r+2}} (\overline{\partial}_M \omega) = 0$$

for any $Z_j \in T_{1,0}(\mathcal{F})$, because $T_{0,1}(\mathcal{F})$ is involutive. Thus

$$\overline{\partial}_M F^r\Omega^{0,m} \subseteq F^r\Omega^{0,m+1}.$$

Q.e.d. Let us set

$$F^r\Omega^{0,\bullet} := \bigoplus_{m=0}^{N} F^r\Omega^{0,m}.$$

PROPOSITION 5.9. *Let $\mathcal{F}$ be a CR foliation on the nondegenerate CR manifold M. Then $\{F^r\Omega^{0,\bullet}\}_{r\geq 0}$ is a decreasing filtration of $\Omega^{0,\bullet}(M)$ by differential ideals. Also* $\dim_{\mathbb{C}} T_{1,0}(\mathcal{F})_x = n$, *$x \in M$, implies that $F^r\Omega^{0,n+r} = \Omega^{0,n+r}(M)$ and* $\dim_{\mathbb{C}} T_{1,0}(M)_x/T_{1,0}(\mathcal{F})_x = k$, *$x \in M$, yields*

$$F^{k+1}\Omega^{0,m} = 0. \tag{5.5}$$

Indeed we have already shown that $\overline{\partial}_M F^r\Omega^{0,\bullet} \subseteq F^r\Omega^{0,\bullet}$. It remains to be shown that

$$\Omega^{0,\bullet}(M) \wedge F^r\Omega^{0,\bullet} \subseteq F^r\Omega^{0,\bullet}.$$

To this end let $\omega = \omega_0 + \cdots + \omega_N$ with $\omega_m \in F^r\Omega^{0,m}$. Then

$$\alpha \wedge \omega_m \in F^r\Omega^{0,m+s},$$

for any $\alpha \in \Omega^{0,s}(M)$. Indeed

$$i_{\overline{Z}_1} \cdots i_{\overline{Z}_{m+s-r+1}} (\alpha \wedge \omega_m) = 0,$$

for any $Z_j \in T_{1,0}(\mathcal{F})$, because at most s of the $\overline{Z}_j$'s enter α, hence there are enough $\overline{Z}_j$'s left to kill ω_m. Hence the desired inclusion follows from

$$\Omega^{0,s}(M) \wedge F^r\Omega^{0,m} \subseteq F^r\Omega^{0,m+s}.$$

To prove (5.5) we need some local considerations. Let $\{T_1, \cdots, T_N\}$ be a local frame of $T_{1,0}(M)$ such that $\{T_1, \cdots, T_n\}$ is a local frame of $T_{1,0}(\mathcal{F})$. Let $\{\theta^1, \cdots, \theta^N\}$ be determined by

$$\theta^i(T_j) = \delta^i_j, \quad \theta^i(T_{\overline{j}}) = 0, \quad \theta^i(T) = 0.$$

Each $\omega \in F^r\Omega^{0,m}$ is locally a sum of monomials of the form

$$\theta^{\overline{\alpha}_1} \wedge \cdots \wedge \theta^{\overline{\alpha}_p} \wedge \theta^{\overline{j}_1} \wedge \cdots \wedge \theta^{\overline{j}_q},$$

$$1 \leq \alpha_1, \cdots, \alpha_p \leq n, \quad n+1 \leq j_1, \cdots, j_q \leq N,$$

with $C^\infty(M)$-coefficients, where $0 \leq p \leq m - r$ and $q := m - p$. If $r = k+1$ then $0 \leq p \leq m - k - 1$, i.e. $q \geq k+1$ hence $\theta^{\overline{j}_1} \wedge \cdots \wedge \theta^{\overline{j}_q} = 0$, and (5.5) is proved.

Let $\mathcal{F}$ be a CR foliation of the nondegenerate CR manifold M. Let θ be a contact form on M. Let $\langle Z, W\rangle := L_\theta(Z, \overline{W})$ and

$$T_{1,0}(\mathcal{F})^\perp := \{Z \in T_{1,0}(M) : \langle Z, W\rangle = 0, \ \text{for any } W \in T_{1,0}(\mathcal{F})\}.$$

An argument of mere linear algebra shows that $T_{1,0}(\mathcal{F})$ is nondegenerate in $(T_{1,0}(M), \langle\ ,\ \rangle)$ and $T_{1,0}(\mathcal{F}) \oplus T_{1,0}(\mathcal{F})^\perp = T_{1,0}(M)$.

PROPOSITION 5.10. *Let $\mathcal{F}$ be a CR foliation of the nondegenerate CR manifold M. Let $\{E_i^{r,s}\}_{i\geq 0}$ be the spectral sequence associated with the filtered differential space $\left(\Omega^{0,\bullet}(M),\ \overline{\partial}_M,\ \{F^r\Omega^{0,\bullet}\}_{r\geq 0}\right)$. Then*

$$E_0^{r,s} \approx Hom\left(\Lambda^s T_{0,1}(\mathcal{F}),\ \Lambda^r[T_{0,1}(\mathcal{F})^\perp]^*\right),$$

$$E_1^{r,0} \approx \Omega^{0,\bullet}(M), \quad E_2^{r,0} \approx H_B^{0,\bullet}(\mathcal{F}),$$

(*isomorphisms of linear spaces*).

Here $T_{0,1}(\mathcal{F})^\perp := \overline{T_{1,0}(\mathcal{F})^\perp} \subset T_{0,1}(M)$. The spectral sequence in Proposition 5.10 is a CR analog to the spectral sequence built by F.W. Kamber & P. Tondeur, [**152**], p. 117-122. See also J.A. Alvarez López, [**5**]. There is however a lack of relationship among the spectral sequence of $\mathcal{F}$ (cf. F.W. Kamber et al., *op. cit.*) and the spectral sequence in Proposition 5.10. A possible solution might consist in the adaptation of M. Riumin's ideas (cf. [**210**]) to the context of tangentially CR foliations. We leave this as an open problem. Let us set

$$Z_i^{r,m} := \{\omega \in F^r\Omega^{0,m} : \overline{\partial}_M \omega \in F^{r+i}\Omega^{0,m+1}\},$$

$$D_i^{r,m} := \left(F^r\Omega^{0,m}\right) \cap \overline{\partial}_M \left(F^{r-i}\Omega^{0,m-1}\right),$$

and

$$E_i^r\Omega^{0,m} := \frac{Z_i^{r,m}}{Z_{i-1}^{r+1,m} + D_{i-1}^{r,m}}.$$

Cf. e.g. [**129**], Vol. III, p. 21. Also, we set $E_i^{r,s} := E_i^r\Omega^{0,r+s}$. Then

$$E_0^{r,s} = \frac{F^r\Omega^{0,r+s}}{F^{r+1}\Omega^{0,r+s}} \approx Hom\left(\Lambda^s\, T_{0,1}(\mathcal{F}),\ \Lambda^r\left[T_{0,1}(\mathcal{F})^\perp\right]^*\right). \tag{5.6}$$

Let us set

$$Z_i^r := \bigoplus_{m=0}^N Z_i^{r,m}, \quad E_i^r := \bigoplus_{m=0}^N E_i^r\Omega^{0,m}.$$

Then $\overline{\partial}_M\, Z_i^r \subset Z_i^{r+i}$ and $\overline{\partial}_M\,\mathrm{Ker}(\pi_i^r) \subset \mathrm{Ker}(\pi_i^{r+i})$, where $\pi_i^r : Z_i^r \to E_i^r$ is the natural projection, hence $\overline{\partial}_M$ induces differentials $d_i^r : E_i^r \to E_i^{r+i}$. The resulting differential $d_0^{r,s} : E_0^{r,s} \to E_0^{r,s}$ corresponds [under the isomorphism (5.6)] to

$$(\overline{\partial}_{\mathcal{CE}}\tilde{\omega})(\overline{Z}_1, \cdots, \overline{Z}_{s+1}) :=$$

$$:= A_{s+1}\left[(Z_1, \cdots, Z_{s+1}) \mapsto i_{\overline{Z}_1}\,\overline{\partial}_M\,\left(\tilde{\omega}(\overline{Z}_2, \cdots, \overline{Z}_{s+1})\right)\right],$$

for any $Z_j \in T_{1,0}(\mathcal{F})$, $1 \le j \le s+1$. Here A_{s+1} is the alternation map (cf. e.g. [**155**], Vol. I, p. 28) and, for any $(0, r+s)$-form ω which is locally (cf. the discussion preceding Proposition 5.10) a sum of monomials of the form $\alpha \wedge \beta$ with $\alpha \in \Lambda^s T_{0,1}(\mathcal{F})^*$ and $\beta \in \Lambda^r[T_{0,1}(\mathcal{F})^\perp]^*$, we set

$$\tilde{\omega}(\overline{Z}_1, \cdots \overline{Z}_s) := i_{\overline{Z}_1} \cdots i_{\overline{Z}_s}\,\omega, \quad Z_j \in T_{1,0}(\mathcal{F}).$$

As the notation suggests, $\overline{\partial}_{\mathcal{CE}}$ is a CR analogue of the Chevalley-Eilenberg differential in [**243**], p. 122. Then

$$E_1^{r,s} \approx H^s\left(Hom\left(\Lambda^\bullet T_{0,1}(\mathcal{F}),\ \Lambda^r[T_{0,1}(\mathcal{F})^\perp]^*\right),\ \overline{\partial}_{CE}\right),$$

hence

$$E_1^{r,0} \approx \Omega_B^{0,r}(\mathcal{F}). \tag{5.7}$$

Since $d_1^{r,0}$ induces, on the right hand side of (5.7), the differential $\overline{\partial}_B$, it follows that $E_2^{r,0} \approx H_B^{0,r}(\mathcal{F})$. Q.e.d.

5.3. The Graham-Lee connection

Let $\varphi \in C^\infty(U)$ be a smooth function defined on an open set $U \subseteq \mathbb{C}^{n+1}$. Let $\Omega = \{z \in U : \varphi(z) < 0\}$ be a smoothly bounded domain i.e. $D\varphi(z) = (\varphi_{z_1}(z), \cdots \varphi_{z_{n+1}}(z)) \neq 0$ for any $z \in \partial\Omega$ and $\partial\Omega = \{z \in U : \varphi(z) = 0\}$. Moreover we assume that Ω is a strictly pseudoconvex domain i.e. $\partial\Omega$ is strictly pseudoconvex as a real hypersurface in U carrying the naturally induced CR structure

$$T_{1,0}(\partial\Omega) = T^{1,0}(U) \cap [T(\partial\Omega) \otimes \mathbb{C}].$$

Here as customary $T^{1,0}(U)$ is the holomorphic tangent bundle over $\mathbb{C}^{n+1}$ i.e. the span of the $\partial/\partial z_j$'s.

Let Δ_φ be the Laplacian of the Kählerian metric on Ω whose Kähler 2-form is $\frac{i}{2}\partial\overline{\partial}\log(-1/\varphi)$. According to C.R. Graham & J.M. Lee, [**124**], if u is a local solution to $\Delta_\varphi u = 0$ which is smooth up to the boundary $\partial\Omega$ then the boundary value f of u must satisfy a compatibility relation $\mathcal{C}_\varphi f = 0$ where $\mathcal{C}_\varphi$ is a differential operator on $\partial\Omega$ of order $2n+2$ (first studied on the Siegel domain and the ball by C.R. Graham, [**125**]-[**126**]). To calculate $\mathcal{C}_\varphi$ one needs to understand the interrelation between the tangential pseudohermitian geometry of the leaves of the foliation $\mathcal{F}$ by level sets of φ and the geometry of the complex ambient space. One key instrument turns out to be a canonical connection on a neighborhood $V \subset \overline{\Omega}$ of the boundary $\partial\Omega$ (the so called *Graham-Lee connection*, cf. Theorem 5.14 below) which induces the Tanaka-Webster connection on each leaf of $\mathcal{F}$. We give an axiomatic description of this connection and hint to how one may recover J.J. Faran's results, cf. [**101**], in this setting (i.e. in the presence of a fixed defining function for the foliation). We merely look at J.J. Faran's third order invariants $h^\alpha{}_{\overline{\beta}}$ and k^α while the problem of recovering J.J. Faran's result dealing with the question whether a given real hypersurface may be a leaf of a Ricci flat foliation (cf. [**101**], p. 403) is left open.

Let us denote by

$$M_\delta = \{z \in \Omega : \varphi(z) = -\delta\}$$

the level sets of φ. For δ sufficiently small M_δ is still a strictly pseudoconvex CR manifold (of CR dimension $n-1$). Therefore, there is a one-sided neighborhood $V \subset \overline{\Omega}$ of $\partial\Omega$ which is foliated by the (strictly pseudoconvex) level sets of φ. Let $\mathcal{F}$ be the relevant foliation and let us denote by $H(\mathcal{F}) \to V$ (respectively by $T_{1,0}(\mathcal{F}) \to V$) the bundle whose portion over M_δ is the Levi distribution $H(M_\delta)$ (respectively the CR structure $T_{1,0}(M_\delta)$). Note that

$$T_{1,0}(\mathcal{F}) \cap T_{0,1}(\mathcal{F}) = (0),$$

$$[\Gamma^\infty(T_{1,0}(\mathcal{F})), \Gamma^\infty(T_{1,0}(\mathcal{F}))] \subseteq \Gamma^\infty(T_{1,0}(\mathcal{F})).$$

Here $T_{0,1}(\mathcal{F}) = \overline{T_{1,0}(\mathcal{F})}$. By a result of J.M. Lee & R. Melrose, [**167**], p. 163

PROPOSITION 5.11. *There is a unique complex vector field ξ of type $(1,0)$ on V such that $\partial\varphi(\xi) = 1$ and ξ is orthogonal to $T_{1,0}(\mathcal{F})$ with respect to $\partial\overline{\partial}\varphi$ i.e. $\partial\overline{\partial}\varphi(\xi, \overline{Z}) = 0$ for any $Z \in T_{1,0}(\mathcal{F})$.*

DEFINITION 5.12. Let $r : V \to \mathbb{R}$ be the C^∞ function defined by $r \equiv 2\,\partial\overline{\partial}\varphi(\xi, \overline{\xi})$. Then r is called the *transverse curvature* of φ. □

The complex vector field ξ and the transverse curvature r are characterized by

$$\xi \rfloor \partial\overline{\partial}\varphi = r\,\overline{\partial}\varphi, \quad \partial\varphi\,(\xi) = 1.$$

Let $\xi = \frac{1}{2}(N - iT)$ be the real and imaginary parts of ξ. Then

$$(d\varphi)(N) = 2, \quad (d\varphi)(T) = 0,$$
$$\theta(N) = 0, \quad \theta(T) = 1,$$
$$\partial\varphi(N) = 1, \quad \partial\varphi(T) = i.$$

In particular T is tangent to (the leaves of) $\mathcal{F}$. $\mathcal{F}$ carries the tangential Riemannian metric g_θ defined by (5.8) below. Note that the pullback of g_θ to each leaf M_δ of $\mathcal{F}$ is the Webster metric of M_δ (associated to the contact form $j_\delta^*\theta$, where $j_\delta : M_\delta \subset V$).

Let $\{W_\alpha : 1 \leq \alpha \leq n-1\}$ be a local frame of $T_{1,0}(\mathcal{F})$ so that $\{W_\alpha, \xi\}$ is a local frame of $T^{1,0}(V)$. Let g_θ be the tensor field given by

$$g_\theta(X, Y) = (d\theta)(X, JY), \quad g_\theta(X, T) = 0, \quad g_\theta(T, T) = 1, \tag{5.8}$$

for any $X, Y \in H(\mathcal{F})$. Then, as mentioned above, g_θ is a tangential Riemannian metric for $\mathcal{F}$ i.e. a Riemannian metric in $T(\mathcal{F}) \to V$. We consider as well

$$L_\theta(Z, \overline{W}) \equiv -i(d\theta)(Z, \overline{W}), \quad Z, W \in T_{1,0}(\mathcal{F}).$$

Note that L_θ and (the $\mathbb{C}$-linear extension of) g_θ coincide on $T_{1,0}(\mathcal{F}) \otimes T_{0,1}(\mathcal{F})$. We set $g_{\alpha\overline{\beta}} = g_\theta(W_\alpha, W_{\overline{\beta}})$. Let $\{\theta^\alpha : 1 \leq \alpha \leq n-1\}$ be the (locally defined) complex 1-forms on V determined by

$$\theta^\alpha(W_\beta) = \delta^\alpha_\beta, \quad \theta^\alpha(W_{\overline{\beta}}) = 0, \quad \theta^\alpha(T) = 0, \quad \theta^\alpha(N) = 0.$$

Then $\{\theta^\alpha, \theta^{\overline{\alpha}}, \theta, d\varphi\}$ is a local frame of $T(V) \otimes \mathbb{C}$ and we may look for $d\theta$ in the form

$$d\theta = B_{\alpha\beta}\,\theta^\alpha \wedge \theta^\beta + B_{\alpha\overline{\beta}}\,\theta^\alpha \wedge \theta^{\overline{\beta}} + B_{\overline{\alpha}\,\overline{\beta}}\,\theta^{\overline{\alpha}} \wedge \theta^{\overline{\beta}} +$$
$$+ (B_\alpha\,\theta^\alpha + B_{\overline{\alpha}}\,\theta^{\overline{\alpha}}) \wedge \theta + (C_\alpha\,\theta^\alpha + C_{\overline{\alpha}}\,\theta^{\overline{\alpha}}) \wedge d\varphi + D\, d\varphi \wedge \theta.$$

As $d\theta = i\partial\overline{\partial}\varphi \in \Omega^{1,1}(U)$ it follows that $B_{\alpha\beta} = 0$, $B_{\overline{\alpha}\,\overline{\beta}} = 0$. Also

$$g_{\alpha\overline{\beta}} = g_\theta(W_\alpha, W_{\overline{\beta}}) = -i(d\theta)(W_\alpha, W_{\overline{\beta}}) = -\frac{i}{2}\,B_{\alpha\overline{\beta}}$$

i.e. $B_{\alpha\overline{\beta}} = 2ig_{\alpha\overline{\beta}}$. Next

$$\frac{1}{2}B_\alpha = (d\theta)(W_\alpha, T) = i\partial\overline{\partial}\varphi(W_\alpha, T) = 0$$

as $T = i(\xi - \overline{\xi})$ (and ξ is orthogonal to $T_{1,0}(\mathcal{F})$ with respect to $\partial\overline{\partial}\varphi$), i.e. $B_\alpha = 0$, $B_{\overline{\alpha}} = 0$. Similarly $C_\alpha = 0$, $C_{\overline{\alpha}} = 0$. Finally

$$D = (d\theta)(N, T) = i\,\partial\overline{\partial}\varphi(N, T) = 2\,\partial\overline{\partial}\varphi(\xi, \overline{\xi}) = r$$

i.e. $D = r$. We obtain the identity

$$d\theta = 2ig_{\alpha\overline{\beta}}\,\theta^\alpha \wedge \theta^{\overline{\beta}} + r\, d\varphi \wedge \theta. \tag{5.9}$$

As an immediate consequence

$$i_T\, d\theta = -\frac{r}{2}\, d\varphi, \tag{5.10}$$

$$i_N\, d\theta = r\,\theta. \tag{5.11}$$

For instance (by (5.9))

$$(d\theta)(X, T) = \frac{r}{2}\{(d\varphi)(X) - (d\varphi)(T)\theta(X)\},$$

for any $X \in T(\mathcal{F})$, hence (as $(d\varphi)(T) = 0$) one derives (5.10). As an application of (5.9) we decompose $[T, N]$ (according to $T(V) \otimes \mathbb{C} = T_{1,0}(\mathcal{F}) \oplus T_{0,1}(\mathcal{F}) \oplus \mathbb{C}T \oplus \mathbb{C}N$). This is a bit trickier, as shown below. By (5.10)

$$\theta([T, N]) = -2(d\theta)(T, N) = r\, d\varphi(N) = 2r.$$

Next

$$\begin{aligned} 2(d\theta)(W_\alpha, [T, N]) &= 2W_\alpha(r) - \theta([W_\alpha, [T, N]]) = \quad \text{(Jacobi's identity)} \\ &= 2W_\alpha(r) + \theta([T, [N, W_\alpha]]) + \theta([N, [W_\alpha, T]]) = \\ &= 2W_\alpha(r) + 2(d\theta)(T, [W_\alpha, N]) - T(\theta([W_\alpha, N])) + \\ &\quad + 2(d\theta)(N, [T, W_\alpha]) - N(\theta([T, W_\alpha])) \end{aligned}$$

hence (by (5.10)-(5.11))

$$(d\theta)(W_\alpha, [T, N]) = W_\alpha(r).$$

We conclude that

$$[T, N] = i\, W^\alpha(r) W_\alpha - i\, W^{\overline{\alpha}}(r) W_{\overline{\alpha}} + 2rT, \tag{5.12}$$

where $W^\alpha(r) = g^{\alpha\overline{\beta}} W_{\overline{\beta}}(r)$ and $W^{\overline{\alpha}}(r) = \overline{W^\alpha(r)}$.

Let ∇ be a linear connection on V. Let us consider the $T(V)$-valued 1-form τ on V defined by

$$\tau(X) = T_\nabla(T, X), \quad X \in T(V),$$

where T_∇ is the torsion tensor field of ∇.

DEFINITION 5.13. We say T_∇ is *pure* if

$$T_\nabla(Z, W) = 0, \;\; T_\nabla(Z, \overline{W}) = 2i L_\theta(Z, \overline{W}) T, \tag{5.13}$$

$$T_\nabla(N, W) = r\, W + i\, \tau(W), \tag{5.14}$$

for any $Z, W \in T_{1,0}(\mathcal{F})$, and

$$\tau(T_{1,0}(\mathcal{F})) \subseteq T_{0,1}(\mathcal{F}), \tag{5.15}$$

$$\tau(N) = -\, J\, \nabla^H r - 2r\, T. \tag{5.16}$$

Here $\nabla^H r$ is defined by $\nabla^H r = \pi_H \nabla r$ and $g_\theta(\nabla r, X) = X(r)$, $X \in T(\mathcal{F})$. Also $\pi_H : T(\mathcal{F}) \to H(\mathcal{F})$ is the projection associated to the direct sum decomposition $T(\mathcal{F}) = H(\mathcal{F}) \oplus \mathbb{R}T$. □

The main purpose of this section is to establish the following

THEOREM 5.14. (C.R. Graham & J.M. Lee, [**124**])
There is a unique linear connection ∇ on V such that i) $T_{1,0}(\mathcal{F})$ *is parallel with respect to* ∇, ii) $\nabla L_\theta = 0$, $\nabla T = 0$, $\nabla N = 0$, *and* iii) T_∇ *is pure.*

DEFINITION 5.15. The linear connection ∇ given by Theorem 5.14 is called the *Graham-Lee connection.* □

Compare to Proposition 1.1 in [**124**], p. 701-702. The axiomatic description in Theorem 5.14 is however new (cf. also Theorem 2 in [**87**]). We first establish

LEMMA 5.16. *Let $\phi : T(\mathcal{F}) \to T(\mathcal{F})$ be the bundle morphism given by $\phi(X) = JX$, for any $X \in H(\mathcal{F})$, and $\phi(T) = 0$. Then*

$$\phi^2 = -I + \theta \otimes T,$$

$$g_\theta(X, T) = \theta(X),$$

$$g_\theta(\phi X, \phi Y) = g_\theta(X, Y) - \theta(X)\theta(Y),$$

for any $X, Y \in T(\mathcal{F})$. Moreover, if ∇ is a linear connection on V satisfying the axioms (i)-(iii) *in Theorem* 5.14 *then*

$$\phi \circ \tau + \tau \circ \phi = 0 \tag{5.17}$$

along $T(\mathcal{F})$. Consequently τ may be computed as

$$\tau(X) = -\frac{1}{2}\phi(\mathcal{L}_T \phi)X, \tag{5.18}$$

for any $X \in H(\mathcal{F})$.

Proof. For any $X \in T(\mathcal{F})$

$$\phi(X) = \phi(\pi_H X + \theta(X)T) = J(\pi_H X) \in H(\mathcal{F}),$$

$$\phi^2(X) = J^2(\pi_H X) = -\pi_H X = -X + \theta(X)T.$$

The second statement in Lemma 5.16 follows from definitions (cf. (5.8)). The third identity follows from

$$g_\theta(\phi X, \phi Y) = (d\theta)(\phi\, \pi_H\, X, \phi^2\, \pi_H\, Y) =$$

$$= g_\theta(\pi_H\, Y, \pi_H\, X) = g_\theta(Y, X) - \theta(X) g_\theta(Y, T).$$

Let us prove (5.17). As $\tau(T_{1,0}(\mathcal{F})) \subseteq T_{0,1}(\mathcal{F})$ (cf. axiom (5.15)) there are complex valued functions $A_\alpha^{\overline{\beta}}$ such that $\tau(W_\alpha) = A_\alpha^{\overline{\beta}} W_{\overline{\beta}}$. Then

$$(\tau \circ \phi + \phi \circ \tau) W_\alpha = i\tau(W_\alpha) + A_\alpha^{\overline{\beta}} \phi(W_{\overline{\beta}}) = 0.$$

It remains that we check (5.18). As $T_{1,0}(\mathcal{F})$ is parallel with respect to ∇ and ∇ is a real operator it follows that $T_{0,1}(\mathcal{F})$ is parallel, hence both $H(\mathcal{F})$ and its complex structure $J_{\mathcal{F}} \equiv J|_{H(\mathcal{F})}$ are parallel. Moreover, as $\nabla T = 0$, it follows that ϕ is parallel, as well. Let $X \in H(\mathcal{F})$. Then (by (5.17))

$$\phi \tau X = -T_\nabla(T, \phi X) = -\nabla_T \phi X + [T, \phi X].$$

Applying ϕ in both sides gives (as $\nabla \phi = 0$)

$$\tau X = -\nabla_T X - \phi[T, \phi X] = -[T, X] - \tau X - \phi[T, \phi X]$$

or

$$2\tau X = -\mathcal{L}_T X - \phi \mathcal{L}_T \phi X.$$

Q.e.d.

Proof of Theorem 5.14. To establish uniqueness, note first that, for any $X = X^{1,0} + X^{0,1} + \theta(X)T \in T(\mathcal{F})$ (with $X^{1,0} \in T_{1,0}(\mathcal{F})$, $X^{0,1} = \overline{X^{1,0}}$) one has (by $\nabla N = 0$)

$$\nabla_N X = [N, X] + T_\nabla(N, X) = \quad (by\ (5.13) - (5.14),\ (5.16))$$

$$= [N, X] + rX^{1,0} + i\tau X^{1,0} + rX^{0,1} - i\tau X^{0,1} + \theta(X)\{J\nabla^H r + 2rT\}$$

that is

$$\nabla_N X = rX + \tau\phi X - [X, N] + \theta(X)\{J\nabla^H r + rT\}, \tag{5.19}$$

for any $X \in T(\mathcal{F})$. In view of (5.18) $\nabla_N X$ is determined. As $\nabla N = 0$, $\nabla T = 0$ it remains that we compute $\nabla_X Z$, for $X \in T(\mathcal{F})$ and $Z \in T_{1,0}(\mathcal{F})$. Note that $\nabla T = 0$, $\nabla L_\theta = 0$ and $\nabla J_{\mathcal{F}} = 0$ yield $\nabla g_\theta = 0$, i.e.

$$X(g_\theta(Y,Z)) = g_\theta(\nabla_X Y, Z) + g_\theta(Y, \nabla_X Z),$$

for any $X, Y, Z \in T(\mathcal{F})$. The well known Christoffel process then leads to

$$\begin{aligned} 2g_\theta(\nabla_X Y, Z) = X(g_\theta(Y,Z)) + Y(g_\theta(X,Z)) - Z(g_\theta(X,Y)) + \\ + g_\theta([X,Y],Z) + g_\theta(T_\nabla(X,Y),Z) + \\ + g_\theta([Z,X],Y) + g_\theta(T_\nabla(Z,X),Y) + \\ + g_\theta(X,[Z,Y]) + g_\theta(X, T_\nabla(Z,Y)). \end{aligned} \tag{5.20}$$

Note that (again by the purity axioms)

$$T_\nabla(X,Y) = 2(d\theta)(X,Y)T + 2(\theta \wedge \tau)(X,Y), \tag{5.21}$$

for any $X, Y \in T(\mathcal{F})$. Indeed (by (5.13))

$$T_\nabla(X,Y) = -2g_\theta(X, \phi Y)T + 2(\theta \wedge \tau)(X,Y).$$

Moreover

$$\begin{aligned} g_\theta(X, \phi Y) = g_\theta(\pi_H X, \phi \pi_H Y) + \theta(X) g_\theta(T, \phi \pi_H Y) = \\ = -(d\theta)(\pi_H X, \pi_H Y) = -(d\theta)(X,Y) + \\ + \theta(X)(d\theta)(T,Y) + \theta(Y)(d\theta)(X,T). \end{aligned}$$

Finally (by (5.10)) $(d\theta)(X,T) = 0$, $X \in T(\mathcal{F})$, and (5.21) is proved. Replacing the torsion terms (from (5.21) into (5.20)) leads to

$$\begin{aligned} 2g_\theta(\nabla_X Z, \overline{W}) = X(g_\theta(Z,\overline{W})) + Z(g_\theta(X,\overline{W})) - \overline{W}(g_\theta(X,Z)) + \\ + g_\theta([X,Z],\overline{W}) + g_\theta([\overline{W},X],Z) + g_\theta(X,[\overline{W},Z]), \end{aligned} \tag{5.22}$$

for any $X \in T(\mathcal{F})$ and $Z, W \in T_{1,0}(\mathcal{F})$, as (by (5.15))

$$\theta(X)\{g_\theta(\tau Z, \overline{W}) - g_\theta(\tau \overline{W}, Z)\} = 0.$$

The uniqueness statement in Theorem 5.14 is proved. The following explicit expressions of (the various components of) ∇ are also available. By (5.13)

$$\nabla_Z \overline{W} = \pi_{0,1}[Z, \overline{W}], \quad Z, W \in T_{1,0}(\mathcal{F}), \tag{5.23}$$

where $\pi_{0,1} : T(\mathcal{F}) \otimes \mathbb{C} \to T_{0,1}(\mathcal{F})$ is the projection. Of course

$$\nabla_{\overline{Z}} W = \overline{\nabla_Z \overline{W}}.$$

Moreover (by $\nabla L_\theta = 0$ and (5.23))

$$L_\theta(\nabla_Z W, \overline{V}) = Z(L_\theta(W, \overline{V})) - L_\theta(W, \pi_{0,1}[Z, \overline{V}]),$$

for any $Z, W, V \in T_{1,0}(\mathcal{F})$, i.e.

$$\nabla_Z W = g^{\alpha\overline{\beta}}\{Z(L_\theta(W, W_{\overline{\beta}})) - L_\theta(W, \pi_{0,1}[Z, W_{\overline{\beta}}])\} W_\alpha \tag{5.24}$$

and

$$\nabla_{\overline{Z}} \overline{W} = \overline{\nabla_Z W}.$$

Next (by (5.14)-(5.15))

$$\nabla_N Z = rZ + \pi_{1,0}[N, Z], \tag{5.25}$$

for any $Z \in T_{1,0}(\mathcal{F})$, and

$$\nabla_N \overline{Z} = \overline{\nabla_N Z}.$$

Finally

$$\nabla_T Z = -\frac{1}{2}\,\phi(\mathcal{L}_T\phi)Z - [Z,T], \tag{5.26}$$

$$\nabla_T \overline{Z} = \overline{\nabla_T Z}, \quad Z \in T_{1,0}(\mathcal{F}).$$

To establish the existence statement in Theorem 5.14 let ∇ be the linear connection on V defined by (5.23)-(5.26) and $\nabla T = 0$, $\nabla N = 0$. Let us check (i)-(iii) in Theorem 5.14. Clearly

$$\nabla_Z W, \ \ \nabla_{\overline{Z}} W, \ \ \nabla_N W \in T_{1,0}(\mathcal{F}),$$

for any $W \in T_{1,0}(\mathcal{F})$, by the very definitions (cf. (5.23)-(5.25)). Moreover (by (5.26))

$$\nabla_T Z = \frac{1}{2}\,(\mathcal{L}_T Z - i\phi \mathcal{L}_T Z)$$

as (by (5.10)) $\mathcal{L}_T Z \in H(\mathcal{F}) \otimes \mathbb{C}$. Therefore

$$\phi \nabla_T Z = i\,\nabla_T Z$$

that is [as $T_{1,0}(\mathcal{F})$ is the eigenspace corresponding to the eigenvalue i of (the $\mathbb{C}$-linear extension to $H(\mathcal{F})$ of) ϕ] $\nabla_T Z \in T_{1,0}(\mathcal{F})$. We conclude that ∇ obeys to (i). Let us check purity. By (5.24)

$$\begin{gathered} L_\theta(\nabla_Z W - \nabla_W Z - [Z,W], \overline{V}) = \\ = Z(L_\theta(W,\overline{V})) - L_\theta(W, \pi_{0,1}[Z,\overline{V}]) - \\ -W(L_\theta(Z,\overline{V})) + L_\theta(Z,\pi_{0,1}[W,\overline{V}]) - L_\theta([Z,W],\overline{V}) = \\ = 3(d^2\theta)(Z,W,\overline{V}) = 0. \end{gathered}$$

Therefore $T_\nabla(Z,W) = 0$. Next (by (5.23) and $\overline{\pi_{0,1}X} = \pi_{1,0}\overline{X}$, $X \in T(\mathcal{F}) \otimes \mathbb{C}$)

$$\begin{gathered} T_\nabla(Z,\overline{W}) = \pi_{0,1}[Z,\overline{W}] - \overline{\pi_{0,1}[W,\overline{Z}]} - [Z,\overline{W}] = \\ = -\theta([Z,\overline{W}])T = 2iL_\theta(Z,\overline{W})T. \end{gathered}$$

Moreover (by (5.25))

$$T_\nabla(N,Z) = \nabla_N Z - [N,Z] = rZ - \pi_{0,1}[N,Z].$$

Also (by (5.26))

$$\tau(Z) = -\frac{1}{2}\phi(\mathcal{L}_T\phi)Z, \quad Z \in T_{1,0}(\mathcal{F}),$$

so that on one hand (5.15) is satisfied, and on the other

$$\begin{gathered} \tau(Z) = -\frac{1}{2}\{i\,\phi[T,Z] + [T,Z]\} = -\pi_{0,1}[T,Z] = \\ = i\left\{\pi_{0,1}[\overline{\xi},Z] - \pi_{0,1}[\xi,Z]\right\} = i\,\pi_{0,1}[\overline{\xi},Z] \end{gathered}$$

i.e.

$$\tau(Z) = i\,\pi_{0,1}[N,Z], \quad Z \in T_{1,0}(\mathcal{F}). \tag{5.27}$$

Here we made use of $T = i(\xi - \overline{\xi})$, $N = \xi + \overline{\xi}$ and $\pi_{0,1}[\xi, Z] = 0$. Then (5.27) yields (5.14). Finally $\nabla T = \nabla N = 0$ and (5.12) yield (5.16) and we conclude that ∇ obeys to (iii). It remains that we check $\nabla L_\theta = 0$. Clearly $\nabla_Z L_\theta = 0$, $Z \in T_{1,0}(\mathcal{F})$ (by (5.23)-(5.24)). Next (by (5.26) and (5.15))

$$(\nabla_T L_\theta)(Z,\overline{W}) = (\mathcal{L}_T L_\theta)(Z,\overline{W}), \quad Z,W \in T_{1,0}(\mathcal{F}),$$

and

$$(\mathcal{L}_T L_\theta)(Z, \overline{W}) = -i\{T(d\theta(Z, \overline{W})) - d\theta([T, Z], \overline{W}) - d\theta(Z, [T, \overline{W}])\} =$$

$$= \frac{i}{2}\{T(\theta([Z, \overline{W}])) - \theta([[T, Z], \overline{W}]) - \theta([Z, [T, \overline{W}]])\} =$$

(by applying the Jacobi identity to the term $\theta([[T, Z], \overline{W}])$)

$$= \frac{i}{2}\{T(\theta([Z, \overline{W}])) + \theta([[Z, \overline{W}], T])\} =$$

$$= i(d\theta)(T, [Z, \overline{W}]) = -\frac{i\, r}{2}\,(d\varphi)([Z, \overline{W}]) = 0$$

(by (5.10) and $[Z, \overline{W}] \in T(\mathcal{F}) \otimes \mathbb{C}$). Hence $\nabla_T L_\theta = 0$. Finally (by (5.25))

$$(\nabla_N L_\theta)(Z, \overline{W}) = -2r g_\theta(Z, \overline{W}) + (\mathcal{L}_N g_\theta)(Z, \overline{W}), \quad Z, W \in T_{1,0}(\mathcal{F}),$$

and $\nabla_N L_\theta = 0$ follows from (5.30) in Lemma 5.17 below.

LEMMA 5.17. *The following identities hold for any $X \in T(\mathcal{F})$*

$$T_\nabla(N, X) = rX + \tau(\phi X) + \theta(X)\{\phi \nabla^H r + rT\}, \tag{5.28}$$

$$[N, \phi X] - \phi[N, X] = 2\tau(X) - \theta(X)\nabla^H r, \tag{5.29}$$

Moreover

$$(\mathcal{L}_N g_\theta)(X, Y) = 2r g_\theta(X, Y) + 2(d\theta)(X, \tau(Y)), \tag{5.30}$$

for any $X, Y \in H(\mathcal{F})$.

Proof. (5.28) follows from (5.14). Let us replace X by ϕX in (5.28)

$$\nabla_N \phi X - [N, \phi X] = r\phi X - \tau(X)$$

and subtract the identity got from (5.28) by applying ϕ to both sides. Since $\nabla\phi = 0$ we obtain (5.29). The proof of (5.30) is a consequence of (5.11), (5.29), and the Jacobi identity

$$(\mathcal{L}_N g_\theta)(X, Y) = N((d\theta)(X, \phi Y)) - (d\theta)([N, X], \phi Y) +$$

$$+2(d\theta)(X, \tau(Y)) - (d\theta)(X, [N, \phi Y]) =$$

$$= -\frac{1}{2}\, N(\theta([X, \phi Y])) + \frac{1}{2}\, \theta([[N, X], \phi Y]) +$$

$$+2(d\theta)(X, \tau(Y)) + \frac{1}{2}\, \theta([X, [N, \phi Y]]) =$$

$$= -\frac{1}{2}\, N(\theta([X, \phi Y])) - \frac{1}{2}\, \theta([[X, \phi Y], N]) - \theta([X, \tau(Y)]) =$$

$$= -(d\theta)(N, [X, \phi Y]) - \theta([X, \tau(Y)]) =$$

$$= -r\, \theta([X, \phi Y]) - \theta([X, \tau(Y)]) =$$

$$= 2r(d\theta)(X, \phi Y) + 2(d\theta)(X, \tau(Y)) =$$

$$= 2r g_\theta(X, Y) + 2(d\theta)(X, \tau(Y)).$$

Q.e.d. Theorem 5.14 is proved.

As to the local calculations, if φ_β^α are the connection 1-forms of the Graham-Lee connection (i.e. $\nabla W_\beta = \varphi_\beta^\alpha \otimes W_\alpha$) then we may look for $d\theta^\alpha$ in the form

$$d\theta^\alpha = B^\alpha_{\beta\gamma}\,\theta^\beta \wedge \theta^\gamma + B^\alpha_{\beta\overline{\gamma}}\,\theta^\beta \wedge \theta^{\overline{\gamma}} + B^\alpha_{\overline{\beta}\,\overline{\gamma}}\,\theta^{\overline{\beta}} \wedge \theta^{\overline{\gamma}} + \tag{5.31}$$

$$+(B^\alpha_\beta\,\theta^\beta + B^\alpha_{\overline{\beta}}\,\theta^{\overline{\beta}}) \wedge \theta + (C^\alpha_\beta\,\theta^\beta + C^\alpha_{\overline{\beta}}\,\theta^{\overline{\beta}}) \wedge d\varphi + D\, d\varphi \wedge \theta.$$

Indeed, applying this identity to the pair (W_β, W_γ) (respectively to $(W_\beta, W_{\overline{\gamma}})$ and (W_β, T)) gives

$$B^\alpha_{\beta\gamma} - B^\alpha_{\gamma\beta} = \varphi^\alpha_\beta(W_\gamma) - \varphi^\alpha_\gamma(W_\beta), \;\; B^\alpha_{\beta\overline{\gamma}} = \varphi^\alpha_\beta(W_{\overline{\gamma}}), \;\; B^\alpha_\beta = \varphi^\alpha_\beta(T).$$

Similarly (applying (5.31) to $(W_{\overline{\beta}}, T)$, (W_β, N) and $(W_{\overline{\beta}}, N)$ respectively)

$$B^\alpha_{\overline{\beta}} = -A^\alpha_{\overline{\beta}}, \;\; C^\alpha_\beta = \frac{1}{2}(\varphi^\alpha_\beta(N) - r\,\delta^\alpha_\beta), \;\; C^\alpha_{\overline{\beta}} = \frac{i}{2}\,A^\alpha_{\overline{\beta}}.$$

Finally (by (5.12))

$$D\,d\varphi(N) = 2(d\theta^\alpha)(N,T) = -\theta^\alpha([N,T]) = i\,W^\alpha(r).$$

Summing up

$$d\theta^\alpha = \theta^\beta \wedge \varphi^\alpha_\beta - i\,\partial\varphi \wedge \tau^\alpha + \frac{i}{2}\,W^\alpha(r)\,d\varphi \wedge \theta + \frac{r}{2}\,d\varphi \wedge \theta^\alpha\,, \tag{5.32}$$

where $\tau^\alpha \equiv A^\alpha_{\overline{\beta}}\,\theta^{\overline{\beta}}$.

Given a linear connection ∇ on V we set $\alpha(X,Y) \equiv \Pi\nabla_X Y$, for any $X, Y \in T(\mathcal{F})$. If ∇ is the Graham-Lee connection then (by the proof of Theorem 5.14) $\alpha = 0$. One may identify, as usual, the normal bundle $\nu(\mathcal{F}) = T(V)/T(\mathcal{F})$ with $\mathbb{R}N$. If $\Pi^\perp : T(V) \to T(\mathcal{F})$ is the projection, let us set $\nabla^{\mathcal{F}} \equiv \Pi^\perp\nabla$. It is easily seen that $\nabla^{\mathcal{F}}$ is the Tanaka-Webster connection of each M_δ (i.e. the pointwise restriction of the Graham-Lee connection to a leaf of $\mathcal{F}$ is the Tanaka-Webster connection of the leaf). In particular $\tau : T(\mathcal{F}) \to T(\mathcal{F})$ is the pseudohermitian torsion of each leaf (hence $g_\theta(\tau X, Y) = g_\theta(X, \tau Y)$, for any $X, Y \in T(\mathcal{F})$).

J.J. Faran determined (cf. [**101**]) a complete system of local invariants under biholomorphic mappings of foliations of V by nondegenerate real hypersurfaces. His study is imitative of the work of S.S. Chern & J.K. Moser, [**65**], and indeed the local invariants of a foliation by real hypersurfaces turn out to be similar to the local invariants of a single real hypersurface. There is however a remarkable difference as one of these invariants is the intrinsic normal direction $N = 2\mathrm{Re}(\xi)$.

PROPOSITION 5.18. (J.J. Faran, [**101**])
The flow along N gives a foliate map, i.e. sending leaves to leaves, whose restriction to each leaf of $\mathcal{F}$ is a contact transformation yet in general not a CR diffeomorphism.

Indeed by (5.9) we have

$$(\mathcal{L}_N\theta)W_\alpha = N(\theta W_\alpha) - \theta[N, W_\alpha] = 2(d\theta)(N, W_\alpha) =$$
$$= 2ih_{\alpha\overline{\beta}}\theta^\alpha \wedge \theta^{\overline{\beta}}(N, W_\alpha) + 4r(d\varphi \wedge \theta)(N, W_\alpha) = 0,$$
$$(\mathcal{L}_N\theta)T = N(\theta T) - \theta[N,T] = 2(d\theta)(N,T) =$$
$$= 4r(d\varphi \wedge \theta)(N,T) = 2rf\varphi(N)\theta(T) = 4r,$$
$$(\mathcal{L}_N\theta)N = 2(d\theta)(N,N) = 0.$$

Summing up we have

$$\mathcal{L}_N\theta = 4r\theta.$$

Q.e.d. On the other hand J.J. Faran has built (cf. *op. cit.*) third order (i.e. which may be calculated at a point by using only the 3-jet of φ at that point) invariants $h^\alpha{}_{\overline{\beta}}$ and k^α and gave their geometric interpretation. Precisely it turns out that

PROPOSITION 5.19. (J.J. Faran, [**101**])
$h^\alpha{}_{\bar\beta}$ measures the failure of the flow on N from being a CR map, while $k^\alpha = 0$ if and only if there is a defining function ρ of Ω such that $\det(\partial^2\rho/\partial z_i\partial\bar z_j) = 0$ i.e. ρ has vanishing transverse curvature.

We shall look at J.J. Faran's invariants starting from the considerations made earlier in this section. As observed in [**124**], R. Graham & J.M. Lee's setting bears the same relationship to J.J. Faran's setting as does S. Webster's (cf. [**250**]) to that of S.S. Chern & J.K. Moser, [**65**]. Using (5.32) one derives

$$(\mathcal{L}_N\theta^\alpha)W_\beta = r\delta^\alpha_\beta - \varphi^\alpha_\beta(N), \quad (\mathcal{L}_N\theta^\alpha)W_{\bar\beta} = -iA^\alpha_{\bar\beta}\,,$$
$$(\mathcal{L}_N\theta^\alpha)T = 2ir^\alpha\,, \quad (\mathcal{L}_N\theta^\alpha)N = 0.$$

Summing up one has

$$\mathcal{L}_N\theta^\alpha = (r\delta^\alpha_\beta - \varphi^\alpha_\beta(N))\theta^\beta - i\tau^\alpha + 2ir^\alpha\theta.$$

In particular

$$\mathcal{L}_N\theta^\alpha \equiv -i\tau^\alpha\,, \mod \theta, \theta^\alpha\,,$$
$$\mathcal{L}_N\theta^\alpha \equiv 2ir^\alpha\theta\,, \mod \theta^\alpha,\ \theta^{\bar\alpha}\,.$$

A comparison with (2.4)-(2.5) in [**101**], p. 401, shows that Faran's third order invariants $h^\alpha{}_{\bar\beta}$ and k^α are essentially $A^\alpha_{\bar\beta}$ and r^α, respectively. Hence

PROPOSITION 5.20. *The flow along N is a CR map when restricted to a leaf of $\mathcal{F}$ if and only if $\tau^\alpha = 0$ i.e. each leaf of $\mathcal{F}$ has vanishing pseudohermitian torsion.*

5.4. Boundary values of Yang-Mills fields

Let $(M, T_{1,0}(M))$ be a compact strictly pseudoconvex CR manifold, of CR dimension n, and θ a contact form on M. Let $(E, \overline\partial_E) \to M$ be a CR-holomorphic vector bundle and h a Hermitian metric in E. Let $\mathcal{C}(E,h)$ be the affine space of all connections D in E such that $Dh = 0$. We consider the functional

$$\mathcal{PYM}(D) = \frac{1}{2}\int_M \|\pi_H R^D\|^2\,\theta\wedge(d\theta)^n. \tag{5.33}$$

Here $\pi_H : \Omega^2(\mathrm{Ad}E) \to \Omega^2(\mathrm{Ad}E)/\mathcal{J}^2_\theta$ is the natural projection and $\mathcal{J}^\bullet_\theta$ the ideal generated by θ in $\Omega^\bullet(\mathrm{Ad}E)$.

DEFINITION 5.21. A *pseudo Yang-Mills field* on M is a critical point of $\mathcal{PYM} : \mathcal{C}(E,h) \to [0,+\infty)$. □

As an application of the Graham-Lee connection we shall show that

THEOREM 5.22. *Let $\Omega = \{z \in U : \varphi(z) < 0\}$ be a smoothly bounded strictly pseudoconvex domain in $\mathbb{C}^n$ and g its Bergman metric. Let $\pi : F \to U$ be a holomorphic vector bundle and h a Hermitian metric on F. Let $D_b \in \mathcal{C}(E,h)$ ($E = \pi^{-1}(\partial\Omega)$) be the boundary values of a Yang-Mills field $D \in \mathcal{C}(F,h)$ on (Ω, g). Assume that $i_T R^{D_b} = 0$. Then D_b is a pseudo Yang-Mills field if and only if $i_N R^D = 0$ on $H(\partial\Omega)$.*

Here T is the characteristic direction of $(\partial\Omega, \theta)$, $\theta \equiv \frac{i}{2}(\overline\partial - \partial)\varphi$, and $H(\partial\Omega)$ is the Levi distribution. Also $N = -JT$ (J is the complex structure on $\mathbb{C}^n$). The proof relies on the explicit relationship among the Levi-Civita connection ∇^g of (Ω, g) and the Graham-Lee connection ∇ of φ.

H. Urakawa has started (cf. [**245**]-[**247**]) a study of Yang-Mills fields on M that is of critical points of the functional

$$\mathcal{YM}(D) = \frac{1}{2}\int_M \|R^D\|^2 \, d\,\mathrm{vol}(g_\theta),$$

where $d\,\mathrm{vol}(g_\theta)$ is the canonical volume form associated to the Webster metric g_θ of (M, θ). See also Chapter 8 of the monograph [**89**]. As it was shown in [**26**] the functionals $\mathcal{YM}$ and $\mathcal{PYM}$ are related. To motivate the definition of $\mathcal{PYM}$ let F_θ be the Fefferman metric of (M, θ). By a result of E. Barletta et al., [**25**], the base map $\phi : M \to N$ corresponding to any smooth S^1-invariant harmonic map $\Phi : C(M) \to N$ from $(C(M), F_\theta)$ into a Riemannian manifold (N, g_N) is locally a subelliptic harmonic map (in the sense of J. Jost & C-J. Xu, [**144**]). Also ϕ is a critical point of the functional

$$E(\phi) = \frac{1}{2}\int_M \mathrm{trace}_{G_\theta}\left(\pi_H \phi^* g_N\right) \theta \wedge (d\theta)^n,$$

where G_θ is the Levi form. Here if B is a bilinear form on $T(M)$ then $\pi_H B$ denotes the restriction of B to $H(M)$, the Levi distribution of $(M, T_{1,0}(M))$. The functional E itself is obtained by integration along the fibre in the ordinary Dirichlet functional

$$\mathbb{E}(\Phi) = \frac{1}{2}\int_{C(M)} \mathrm{trace}_{F_\theta}(\Phi^* g_N)\, d\mathrm{vol}(F_\theta),$$

where $\Phi = \phi \circ \pi$. Then perhaps subelliptic harmonic maps (rather than harmonic maps, with respect to the Webster metric) are the natural objects of study in CR geometry. Subelliptic harmonic maps were discussed in Chapter 4 of the monograph [**89**]. Another example of the sort is the *CR Yamabe problem* i.e. given a contact form θ on M such that G_θ is positive definite find a contact form $\hat{\theta} = e^u \theta$, $u \in C^\infty(M)$, such that the pseudohermitian scalar curvature $\hat{\rho}$ of $(M, \hat{\theta})$ is a constant λ. By a result of J.M. Lee, [**168**], the scalar curvature $K : C(M) \to \mathbb{R}$ of $(C(M), F_\theta)$ is S^1-invariant and the corresponding base function $\pi_* K : M \to \mathbb{R}$ is, up to a constant, the pseudohermitian scalar curvature ρ of (M, θ) i.e. precisely

$$\pi_* K = \frac{2n+1}{n+1}\rho.$$

Therefore the CR Yamabe problem is nothing but the Yamabe problem for the Fefferman metric and the relevant equation (the Yamabe equation on $(C(M), F_\theta)$) projects on

$$c_n \Delta_b u + \rho u = \lambda u^{p-1}$$

(the so caled *CR Yamabe equation*), a nonlinear subelliptic equation on M (which may be analyzed with the techniques in [**107**], cf. D. Jerison & J.M. Lee, [**142**]-[**143**], and N. Gamara & R. Yacoub, [**114**], for a complete solution to the CR Yamabe problem). The solution to the CR Yamabe problem is discussed in detail in Chapter 3 of [**89**]. The common feature of the two examples above is that both provide natural objects on M, as projections of (S^1-invariant) geometric quantities on $C(M)$, associated to the Fefferman metric. A more refined statement is that both examples lead to nonlinear subelliptic problems on M. This has been already emphasized for the CR Yamabe problem. As to the example of S^1-invariant harmonic maps $\Phi : C(M) \to N$, the base map is a solution to

$$\Delta_b \phi^i + g^{\alpha\overline{\beta}} T_\alpha(\phi^j) T_{\overline{\beta}}(\phi^k)((\Gamma_N)^i_{jk} \circ \phi) = 0,$$

where $(\Gamma_N)^i_{jk}$ are the Christoffel symbols of the second kind of g_N. On the same line of thought we may state the following

THEOREM 5.23. (E. Barletta et al, [**26**])
Let M be a compact strictly pseudoconvex CR manifold, of CR dimension n. Let θ be a contact form on M with G_θ positive definite. Let $(E, \overline{\partial}_E) \to M$ be a CR-holomorphic vector bundle and h a Hermitian metric in E. i) There is a constant c_n depending only on the dimension and the orientation of M such that

$$c_n \, \mathcal{YM}(D) = \mathcal{PYM}(D) + 2 \int_M \|i_T R^D\|^2 \, \theta \wedge (d\theta)^n, \quad D \in \mathcal{C}(E,h). \tag{5.34}$$

Consequently, given a Hermitian connection D in E whose curvature R^D is of type $(1,1)$, D is a pseudo Yang-Mills field on M if and only if D is the Tanaka connection of $(E, \overline{\partial}_E, h)$. ii) Let

$$\widehat{\mathcal{YM}}(\mathbb{D}) = \frac{1}{2} \int_{C(M)} \langle R^{\mathbb{D}}, R^{\mathbb{D}} \rangle \, d\,\mathrm{vol}(F_\theta)$$

be the Yang-Mills functional on $C(M)$, for $\mathbb{D} \in \mathcal{C}(\pi^* E, \pi^* h)$. Then

$$\widehat{\mathcal{YM}}(\pi^* D) = 2\pi \mathcal{PYM}(D), \quad D \in \mathcal{C}(E,h). \tag{5.35}$$

Consequently, if $\pi^* D$ is a Yang-Mills field on $(C(M), F_\theta)$ then D is a pseudo Yang-Mills field on M. Viceversa, let D be a pseudo Yang-Mills field on M such that $i_T R^D = 0$. Then $\pi^* D$ is a Yang-Mills field on $C(M)$ if and only if

$$(R^{\alpha\overline{\beta}} - \frac{\rho}{2(n+1)} g^{\alpha\overline{\beta}}) R^D(T_\alpha, T_{\overline{\beta}}) u = 0, \tag{5.36}$$

for some local frame $\{T_\alpha : 1 \leq \alpha \leq n\}$ of $T_{1,0}(M)$ at any point $x \in M$, and

$$\Lambda_\theta R^D = 0. \tag{5.37}$$

In particular, if M is (pseudohermitian) Ricci flat then the pullback $\pi^* D$ of the canonical Tanaka connection D of (E,h) is a Yang-Mills field.

The proof of Theorem 5.23 will not be given here. Its main ingredients are a local coordinate calculation of the Fefferman metric of (M,θ), the explicit relationship among the Levi-Civita connection $\nabla^{C(M)}$ of $(C(M), F_\theta)$ and the Tanaka-Webster connection ∇ of (M,θ), and Theorem 2.3 in [**245**], p. 551. We may also state

THEOREM 5.24. (E. Barletta et al., [**26**])
Let D be a pseudo Yang-Mills field and $D^t = D + A^t$, $|t| < \epsilon$, a smooth variation of D whose first order part $\varphi \equiv \{dA^t/dt\}_{t=0}$ satisfies $i_T \varphi = 0$ and $\delta_b^D \varphi = 0$. Then

$$\frac{d^2}{dt^2} \{\mathcal{PYM}(D^t)\}_{t=0} = \int_M \langle \mathcal{S}_b^D(\varphi), \varphi \rangle \, \theta \wedge (d\theta)^n \tag{5.38}$$

where $\mathcal{S}_b^D(\varphi) \equiv \Delta_b^D \varphi + \mathcal{R}_b^D(\varphi)$ and $\Delta_b^D \varphi \equiv d_b^D \delta_b^D \varphi + \delta_b^D d_b^D \varphi$ is the generalized sublaplacian. The operator $\mathcal{S}_b^D : \Omega^{0,1}(\mathrm{Ad}(E)) \to \Omega^{0,1}(\mathrm{Ad}(E))$ is subelliptic of order $1/2$.

The proof of Theorem 5.24 also falls beyond the scope of this book. As $\mathcal{R}_b^D$ is a zero order operator the crucial point in the proof of Theorem 5.24 is to show that

$$(\Delta_b^D \varphi) \otimes e_j = 2\{\Box_b \varphi^i_j + \tag{5.39}$$

$$+ (n-1)(\nabla_T \varphi^i_j + \varphi^i_j \circ \tau) \circ J\} \otimes e_i + \text{lower order terms},$$

for any $\varphi \in \Omega^{0,1}(\mathrm{Ad}(E))$, $\varphi e_j = \varphi^i_j \otimes e_i$, and then exploit the subellipticity of the Kohn-Rossi operator $\Box_b$ on scalar $(0,1)$-forms.

The scope of this section is to give a proof of Theorem 5.22. We start by preparing the needed material on CR-holomorphic vector bundles and relative connection theory. Let $E \to M$ be a complex vector bundle over a CR manifold.

DEFINITION 5.25. A *pre-$\overline{\partial}$-operator* is a first order differential operator

$$\overline{\partial}_E : \Gamma^\infty(E) \to \Gamma^\infty(T_{0,1}(M)^* \otimes E)$$

such that

$$\overline{\partial}_E(fu) = f\overline{\partial}_E u + (\overline{\partial}_b f) \otimes u,$$

for any $f \in C^\infty(M)$ and any $u \in \Gamma^\infty(E)$. $\Box$

DEFINITION 5.26. A pair $(E, \overline{\partial}_E)$ consisting of a complex vector bundle and a pre-$\overline{\partial}$-operator is a *CR-holomorphic* vector bundle if $\overline{\partial}_E$ satisfies the *integrability condition*

$$[\overline{Z}, \overline{W}] \cdot u = \overline{Z} \cdot \overline{W} \cdot u - \overline{W} \cdot \overline{Z} \cdot u,$$

for any $u \in \Gamma^\infty(E)$, $Z, W \in T_{1,0}(M)$. Here $\overline{Z} \cdot u$ is short for $(\overline{\partial}_E u)\overline{Z}$. $\Box$

Let $(E, \overline{\partial}_E) \to M$ be a CR-holomorphic vector bundle. Let h be a Hermitian metric in E. Let $\mathcal{C}(E, h)$ be the affine space of all connections D in E such that $Dh = 0$ i.e.

$$X(h(u, v)) = h(D_X u, v) + h(u, D_{\overline{X}} v),$$

for any $X \in T(M) \otimes \mathbb{C}$ and any $u, v \in \Gamma^\infty(E)$.

DEFINITION 5.27. A connection $D \in \mathcal{C}(E, h)$ is *Hermitian* if $D^{0,1} = \overline{\partial}_E$. Here $D^{0,1}u$ is the restriction of Du to $T_{0,1}(M)$. $\Box$

Let $\mathrm{Ad}(E) \to M$ be the subbundle of $\mathrm{End}(E) \to M$ consisting of all skew-symmetric endomorphisms S, i.e. $h(Su, v) + h(u, Sv) = 0$, for any $u, v \in \Gamma^\infty(E)$. By a result in [**90**], p. 43, given a contact form θ and an endomorphism $S \in \Gamma^\infty(\mathrm{Ad}(E))$ there is a unique Hermitian connection $D = D(h, \theta, S)$ in E (the *canonical S-connection*) such that

$$\Lambda_\theta \, R^D = 2nS. \tag{5.40}$$

Here $R^D = D \circ D : \Omega^0(E) \to \Omega^2(E)$ is the curvature 2-form of D. Also we set $\Omega^k(E) = \Gamma^\infty(\Lambda^k T^*(M) \otimes E)$, $k \geq 0$. If $F \to M$ is a vector bundle and $\varphi \in \Gamma^\infty(T^*(M) \otimes T^*(M) \otimes F)$ the trace $\Lambda_\theta \varphi$ of φ is given by

$$i(\Lambda_\theta \varphi)_x = \sum_{\alpha=1}^{n} \varphi(Z_\alpha, Z_{\overline{\alpha}})_x \,,$$

where $\{Z_\alpha\}$ is a (local) orthonormal (i.e. $L_\theta(Z_\alpha, Z_{\overline{\beta}}) = \delta_{\alpha\beta}$) frame of $T_{1,0}(M)$ on $U \ni x$. Therefore $\Lambda_\theta \varphi \in \Gamma^\infty(F)$. When $S = 0$ the canonical S-connection is the *Tanaka connection* $D(h, \theta, 0)$ in $E \to M$ (cf. [**235**]). $D(h, \theta, 0)$ is a CR invariant.

Let $D \in \mathcal{C}(E, h)$ be a Hermitian connection such that its curvature R^D is a form of type $(1,1)$. By a result in [**245**], D is a Yang-Mills field if and only if D is the Tanaka connection $D(h, \theta, 0)$. In general, canonical S-connections solve the inhomogeneous Yang-Mills equations $\delta^D R^D = f$, in the presence of suitable compatibility conditions satisfied by f (cf. Theorem 2 in [**90**], p. 44-45).

Let $\Omega = \{z \in U : \varphi(z) < 0\}$ be a smoothly bounded strictly pseudoconvex domain as above. Let $\pi : F \to U$ be a holomorphic vector bundle.

PROPOSITION 5.28. *The portion $E = \pi^{-1}(\partial\Omega)$ of F over the boundary of Ω is CR-holomorphic.*

Indeed, as F is holomorphic, there is a natural differential operator

$$\overline{\partial}_F : \Gamma^\infty(F) \to \Gamma^\infty(T^{0,1}(U)^* \otimes F)$$

where $T^{0,1}(U)$ is the anti-holomorphic tangent bundle over U. Given $u \in \Gamma^\infty(E)$ let $\tilde{u} \in \Gamma^\infty(F)$ be a C^∞ extension of u as a cross-section in F and let us set

$$\left(\overline{\partial}_E u\right)_z = \left(\overline{\partial}_F \tilde{u}\right)_z, \quad z \in \partial\Omega.$$

The definition of $\left(\overline{\partial}_E u\right)_z$ does not depend upon the choice of extension $\tilde{u}$ of u because

$$(\overline{\partial} f)|_{T_{0,1}(\partial\Omega)} = \overline{\partial}_b(f|_{\partial\Omega})$$

for any C^∞ function $f : U \to \mathbb{C}$. Let

$$\{\Phi_\alpha : \pi^{-1}(\Omega_\alpha) \to \Omega_\alpha \times \mathbb{C}^m : \alpha \in I\}$$

be a trivialization atlas for F and

$$G_{\beta\alpha} : \Omega_\beta \cap \Omega_\alpha \to \mathrm{GL}(m, \mathbb{C})$$

the corresponding transition functions. Let us set $U_\alpha = \Omega_\alpha \cap \partial\Omega$ and $g_{\beta\alpha} = G_{\beta\alpha}|_{U_\alpha \cap U_\beta}$. As $G_{\beta\alpha}$ are holomorphic, it follows that $E \to \partial\Omega$ is a peculiar type of CR-holomorphic vector bundle in that its transition functions $g_{\beta\alpha}$ are matrix valued CR functions on $\partial\Omega$. Q.e.d.

Let $K(\zeta, z)$ be the Bergman kernel of Ω. By a classical result in [**104**]

THEOREM 5.29. (C. Fefferman, [**104**])
Let $\Omega = \{\varphi < 0\} \subset \mathbb{C}^n$ be a smoothly bounded strictly pseudoconvex domain. Then

$$K(\zeta, z) = c_\Omega |\nabla\varphi(z)|^2 \cdot \det L_\varphi(z) \cdot \Psi(\zeta, z)^{-(n+1)} + H(\zeta, z), \tag{5.41}$$

where $H \in C^\infty(\overline{\Omega} \times \overline{\Omega} \setminus \Delta)$, Δ is the diagonal of $\partial\Omega \times \partial\Omega$, and H satisfies the estimate

$$|H(\zeta, z)| \leq c'_\Omega |\Psi(\zeta, z)|^{-(n+1)+1/2} \cdot |\log|\Psi(\zeta, z)||. \tag{5.42}$$

The formula (5.41) is called the *Fefferman asymptotic expansion formula* for the Bergman kernel. Here L_φ is the restriction of $\partial\overline{\partial}\varphi$ to $T_{1,0}(\partial\Omega) \otimes T_{0,1}(\partial\Omega)$. Also we set

$$\Psi(\zeta, z) = (F(\zeta, z) - \varphi(z))\chi(|\zeta - z|) + (1 - \chi(|\zeta - z|))|\zeta - z|^2$$

where

$$F(\zeta, z) = -\sum_{j=1}^n \frac{\partial\varphi}{\partial z^j}(z)(\zeta^j - z^j) - \frac{1}{2}\sum_{j,k=1}^n \frac{\partial^2\varphi}{\partial z^j \partial z^k}(z)(\zeta^j - z^j)(\zeta^k - z^k)$$

and $\chi(t)$ is a C^∞ cut-off function with $\chi(t) = 1$ for $|t| < \epsilon_0/2$ and $\chi(t) = 0$ for $|t| \geq 3\epsilon_0/4$. As a consequence of (5.41)

$$K(z, z)^{-1/(n+1)} = |\varphi(z)| \left(\Phi(z) + H(z, z)|\varphi(z)|^{n+1}\right)^{-1/(n+1)}$$

where $\Phi(z) \equiv c_\Omega |\nabla\varphi(z)|^2 \det L_\varphi(z)$ stays finite near $\partial\Omega$ and (by (5.42))

$$|H(z, z)|\, |\varphi(z)|^{n+1} \leq c'_\Omega |\varphi(z)|^{1/2} |\log|\varphi(z)|| \to 0, \quad \textit{as } z \to \partial\Omega.$$

Therefore $K(z,z)^{-1/(n+1)}$ vanishes at $\partial\Omega$. Also, as $\Phi(z) \neq 0$ near the boundary, $\nabla K(z,z)^{-1/(n+1)} \neq 0$ along $\partial\Omega$, hence

COROLLARY 5.30. *$K(z,z)^{-1/(n+1)}$ is a defining function for Ω.*

For the rest of this section we assume that $\varphi(z) \equiv -K(z,z)^{-1/(n+1)}$ and set $\theta \equiv \frac{i}{2}(\overline{\partial} - \partial)\varphi$. Then $d\theta = i\,\partial\overline{\partial}\varphi$. Let us differentiate

$$\log|\varphi| = -(1/(n+1))\log K$$

(where K is short for $K(z,z)$) so that to obtain

$$\frac{1}{\varphi}\,\overline{\partial}\varphi = -\frac{1}{n+1}\,\overline{\partial}\log K.$$

Applying the operator $i\,\partial$ leads to

$$\frac{1}{\varphi}\,d\theta - \frac{i}{\varphi^2}\,\partial\varphi\wedge\overline{\partial}\varphi = -\frac{i}{n+1}\,\partial\overline{\partial}\log K. \tag{5.43}$$

We shall need the *Bergman metric*

$$g = \frac{\partial^2 \log K}{\partial z^j \partial \overline{z}^k}\,dz^j \odot d\overline{z}^k\,.$$

As well known, g is a Kähler metric on Ω (Kähler-Einstein when Ω is homogeneous). Here $\odot$ denotes the symmetric tensor product, i.e. $\alpha\odot\beta = \frac{1}{2}(\alpha\otimes\beta + \beta\otimes\alpha)$. Let us set

$$\omega(X,Y) = g(X,JY),$$

the Kähler 2-form of (Ω, J, g), where J is the underlying complex structure. Then

$$\omega = -i\,\partial\overline{\partial}\log K$$

and (5.43) may be written

PROPOSITION 5.31. *Let $\Omega\subset\mathbb{C}^n$ be a smoothly bounded strictly pseudoconvex domain and $\varphi(z) = -K(z,z)^{-1/(n+1)}$, $z\in\Omega$. Then the Bergman metric g of Ω is expressed by*

$$g(X,Y) = \frac{n+1}{\varphi}\{\frac{i}{\varphi}\,(\partial\varphi\wedge\overline{\partial}\varphi)(X,JY) - d\theta(X,JY)\}, \tag{5.44}$$

for any $X, Y\in\mathcal{X}(\Omega)$.

Let $\mathcal{F}$ be the foliation of V by level sets

$$M_\delta = \{z\in\Omega : \varphi(z) = -\delta\}$$

as in the previous section. As a consequence of (5.44) and of $JT = -N$

COROLLARY 5.32. *The Bergman metric g of Ω is expressed by*

$$g(X,Y) = -\frac{n+1}{\varphi}\,g_\theta(X,Y), \quad X,Y\in H(\mathcal{F}). \tag{5.45}$$

$$g(X,T) = 0, \;\; g(X,N) = 0, \quad X\in H(\mathcal{F}), \tag{5.46}$$

$$g(T,N) = 0, \;\; g(T,T) = g(N,N) = \frac{n+1}{\varphi}\left(\frac{1}{\varphi} - r\right). \tag{5.47}$$

In particular $1 - r\varphi > 0$ everywhere in Ω.

Using (5.45)-(5.47) we may relate the Levi-Civita connection ∇^g of (V,g) to the Graham-Lee connection ∇. By (5.45) (as $X(\varphi)=0$, $X\in T(\mathcal{F})$)

LEMMA 5.33.

$$g(\nabla^g_X Y, Z) = g(\nabla_X Y, Z), \quad X,Y,Z \in H(\mathcal{F}). \tag{5.48}$$

Note that any tangent vector field $X \in T(V)$ decomposes as

$$X = \pi_H X + \theta(X)T + \frac{1}{2}(d\varphi)(X)N,$$

where $\pi_H : T(V) \to H(\mathcal{F})$ is the projection. Moreover it is easy to see that $\theta([T,X]) = 0$ for any $X \in H(\mathcal{F})$. Also $[T,X] \in T(\mathcal{F})$ hence $[T,X] \in H(\mathcal{F})$ for any $X \in H(\mathcal{F})$. Taking into account the identity

$$2g(\nabla^g_X Y, Z) = X(g(Y,Z)) + Y(g(X,Z)) - Z(g(X,Y)) + \tag{5.49}$$
$$+ g([X,Y],Z) + g([Z,X],Y) + g(X,[Z,Y]),$$

for any $X,Y,Z \in T(V)$, one has (by (5.46))

$$2g(\nabla^g_X Y, T) = -T(g(X,Y)) +$$
$$+ g([X,Y],T) + g([T,X],Y) + g(X,[T,Y]) =$$
$$= \frac{n+1}{\varphi}\,\{T(g_\theta(X,Y)) - g_\theta([T,X],Y) - g_\theta(X,[T,Y])\} +$$
$$+ \frac{n+1}{\varphi}\left(\frac{1}{\varphi} - r\right)\theta([X,Y]),$$

for any $X,Y \in H(\mathcal{F})$. Since $\nabla_X Y \in H(\mathcal{F})$ it follows that

$$T(g_\theta(X,Y)) - g_\theta([T,X],Y) - g_\theta(X,[T,Y]) = 2g_\theta(\tau X, Y).$$

Note that one makes use of the fact that $\tau : H(\mathcal{F}) \to H(\mathcal{F})$ is self-adjoint i.e. $g_\theta(\tau X, Y) = g_\theta(X, \tau Y)$, $X,Y \in H(\mathcal{F})$. Hence

$$g(\nabla^g_X Y, T) = -g(\tau X, Y) - \frac{n+1}{\varphi}\left(\frac{1}{\varphi} - r\right)(d\theta)(X,Y)$$

or

LEMMA 5.34.

$$g(\nabla^g_X Y, T) = -g(\tau X, Y) - \left(\frac{1}{\varphi} - r\right) g(X, \phi Y), \tag{5.50}$$

for any $X,Y \in H(\mathcal{F})$.

Exploiting again $\nabla^g g = 0$ (and $g([X,Y],N) = 0$) we get

$$2g(\nabla^g_X Y, N) = -(\mathcal{L}_N g)(X,Y), \quad X,Y \in H(\mathcal{F}).$$

Hence

$$2g(\nabla^g_X Y, N) = -\frac{n+1}{\varphi^2}\,N(\varphi)\,g_\theta(X,Y) + \frac{n+1}{\varphi}\,(\mathcal{L}_N g_\theta)(X,Y) =$$
$$= 2\left(\frac{1}{\varphi} - r\right) g(X,Y) + \frac{2(n+1)}{\varphi}\,(d\theta)(X,\tau Y)$$

that is

LEMMA 5.35.

$$g(\nabla^g_X Y, N) = \left(\frac{1}{\varphi} - r\right) g(X,Y) + g(X, \phi\,\tau\,Y), \tag{5.51}$$

for any $X, Y \in H(\mathcal{F})$.

Note that (5.51) may be also derived from (5.50) by using the fact that g is a Kähler metric. Indeed

$$g(\nabla^g_X Y, N) = g(J\nabla^g_X Y, JN) = g(\nabla^g_X JY, T) =$$
$$= -g(\tau X, \phi Y) - \left(\frac{1}{\varphi} - r\right) g(X, \phi^2 Y),$$

etc. For further use, let us also retain that

$$(\mathcal{L}_N g)(X,Y) = -2\left(\frac{1}{\varphi} - r\right) g(X,Y) - 2g(X, \phi\tau Y), \tag{5.52}$$

for any $X, Y \in H(\mathcal{F})$. At this point, the identities (5.48) and (5.50)-(5.51) lead to

PROPOSITION 5.36.

$$\begin{aligned}\nabla^g_X Y = \nabla_X Y + \\ + \left\{\frac{\varphi}{1-\varphi r}\, g_\theta(\tau X, Y) + g_\theta(X, \phi Y)\right\} T - \\ - \left\{g_\theta(X,Y) + \frac{\varphi}{1-\varphi r}\, g_\theta(X, \phi\,\tau\,Y)\right\} N,\end{aligned} \tag{5.53}$$

for any $X, Y \in H(\mathcal{F})$.

To compute $\nabla^g_X T$ we use (5.53) and

$$g(\nabla^g_X T, Y) = -g(T, \nabla^g_X Y)$$

so that

$$g(\nabla^g_X T, Y) = g(\tau X, Y) + \left(\frac{1}{\varphi} - r\right) g(X, \phi Y). \tag{5.54}$$

The component along T is $\frac{1}{2} X(\|T\|^2)$ hence

$$g(\nabla^g_X T, T) = -\frac{n+1}{2\varphi} X(r). \tag{5.55}$$

Moreover

$$2g(\nabla^g_X T, N) = g(X, [N,T]) = -g(X, \phi\nabla^H r)$$

that is

$$g(\nabla^g_X T, N) = -\frac{n+1}{2\varphi} (\phi X)(r). \tag{5.56}$$

Summing up (by (5.54)-(5.56))

PROPOSITION 5.37.

$$\nabla^g_X T = \tau X - \left(\frac{1}{\varphi} - r\right)\phi X - \frac{\varphi}{2(1-r\varphi)}\left\{X(r)T + (\phi X)(r) N\right\}, \tag{5.57}$$

for any $X \in H(\mathcal{F})$.

Again by (5.53) and

$$g(\nabla^g_X N, Y) = -g(N, \nabla^g_X Y)$$

we get

$$g(\nabla^g_X N, Y) = -\left(\frac{1}{\varphi} - r\right) g(X, Y) - g(X, \phi\tau Y). \tag{5.58}$$

Next

$$g(\nabla^g_X N, T) = \frac{n+1}{2\varphi}(\phi X)(r). \tag{5.59}$$

Finally the component along N is $\frac{1}{2}X(\|N\|^2)$ hence

$$g(\nabla^g_X N, N) = -\frac{n+1}{2\varphi} X(r). \tag{5.60}$$

Summing up (by (5.58)-(5.60))

PROPOSITION 5.38.

$$\nabla^g_X N = -\left(\frac{1}{\varphi} - r\right) X + \tau\phi X + \tag{5.61}$$

$$+\frac{\varphi}{2(1-r\varphi)}\{(\phi X)(r)\, T - X(r)\, N\},$$

for any $X \in H(\mathcal{F})$.

We wish to compute $\nabla^g_T X$. To this end (by $\nabla g_\theta = 0$)

$$2g_\theta(\nabla_T X, Y) = T(g_\theta(X, Y)) + g_\theta([T, X], Y) + g_\theta([Y, T], X) +$$

$$+ g_\theta(T, [Y, X]) + g_\theta(\tau X, Y) - g_\theta(\tau Y, X) - 2(d\theta)(X, Y)$$

yielding (upon multiplication by $-(n+1)/\varphi$)

$$T(g(X, Y)) + g([T, X], Y) + g([Y, T], X) = 2g(\nabla_T X, Y).$$

Therefore (by $\nabla^g g = 0$)

$$2g(\nabla^g_T X, Y) = T(g(X, Y)) + g([T, X], Y) + g([Y, T], X) +$$

$$+ g(T, [Y, X]) = 2g(\nabla_T X, Y) - \theta([X, Y])\|T\|^2$$

or

$$g(\nabla^g_T X, Y) = g(\nabla_T X, Y) + \left(\frac{1}{\varphi} - r\right) g(X, \phi Y). \tag{5.62}$$

Similar to the above

$$g(\nabla^g_T X, T) = -\frac{n+1}{2\varphi} X(r), \tag{5.63}$$

$$g(\nabla^g_T X, N) = -\frac{n+1}{2\varphi}(\phi X)(r). \tag{5.64}$$

Collecting the information in (5.62)-(5.64) we have proved

PROPOSITION 5.39.

$$\nabla^g_T X = \nabla_T X - \left(\frac{1}{\varphi} - r\right)\phi X - \tag{5.65}$$

$$-\frac{\varphi}{2(1-r\varphi)}\{X(r)T + (\phi X)(r)N\},$$

for any $X \in H(\mathcal{F})$.

Let us compute $\nabla^g_N X$. We have

$$2g(\nabla^g_N X, Y) = N(g(X,Y)) + g([N,X],Y) + g([Y,N],X) =$$

$$= 2g([N,X],Y) + (\mathcal{L}_N g)(X,Y).$$

Using (5.52) and

$$[N,X] = \nabla_N X - rX - \tau(\phi X)$$

one shows that

$$g(\nabla^g_N X, Y) = g(\nabla_N X, Y) - \frac{1}{\varphi}\, g(X,Y). \tag{5.66}$$

Calculations similar to the above also furnish

$$g(\nabla^g_N X, T) = \frac{n+1}{2\varphi}\,(\phi X)(r), \tag{5.67}$$

$$g(\nabla^g_N X, N) = -\frac{n+1}{2\varphi}\, X(r). \tag{5.68}$$

Using (5.66)-(5.68) we may now conclude that

PROPOSITION 5.40.

$$\nabla^g_N X = \nabla_N X - \frac{1}{\varphi} X + \tag{5.69}$$

$$+\frac{\varphi}{2(1-r\varphi)}\{(\phi X)(r)T - X(r)N\},$$

for any $X \in H(\mathcal{F})$.

Moreover (omitting the details)

$$g(\nabla^g_N T, X) = -\frac{n+1}{2\varphi}\,(\phi X)(r),$$

$$g(\nabla^g_N T, T) = -\frac{n+1}{2\varphi}\left\{N(r) + \frac{4}{\varphi^2} - \frac{2r}{\varphi}\right\},$$

$$g(\nabla^g_N T, N) = -\frac{n+1}{2\varphi}\, T(r),$$

so that

PROPOSITION 5.41.

$$\nabla^g_N T = -\frac{1}{2}\,\phi\,\nabla^H r - \tag{5.70}$$

$$-\frac{\varphi}{2(1-r\varphi)}\left\{\left(N(r) + \frac{4}{\varphi^2} - \frac{2r}{\varphi}\right)T + T(r)N\right\}.$$

Similarly we find

PROPOSITION 5.42.

$$\nabla^g_T N = \frac{1}{2}\phi\nabla^H r - \tag{5.71}$$
$$-\frac{\varphi}{2(1-r\varphi)}\left\{\left(N(r)+\frac{4}{\varphi^2}-\frac{6r}{\varphi}+4r^2\right)T+T(r)N\right\},$$

$$\nabla^g_T T = -\frac{1}{2}\,\nabla^H r - \tag{5.72}$$
$$-\frac{\varphi}{2(1-r\varphi)}\left\{T(r)T-\left(N(r)+\frac{4}{\varphi^2}-\frac{6r}{\varphi}+4r^2\right)N\right\},$$

$$\nabla^g_N N = -\frac{1}{2}\,\nabla^H r + \tag{5.73}$$
$$+\frac{\varphi}{2(1-r\varphi)}\left\{T(r)T-\left(N(r)+\frac{4}{\varphi^2}-\frac{2r}{\varphi}\right)N\right\}.$$

Let us consider a holomorphic vector bundle $\pi : F \to U$, carrying the Hermitian metric h, and set $E_\delta = \pi^{-1}(M_\delta)$ (the portion of F over a leaf of $\mathcal{F}$). A connection $D \in \mathcal{C}(F,h)$ induces a connection $D^\delta \in \mathcal{C}(E_\delta, h_\delta)$ (where $h_{\delta,z} = h_z$, $z \in M_\delta$). D^δ is most easily described with respect to a local trivialization $\Phi : \pi^{-1}(O) \to O \times \mathbb{C}^m$ of F, for some open subset $O \subseteq U$. Let us set

$$\sigma_i(z) = \Phi^{-1}(z, e_i), \quad z \in O, \quad 1 \le i \le m,$$

where $\{e_1, \cdots, e_m\}$ is the canonical linear basis in $\mathbb{C}^m$. If $u_i \equiv \sigma_i|_{O\cap M_\delta}$ then D^δ is given by

$$(D^\delta_X u)_z = X(f^i)_z u_i(z) + f^i(z)(D_{(d\,j_\delta)X}\sigma_i)_z\,, \quad z \in O \cap M_\delta,$$

for any section $u = f^i u_i$, $f^i \in C^\infty(O \cap M_\delta)$, and any $X \in \mathcal{X}(M_\delta)$. It is easily shown that the definition of $(D^\delta_X u)_z$ doesn't depend upon the local trivialization chart Φ at z i.e. if

$$g = [g_{ij}] : O \cap O' \to \mathrm{GL}(m, \mathbb{C}), \quad g(z) = \Phi'_z \circ \Phi_z^{-1}\,,$$

are the transition functions of F then $(D^\delta_X u)_z$ is invariant under the transformation $\sigma_j(z) = g^i_j(z)\sigma'_i(z))$. Let $R^D \in \Omega^2(\mathrm{Ad}(F))$ and ω^i_j be the curvature tensor field and connection 1-forms of D i.e.

$$D\sigma_j = \omega^i_j \otimes \sigma_i\,,$$

so that

$$R^D\sigma_j = 2(d\omega^i_j - \omega^i_k \wedge \omega^k_j) \otimes \sigma_i\,.$$

Also let $R^\delta \in \Omega^2(\mathrm{Ad}(E_\delta))$ and $(\omega_\delta)^i_j$ be the curvature tensor field and the connection 1-forms of D^δ, respectively. Then $(\omega_\delta)^i_j = j^*_\delta \omega^i_j$ yields

$$R^\delta u_i = (j^*_\delta R^D)\sigma_i\,, \quad 1 \le i \le m. \tag{5.74}$$

Let $E \to M$ be a vector bundle and D a connection in E. We shall need the differential operator $d^D : \Omega^k(E) \to \Omega^{k+1}(E)$ given by

$$(d^D\varphi)(X_1, \cdots, X_{k+1}) = \sum_{i=1}^{k+1}(-1)^{i+1} D_{X_i}(\varphi(X_1, \cdots, \hat{X}_i, \cdots, X_{k+1})) +$$
$$+ \sum_{1\le i<j\le k+1} (-1)^{i+j}\varphi([X_i, X_j], X_1, \cdots, \hat{X}_i, \cdots, \hat{X}_j, \cdots, X_{k+1})$$

for any $\varphi \in \Omega^k(E)$ and any $X_i \in T(M)$, $1 \le i \le k$. Here a hat indicates, as usual, the suppression of a term.

Let $D \in \mathcal{C}(E,h)$ and let us denote by the same symbol the connection induced by D in $\mathrm{Ad}(E) \to M$.

DEFINITION 5.43. The operator δ^D is the formal adjoint of

$$d^D : \Omega^1(\mathrm{Ad}(E)) \to \Omega^2(\mathrm{Ad}(E))$$

with respect to the inner product

$$(\varphi, \psi) = \int_M \langle \varphi, \psi \rangle \, \theta \wedge (d\theta)^n \, , \quad \varphi, \psi \in \Omega^k(E). \tag{5.75}$$

□

As an illustration of the use of δ^D let us derive the Euler-Lagrange equations of the variational principle $\delta\, \mathcal{PYM}(D) = 0$. To this end let $\varphi \in \Omega^1(\mathrm{Ad}(E))$. A standard calculation shows that

$$R^{D+t\varphi} = R^D + t\, d^D\varphi + t^2\, [\varphi \wedge \varphi]$$

where

$$[\varphi \wedge \psi]_{X,Y} := [\varphi_X, \psi_Y] - [\varphi_Y, \psi_X],$$

for any $X, Y \in T(M)$, $\varphi, \psi \in \Omega^1(\mathrm{Ad}(E))$, hence

$$\|\pi_H\, R^{D+t\varphi}\|^2 = \|\pi_H\, R^D\|^2 + 2t\, \langle \pi_H\, R^D \, , \, \pi_H\, d^D\varphi \rangle + O(t^2)$$

and

$$\frac{d}{dt}\{\mathcal{PYM}(D + t\varphi)\}_{t=0} = \frac{1}{2} \int_M \frac{d}{dt}\{\|\pi_H R^{D+t\varphi}\|^2\}_{t=0}\, \theta \wedge (d\theta)^n =$$

$$= \int_M \langle \pi_H\, R^D \, , \, d^D\varphi \rangle \, \theta \wedge (d\theta)^n = \int_M \langle \delta^D \pi_H R^D \, , \, \varphi \rangle \, \theta \wedge (d\theta)^n.$$

Then $\frac{d}{dt}\{\mathcal{PYM}(D + t\varphi)\}_{t=0} = 0$ yields

$$\delta^D\, \pi_H\, R^D = 0. \tag{5.76}$$

These are the *pseudo Yang-Mills equations* and a solution D to (5.76) is a *pseudo Yang-Mills field.* Let $D \in \mathcal{C}(E,h)$ such that $i_T R^D = 0$. Then (by (5.76)) D is a pseudo Yang-Mills field if and only if D is a Yang-Mills field, and the last statement in part i) of Theorem 5.23 follows from Theorem 2.3 in [**245**], p. 551.

DEFINITION 5.44. Let us consider the operator $\delta_b^D : \Omega^{k+1}(E) \to \Omega^k(E)$ given by

$$(\delta_b^D \varphi)(X_1, \cdots, X_k) = -\sum_{a=1}^{2n} (D_{E_a}\varphi)(E_a, X_1, \cdots, X_k),$$

for any $\varphi \in \Omega^{k+1}(E)$ and $X_i \in T(M)$, $1 \le i \le k$, where $\{E_a : 1 \le a \le 2n\}$ is a local G_θ-orthonormal frame of $H(M)$. □

Clearly if $\varphi \in \Omega^k_H(E)$ then $\delta^D\varphi = \delta_b^D\varphi$ and $i_T\delta_b^D\varphi = 0$. Consequently if $i_T R^D = 0$ then the equations (5.76) may also be written

$$\delta_b^D R^D = 0. \tag{5.77}$$

Let now $\{W_\alpha\}$ be a local orthonormal ($g_\theta(W_\alpha, W_{\overline{\beta}}) = \delta_{\alpha\beta}$) frame of $T_{1,0}(\mathcal{F})$ and let us set

$$E_\alpha \equiv \sqrt{-\frac{\varphi}{n+1}}\, W_\alpha\,, \quad 1 \le \alpha \le n-1, \quad E_n \equiv \sqrt{\frac{2f\varphi}{n+1}}\, \xi\,,$$

where $f \equiv \varphi/(1 - r\varphi)$. Then given a connection D in $F \to U$, for any $X \in H(M_\delta)$

$$(\delta^D R^D)X = -\sum_{a=1}^{n}\{(D_{E_a} R^D)(E_{\overline{a}}, X) + (D_{E_{\overline{a}}} R^D)(E_a, X)\} =$$

$$= \frac{\varphi}{n+1}\sum_{\alpha=1}^{n-1}\{(D_{W_\alpha} R^D)(W_{\overline{\alpha}}, X) + (D_{W_{\overline{\alpha}}} R^D)(W_\alpha, X)\} -$$
$$-\frac{2f\varphi}{n+1}\{(D_\xi R^D)(\overline{\xi}, X) + (D_{\overline{\xi}} R^D)(\xi, X)\}$$

and

$$\sum_{\alpha=1}^{n-1}(D_{W_\alpha} R^D)(W_{\overline{\alpha}}, X)\sigma_j =$$
$$= \sum_\alpha\{D_{W_\alpha}\left(R^D(W_{\overline{\alpha}}, X)\sigma_j\right) - R^D(W_{\overline{\alpha}}, X)D_{W_\alpha}\sigma_j -$$
$$-R^D(\nabla^g_{W_\alpha} W_{\overline{\alpha}}, X)\sigma_j - R^D(W_{\overline{\alpha}}, \nabla^g_{W_\alpha} X)\sigma_j\} = \quad (by\ (5.53))$$
$$= \sum_\alpha\{(D^\delta_{W_\alpha} R^\delta)(W_{\overline{\alpha}}, X)u_j -$$
$$-[f\, g_\theta(\tau\, W_\alpha, W_{\overline{\alpha}}) + g_\theta(W_\alpha, \phi\, W_{\overline{\alpha}})]R^D(T, X)\sigma_j +$$
$$+[g_\theta(W_\alpha, W_{\overline{\alpha}}) + f\, g_\theta(W_\alpha, \phi\,\tau\, W_{\overline{\alpha}})]R^D(N, X)\sigma_j -$$
$$-[f\, g_\theta(\tau\, W_\alpha, X) + g_\theta(W_\alpha, \phi X)]R^D(W_{\overline{\alpha}}, T)\sigma_j +$$
$$+\,[g_\theta(W_\alpha, X) + f\, g_\theta(W_\alpha, \phi\,\tau\, X)]R^D(W_{\overline{\alpha}}, N)\sigma_j\}\,.$$

Therefore (by the purity property of T_∇)

$$\sum_\alpha(D_{W_\alpha} R^D)(W_{\overline{\alpha}}, X) = \sum_\alpha(D^\delta_{W_\alpha} R^\delta)(W_{\overline{\alpha}}, X)u_j -$$
$$+i(n-1)R^D(T, X)\sigma_j + (n-1)R^D(N, X)\sigma_j -$$
$$-f\, R^D(\pi_{0,1}\,\tau\, X, T) - R^D(\pi_{0,1}\,\phi\, X, T) +$$
$$+R^D(\pi_{0,1}\, X, N) + f\, R^D(\pi_{0,1}\,\phi\,\tau\, X, N).$$

We obtain

$$\sum_\alpha\{(D_{W_\alpha} R^D)(W_{\overline{\alpha}}, X) + (D_{W_{\overline{\alpha}}} R^D)(W_\alpha, X)\} = \tag{5.78}$$
$$-(\delta_b^{D^\delta} R^\delta)_X\, u_j + \{R^D(N, (2n-3)X - f\,\phi\,\tau\, X) +$$
$$+R^D(T, \phi\, X + f\,\tau\, X)\}\sigma_j\,.$$

Moreover

$$\{(D_\xi R^D)(\overline{\xi}, X) + (D_{\overline{\xi}} R^D)(\xi, X)\}\sigma_j =$$
$$= \frac{1}{2}\{D_N\left(R^D(N, X)\sigma_j\right) + D_T\left(R^D(T, X)\sigma_j\right) -$$
$$-R^D(N, X)D_N\sigma_j - R^D(T, X)D_T\sigma_j -$$
$$-R^D(\nabla^g_N N, X)\sigma_j - R^D(\nabla^g_T T, X)\sigma_j -$$

$$-R^D(N, \nabla^g_N X)\sigma_j - R^D(T, \nabla^g_T X)\sigma_j\}.$$

Substitution from (5.65), (5.69) and (5.72)-(5.73) gives

$$-R^D(\nabla^g_N N, X) - R^D(\nabla^g_T T, X) - R^D(N, \nabla^g_N X) - R^D(T, \nabla^g_T X) =$$
$$= R^D(\nabla^H r, X) - R^D(T, \nabla_T X) - R^D(N, \nabla_N X) +$$
$$+ \frac{1}{f}\, R^D(T, \phi X) + f\,(\phi X)(r)\, R^D(T, N) + \left(\frac{1}{\varphi} + 2r\right) R^D(N, X).$$

We conclude that

$$\{(D_\xi R^D)(\overline{\xi}, X) + (D_{\overline{\xi}} R^D)(\xi, X)\}\sigma_j = \tag{5.79}$$
$$= \frac{1}{2}\{(D_N\, i_N R^D)X + (D_T\, i_T R^D)X + R^D(\nabla^H r, X) +$$
$$+ \frac{1}{f}\, R^D(T, \phi X) + f\,(\phi X)(r)\, R^D(T, N) + \left(\frac{1}{\varphi} + 2r\right) R^D(N, X)$$

(the covariant derivatives in the right hand member of (5.79) are defined with respect to D and ∇). Finally (by (5.78)-(5.79))

$$(\delta^D R^D)_X\, \sigma_j = \frac{\varphi}{n+1}\left\{-(\delta_b^{D^\delta} R^\delta)_X\, u_j + \right.$$
$$+ [R^D(N, (2n-3)X - f\,\phi\,\tau\, X) + R^D(T, \phi\, X + f\,\tau\, X)]\sigma_j\} -$$
$$- \frac{f\,\varphi}{n+1}\{(D_N\, i_N R^D)X + (D_T\, i_T R^D)X + R^D(\nabla^H r, X) +$$
$$\left. + R^D(T\,,\, \frac{1}{f}\,\phi\, X + f\,(\phi\, X)(r)\, N) + \left(\frac{1}{\varphi} + 2r\right) R^D(N, X)\right\}\sigma_j\,.$$

Assume that D is a Yang-Mills field on (Ω, g), i.e. $\delta^D R^D = 0$ in Ω. Then, for $\varphi \to 0$ (as r and $\nabla^H r$ stay finite near $\partial\Omega$, cf. [**124**], p. 164)

$$(\delta_b^{D_b} R^{D_b})_X u_j = 2(n-2)R^D(N, X)\sigma_j$$

where $D_b \equiv D^0$ is the boundary values of D. Therefore, if $i_T R^{D_b} = 0$ then D_b is a pseudo Yang-Mills field on $\partial\Omega$ if and only if $i_N R^D = 0$ on $H(\partial\Omega)$. Theorem 5.22 is proved. With the same techniques we may show that

COROLLARY 5.45. *Let $D \in \mathcal{C}(F, h)$ be a Yang-Mills field on (Ω, g) such that $i_N R^D = 0$. Then the boundary values D_b of D satisfy $\Lambda_\theta R^{D_b} = 0$.*

Corollary 5.45 shows that the axiom (5.40) (with $S = 0$) in the description of the Tanaka connection, as well as (5.37) in Theorem 5.23, are rather natural occurrences. The proof is

$$0 = (\delta^D R^D)_T \sigma_j =$$
$$= \frac{\varphi}{n+1} \sum_\alpha \{(D_{W_\alpha} R^D)(W_{\overline{\alpha}}, T) + (D_{W_{\overline{\alpha}}} R^D)(W_\alpha, T)\}\sigma_j -$$
$$- \frac{2f\varphi}{n+1}\{D_N(R^D(N, T)\sigma_j) - R^D(N, T)D_N\sigma_j -$$
$$- R^D(\nabla^g_N N, T) - R^D(N, \nabla^g_N T)\sigma_j\}$$

or (by (5.53), (5.57), (5.70) and (5.73))

$$0 = \varphi\{(\delta^{D^\delta} R^\delta)_T\, u_j + 2(n-1)R^D(T, N)\sigma_j\} +$$
$$+ \frac{1}{2}\,\varphi\,\{f\,[R^D(T, \nabla^H r) + R^D(N, \phi\,\nabla^H r)] + \mathit{trace}\,\pi_H R^D(\cdot\,,\,\tau\,\cdot)\}\sigma_j -$$

$$+(n+1)f\{2(D_N\, i_N R^D)T + R^D(N, \phi\nabla^H r) - R^D(T, \nabla^H r)\}\sigma_j -$$
$$-\frac{2\varphi}{f}\,\Lambda_\theta R^D\sigma_j - (n+1)f^2\left\{N(r) + \frac{4}{\varphi^2} - \frac{2r}{\varphi}\right\} R^D(T,N)\sigma_j\,.$$

When $\varphi \to 0$ one observes $\varphi/f \to 1$ and $f^2/\varphi^2 \to 1$ hence

$$\left(\Lambda_\theta R^{D_b}\right) u_j = -2(n+1)R^D(T,N)\sigma_j\,.$$

Q.e.d.

A similar method for domains in a strictly pseudoconvex CR manifolds was devised by R.K. Hladky, [**138**]. Roughly speaking such domains are foliated by compact CR submanifolds of real codimension 2 (see also D. Ciampa, [**68**]) This is useful in the study of the $\overline{\partial}_b$-Neumann problem[2] and leads to a new approach[3] to finding estimates based on decomposing the operator into its tangential and transverse parts with respect to the foliation. Estimates for the tangential parts follow from results about the tangential Cauchy-Riemann complex on compact CR manifolds, while estimates for the transverse part reduce to elliptic estimates in the plane. R.K. Hladky used (cf. *op. cit.*) this approach for certain domains in the Heisenberg group (including the unit Heisenberg ball) to establish sharp regularity and Fredholm theorems for the $\overline{\partial}_b$-Neumann problem.

5.5. Flows

Let $(M, T_{1,0}(M))$ be a nondegenerate CR manifold and θ a contact form on M. Let $\epsilon \in \{\pm 1\}$ and let X be a tangent vector field on M so that $g_\theta(X,X) = \epsilon$ everywhere on M. Let $\mathcal{F}$ be the flow determined by X i.e. the foliation whose leaves are the maximal integral curves of X. Let $\chi_\mathcal{F}$ be defined by

$$\chi_\mathcal{F}(Y) = g_\theta(Y, X), \quad Y \in T(M).$$

Then $\chi_\mathcal{F} \in \Omega^1(M)$ is the characteristic form of $\mathcal{F}$ on (M, g_θ). Next

$$\kappa(Z) = \epsilon g_\theta(\nabla_X X, Z).$$

Yet $2g_\theta(\nabla_X X, X) = X(\epsilon) = 0$ yields $\nabla_X X \in P^\perp$ hence

$$t = \epsilon\nabla_X X. \tag{5.80}$$

THEOREM 5.46. *Let $(M, T_{1,0}(M))$ be a nondegenerate CR manifold, θ a contact form on M, and T the characteristic direction of (M,θ). Let $\mathcal{F}$ be the flow on M defined by T. Then*

i) $\mathcal{F}$ is totally geodesic in (M, g_θ),

ii) The orbits of T are autoparallel curves of ∇,

iii) $\mathcal{L}_T\chi_\mathcal{F} = 0$,

iv) g_P is invariant under flows of vector fields lying in the Levi distribution of M,

v) $d\chi_\mathcal{F} \in F^2\Omega^2(M)$.

Here $F^r\Omega^m(M)$ is the filtration (of the de Rham complex of M) given by (9.10) in [**243**], p. 120. Theorem 5.46 follows from

[2]The analog within CR geometry of the $\overline{\partial}$-Neumann problem, cf. [**106**].

[3]Cf. J.M. Lee, *Foliations of CR manifolds and estimates for tangential Cauchy-Riemann complex*, unpublished lecture, University of California, Department of Mathematics, Analysis Seminar, February 20, 2004.

PROPOSITION 5.47. *Let* $\epsilon \in \{\pm 1\}$ *and* $X \in T(M)$ *so that* $g_\theta(X,X) = \epsilon$. *Let* $\mathcal{F}$ *be the flow defined by* X. *The following statements are equivalent*

i) $\kappa = 0$.

ii) *The orbits of* X *are autoparallel curves o* ∇.

iii) *The following identity holds*

$$\mathcal{L}_X \chi_{\mathcal{F}} = \theta(X)(2JX + \tau X)^\flat - A(X,X)\theta.$$

iv) *For any* $Z \in P^\perp$

$$(\mathcal{L}_Z g_P)(X,X) = 2A(X,X)\theta(Z) - 2\theta(X) g_\theta((2J+\tau)X, Z).$$

Here $\flat$ denotes lowering of indices by g_θ i.e. $X^\flat(Y) = g_\theta(X,Y)$ for any $X, Y \in T(M)$. The equivalence i) $\Longleftrightarrow$ ii) follows from (5.80). Moreover (again by (5.80))

$$\kappa(Z) = -\epsilon g_\theta(X, \nabla_X Z)$$

hence

$$(\mathcal{L}_X \chi_{\mathcal{F}})Z = \epsilon \kappa(Z) + g_\theta(T_\nabla(X,Z), X) \tag{5.81}$$

for any $Z \in T(M)$. Finally

$$g_\theta(T_\nabla(X,Z), X) = -A(X,X)\theta(Z) + \theta(X) g_\theta((2J+\tau)X, Z) \tag{5.82}$$

for any $Z \in T(M)$. Then (5.81)-(5.82) yield *i)* $\Longleftrightarrow$ *iii)*. Next (by (2.32) and (5.82))

$$(\mathcal{L}_Z g_P)(X,X) = -2\epsilon\kappa(Z) + 2A(X,X)\theta(Z) - 2\theta(X) g_\theta((2J+\tau)X, Z)$$

hence i) $\Longleftrightarrow$ iv). Q.e.d. All that remains to be checked is v) in Theorem 5.46. This follows from

$$F^2\Omega^2(M) = \{\omega \in \Omega^2(M) : i_X \omega = 0\}$$

and the fact that $\kappa = 0$ if and only if

$$i_X d\chi_{\mathcal{F}} = A(X,X)\theta - \theta(X)(2JX + \tau X)^\flat .$$

Next we shall restrict our attention to flows defined by infinitesimal pseudohermitian transformations.

DEFINITION 5.48. A C^∞ diffeomorphism $f : M \to M$ is a *pseudohermitian transformation* of (M, θ) if 1) f is a CR map, and 2) $f^*\theta = \theta$. □

Let $\mathrm{Psh}(M,\theta)$ be the group of all pseudohermitian transformations of (M,θ). It has been studied by S. Webster (cf. Theorem 1.2 in [**250**]) and E. Musso (cf. Theorem 4.10 in [**186**]). Let $U(M,\theta) \to M$ be the principal $\mathrm{U}(r,s)$-subbundle of $L(T(M)) \to M$ consisting of all linear frames of the form

$$u = (x, \{X_\alpha, J_x X_\alpha, T(x)\})$$

with

$$g_{\theta,x}(X_i, X_j) = \epsilon_i \delta_{ij}\,, \quad 1 \le i,j \le 2n,$$

where $X_{\alpha+n} = J_x X_\alpha$, $X_\alpha \in H(M)_x$, $1 \le \alpha \le n$. We may state

PROPOSITION 5.49. *A* C^∞ *diffeomorphism* f *of* M *is a pseudohermitian transformation of* (M,θ) *if and only if the induced transformation of* $L(T(M))$ *maps* $U(M,\theta)$ *into itself. Any fibre preserving transformation of* $U(M,\theta)$ *which leaves the canonical form of* $U(M,\theta)$ *invariant is induced by a pseudohermitian transformation* $f \in \mathrm{Psh}(M,\theta)$.

The proof of Proposition 5.49 follows from the fact that $f \in \mathrm{Psh}(M, \theta)$ if and only if f is a C^∞ diffeomorphism, $f^*\theta = \theta$, and $f^*\theta^\alpha = U^\alpha_\beta \theta^\beta$, for any local frame $\{\theta^\alpha\}$ of $T_{1,0}(M)^*$ and some (locally defined) C^∞ functions U^α_β on M (it mimics the proof of Proposition 3.1 in [**155**], vol. I, p. 236).

DEFINITION 5.50. A tangent vector field X on M is an *infinitesimal pseudohermitian transformation* of (M, θ) if its local 1-parameter group of local transformations consists of local pseudohermitian transformations of (M, θ). □

Let $\mathbf{i}(M, \theta)$ be the set of all infinitesimal pseudohermitian transformations of (M, θ). By analogy with Proposition 3.2 in [**155**], vol. I, p. 237, one may establish

PROPOSITION 5.51. *For a vector field X tangent to a nondegenerate CR manifold M, the following statements are equivalent*
i) $X \in \mathbf{i}(M, \theta)$.
ii) *The natural lift of X to $L(T(M))$ is tangent to $U(M, \theta)$ at each point of $U(M, \theta)$.*
iii) $\mathcal{L}_X \theta = 0$ *and* $\mathcal{L}_X \theta^\alpha = V^\alpha_\beta \theta^\beta$ *for any local frame $\{\theta^\alpha\}$ of $T_{1,0}(M)^*$ and some local C^∞ functions V^α_β on M.*

By iii) in Proposition 5.51, $\mathbf{i}(M, \theta)$ is a Lie algebra, as

$$\mathcal{L}_{[X,Y]} = [\mathcal{L}_X, \mathcal{L}_Y].$$

Any pseudohermitian transformation of (M, θ) preserves the Webster metric g_θ hence

$$\mathrm{Psh}(M, \theta) \subseteq \mathrm{Iso}(M, g_\theta). \tag{5.83}$$

Let $\epsilon \in \{\pm 1\}$ and $X \in \mathbf{i}(M, \theta)$ such that $g_\theta(X, X) = \epsilon$. Let $\mathcal{F}$ be the flow determined by X. Then (by (5.83)) $[X, Z] \in P^\perp$ for any $Z \in P^\perp$. Consequently

$$\mathcal{L}_X \chi_{\mathcal{F}} = 0 \tag{5.84}$$

Let us assume for the rest of this section that M is compact. Then (by Theorem 1.2 in [**250**], p. 31) $\mathrm{Psh}(M, \theta)$ is compact. Let G be the closure in $\mathrm{Psh}(M, \theta)$ of the 1-parameter group of transformations obtained by integrating X. Then G is compact and abelian hence a torus. Let $\Omega^k(M)^G$ denote the space of G-invariant k-forms on M. As a corollary of (5.84) there is a short exact sequence

$$0 \to \Omega^k_B(\mathcal{F}) \hookrightarrow \Omega^k(M)^G \xrightarrow{i_X} \Omega^{k-1}_B(\mathcal{F}) \to 0 \tag{5.85}$$

hence one may conclude (as in [**243**], p. 139) that

PROPOSITION 5.52. *$H^k_B(\mathcal{F})$, $0 \leq k \leq 2n$, are finite dimensional and zero for $k > 2n$.*

By Proposition 2.2 in [**250**], p. 33

PROPOSITION 5.53. *If $\tau = 0$ then $T \in \mathbf{i}(M, \theta)$. In particular if $X = T$ then the connecting homomorphism $\Delta : H^{k-1}_B(\mathcal{F}) \to H^{k+1}_B(\mathcal{F})$ in the long exact cohomology sequence associated with* (5.85) *is given by*

$$\Delta[\omega] = [(d\theta) \wedge \omega], \quad [\omega] \in H^{k-1}_B(\mathcal{F}).$$

As a corollary of G. Gigante's Theorem 5.8 and of Theorem 10.13 in [**243**], p. 139,

COROLLARY 5.54. *Let M be a nondegenerate CR manifold and θ a contact form on M. Let T be the characteristic direction of $d\theta$. If $X = T$ then the map $(i_T)_* : H^1(M, \mathbb{R}) \to H^0_B(\mathcal{F})$ is injective.*

5.6. Monge-Ampère foliations

This section is based on results in [**84**] and is about foliations of a strictly pseudoconvex CR manifold which are CR analogues of the Monge-Ampére foliations of E. Bedford & M. Kalka, [**30**], and P.M. Wong, [**256**]. These are defined by equations of the form

$$(\partial_M \overline{\partial}_M f)^p \neq 0, \quad (\partial_M \overline{\partial}_M f)^{p+1} = 0,$$

for some $0 \leq p \leq n$ (where n is the CR dimension of the given CR manifold M) and some $f \in C^\infty(M) \otimes \mathbb{C}$ whose real and imaginary parts are $\overline{\partial}_M$-plurisubharmonic (cf. Theorem 5.61). If $\mathcal{F}$ is such a tangential Monge-Ampére foliation then each leaf N of $\mathcal{F}$ is a CR manifold and the inclusion $\iota : N \subset M$ is a pseudohermitian immersion. By a result in this section $\mathcal{F}$ must be harmonic.

Let M be a nondegenerate CR manifold, θ a contact form on M, and T the characteristic direction of $d\theta$.

DEFINITION 5.55. A complex valued differential 2-form ω on M is of *type* $(1,1)$ if $\omega(Z, W) = 0$ whenever Z, W are of the same type i.e. both in $T_{1,0}(M)$ or both in $T_{0,1}(M)$). $\square$

Let $\Lambda^{1,1}(M)$ be the bundle of all $(1,1)$-forms on M.

DEFINITION 5.56. Let us consider the differential operator

$$\partial_M : \Gamma^\infty(\Lambda^{0,1}(M)) \to \Gamma^\infty(\Lambda^{1,1}(M))$$

defined as follows. Let σ be a $(0,1)$-form. Then $\partial_M \sigma$ is the unique $(1,1)$-form on M which agrees with $d\sigma$ when restricted to $T_{1,0}(M) \otimes T_{0,1}(M)$ and such that $T \rfloor \partial_M \sigma = 0$. $\square$

If $\sigma = \sigma_{\bar{\alpha}} \theta^{\bar{\alpha}} + \sigma_0 \theta$ is an arbitrary $(0,1)$-form then

$$\partial_M \sigma = (\sigma_{\bar{\beta},\alpha} + i h_{\alpha\bar{\beta}} \sigma_0) \theta^\alpha \wedge \theta^{\bar{\beta}}.$$

In particular, if $u \in C^\infty(M)$ and $\sigma = \overline{\partial}_M u$ then

$$\partial_M \overline{\partial}_M u = u_{\bar{\beta},\alpha} \theta^\alpha \wedge \theta^{\bar{\beta}}.$$

A coma denotes the covariant derivative with respect to the Tanaka-Webster connection of (M, θ) i.e. $u_{A,B} = (\nabla_{T_B} du) T_A$.

DEFINITION 5.57. u is called $\overline{\partial}_M$-*pluriharmonic* if $\partial_M \overline{\partial}_M u = 0$. Also u is $\overline{\partial}_M$-*plurisubharmonic* if $T(u) = 0$ and $i \partial_M \overline{\partial}_M u \geq 0$. $\square$

Throughout the present section we adopt the following more specialized concept of a tangentially CR foliation. Besides from requesting that the leaves of the given foliation $\mathcal{F}$ be CR immersed we wish that the given pseudohermitian structures be preserved and that the leaves of $\mathcal{F}$ occupy a precise position with respect to the preferred direction of the ambient space i.e. its characteristic direction (relative to a fixed choice of contact form). We need to collect a few notions of CR geometry.

DEFINITION 5.58. Let M be a nondegenerate CR manifold of CR codimension 1 and let θ be a contact form on M. Let T be the characteristic direction of $d\theta$. Let N be another CR manifold of hypersurface type and $f : N \to M$ a CR immersion i.e. a C^∞ immersion and a CR map. Let θ_N be a pseudohermitian structure on N. We say f is *isopseudohermitian* if $f^*\theta = \theta_N$. An isopseudohermitian immersion $f : N \to M$ is a *pseudohermitian immersion* if $(d_x f)T_x(N)$ is nondegenerate in $(T_{f(x)}(M), g_{\theta,x})$ for any $x \in N$ and $f(N)$ is tangent to T i.e. $T^\perp = 0$ where $T^\perp$ is the normal component of T with respect to the decomposition

$$T_{f(x)}(M) = [(d_x f)T_x(N)] \oplus E(f)_x\,, \quad x \in N,$$

and $E(f) \to N$ is the normal bundle of the immersion f with respect to the Webster metric g_θ of (M, θ). □

A study of the geometry of pseudohermitian immersions has been started in [**84**] and [**20**]. See also [**89**]. Let M be a strictly pseudoconvex CR manifold of CR dimension $N = n + k$ and θ a contact form on M such that the Levi form G_θ is positive definite. Given a CR immersion $f : N \to M$ of a $(2k+1)$-dimensional CR manifold N of CR dimension k we may consider $\theta_N := f^*\theta$. Then θ_N is a contact form on N and the Levi form G_{θ_N} is positive definite. Also f is an isopseudohermitian immersion of (N, θ_N) into (M, θ). Given a tangentially CR foliation $\mathcal{F}$ of (M, θ) each leaf $N \in M/\mathcal{F}$ is thought of as endowed with the induced contact form $\iota^*\theta$ where $\iota : N \to M$ is the inclusion.

DEFINITION 5.59. Let M be a strictly pseudoconvex CR manifold. A *pseudohermitian foliation* of M is a codimension $2p$ foliation $\mathcal{F}$ of M such that 1) each leaf N of $\mathcal{F}$ is a nondegenerate CR manifold of CR dimension $n - p$, and 2) the inclusion $\iota : N \to M$ is a pseudohermitian immersion. □

DEFINITION 5.60. Let $\mathcal{F}$ be a tangentially CR foliation of a nondegenerate CR manifold M. A real valued function $u \in C^\infty(M)$ is said to be *$\overline{\partial}_M$-pluriharmonic on $\mathcal{F}$* if for any leaf N of $\mathcal{F}$ the restriction $u \circ \iota$ is $\overline{\partial}_N$-pluriharmonic on N. □

We may state the following

THEOREM 5.61. ([**84**])
Let $(M, T_{1,0}(M))$ be a strictly pseudoconvex CR manifold of CR dimension $n \geq 2$ with a fixed contact structure θ so that the Levi form L_θ is positive definite and the Tanaka-Webster connection of (M, θ) has vanishing pseudohermitian torsion $(\tau = 0)$. Let $f \in C^\infty(M) \otimes \mathbb{C}$ such that

$$(\partial_M \overline{\partial}_M f)^p \neq 0, \quad (\partial_M \overline{\partial}_M f)^{p+1} = 0,$$

everywhere on M, for some $0 \leq p \leq n$. Let $f = u + iv$ with u and v real. If both u, v are $\overline{\partial}_M$-plurisubharmonic then there is a harmonic foliation $\mathcal{F}$ of M, of codimension $2p$, such that $\mathcal{F}$ is pseudohermitian and

i) *u, v are $\overline{\partial}_M$-pluriharmonic on $\mathcal{F}$.*

ii) *If $M = \mathbb{H}_n$ (the Heisenberg group) then $u_\alpha, v_\alpha \in CR^\infty(\mathcal{F})$ for $1 \leq \alpha \leq n$ (where $u_\alpha = T_\alpha(u)$ and $v_\alpha = T_\alpha(v)$).*

To prove Theorem 5.61 we need the following

LEMMA 5.62. *Let M be a nondegenerate CR manifold and consider a real valued function $u \in C^\infty(M)$. Then $i\partial_M \overline{\partial}_M u$ is a real form if and only if $T(u) = 0$. Moreover let us assume that $n \geq 2$. If* i) *$\tau = 0$ and* ii) *$dT(u) = T^2(u)\theta$ then*

$\partial_M \overline{\partial}_M u$ is closed. Conversely, if $\partial_M \overline{\partial}_M u$ is closed and $T_\alpha(u)_x \neq 0$ for any $x \in M$ and some local frame $\{T_\alpha\}$ in $T_{1,0}(M)$ (and thus for all) then i)-ii) *must hold.*

Proof. Let us define $\nabla^2 u$ by setting $(\nabla^2 u)(X,Y) = (\nabla_X du)Y$ for any $X, Y \in T(M)$, where ∇ is the Tanaka-Webster connection. One has

$$(\nabla^2 u)(X,Y) = (\nabla^2 u)(Y,X) + \Omega_\theta(X,Y)T(u), \tag{5.86}$$

$$(\nabla^2 u)(X,T) = (\nabla^2 u)(T,X) + \tau(X)(u), \tag{5.87}$$

for any $X, Y \in H(M)$. Thus $\nabla^2 u$ is symmetric if and only if $T(u) = 0$ and $\tau = 0$. By (5.86) we obtain the commutation relations

$$u_{\beta,\alpha} = u_{\alpha,\beta}\,, \quad u_{\bar\beta,\bar\alpha} = u_{\bar\alpha,\bar\beta}\,, \tag{5.88}$$

$$u_{\bar\beta,\alpha} = u_{\alpha,\bar\beta} - i h_{\alpha\bar\beta} T(u). \tag{5.89}$$

Then $\overline{u_{\bar\alpha,\beta}} = u_{\alpha,\bar\beta}$ (as u is real) and (5.89) yield

$$\overline{i\partial_M \overline{\partial}_M u} = i\partial_M \overline{\partial}_M u + iT(u)d\theta. \tag{5.90}$$

To prove the second statement in Lemma 5.62 let us set $\omega = \partial_M \overline{\partial}_M u$. Then

$$(d\omega)(T_\alpha, T_\beta, T_\gamma) = (d\omega)(T_{\bar\alpha}, T_{\bar\beta}, T_{\bar\gamma}) = 0, \tag{5.91}$$

$$6(d\omega)(T_\alpha, T_\beta, T_{\bar\gamma}) = i(h_{\alpha\bar\gamma} A^{\bar\rho}_{\beta} - h_{\beta\bar\gamma} A^{\bar\rho}_{\alpha}) u_{\bar\rho}\,, \tag{5.92}$$

$$6(d\omega)(T_{\bar\alpha}, T_{\bar\beta}, T_\gamma) = i h_{\gamma\bar\beta}\{T_{\bar\alpha}(u_0) - A^{\rho}_{\bar\alpha} u_\rho\} - i h_{\gamma\bar\alpha}\{T_{\bar\beta}(u_0) - A^{\rho}_{\bar\beta} u_\rho\}, \tag{5.93}$$

where $u_\alpha = T_\alpha(u)$, $u_{\bar\alpha} = T_{\bar\alpha}(u)$ and $u_0 = T(u)$. Clearly i)-ii) yield $d\omega = 0$ (by (5.91)-(5.93)). Conversely, let us assume that ω is closed. As $u_{\bar\rho} \neq 0$ everywhere on M, by (5.92) we obtain

$$h_{\alpha\bar\gamma} A^{\bar\rho}_{\beta} - h_{\beta\bar\gamma} A^{\bar\rho}_{\alpha} = 0.$$

Contraction with $h^{\bar\gamma\alpha}$ leads to $(n-1)A^{\bar\rho}_{\beta} = 0$ so that i) holds. Similarly (5.93) and $A^{\alpha}_{\bar\beta} = 0$ yield $T_{\bar\alpha}(u_0) = 0$, that is ii) holds.

Let us consider the q-forms ω_j , $1 \leq j \leq k$ on M and let us set

$$\mathrm{Ann}(\omega_1, \cdots, \omega_k)_x = \{X \in T_x(M) : X \rfloor \omega_{j,x} = 0, \ \forall 1 \leq j \leq k\}.$$

The resulting distribution does not have constant dimension, in general. If the q-forms ω_j are closed then $\mathrm{Ann}(\omega_1, \cdots, \omega_k)$ is involutive. Let ω be a $(1,1)$-form with $T \rfloor \omega = 0$. Locally

$$\omega = \frac{i}{2} a_{\alpha\bar\beta} \theta^\alpha \wedge \theta^{\bar\beta}. \tag{5.94}$$

Then ω is real if and only if $a_{\alpha\bar\beta} = \overline{a_{\beta\bar\alpha}}$ i.e. $A = (a_{\alpha\bar\beta})$ is hermitian. If A is hermitian then there is $U \in \mathrm{U}(n)$ such that $U^{-1}AU = \mathrm{diag}(\lambda_1, \cdots, \lambda_n)$, where $\lambda_j \in \mathbb{R}$. We define a vector $(\alpha_1, \cdots, \alpha_n)$ of $(1,0)$-forms on M by $(\alpha_1, \cdots, \alpha_n) = (\theta^1, \cdots, \theta^n)U$. Therefore any real $(1,1)$-form ω on M satisfying $T \rfloor \omega = 0$ may be written

$$\omega = \frac{i}{2} \sum_{\sigma=1}^{n} \lambda_\sigma \alpha_\sigma \wedge \overline{\alpha_\sigma}\,. \tag{5.95}$$

We shall need

LEMMA 5.63. *Let* ω_j, $1 \leq j \leq k$, *be the real* $(1,1)$*-forms on* M *such that* $T \rfloor \omega_j = 0$, $1 \leq j \leq k$. *Then* $\text{Ann}(\omega_1, \cdots, \omega_k)$ *is* J*-invariant.*

Proof. Let ω be given by (5.94). Let us set $\beta_\rho = |\lambda_\rho|^{1/2}\alpha_\rho$ for $1 \leq \rho \leq n$. The (5.95) becomes

$$\omega = \frac{i}{2}\sum_{\rho=1}^{n} \text{sign}(\lambda_\rho)\beta_\rho \wedge \overline{\beta}_\rho . \tag{5.96}$$

Let us set $b_\rho = \text{Re}(\beta_\rho)$. Then (5.96) may be written

$$\omega = -\sum_{\rho=1}^{n} \text{sign}(\lambda_\rho) b_\rho \wedge J^* b_\rho . \tag{5.97}$$

Here if α is a 1-form on M then $(J^*\alpha)X = \alpha(JX)$ and J is the complex structure of the Levi distribution (regarded as a $(1,1)$-tensor field on M with $JT = 0$). Then $J^*\omega = \omega$. To prove Lemma 5.63 one applies this fact to the $(1,1)$-forms ω_j.

LEMMA 5.64. *Let* ω *be given by* (5.94). *If* $\omega^p \neq 0$ *and* $\omega^{p+1} = 0$ *everywhere on* M, *for some* $0 \leq p \leq n$, *then* $\dim_{\mathbb{R}} \text{Ann}(\omega)_x = 2(n-p)+1$, *for any* $x \in M$.

Proof. Let $X = Z^\alpha T_\alpha + Z^{\bar\alpha} T_{\bar\alpha} + Z^0 T$ be a real tangent vector. Then $X \in \text{Ann}(\omega)$ if and only if $a_{\alpha\bar\beta} Z^\alpha = 0$. Thus

$$\text{Ann}(\omega)_x \approx \{(Z^1, \cdots, Z^n) \in \mathbb{C}^n : a_{\alpha\bar\beta}(x) Z^\alpha = 0\} \oplus \mathbb{R}T(x)$$

(a $\mathbb{R}$-linear isomorphism). Thus

$$\dim_{\mathbb{R}} \text{Ann}(\omega)_x = 2(n - \text{rank}(a_{\alpha\bar\beta}(x))) + 1.$$

Finally it is easy to see that $\omega^p \neq 0$ if and only if $\text{rank}(a_{\alpha\bar\beta}) = p$.

The condition that ω in Lemma 5.64 is real is necessary, as shown by the following

EXAMPLE 5.65. Let us consider the Heisenberg group $\mathbb{H}_2$ with the natural coordinates (z_1, z_2, t). Let us set $\theta^\alpha = dz^\alpha$, $\alpha = 1, 2$. Then

$$\omega = \theta^1 \wedge \theta^{\bar 1} + \theta^2 \wedge \theta^{\bar 2} + 2i\text{Re}(\theta^1 \wedge \theta^{\bar 2})$$

is a $(1,1)$-form on $\mathbb{H}_2$ such that $T \rfloor \omega = 0$ and $\omega \neq 0$, $\omega^2 = 0$ (where $T = \partial/\partial t$). Were Lemma 5.64 to apply (with $p = 1$) we would get $\dim_{\mathbb{R}} \text{Ann}(\omega)_x = 3$, $x \in \mathbb{H}_2$. Yet this is false (one actually has $\text{Ann}(\omega)_x = \mathbb{R}T(x)$) as ω is not real. □

DEFINITION 5.66. Let ω be given by (5.94). Then ω is *nonnegative* (and we write $\omega \geq 0$) if $a_{\alpha\bar\beta}\xi^\alpha\overline{\xi^\beta} \geq 0$ for any $(\xi^1, \cdots, \xi^n) \in \mathbb{C}^n$. □

If $\omega \geq 0$ then $\lambda_\rho \geq 0$, $1 \leq \rho \leq n$.

Let ω_j, $j = 1, 2$, be two real $(1,1)$-forms on M such that $T \rfloor \omega_j = 0$. By applying (5.97) to both ω_j there exist $\lambda_\rho, \mu_\rho \in \mathbb{R}$ and $a_\rho, b_\rho \in \Gamma^\infty(T^*(M))$ such that

$$\omega_1 = -\sum_{\rho=1}^{n} \text{sign}(\mu_\rho) a_\rho \wedge J^* a_\rho , \quad \omega_2 = -\sum_{\rho=1}^{n} \text{sign}(\lambda_\rho) b_\rho \wedge J^* b_\rho .$$

Let us assume $\omega_j \geq 0$ so that $\text{sign}(\lambda_\rho) = \text{sign}(\mu_\rho) = +1$.

LEMMA 5.67. *The following statements are equivalent*
i) $\dim_{\mathbb{R}} \mathrm{Span}(a_\rho, J^* a_\rho, b_\rho, J^* b_\rho)_x = 2p$, *for any* $x \in M$.
ii) $(\omega_1 + i\omega_2)^p \neq 0$ *and* $(\omega_1 + i\omega_2)^{p+1} = 0$ *everywhere on* M.

Proof. i) $\Longrightarrow$ ii)
Let us set $S = \mathrm{Span}(a_\rho, J^* a_\rho, b_\rho, J^* b_\rho)$ and $A = \mathrm{Ann}(a_\rho, J^* a_\rho, b_\rho, J^* b_\rho)$. Then $\dim_{\mathbb{R}} S_x = 2n + 1 - \dim_{\mathbb{R}} A_x$ so that $\dim_{\mathbb{R}} A_x = 2(n-p)+1$. Note that $T \in A$. Let L_x be the orthogonal complement (with respect to $g_{\theta,x}$) of A_x in $T_x(M)$. It is our purpose to show that

$$(\omega_1 + i\omega_2)^p \neq 0 \tag{5.98}$$

everywhere on M. As L is J-invariant and $T(x) \notin L_x$, J restricts to a complex structure on L so that we may decompose (pointwise) $L^* \otimes \mathbb{C} = L^{1,0} \oplus L^{0,1}$ where $L^{1,0} = \mathrm{Eigen}(i)$ (with respect to J^*) and $L^{0,1} = \overline{L^{1,0}}$. Let $\{\gamma_1, \cdots, \gamma_p, \overline{\gamma_1}, \cdots, \overline{\gamma_p}\}$ be a frame of $L^* \otimes \mathbb{C}$ with $\gamma_j \in L^{1,0}$. Then $a_\rho = a^j_\rho \gamma_j + \overline{a^j_\rho}\, \overline{\gamma_j}$ for some $a^j_\rho \in C^\infty(M) \otimes \mathbb{C}$. Then

$$J^* a_\rho = i(a^j_\rho \gamma_j - \overline{a^j_\rho}\, \overline{\gamma_j})$$

so that

$$a_\rho \wedge J^* a_\rho = -2i a^j_\rho \overline{a^k_\rho} \gamma_j \wedge \overline{\gamma_k}\,.$$

Thus ω_1 may be written

$$\omega_1 = 2i \sum_{\rho=1}^{n} a^j_\rho \overline{a^k_\rho} \gamma_j \wedge \overline{\gamma_k}\,.$$

One may use the similar expression of ω_2 so that to yield

$$\omega_1 + i\omega_2 = i(r^{j\bar{k}} + i s^{j\bar{k}}) \gamma_j \wedge \overline{\gamma}_k \tag{5.99}$$

where $r^{j\bar{k}} = 2\sum_{\rho=1}^n a^j_\rho \overline{a^k_\rho}$, $s^{j\bar{k}} = 2\sum_{\rho=1}^n b^j_\rho \overline{b^k_\rho}$. Let us set $R = (r^{j\bar{k}})$, $S = (s^{j\bar{k}})$. Then R, S are hermitian symmetric. Also $R \geq 0$, $S \geq 0$ (for instance $r^{j\bar{k}} \xi_j \overline{\xi_k} = 2\sum_\rho |b^j_\rho \xi_j|^2 \geq 0$). By (5.99)

$$(\omega_1 + i\omega_2)^p = p! i^p \det(R + iS) \gamma_1 \wedge \overline{\gamma_1} \wedge \cdots \wedge \gamma_p \wedge \overline{\gamma_p} \tag{5.100}$$

so that checking (5.98) amounts to showing that $R + iS$ is nonsingular. Note that $R + iS$ may be thought of as a bundle map

$$R + iS : L \to L^* \otimes \mathbb{C}.$$

Indeed, let $\{Z^1, \cdots, Z^p, \overline{Z^1}, \cdots, \overline{Z^p}\}$ be a frame of $L \otimes \mathbb{C} = L_{1,0} \oplus L_{0,1}$ such that $Z^j \in L_{1,0} = \mathrm{Eigen}(i)$, where $L_{0,1} = \overline{L_{1,0}}$, dual to $\{\gamma_1, \cdots, \gamma_p, \overline{\gamma}_1, \cdots, \overline{\gamma_p}\}$. Then let us think of R as the bundle map

$$R : L \to L^*, \quad R Z^j = r^{j\bar{k}} \overline{\gamma_k}\,, \quad R\overline{Z^k} = r^{j\bar{k}} \gamma_j\,.$$

Let $X = \xi_j Z^j + \overline{\xi_j}\, \overline{Z^j} \in L$. Let us assume that $X \in \mathrm{Ker}(R + iS)$. Then

$$\xi_j \overline{\xi_k} (r^{j\bar{k}} + i s^{j\bar{k}}) = 0. \tag{5.101}$$

Let us take the complex conjugate of (5.101) and use the hermitian symmetry of R, S. Next, let us add (respectively subtract) the resulting identity to (5.101). This procedure yields

$$r^{j\bar{k}} \xi_j \overline{\xi_k} = 0, \quad s^{j\bar{k}} \xi_j \overline{\xi_k} = 0.$$

At this point one may use the positivity of R, S. For instance

$$0 = r^{j\bar{k}} \xi_j \overline{\xi_k} = 2 \sum_{\rho=1}^{n} |a^j_\rho \xi_j|^2$$

so that $a^j_\rho \xi_j = 0$. Consequently $a_\rho X = 0$, $(J^* a_\rho)X = 0$. Similarly one may exploit $S \geq 0$ to show that $b_\rho = 0$, $(J^* b_\rho)X = 0$. Therefore $X \in A$. But the sum $A_x + L_x$ is direct so that $X = 0$. Thus $\operatorname{Ker}(R + iS) = 0$ so that (5.98) holds good. The fact that $(\omega_1 + i\omega_2)^{p+1} = 0$ is a consequence of (5.100). The proof of ii) $\Longrightarrow$ i) is similar to the above and therefore omitted.

REMARK 5.68. Let $(M, T_{1,0}(M))$ be a strictly pseudoconvex CR manifold of CR dimension n and θ a contact 1-form on M so that L_θ is positive definite. Let ω be a real closed $(1,1)$-form on M such that $T \rfloor \omega = 0$. From the discussion above it follows that if, for some $2 \leq p \leq n$, one has $\omega^p = 0$, $\omega^{p-1} \neq 0$ everywhere on M then $\operatorname{Ann}(\omega)$ is an integrable $(2n - 2p + 3)$-dimensional distribution which defines a pseudohermitian foliation $\mathcal{F}_p(\omega)$ whose leaves are CR manifolds of CR dimension $n - p + 1$. □

Let us look at two particular cases.

EXAMPLE 5.69. If $\omega = d\theta = i h_{\alpha\bar{\beta}} \theta^\alpha \wedge \theta^{\bar{\beta}}$ then $p = n$, that is $\omega^{n+1} = 0$ (as a $(2n+2)$-form) and $\omega^n \neq 0$ (by nondegeneracy). Consequently $\operatorname{Ann}(\omega)$ is the 1-dimensional distribution spanned by T (and the leaves of $\mathcal{F} = \mathcal{F}_n(d\theta)$ are the maximal integral curves of T). □

EXAMPLE 5.70. Let $u \in C^\infty(M)$ such that $T(u) = 0$. Assume that $\tau = 0$. Let $\omega = i\partial_M \overline{\partial}_M u$. It is a closed $(1,1)$-form by Lemma 5.62. Assume that u satisfies

$$(\partial_M \overline{\partial}_M u)^{n-1} \neq 0, \quad (\partial_M \overline{\partial}_M u)^n = 0$$

everywhere on M. The last equation is termed the *homogeneous tangential Monge-Ampère equation.* Then

$$\mathcal{F}_n(u) = \mathcal{F}_n(i\partial_M \overline{\partial}_M u)$$

is a pseudohermitian foliation on M whose leaves are 3-dimensional CR manifolds. In analogy with [**256**] we may call $\mathcal{F}_n(u)$ the *tangential Monge-Ampère foliation.* □

We may now prove Theorem 5.61. Let $\omega_1 = i\partial_M \overline{\partial}_M u$, $\omega_2 = i\partial_M \overline{\partial}_M v$. Then ω_j are real $(1,1)$-forms and (by hypothesis) $\omega_j \geq 0$. We may apply Lemma 5.62 to conclude that both ω_j are closed. Then $\operatorname{Ann}(\omega_1, \omega_2)$ is involutive. As $\omega_1 + i\omega_2 = i\partial_M \overline{\partial}_M f$ one may apply Lemma 5.67 to yield (with the notations there) $\dim_{\mathbb{R}} A_x = 2(n-p)+1$. As $\omega_j \geq 0$ (cf. also the proof of Lemma 5.67) $\operatorname{Ann}(\omega_1, \omega_2) = A$. Thus $\operatorname{Ann}(\omega_1, \omega_2)$ has constant dimension. By Frobenius theorem $\operatorname{Ann}(\omega_1, \omega_2)$ defines a foliation $\mathcal{F}$ of M of codimension $2p$. Let N be a leaf of $\mathcal{F}$ i.e. a maximal connected integral manifold of $\operatorname{Ann}(\omega_1, \omega_2)$. One has (by the very definition of ∂_M) that

$$T(x) \in \operatorname{Ann}(\omega_1, \omega_2)_x = T_x(N), \quad x \in N.$$

Thus T is tangent to N. Let $H(N)_x$ be the orthogonal complement of $\mathbb{R}T(x)$ in $T_x(N)$ (with respect to $g_{\theta,x}$). Then

$$H(N)_x \subset H(M)_x, \quad x \in N.$$

By Lemma 5.63 it follows that $\operatorname{Ann}(\omega_1, \omega_2)$ is J-invariant so that J restricts to a complex structure in $H(N)$. Thus N is a CR manifold of CR dimension $n - p$ (and

$H(N)$ is its Levi distribution). Let us endow N with the induced pseudohermitian structure $\iota^*\theta$ where $\iota : N \to M$ is the inclusion. Then N is seen to be strictly pseudoconvex and ι is isopseudohermitian. Thus we may apply Theorem 7 of [**84**], p. 189, to conclude that $\mathcal{F}$ is harmonic i.e. each leaf of $\mathcal{F}$ is minimal in (M, g_θ).

Let N be a leaf of $\mathcal{F}$. We wish to show that $u \circ \iota$ is $\overline{\partial}_N$-pluriharmonic. We need the following

LEMMA 5.71. *Let M, A be two strictly pseudoconvex CR manifolds and $f : M \to A$ a pseudohermitian immersion. Let $u \in C^\infty(A)$ and σ a $(0,1)$-form on A. Then*

$$\overline{\partial}_M(u \circ f) = f^*\overline{\partial}_A u, \quad \overline{\partial}_M(f^*\sigma) = f^*\overline{\partial}_A \sigma.$$

The proof of Lemma 5.71 follows from the axiomatic description of $\overline{\partial}_M$. We may use Lemma 5.71 to prove i) in Theorem 5.61. Indeed

$$\partial_N \overline{\partial}_N (u \circ \iota) = \iota^* \partial_M \overline{\partial}_M u = -i\iota^*\omega_1 = 0$$

as

$$T_x(N) = \mathrm{Ann}(\omega_1, \omega_2)_x \subseteq \mathrm{Ann}(\omega_1)_x\,, \quad x \in N.$$

To prove ii) let $u_\alpha = T_\alpha(u)$. Let N be a leaf of $\mathcal{F}$ and $Z \in T_{1,0}(N) \subseteq T_{1,0}(M)$. Then $Z = Z^\alpha T_\alpha = X - iJX$, for some $X \in H(N)$. Now on one hand $T_x(N) \subseteq \mathrm{Ann}(\omega_1)$ and on the other

$$J_x T_x(N) = J_x \mathrm{Ann}(\omega_1, \omega_2)_x \subseteq \mathrm{Ann}(\omega_1, \omega_2)_x \subseteq \mathrm{Ann}(\omega_1)_x\,, \quad x \in N,$$

so that $Z \in \mathrm{Ann}(\omega_1)$. Then

$$0 = Z \rfloor \omega_1 = iZ \rfloor \partial_M \overline{\partial}_M u = iZ \rfloor (u_{\bar\beta,\alpha}\theta^\alpha \wedge \theta^{\bar\beta}) = \frac{i}{2} u_{\bar\beta,\alpha} Z^\alpha \theta^{\bar\beta}\,.$$

Next

$$\overline{Z}(u_\alpha) = \overline{Z^\beta} T_{\bar\beta}(u_\alpha) = \overline{Z^\beta} u_{\alpha,\bar\beta} + \Gamma^\rho_{\bar\beta\alpha} \overline{Z^\beta} u_\rho$$

so that

$$\overline{Z}(u_\alpha) = \Gamma^\rho_{\bar\beta\alpha} \overline{Z^\beta} u_\rho\,.$$

Finally if $M = \mathbb{H}_n$ then u_α is CR-holomorphic on N (as $\Gamma^\rho_{\bar\beta\alpha} = 0$). The proof of Theorem 5.61 is complete.

We end this section by formulating a few open problems directly related to the material previously developed.

1) Study the geometry of the second fundamental form of the foliation by level sets of $f(z, u + iv) = \alpha(|z|^2 - v)e^u$, in the spirit of [**243**], p. 81-82 and p. 107, and related to the solutions to the Lewy equation $\partial u/\partial \bar z - 2z\beta(\rho)\, \partial u/\partial \overline{w} = f$ (see also [**128**]).

2) Compute the basic Kohn-Rossi cohomology $H^{0,\bullet}_B(\mathcal{F}_n(u))$ of a tangential Monge-Ampère foliation $\mathcal{F}_n(u)$, of a nondegenerate CR manifold M of CR dimension n, determined by a function $u \in C^\infty(M)$ satisfying $(\partial_M \overline{\partial}_M u)^{n-1} \neq 0$ and $(\partial_M \overline{\partial}_M u)^n = 0$.

3) Given a 3-dimensional conformal manifold $(M, [g])$, $N := \{\tau \in T^*(M) \otimes \mathbb{C} : g^*(\tau, \tau) = 0,\ \tau \neq 0\}/\mathbb{C}_*$ is a 5-dimensional nondegenerate CR manifold with the CR structure determined by the Hamilton form of $T^*(M) \otimes \mathbb{C}$ (cf. Section 2.7 in this book or [**165**], p. 608). The fibers of the projection $\pi : N \to M$ are the leaves of a smooth foliation $\mathcal{F}$ of N by $\mathbb{C}P^1$'s. Study the geometry of the

(pseudohermitian analogue of the) second fundamental form of $\mathcal{F}$ in N. Apply the results to R. Penrose's foliated CR manifold (cf. [**199**]) associated to a totally umbilical spacelike hypersurface of a 4-dimensional Lorentzian manifold.

4) Study the $d_{\mathcal{F}}$-cohomology and compute the Atiyah class $A(M, \mathcal{F})$ (cf. [**179**], p. 58-61) for a foliation $\mathcal{F}$ of a CR manifold M.

An array of results in complex analysis, regarding the complex Monge-Ampère equation on a complex manifold, suggest analogous CR notions, and await for a meaningful generalization. Let us recall (cf. W. Stoll, [**231**]) that a *strictly parabolic manifold* is a n-dimensional complex manifold M together with a strictly plurisubharmonic exaustion function $f : M \to [0, +\infty)$ such that $u = \log f$ satisfies the complex Monge-Ampère equation $(dd^c u)^n = 0$ on $M^* := M \setminus f^{-1}(0)$. Here $d^c = (1/(4i))(\partial - \overline{\partial})$. In the same paper W. Stoll conjectured that the only strictly parabolic manifold is, up to a biholomorphism, the pair $M = \mathbb{C}^n$ and $f(z) = |z|^2$. The conjecture was solved by W. Stoll himself (cf. [**232**]) and by D. Burns, [**57**], whose approach we briefly recall. Let u be a solution to $(dd^c u)^n = 0$ and let $\mathcal{F}$ be the foliation of M^* associated to u i.e. the foliation $\mathcal{F}$ tangent to the distribution defined by the Pfaffian equation $dd^c u = 0$ (the leaves of $\mathcal{F}$ are the integral manifolds of $\mathrm{Ker}(dd^c u)$). When f is strictly plurisubharmonic it may be shown (cf. [**57**]) that the leaves of $\mathcal{F}$ are totally geodesic in (M^*, g_f), where $g_f = \partial^2 f / \partial z^j \partial \overline{z}^k \, dz^j \odot d\overline{z}^k$ is the Kähler metric of potential f. The precise statement of the solution to Stoll's conjecture is

THEOREM 5.72. (D. Burns, [**57**])
Let M be a connected complex manifold, of complex dimension n, and $f : M \to [0, R^2)$ a C^∞ strictly plurisubharmonic exaustion function $(0 < R \leq +\infty)$ such that $u = \log f$ satisfies $(dd^c u)^n = 0$. Then there exists a biholomorphic map $\Phi : B(0, R) \to M$ such that $\Phi^* f = |z|^2$.

Here $B(0, R) = \{z \in \mathbb{C}^n : |z| < R\}$. The proof of Theorem 5.72 relies on a lemma of W. Stoll (cf. [**231**]) according to which for f as in Theorem 5.72 there is a unique point $O \in M$ such that $f^{-1}(0) = \{O\}$. The construction of Φ exploits the exponential mapping of the Kähler metric g_f. The main argument is to compare the behavior of the geodesics on the leaves of $\mathcal{F}$ to their behavior in M, using the fact that each leaf of $\mathcal{F}$ is totally geodesic together with a technique of Jacobi fields. All these notions admit well defined CR analogs and the study of the geometric properties of the (CR analog to the) foliation $\mathcal{F}$ is interesting in its own right. Proving a CR analog to Theorem 5.72 above is an open question. See also E. Bedford & D. Burns, [**31**], P-M. Wong, [**256**], T. Duchamp & M. Kalka, [**92**].

CHAPTER 6

Transversally CR foliations

The geometry of foliations of C^∞ manifolds possessing a specified additional transverse structure, e.g. a Riemannian (cf. B. Reinhart, [**204**]) or conformal (cf. L. Conlon, [**70**]) or contact (cf. W. L. Ting, [**240**]) structure, etc., has enjoyed a large interest during recent years. Cf. also C. Godbillon, [**121**], p. 141-143. These structures are supported by the normal (or transverse) bundle of the foliation and are requested to satisfy a holonomy invariance condition (for instance a Riemannian foliation is one whose normal bundle carries a Riemannian metric invariant under sliding along the leaves (cf. P. Tondeur, [**243**], p. 51)) and A. Haefliger's (cf. [**130**]), by now classical, approach to foliations (as Γ-foliations) provides the elegant means for a unified treatment. In this setting a Riemannian foliation is a $\Gamma(N)$-foliation, where $\Gamma(N)$ is the pseudogroup of all local isometries of a (model) Riemannian manifold N. Particularly deep results (e. g. the finite dimensionality of the basic cohomology of a Riemannian foliation, cf. A. El Kacimi-Alaoui & V. Sergiescu & G. Hector, [**147**]) were obtained by reducing the study of Riemannian foliations to the study of transversally parallelizable (cf. P. Molino, [**178**]) and transversally Lie (cf. F. Fedida, [**103**]) foliations. See also N. Abe, cf. [**1**], [**2**], for a study of characteristic classses of Γ-foliations. Also see S. Nishikawa & P. Tondeur & L. Vanhecke, [**191**], for an attempt to describe Riemannian foliations by spectral invariants.

In this spirit, the purpose of the present chapter is to study the geometry of $\Gamma^r_{CR}(N)$-foliations (with $r = \infty$ or $r = \omega$) where $\Gamma^r_{CR}(N)$ is the pseudogroup of all local CR automorphisms of class C^r of a given (model) CR manifold N. These are called *CR foliations* and CR manifolds correspond to the case of the trivial CR foliation by points.

Our motivation comes from the theory of complex (and Levi) foliations on CR manifolds (cf. e. g. E. M. Chirka, [**67**], p. 150). A complex foliation $\mathcal{F}$ of a CR manifold $(M, H(M))$ (where $H(M)$ is its Levi distribution) is one whose tangent bundle P is a complex subbundle of $H(M)$ and whose foliated charts restricted to plaques give biholomorphisms. The quotient $H(M)/P$ carries a natural complex structure J and in cases of interest (cf. our Theorem 6.27) J is parallel with respect to the Bott connection of $\mathcal{F}$, so that $\mathcal{H} = \text{Eigen}(i)$ (the eigenbundle of J corresponding to the eigenvalue $i = \sqrt{-1}$) is a *transverse almost CR structure.* Moreover $\mathcal{H}$ is *integrable* (for any $x \in M$ there is an open neighborhood U and an *admissible* frame $\{\zeta_\alpha\}$ of $\mathcal{H}$ on U, that is each ζ_α is a transverse vector field and $[\zeta_\alpha, \zeta_\beta] \in \mathcal{H}$) in most examples at hand.

To further motivate our line of thought, let us recall (cf. [**67**], p. 155, or S. I. Pinchuk & S. I. Tsyganov, [**202**]) that

DEFINITION 6.1. A complex foliation $\mathcal{F}$ of complex dimension k of a CR manifold M is *CR-straightenable* if there are a domain $\Omega \subset \mathbb{C}^k$, a CR manifold N, and a CR diffeomorphism $\varphi : \Omega \times N \to M$ such that $\varphi(\Omega \times \{p\})$ is a leaf of $\mathcal{F}$, for any $p \in N$. □

If $\mathcal{F}$ is CR-straightenable then the transverse geometry of $\mathcal{F}$ is modelled on N i.e. $\mathcal{F}$ is a $\Gamma^{\infty}_{CR}(N)$-foliation. See also M. Freeman, [**113**].

In the end, it is worth mentioning that the notion of (transversally) CR foliation is implicit in [**67**], p. 157. There, a CR foliation is a foliation $\mathcal{F}$ of a CR manifold M such that for any defining local submersion $f : U \to U'$ (i. e. the leaves of $\mathcal{F}_U$ are the fibres of f) the local quotient manifold U' is a CR manifold, f is a CR map, and $df : H(U) \to H(U')$ is onto. Such $\mathcal{F}$ carries a transverse CR structure. Yet, on one hand $\Gamma^{\infty}_{CR}(N)$-foliations make sense on arbitrary C^{∞} manifolds (not just on CR manifolds). On the other, the requirement that $f : U \to U'$ be CR is somewhat misleading. Indeed this yields a tangential CR structure (so that each leaf becomes a CR submanifold of M) thus prompting the choice of terminology (CR foliations) in [**67**] (the transverse CR structure is not looked at there). Also (at least in the CR codimension one case) the Levi form of M must have a nontrivial kernel. Our point of view is, of course, that the tangential CR structure of $\mathcal{F}$ is only incidental, and that E. Chirka's approach to CR foliations (requiring that the local quotient manifolds possess some G-structure) is just the typical manner (cf. e. g. Proposition 2.6 in [**179**], p. 51-52) of assigning a transverse G-structure to $\mathcal{F}$.

Let N be a CR manifold. The group $\mathrm{Aut}_{CR}(N)$ of all global CR automorphisms of N is a Lie transformation group (by a result of S. S. Chern & J. Moser, [**65**]). We discuss foliations defined by suspension of a homomorphism $h : \pi_1(B, x_0) \to \mathrm{Aut}_{CR}(N)$. These turn out to be CR foliations with nontrivial holonomy, cf. Theorem 6.14.

When the normal bundle of the given CR foliation has odd real rank we develop a foliated analogue of S.M. Webster's (cf. [**250**]) pseudohermitian geometry, cf. our Theorems 6.15 and 6.20. We introduce notions such as *transverse pseudohermitian structrure*, *transverse Levi form*, and *transverse Webster metric* g_θ, as well as notions of (transverse) nondegeneracy and strict pseudoconvexity. In the nondegenerate case, the transverse Webster metric g_θ is a transverse metric (in the sense of [**179**], p. 77) for $\mathcal{F}$ and thus there is a bundle-like semi-Riemannian metric g on M inducing g_θ. Our main result in this direction is that there is an adapted connection ∇ in the normal bundle of the given nondegenerate CR foliation $\mathcal{F}$ which parallelizes both the transverse Levi form and the complex structure in the transverse Levi distribution (∇ is unique under some assumption on its torsion, cf. Theorem 6.39). In addition, ∇ does not depend upon the choice of bundle-like semi-Riemannian metric g (inducing the transverse Webster metric) used in its construction. For the case of a CR foliation by points ∇ is the Tanaka-Webster connection.

We show that any CR foliation comes equipped with a natural differential operator $\overline{\partial}_Q$ (a foliated analogue of the tangential Cauchy-Riemann operator in complex analysis) acting on transverse $(0, k)$-forms. We look at the cohomology of the resulting $\overline{\partial}_Q$-complex. For the case of a simple CR foliation defined by a submersion this cohomology turns out to be the Kohn-Rossi cohomology of the base CR manifold (cf. Theorems 6.30 and 6.35).

Let $\mathcal{F}$ be a CR foliation of type (n, k) of M and $\mathcal{H}$ its transverse CR structure. We introduce a concept of *embedding* of $(M, \mathcal{H})$. This is essentially an immersion

$\psi : M \to \mathbb{C}^N$ for some $N > n + k$ which induces a bundle monomorphism G of the normal bundle into $T(\mathbb{C}^{n+k})$ such that G maps $\mathcal{H}$ into the holomorphic tangent bundle over $\mathbb{C}^{n+k}$. Any real analytic transverse CR structure is shown (cf. our Theorem 6.44) to be locally embeddable.

6.1. Transversally CR foliations

Let M be a C^∞ manifold and $\mathcal{F}$ a codimension q foliation ($q = 2n + k$, $k \geq 1$) of class C^∞ of M. Let $P = T(\mathcal{F}) \subset T(M)$ be the tangent bundle of $\mathcal{F}$. Let $Q = \nu(\mathcal{F}) = T(M)/P$ be the normal (or transverse) bundle of $\mathcal{F}$ and $\Pi : T(M) \to Q$ the natural bundle epimorphism. Let $\overset{\circ}{\nabla}$ be the Bott connection of $(M, \mathcal{F})$. The following notions are central for the rest of the present chapter.

DEFINITION 6.2. Let $\mathcal{H} \subset Q \otimes \mathbb{C}$ be a complex subbundle of complex rank n such that 1) $\mathcal{H} \cap \overline{\mathcal{H}} = \{0\}$. Then $H := \mathrm{Re}\{\mathcal{H} \oplus \overline{\mathcal{H}}\}$ is the *transverse Levi distribution*. We call $\mathcal{H}$ a *transverse almost CR structure* (of *transverse CR dimension* n) if 2) H is parallel with respect to the Bott connection of $\mathcal{F}$ i.e.

$$\overset{\circ}{\nabla}_X \Gamma^\infty(H) \subseteq \Gamma^\infty(H), \quad X \in \Gamma^\infty(P),$$

and 3) $\mathcal{L}_X J = 0$ for any $X \in \Gamma^\infty(P)$. Here $J : H \to H$ is the complex structure given by

$$J(\alpha + \overline{\alpha}) = i(\alpha - \overline{\alpha}), \quad \alpha \in \Gamma^\infty(\mathcal{H}). \tag{6.1}$$

Also $i = \sqrt{-1}$. □

Lie derivatives are defined with respect to $\overset{\circ}{\nabla}$ e.g.

$$(\mathcal{L}_X J)s = \overset{\circ}{\nabla}_X (Js) - J \overset{\circ}{\nabla}_X s, \quad s \in \Gamma^\infty(Q).$$

Also, if $\omega \in \Gamma^\infty(\Lambda^k Q^*)$ and $s_1, \cdots, s_k \in \Gamma^\infty(Q)$ then we set

$$(\mathcal{L}_X \omega)(s_1, \cdots, s_k) = X(\omega(s_1, \cdots, s_k)) -$$

$$- \sum_{j=1}^{k} \omega(s_1, \cdots, s_{j-1}, \overset{\circ}{\nabla}_X s_j, s_{j+1}, \cdots, s_k)$$

for any $X \in \Gamma^\infty(P)$.

Let $V(\mathcal{F}) = V(M, \mathcal{F}) \subset \mathcal{X}(M)$ be the Lie subalgebra of all foliate vector fields (or infinitesimal automorphisms of $\mathcal{F}$) and $\ell(\mathcal{F}) = \ell(M, \mathcal{F}) \subset \Gamma^\infty(Q)$ the subset of all transverse vector fields. Let $\Gamma^\infty_B(Q)$ consist of all $s \in \Gamma^\infty(Q)$ with $\mathcal{L}_X s = 0$ for any $X \in \Gamma^\infty(P)$. Note that

$$\Gamma^\infty_B(Q) = \ell(\mathcal{F}) \tag{6.2}$$

(so that the Lie bracket $[s, r]$ of any $s, r \in \Gamma^\infty_B(Q)$ is well defined). Indeed if $s \in \Gamma^\infty(Q)$ is invariant under sliding along the leaves then each $Y \in \mathcal{X}(M)$ with $\Pi Y = s$ is foliate (as $\Pi[X, Y] = \overset{\circ}{\nabla}_X \Pi Y = \mathcal{L}_X s = 0$) so that s is a transverse vector field, and conversely.

DEFINITION 6.3. A transverse almost CR structure $\mathcal{H} \subset Q \otimes \mathbb{C}$ is termed *integrable* if for any $x \in M$ there is an open neighborhood $U \subseteq M$, $x \in U$, and there is a frame $\{\zeta_1, \cdots, \zeta_n\}$ of $\mathcal{H}$ on U such that $\zeta_\alpha \in \Gamma^\infty_B(Q \otimes \mathbb{C})$ and $[\zeta_\alpha, \zeta_\beta] \in \Gamma^\infty(\mathcal{H})$ for any $1 \leq \alpha, \beta \leq n$. Such a (local) frame of $\mathcal{H}$ is termed *admissible*. An integrable transverse almost CR structure is referred to as a *transverse CR structure* on $(M, \mathcal{F})$. □

PROPOSITION 6.4. *Let $\mathcal{H} \subset Q \otimes \mathbb{C}$ be a transverse almost CR structure such that*

$$[\Gamma_B^\infty(\mathcal{H}), \Gamma_B^\infty(\mathcal{H})] \subseteq \Gamma_B^\infty(\mathcal{H}).$$

Then $\mathcal{H}$ is integrable.

Proof. Let $x_0 \in M$ and $f : U \to U'$ a local defining submersion of $\mathcal{F}$ with $x_0 \in U$. We claim that U' is a CR manifold. Indeed let $y \in U'$ and $x \in U$ such that $f(x) = y$. Let us set

$$T_{1,0}(U')_y = (d_T f)_x \mathcal{H}_x \, .$$

The definition of $T_{1,0}(U')_y$ does not depend upon the choice of $x \in f^{-1}(y)$. To check this statement, let $x' \in f^{-1}(y)$ be another point over y and

$$C : [0,1] \to f^{-1}(y), \quad C(0) = x, \quad C(1) = x'.$$

a curve in the fibre over y, joining x and x'. Next let $s(t)$ be the solution to the Cauchy problem

$$\begin{cases} \mathring{\nabla}_{dC/dt} s = 0, \\ s(0) = z \in Q_x. \end{cases}$$

Let $\tau_C : Q_x \to Q_{x'}$ be the linear isomorphism given by $\tau_C(z) = s(1)$. Then

$$(d_T f)_{x'} \circ \tau_C = (d_T f)_x$$

(this proves that $T_{1,0}(U')_y$ is well defined because $\tau_C(\mathcal{H}_x) = \mathcal{H}_{x'}$ (as $\mathcal{H}$ is parallel with respect to $\mathring{\nabla}$)). Indeed let $Y \in T(M)$ such that $\Pi Y = s$. Then

$$\Pi \left[\frac{dC}{dt}, Y \right] = 0 \, , \; (df) \, \frac{dC}{dt} = 0$$

hence

$$\left[\frac{dC}{dt}, Y \right] \in T(\mathcal{F}) = \mathrm{Ker}(df).$$

Let $h : U' \to \mathbb{R}$ be a C^∞ function. Then $(df)[dC/dt, Y] = 0$ yields

$$0 = [\frac{dC}{dt}, Y](h \circ f) = \frac{dC}{dt}(Y(h \circ f)) - Y(\frac{dC}{dt}(h \circ f))$$

i.e. $(dC/dt)(Y(h \circ f)) = 0$. Finally

$$0 = \frac{d}{dt}\{Y(h \circ f)\} = \frac{d}{dt}\{(f_* Y)h\}$$

so that

$$(d_{C(0)} f) Y(0) = (d_{C(1)} f) Y(1).$$

Q.e.d. Next $T_{1,0}(U')$ is integrable because $\Gamma^\infty(T_{1,0}(U') \approx \Gamma_B^\infty(\mathcal{H})$. Finally one may lift to U a local frame of $T_{1,0}(U')$ at $f(x_0)$.

DEFINITION 6.5. A Γ-*foliation of codimension q and class C^∞* on M consists of the following data i) an open covering $\{U_i\}_{i \in I}$ of M, ii) an additional C^∞ manifold N and a pseudogroup Γ of local transformations of N, iii) for each $i \in I$ a C^∞ submersion $f_i : U_i \to N$, iv) for any $i, j \in I$ (with $U_{ji} = U_i \cap U_j \neq \emptyset$) an element $\gamma_{ji} \in \Gamma$ so that $f_j = \gamma_{ji} \circ f_i$ on U_{ji}. □

Cf. [**130**]. Given a Γ-foliation as in Definition 6.5 let $x \in M$ and $i \in I$ with $x \in U_i$ and let us set

$$P_x = \text{Ker}(d_x f_i).$$

Then $P \subset T(M)$ is a well defined (by iv)) integrable distribution so that any Γ-foliation gives rise to a foliation $\mathcal{F}$ of M. Conversely, by the classical Frobenius theorem any foliation $\mathcal{F}$ of M gives rise to a Γ-foliation with $N = \mathbb{R}^q$ and Γ the pseudogroup of all local C^∞ diffeomorphisms of $\mathbb{R}^q$.

Let N be a $(2n+k)$-dimensional C^∞ manifold. Let $T_{1,0}(N)$ be a CR structure of CR dimension n on N. Let $\Gamma^\infty_{CR}(N)$ be the pseudogroup of all local CR automorphisms of $(N, T_{1,0}(N))$, of class C^∞.

DEFINITION 6.6. Let $\mathcal{F}$ be a $\Gamma^\infty_{CR}(N)$-foliation of M. Then $\mathcal{F}$ is said to be a *(transversally) CR foliation* of *transverse CR dimension* n and *transverse CR codimension* k. □

Let $\mathcal{F}$ be a CR foliation and $x \in U_i$. The differential $d_x f_i : T_x(M) \to T_{f_i(x)}(N)$ descends to a $\mathbb{R}$-linear isomorphism $F_{i,x} : Q_x \to T_{f_i(x)}(N)$ with $F_{i,x} \circ \Pi_x = d_x f_i$. Let us set

$$H_x = F^{-1}_{i,x} H(N)_{f_i(x)}\,.$$

As $\gamma_{ji} \in \Gamma^\infty_{CR}(N)$ one has in particular $(d\gamma_{ji})H(N) = H(N)$ so that H_x is well defined. It carries the complex structure J_x given by

$$J_x = F^{-1}_{i,x} \circ J_{N, f_i(x)} \circ F_{i,x}\,.$$

Once again, as $\gamma_{ji} \in \Gamma^\infty_{CR}(N)$, in particular $(d\gamma_{ji}) \circ J_N = J_N \circ (d\gamma_{ji})$ so that J_x is well defined.

DEFINITION 6.7. H is referred to as the *transverse Levi distribution* of $(M, \mathcal{F})$. □

We may state the following

THEOREM 6.8. *Let $(N, T_{1,0}(N))$ be a CR manifold of class C^∞ and type (n,k). Let $\mathcal{F}$ be a CR foliation of M whose transverse geometry is modelled on $(N, T_{1,0}(N))$. Let H be the transverse Levi distribution of $\mathcal{F}$. Let us extend J to $H \otimes \mathbb{C}$ by $\mathbb{C}$-linearity and set $\mathcal{H} = \text{Eigen}(i)$. Then $\mathcal{H}$ is a transverse CR structure of transverse CR dimension n (and $F_{i,x}(\mathcal{H}_x) = T_{1,0}(N)_{f_i(x)}$ for any $x \in U_i$).*

Proof. Let $X \in \Gamma^\infty(P)$. Let $(\varphi_t)_{|t|<\epsilon}$ be the local 1-parameter group of local transformations of M got by locally integrating X. Let $x \in M$ and $C_x : (-\epsilon, \epsilon) \to M$ given by $C_x(t) = \varphi_t(x)$, $|t| < \epsilon$. Then $X_{C_x(t)} = \dot{C}_x(t)$, $|t| < \epsilon$. Let $i \in I$ with $x \in U_i$. Then $C_x(t) \in U_i$ for any $|t| < \epsilon_i$ and some $0 < \epsilon_i \leq \epsilon$. As X restricted to U_i is vertical (i.e. tangent to the fibres of f_i) it follows by a standard argument that $f_i \circ C_x = const.$ (with respect to t). Thus

$$f_i \circ \varphi_t = f_i \;\;,\;\; |t| < \epsilon_i\,. \tag{6.3}$$

Let $s \in \Gamma^\infty(H)$ and $Y_s \in \mathcal{X}(M)$ with $\Pi Y_s = s$. Then $(d_x f_i) Y_s(x) = F_{i,x} s(x) \in H(N)_{f_i(x)}$. Finally

$$[X, Y_s]_x = \lim_{t \to 0} \frac{1}{t} \left\{ Y_s(x) - (d_{\varphi_t(x)} \varphi_{-t}) Y_s(\varphi_t(x)) \right\}$$

so that (by (6.3))

$$F_{i,x}(\mathring{\nabla}_X s)_x = \lim_{t \to 0} \frac{1}{t} \left\{ (d_x f_i) Y_s(x) - (d_{\varphi_t(x)} f_i) Y_s(\varphi_t(x)) \right\} \in H(N)_{f_i(x)}\,.$$

Thus H is parallel with respect to the Bott connection of $\mathcal{F}$. Next we conduct the following calculation

$$(\mathcal{L}_X J)_x s(x) = (\mathcal{L}_X J_s)_x - (J\mathcal{L}_X s)_x = \Pi_x[X, Y_{Js}]_x - J_x \Pi_x [X, Y_s]_x =$$

$$= \lim_{t\to t} \frac{1}{t}\left\{J_x \Pi_x (d_{\varphi_t(x)}\varphi_{-t}) Y_s(\varphi_t(x)) - \Pi_x (d_{\varphi_t(x)}\varphi_{-t}) Y_{Js}(\varphi_t(x))\right\}$$

for any $s \in \Gamma^\infty(H)$ and some $Y_s, Y_{Js} \in \mathcal{X}(M)$ with $\Pi Y_s = s$, $\Pi Y_{Js} = Js$. Note that P is φ_t-invariant. Thus $d_x\varphi_t : T_x(M) \to T_{\varphi_t(x)}(M)$ induces a $\mathbb{R}$-linear isomorphism $L_{x,t} : Q_x \to Q_{\varphi_t(x)}$ which maps H_x onto $H_{\varphi_t(x)}$. As a consequence of definitions

$$L_{\varphi_t(x),-t} \circ \Pi_{\varphi_t(x)} = \Pi_x \circ (d_{\varphi_t(x)}\varphi_t). \tag{6.4}$$

Thus (by (6.4))

$$(\mathcal{L}_X J)_x s(x) = \lim_{t\to 0} \frac{1}{t}\left\{J_x L_{\varphi_t(x),-t} - L_{\varphi_t(x),-t} J_{\varphi_t(x)}\right\} s(\varphi_t(x)) =$$

$$= \lim_{t\to 0} \frac{1}{t}\Big\{F_{i,x}^{-1} J_{N,f_i(x)} F_{i,x} L_{\varphi_t(x),-t} -$$

$$- L_{\varphi_t(x),-t} F^{-1}_{i,\varphi_t(x)} J_{N,f_i(\varphi_t(x))} F_{i,\varphi_t(x)}\Big\} s(\varphi_t(x)) = 0$$

where the last equality follows from the commutativity of the diagram

$$\begin{array}{ccccc} & H_{\varphi_t(x)} & \stackrel{L_{\varphi_t(x),-t}}{\longrightarrow} & H_x & \\ F_{i,\varphi_t(x)} & \downarrow & & \downarrow & F_{i,x} \\ & H(N)_{f_i(\varphi_t(x))} & = & H(N)_{\varphi_t(x)} & \end{array}$$

Thus J (the complex structure of the transverse Levi distribution) is invariant under sliding along the leaves. See also [**179**], p. 24.

Let $x_0 \in U_i$ and set $p_0 = f_i(x_0) \in N$. There is an open set $V \subset N$, $p_0 \in V$, and a frame $\{Z_1, \cdots, Z_n\}$ of $T_{1,0}(N)$ on V. Let us set $U = f_i^{-1}(V) \subset U_i$ and $\zeta_\alpha(x) = F_{i,x}^{-1} Z_\alpha(f_i(x))$ for any $x \in U$. Let $X \in \Gamma^\infty(P)$ and let T_α be a complex tangent vector field on U such that $\Pi T_\alpha = \zeta_\alpha$. Then

$$(\mathcal{L}_X \zeta_\alpha)_x = \Pi_x [X, T_\alpha]_x =$$

$$= \lim_{t\to 0}\frac{1}{t}\left\{\zeta_\alpha(x) - \Pi_x(d_{\varphi_t(x)}\varphi_t) T_\alpha(\varphi_t(x))\right\} = 0$$

so that $\zeta_\alpha \in \Gamma^\infty_B(Q\otimes\mathbb{C})$ i.e. T_α is foliate. Thus (by (6.2)) the Lie bracket $[\zeta_\alpha, \zeta_\beta]$ is well defined. On the other hand Z_α and T_α are f_i-related, and thus $\{\zeta_1, \cdots, \zeta_n\}$ follows to be an admissible frame of $\mathcal{H}$ on U. Theorem 6.8 is completely proved.

Before going any further we wish to give an example of a CR foliation which is imitative of the CR structure of a real hypersurface in a complex ambient space (cf. Proposition 6.11 below).

DEFINITION 6.9. By a *transversally Kählerian* foliation $\mathcal{F}$ on a C^∞ manifold M we mean a foliation satisfying the following conditions i) $\mathcal{F}$ is Riemannian, with a bundle-like metric g on M inducing the holonomy invariant metric g_Q on $Q = \nu(\mathcal{F}) \approx P^\perp$, ii) there is a holonomy invariant almost complex structure $J : Q \to Q$ (where $q = \text{cod}(\mathcal{F}) = 2n$), with respect to which g_Q is Hermitian, i.e. $g_Q(Jr, Js) = g_Q(r, s)$ for any $r, s \in Q$, and iii) if ∇ is the unique metric and torsion free connection in Q (cf. e.g. [**243**], p. 48) then ∇ is almost complex, i.e. $\nabla J = 0$. □

The notion is due to S. Nishikawa & P. Tondeur, cf. [**189**], where they study the Lie algebra of transversally holomorphic fields of a harmonic transversally Kählerian foliation of a closed orientable manifold (generalizing to the foliation context a result by A. Lichnerowicz, [**173**]). For instance we have the following

EXAMPLE 6.10. Let $\mathbb{C}H^n$ be the complex hyperbolic space i.e. the quotient space of the anti de Sitter space H_1^{2n+1} under the canonical circle action, cf. [**155**], vol. II, p. 282. Here

$$H_1^{2n+1} = \{z = (z_0, \cdots, z_n) \in \mathbb{C}^{n+1} : (z, z) = -1\},$$

$$(z, w) = -z_0 \overline{w}_0 + \sum_{j=1}^{n} z_j \overline{w}_j \, .$$

The induced metric on H_1^{2n+1} has signature $(1, 2n)$ and constant sectional curvature -1. The canonical S^1-action on H_1^{2n+1} gives rise to a fibration

$$S^1 \to H_1^{2n+1} \longrightarrow \mathbb{C}H^n$$

and the corresponding simple foliation $\mathcal{F}$ (tangent to the vertical distribution) is a harmonic transversally Kählerian foliation (actually, each leaf of $\mathcal{F}$ is totally geodesic in H_1^{2n+1}). □

See also L.A. Cordero & R. Wolak, [**72**] (discussing the basic cohomology of transversally Kählerian foliations by means of a basic version of the Frölicher spectral sequence) and Chapter V of the monograph [**255**] (on transversally Hermitian and transversally Kählerian foliations).

PROPOSITION 6.11. *Let $\mathcal{F}$ be a transversally Kählerian foliation of a C^∞ manifold M. Let M' be an orientable real hypersurface of M. If M' is a saturated subset of $(M, \mathcal{F})$ then the induced foliation $\mathcal{F}'$ on M' is a transversally CR foliation.*

Proof. Let g be a bundle-like Riemannian metric on M inducing the Hermitian metric g_Q on (Q, J) as above. Let U be a unit normal field on M' and let us set

$$\xi = \Pi U \in \Gamma^\infty(Q'^\perp)$$

where $Q'^\perp$ is the orthogonal complement of $Q' = \nu(\mathcal{F}')$ in (Q, g_Q). Then $J\xi \in Q'$. Indeed this follows from $g_Q(J\xi, \xi) = 0$. Let H_x be the orthogonal complement of $\mathbb{R}J_x\xi_x$ in Q'_x, for any $x \in M'$. Then $JH = H$. Indeed given $s \in H$ one has

$$g_Q(Js, J\xi) = g_Q(s, \xi) = 0.$$

Clearly ξ is holonomy invariant. Next we should check that H is parallel with respect to the Bott connection $\mathring{\nabla}'$ of $\mathcal{F}'$, and that $\mathcal{L}_X J' = 0$ for any $X \in T(\mathcal{F}')$ where J' is induced by J on H. For instance

$$g_{Q'}(\mathring{\nabla}'_X s, J\xi) = g_Q(\mathring{\nabla}_X s, J\xi) = g_Q(\nabla_X s, J\xi) =$$

$$= -g_Q(\nabla_X Js, \xi) = -g_Q(\mathring{\nabla}_X Js, \xi) = -g_Q(\mathring{\nabla}'_X Js, \xi) = 0$$

for any $X \in T(\mathcal{F}')$ and $s \in H$. Finally it suffices to show that

$$[\Gamma^\infty_B(\mathcal{H}), \Gamma^\infty_B(\mathcal{H})] \subseteq \Gamma^\infty_B(\mathcal{H}).$$

Indeed let $\alpha = u - iJu$ and $\beta = v - iJv$ with $u, v \in H$. Then

$$[\alpha, \beta] = [u, v] - [Ju, Jv] - i\{[u, Jv] + [Ju, v]\}.$$

Then on one hand the fact that ∇ is torsion free yields

$$g_Q([u, v] - [Ju, Jv], J\xi) = 0$$

hence $[u, v] - [Ju, Jv] \in H$, and on the other $\nabla J = 0$ yields

$$[Ju, Jv] - [u, v] - J\{[Ju, v] + [u, Jv]\} = 0.$$

Q.e.d.

One may look at transverse (almost) CR structures as transverse G-structures, as well. Let $\mathcal{F}$ be a codimension q foliation of M. Let

$$B_T^1 = B_T^1(M, \mathcal{F})$$

be the (total space of the) principal $\mathrm{GL}(q, \mathbb{R})$-bundle of transverse frames and $G \subset \mathrm{GL}(q, \mathbb{R})$ a Lie subgroup.

DEFINITION 6.12. A transverse G-structure $E \subset B_T^1$ is *locally flat* if for any $x \in M$ there is a foliated coordinate chart $(U, x^1, \cdots, x^p, y^1, \cdots, y^q)$ at x such that $\sigma_T(U) \subset E$. Here $\sigma_T : U \to B_T^1$ is the natural field of transverse frames

$$\sigma_T(x) = (x, \{(\Pi \frac{\partial}{\partial y^j})_x : 1 \leq j \leq q\}), \quad x \in U.$$

□

Let $\mathcal{H}$ be a transverse almost CR structure of type (n, k) and (H, J) the corresponding transverse Levi distribution. Let E_x consist of all $\mathbb{R}$-linear isomorphisms $z : \mathbb{R}^q \to Q_x$ with $z(e_\alpha) \in H_x$ and $J_x z(e_\alpha) = z(e_{\alpha+n})$. Here $q = 2n + k$ while $\{e_\alpha, e_{\alpha+n}, e_{j+2n}\}$ denotes the canonical linear basis in $\mathbb{R}^q$ (with $1 \leq \alpha \leq n$, $1 \leq j \leq k$). Let $G \subset \mathrm{GL}(q, \mathbb{R})$ consist of all nonsingular matrices of the form

$$\begin{bmatrix} g^\alpha_\beta & \Omega^\alpha_\beta & u^\alpha_j \\ -\Omega^\alpha_\beta & g^\alpha_\beta & v^\alpha_j \\ 0 & 0 & w^i_j \end{bmatrix}$$

THEOREM 6.13. *E is a transverse G-structure. If E is locally flat then $\mathcal{H}$ is integrable.*

Proof. Let $z \in E$ and $x = p_T(z) \in M$. Let $U \subset M$ be a simple open subset, $x \in U$, and $f : U \to N$ a local defining submersion (so that $T(\mathcal{F})_{|U} = \mathrm{Ker}(df)$). Let $B^1(N)$ be the bundle of linear frames tangent to N and $f_T : B_T^1(U, \mathcal{F}_U) \to B^1(N)$ the natural bundle map (cf. also [**179**], p. 47). Then f_T is a local defining submersion for $\mathcal{F}_T^1$ (so that $P^1_{T,z} = Ker(d_z f_T)$ for any $z \in B_T^1(U, \mathcal{F}_U)$). Let $X \in \mathcal{X}(p_T^{-1}(U))$ and $(\varphi_t)_{|t|<\epsilon}$ the local 1-parameter group of local transformations of B_T^1 induced by X. Then $f_T(C_z(t)) = f_T(z)$ where $C_z(t) = \varphi_t(z)$, $|t| < \epsilon$. Let $z = (x, \{s_j\}) \in E$ and $C_z(t) = (x(t), \{s_j(t)\})$. Then $s_\alpha(t) = F^{-1}_{x(t)} F_x s_\alpha \in H_{x(t)}$ (as $F_x s_\alpha \in H(N)_x$). Next $f(x(t)) = f(x)$ and $J_{x(t)} s_\alpha(t) = s_{\alpha+n}(t)$ (by the very definition of J) so that C_z lies in E, i.e. X is tangent to E. If E is locally flat then for any $x \in M$ there is a foliated coordinate chart $(U, x^1, \cdots, x^p, y^1, \cdots, y^q)$ at x such that $\zeta_\alpha = \Pi(\partial/\partial y^\alpha - i\partial/\partial y^{\alpha+n})$ is a frame of $\mathcal{H}$ on U, $1 \leq \alpha \leq n$. Clearly $\partial/\partial y^\alpha - i\partial/\partial y^{\alpha+n} \in L(\mathcal{F}) \otimes \mathbb{C}$ and $[\zeta_\alpha, \zeta_\beta] = 0$ so that $\{\zeta_\alpha\}$ is admissible. Q.e.d.

6.2. CR foliations built by suspension

Let B, T be two compact connected manifolds and

$$h : \pi_1(B, x_0) \to \mathrm{Diff}(T)$$

a group homomorphism. Here $\pi_1(B, x_0)$ is the first homotopy group of B (with base point $x_0 \in B$) and $\mathrm{Diff}(T)$ denotes the group of all global diffeomorphisms of T. Set $G = h(\pi_1(B, x_0))$. We may state the following

THEOREM 6.14. *Let $\mathcal{F}$ be a CR foliation (whose transverse geometry is modelled on the CR manifold $(N, T_{1,0}(N))$ of the C^∞ manifold M. If $\mathcal{F}$ is defined by suspension of $h : \pi_1(B, x_0) \to \mathrm{Diff}(T)$ then T is a CR manifold of type (n, k) and G a group of global CR automorphisms of T. Conversely, if $\mathrm{Aut}_{CR}(T)$ is the group of all global CR automorphisms of a CR manifold $(T, T_{1,0}(T))$ then the foliation $\mathcal{F}$ defined by suspension of $h : \pi_1(B, x_0) \to \mathrm{Aut}_{CR}(T)$ is a CR foliation.*

Proof. Let $\hat{p} : \hat{B} \to B$ be the universal cover of B. Set $\tilde{M} = \hat{B} \times T$. There is a natural action of $\pi_1(B, x_0)$ on $\tilde{M}$ given by

$$(\hat{x}, y) \cdot [\gamma] = (\hat{x} \cdot [\gamma], h([\gamma]^{-1})(y))$$

for any $\hat{x} \in \hat{B}$, $y \in T$ and $[\gamma] \in \pi_1(B, x_0)$. Let

$$\rho : \tilde{M} \to \tilde{M}/\pi_1(B, x_0)$$

be the canonical projection. Let $\tilde{\mathcal{F}}$ be the (simple) foliation of $\tilde{M}$ whose leaves are the fibres of

$$p_2 : \tilde{M} \to T \ , \ \ p_2(\hat{x}, y) = y$$

The assumption on $(M, \mathcal{F})$ in Theorem 6.14 amounts to

$$M = \tilde{M}/\pi_1(B, x_0)$$

and $\tilde{\mathcal{F}} = \rho^* \mathcal{F}$, that is $\tilde{\mathcal{F}}$ is the pullback of $\mathcal{F}$ by ρ, cf. [**131**] and [**179**], p. 28. Then

$$T(\tilde{\mathcal{F}})_{\tilde{x}} = (d_{\tilde{x}}\rho)^{-1} T(\mathcal{F})_{\rho(\tilde{x})}$$

for any $\tilde{x} \in \tilde{M}$. Consequently $d_{\tilde{x}}\rho$ descends to an isomorphism

$$\Pi_{\tilde{x}} : \nu(\tilde{\mathcal{F}})_{\tilde{x}} \to \nu(\mathcal{F})_{\rho(\tilde{x})}.$$

Let H be the transverse Levi distribution of $\mathcal{F}$ and J its complex structure. Let us set

$$\tilde{H}_{\tilde{x}} = \Pi_{\tilde{x}}^{-1} H_{\rho(\tilde{x})}, \quad \tilde{J}_{\tilde{x}} = \Pi_{\tilde{x}}^{-1} \circ J_{\rho(\tilde{x})} \circ \Pi_{\tilde{x}}.$$

Then $(\tilde{H}, \tilde{J})$ makes $\mathcal{F}$ into a (transversally) CR foliation. This may be seen in yet another way. Let $\{U_i, f_i, \gamma_{ji}\}_{i,j \in I}$ be the data defining $\mathcal{F}$ (as a $\Gamma^\infty_{CR}(N)$-foliation, for some given model CR manifold N). We may assume w.l.o.g. that $U_i = \rho(\hat{V}_i \times T)$ where $\hat{V}_i = \hat{p}^{-1}(V_i)$ for some simply connected open subset $V_i \subset B$. Let $\tilde{U}_i = \hat{V}_i \times T$ and define $\tilde{f}_i : \tilde{U}_i \to N$ by $\tilde{f}_i = f_i \circ \rho$. Then the data $\{\tilde{U}_i, \tilde{f}_i, \gamma_{ji}\}_{i,j \in I}$ determines $\tilde{\mathcal{F}}$ (that is $\mathcal{F}$ is a $\Gamma^\infty_{CR}(N)$-foliation of $\tilde{M}$).

The differential $d_{\tilde{x}} p_2$ descends to an isomorphism

$$A_{\tilde{x}} : \nu(\tilde{\mathcal{F}})_{\tilde{x}} \to T_y(T)$$

for any $\tilde{x} = (\hat{x}, y) \in \tilde{M}$. Let $y \in T$ and $\tilde{x} \in p_2^{-1}(y)$. Let us set by definition

$$H(T)_y = A_{\tilde{x}} \tilde{H}_{\tilde{x}}.$$

As $\tilde{H}$ is invariant under sliding along the leaves of $\tilde{\mathcal{F}}$ (and $p_2^{-1}(y)$ is a leaf of $\tilde{\mathcal{F}}$) it follows that $H(T)_y$ is well defined i.e. its definition does not depend upon the choice of $\tilde{x} \in p_2^{-1}(y)$. Similar considerations apply to the complex structure $J_{T,y}$ given by

$$J_{T,y} = A_{\tilde{x}} \circ \tilde{J}_{\tilde{x}} \circ A_{\tilde{x}}^{-1}$$

for any $y \in T$ and some $\tilde{x} \in p_2^{-1}(y)$. Thus T becomes a CR manifold and $H(T)$ is its Levi distribution.

Let $g = h([\gamma]) \in G$. Let us consider $L_g : T \to T$ given by $L_g(y) = g^{-1}(y)$. Then

$$p_2 \circ R_{[\gamma]} = L_g \circ p_2 \tag{6.5}$$

where $R_{[\gamma]} : \tilde{M} \to \tilde{M}$ is the right translation with $[\gamma] \in \pi_1(B, x_0)$. If $X \in T(\tilde{\mathcal{F}})_{\tilde{x}} = \mathrm{Ker}(d_{\tilde{x}} p_2)$ then (by (6.5)) we obtain

$$(d_{\tilde{x}\cdot[\gamma]}\, p_2) \circ (d_{\tilde{x}} R_{[\gamma]})X = 0$$

that is $T(\tilde{\mathcal{F}})$ is $\pi_1(B, x_0)$-invariant. Thus $d_{\tilde{x}} R_{[\gamma]}$ descends to an isomorphism

$$B_{\tilde{x}\,,\,[\gamma]} : \nu(\tilde{\mathcal{F}})_{\tilde{x}} \to \nu(\tilde{\mathcal{F}})_{\tilde{x}\cdot[\gamma]}.$$

Furthermore

$$(d_y L_g) \circ A_{\tilde{x}} = A_{\tilde{x}\cdot[\gamma]} \circ B_{\tilde{x}\,,\,[\gamma]}$$

so that

$$(d_y L_g)\, H(T)_y = (d_y L_g)\, A_{\tilde{x}}\, \tilde{H}_{\tilde{x}} =$$
$$= A_{\tilde{x}\cdot[\gamma]}\, B_{\tilde{x}\,,\,[\gamma]}\, \tilde{H}_{\tilde{x}} = A_{\tilde{x}\cdot[\gamma]}\, B_{\tilde{x}\,,\,[\gamma]}\, \Pi_{\tilde{x}}^{-1}\, H_{\rho(\tilde{x})}.$$

Finally, we may use

$$\Pi_{\tilde{x}} = \Pi_{\tilde{x}\cdot[\gamma]} \circ B_{\tilde{x}\,,\,[\gamma]}$$

to conclude that

$$(d_y L_g) H(T)_y = H(T)_{g^{-1}y}\,.$$

Similarly dL_g commutes with J_T. Thus $L_g \in \mathrm{Aut}_{CR}(T)$. Q.e.d.

Given a CR foliation $\mathcal{F}$ defined by suspension of $h : \pi_1(B, x_0) \to \mathrm{Aut}_{CR}(T)$ one may attempt (in analogy with the case of Riemannian foliations, cf. [**179**], p. 97-99) to describe the closure of the leaf L passing through a point of the fibre $p^{-1}(x_0) \approx T$, under the assumption that $\mathrm{Aut}_{CR}(T)$ is compact. We leave this as an open problem.

6.3. Transverse pseudohermitian geometry

The basic ideas of pseudohermitian geometry carry over to the context of CR foliations as follows. Let $(\mathcal{F}, H, J)$ be a CR foliation whose transverse geometry is modelled on the CR manifold N of hypersurface type. On each U_i one may consider the 1-form $\theta_{T,i}$ given by $\theta_{T,i} = f_i^* \theta_N$. Next let us define $\theta_i \in \Gamma^\infty(Q^*)$ by setting $\theta_{i,x} \circ \Pi_x = (\theta_{T,i})_x$ for any $x \in U_i$. Clearly $H_x = \mathrm{Ker}(\theta_{i,x})$ for any $x \in U_i$. Let $j \in I$ with $U_{ji} \neq 0$. As $\gamma_{ji} \in \Gamma^\infty_{CR}(N)$ it follows in particular that γ_{ji} is a contact transformation. Therefore

$$\theta_j = (\lambda_{ji} \circ f_i)\theta_i \tag{6.6}$$

on U_{ji} for some nowhere vanishing C^∞ functions $\lambda_{ji} : f_i(U_{ji}) \to \mathbb{R}$. To investigate the properties of (U_i, θ_i) we only need to look at the case of a simple foliation (defined by a submersion). We may state

THEOREM 6.15. *Let $f : M \to N$ be a submersion with connected fibres from a manifold M onto an orientable CR manifold $(N, T_{1,0}(N))$ (of hypersurface type) on which a pseudohermitian structure θ_N has been fixed. Let $\mathcal{F}$ be the foliation of M tangent to the vertical bundle of f. Let $\theta_T = f^*\theta_N$ and $\theta \in \Gamma^\infty(Q^*)$ given by $\theta \circ \Pi = \theta_T$. Then $\mathcal{L}_X\theta = 0$ for any $X \in \Gamma^\infty(P)$ i.e. $\theta \in \Gamma_B^\infty(Q^*)$. Assume that $(N, T_{1,0}(N))$ is nondegenerate and let us set $\xi_x = F_x^{-1}(\xi_{N,f(x)})$ where $F_x : Q_x \to T_{f(x)}(N)$ is given by $F_x \circ \Pi_x = d_x f$ for any $x \in M$. Then $\xi \in \Gamma_B^\infty(Q)$ and*

$$\theta(\xi) = 1, \quad \xi \rfloor d_Q\theta = 0. \tag{6.7}$$

Throughout $\Gamma_B^\infty(\Lambda^k Q^*)$ consists of all $\omega \in \Gamma^\infty(\Lambda^k Q^*)$ with $\mathcal{L}_X\omega = 0$ for any $X \in \Gamma^\infty(P)$. Also ξ_N is the characteristic direction of $d\theta_N$. We shall prove Theorem 6.15 later on.

Let N be a CR manifold and θ_N a pseudohermitian structure on N. Clearly if $\mathcal{F}$ is a Γ-foliation of M where $\Gamma \subset \Gamma_{CR}^\infty(N)$ is the subpseudogroup of all local pseudohermitian (with respect to θ_N, cf. Definition 5.48) CR automorphisms of N, then the local sections θ_i glue up to a global section $\theta \in \Gamma_B^\infty(Q^*)$. We are led to the following general considerations. Let $(\mathcal{F}, H, J)$ be a CR foliation of transverse CR codimension $k = 1$.

DEFINITION 6.16. A globally defined nowhere vanishing section $\theta \in \Gamma_B^\infty(Q^*)$ is a *transverse pseudohermitian structure* if $H = \text{Ker}(\theta)$. □

By Theorem 6.15 any simple foliation given by a C^∞ submersion onto a CR hypersurface N carries a transverse pseudohermitian structure induced by a fixed pseudohermitian structure on N.

Let $\Omega_B^k(\mathcal{F}) = \Omega_B^k(M, \mathcal{F})$ be the $\Omega_B^0(\mathcal{F})$-module of basic k-forms, where $\Omega_B^0(\mathcal{F}) = \Omega_B^0(M, \mathcal{F})$ is the ring of all basic C^∞ functions $\lambda : M \to \mathbb{R}$.

Let $\mathcal{F}$ be a CR foliation and θ a transverse pseudohermitian structure. Let $\theta_T = \theta \circ \pi$ be the corresponding basic 1-form. Set $D = \text{Ker}(\theta_T)$. Then $H = D/P$. Next let us set

$$K_x = \{\alpha \in Q_x^* : \text{Ker}(\alpha) \supseteq H_x\}$$

for any $x \in M$. This furnishes a real line subbundle $K \subset Q^*$ and any transverse pseudohermitian structure may be viewed as a globally defined nowhere zero section in K. Thus if $\hat{\theta}$ is another transverse pseudohermitian structure then $\hat{\theta} = \lambda\theta$ for some nowhere vanishing basic function λ on M. Indeed

$$\mathcal{L}_X\hat{\theta} = X(\lambda)\theta + \lambda\mathcal{L}_X\theta$$

so that $X(\lambda) = 0$ for any $X \in \Gamma^\infty(P)$ i.e. $\lambda \in \Omega_B^0(\mathcal{F})$.

With each $\omega \in \Gamma^\infty(\Lambda^k Q^*)$ we associate a differential k-form $\omega_T = \Phi_k\omega$ on M given by

$$\omega_T(Y_1, \cdots, Y_k) = \omega(\Pi Y_1, \cdots, \Pi Y_k)$$

for any $Y_1, \cdots, Y_k \in \mathcal{X}$. The map Φ_k yields a $\mathbb{R}$-linear isomorphism

$$\Gamma_B^\infty(\Lambda^k Q^*) \approx \Omega_B^k(\mathcal{F}).$$

Indeed $X \rfloor \omega_T = 0$ for any $X \in \Gamma^\infty(P)$ (as a consequence of definitions). Also

$$(X \rfloor d\omega_T)(Y_1, \cdots, Y_k) = (\mathcal{L}_X\omega)(\Pi Y_1, \cdots, \Pi Y_k) = 0$$

so that ω_T is a basic k-form. Finally Φ_k admits the inverse Ψ_k given by

$$(\Psi_k\beta)(s_1, \cdots, s_k) = \beta(Y_{s_1}, \cdots, Y_{s_k})$$

for any $s_j \in \Gamma^\infty(Q)$ and some $Y_{s_j} \in \mathcal{X}(M)$ with $\Pi Y_{s_j} = s_j$, $1 \leq j \leq k$. As β is basic $\Psi_k\beta$ is well defined. Also

$$(\mathcal{L}_X \Psi_k \beta)(s_1, \cdots, s_k) = (d\beta)(X, Y_{s_1}, \cdots, Y_{s_k}) = 0$$

and Ψ_k is well defined. We shall need the differential operator

$$d_Q = \Psi_{k+1} \circ d \circ \Phi_k : \Gamma_B^\infty(\Lambda^k Q^*) \to \Gamma_B^\infty(\Lambda^{k+1} Q^*).$$

DEFINITION 6.17. Let $\mathcal{F}$ be a CR foliation and θ a transverse pseudohermitian structure. The *transverse Levi form* G_θ is defined by

$$G_\theta(s, r) = (d_Q\theta)(s, Jr) \tag{6.8}$$

for any $s, r \in \Gamma^\infty(H)$. □

Then $G_{\lambda\theta} = \lambda G_\theta$. We need the following

LEMMA 6.18. *For any $\alpha, \beta \in \Gamma^\infty(\mathcal{H})$:*

$$(d_Q\theta)(\alpha, \beta) = (d_Q\theta)(\overline{\alpha}, \overline{\beta}) \tag{6.9}$$

That is G_θ (as a real $(0,2)$-tensor field) is symmetric and $G_\theta(Js, Jr) = G_\theta(s, r)$ for any $s, r \in \Gamma^\infty(H)$.

Proof. By Theorem 6.8, for any $x \in M$ there is $U \subseteq M$ open, $x \in U$, and there is an admissible frame $\{\zeta_1, \cdots, \zeta_n\}$ of $\mathcal{H}$ on U. For each $1 \leq \alpha \leq n$ there is $Y_\alpha \in L(\mathcal{F}) \otimes \mathbb{C}$ such that $\pi Y_\alpha = \zeta_\alpha$. Thus

$$(d_Q\theta)(\zeta_\alpha, \zeta_\beta) = (d\theta_T)(Y_\alpha, Y_\beta) = -\frac{1}{2}\theta_T([Y_\alpha, Y_\beta]) = -\frac{1}{2}\theta[\zeta_\alpha, \zeta_\beta]) = 0$$

and Lemma 6.18 is proved.

DEFINITION 6.19. A CR foliation $(\mathcal{F}, \mathcal{H})$ is *nondegenerate* if G_θ is nondegenerate for some transverse pseudohermitian structure θ. □

Any simple foliation given by a C^∞ submersion onto a nondegenerate CR manifold of hypersurface type is itself nondegenerate. Any nondegenerate CR foliation is a contact foliation, in the sense of W-L. Ting, [**240**]. We establish the following

THEOREM 6.20. *For any transversally orientable nondegenerate CR foliation $(\mathcal{F}, \mathcal{H})$ on which a transverse pseudohermitian structure has been fixed, there is a globally defined nowhere vanishing section $\xi \in \Gamma^\infty(Q)$ such that $\xi \rfloor \theta = 1$ and $\xi \rfloor d_Q\theta = 0$. Such ξ is unique and invariant under sliding along the leaves.*

Proof. Let $\mathrm{Null}(d_Q\theta)$ be the null space bundle of $d_Q\theta$ i.e. $z \in \mathrm{Null}(d_Q\theta)_x$ if and only if $z \in Q_x$ and $z \rfloor (d_Q\theta)_x = 0$. Then

$$\mathrm{rank}_{\mathbb{R}} \mathrm{Null}(d_Q\theta) = 1$$

by the nondegeneracy of $d_Q\theta$ on H. Note that

$$\mathrm{Null}(d_Q\theta) \approx Q/H \tag{6.10}$$

(a vector bundle isomorphism). To check (6.10) it suffices to show that the map

$$z \mapsto z + H_x\,, \quad z \in \mathrm{Null}(d_Q\theta)_x\,,$$

is a bundle monomorphism. This follows from the fact that $\mathrm{Null}(d_Q\theta) \cap H = \{0\}$.

As Q is orientable and H oriented by its complex structure it follows that Q/H admits a globally defined nowhere zero section S (M is assumed to be connected). Next there is $\gamma \in \Gamma^\infty(Q)$ such that $S(x) = \gamma(x) + H_x$ for any $x \in M$. Let us set

$\lambda = \theta(\gamma) \in \Omega^0(M)$. Then λ is nowhere zero. Indeed if $\lambda(x) = 0$ for some $x \in M$ then $\gamma(x) \in \text{Ker}(\theta_x) = H_x$ i.e. $S(x) = 0$, a contradiction. Let us set $s = (1/\lambda)\gamma$. By (6.10) there is $\xi \in \Gamma^\infty(\text{Null}(d_Q\theta))$ such that $\xi(x) + H_x = s(x) + H_x$. Consequently $\xi(x) - s(x) \in H_x = \text{Ker}(\theta_x)$ i.e. $\theta(\xi) = \theta(s) = 1$. To prove the last statement in Theorem 6.20 note that $\mathcal{L}_X\theta = 0$ and $\theta(\xi) = 1$ yield $0 = (\mathcal{L}_X\theta)\xi = X(\theta\xi) - \theta(\mathcal{L}_X\xi)$ that is $\mathcal{L}_X\xi \in \Gamma^\infty(H)$ for any $X \in \Gamma^\infty(P)$. Next for any $s \in \Gamma^\infty(H)$ we have

$$0 = (\mathcal{L}_X d_Q\theta)(\xi, s) =$$

$$= X((d_Q\theta)(\xi, s) - (d_Q\theta)(\mathcal{L}_X\xi, s) - (d_Q\theta)(\xi, \mathcal{L}_X s)$$

so that (by the nondegeneracy of $d_Q\theta$ on H) we get

$$\mathcal{L}_X\xi = 0 \tag{6.11}$$

for any $X \in \Gamma^\infty(P)$, i.e. $\xi \in \Gamma^\infty_B(Q)$.

DEFINITION 6.21. The transverse vector field ξ in Theorem 6.20 is referred to as the *characteristic direction* of $(\mathcal{F}, \mathcal{H}, \theta)$. □

Note that

$$Q = H \oplus \mathbb{R}\xi. \tag{6.12}$$

By taking into account (6.12) we may extend the transverse Levi form G_θ to a semi-Riemannian holonomy invariant bundle metric g_θ in Q by setting $g_\theta(s,r) = G_\theta(s,r)$, $g_\theta(s,\xi) = 0$ and $g_\theta(\xi,\xi) = 1$, for any $s, r \in \Gamma^\infty(H)$. The holonomy invariance of g_θ follows from i) the fact that H is parallel (with respect to the Bott connection of $\mathcal{F}$), ii) $\mathcal{L}_X d_Q\theta = 0$ for any $X \in \Gamma^\infty(P)$, and from (6.11).

DEFINITION 6.22. g_θ is called the *transverse Webster metric* of $(M, (\mathcal{F}, \mathcal{H}, \theta))$. □

The transverse Levi form may be viewed as a Hermitian form on $H \otimes \mathbb{C}$ i.e. let L_θ be given by

$$L_\theta(\alpha, \overline{\beta}) = -i(d_Q\theta)(\alpha, \overline{\beta})$$

$$L_\theta(\alpha, \beta) = L_\theta(\overline{\alpha}, \overline{\beta}) = 0, \quad L_\theta(\overline{\alpha}, \beta) = \overline{L_\theta(\alpha, \overline{\beta})},$$

for any $\alpha, \beta \in \Gamma^\infty(\mathcal{H})$. Then L_θ and the $\mathbb{C}$-linear extension of G_θ to $H \otimes \mathbb{C}$ coincide.

DEFINITION 6.23. A CR foliation $(\mathcal{F}, \mathcal{H})$ is *strictly pseudoconvex* if

$$(L_\theta)_x(\sigma, \overline{\sigma}) > 0, \quad \sigma \in \mathcal{H}_x \setminus \{0\}, \quad x \in M,$$

for some transverse pseudohermitian structure θ. □

If this is the case then $(\mathcal{F}, g_\theta)$ is a Riemannian foliation.

Let $(\mathcal{F}, \mathcal{H})$ be a nondegenerate CR foliation carrying the transverse pseudohermitian structure θ. Let $\{\zeta_\alpha\} \subset \Gamma^\infty_B(\mathcal{H})$ be an admissible frame of $\mathcal{H}$ on $U \subseteq M$. Set $h_{\alpha\overline{\beta}} = L_\theta(\zeta_\alpha, \zeta_{\overline{\beta}})$ where $\zeta_{\overline{\alpha}} = \overline{\zeta_\alpha}$. Then $h_{\alpha\overline{\beta}} : U \to \mathbb{C}$ are basic functions. Let $\lambda_\alpha(x)$ be the eigenvalues of $[h_{\alpha\overline{\beta}}(x)]$, $x \in U$. As $[h_{\alpha\overline{\beta}}]$ is Hermitian each λ_α is $\mathbb{R}$-valued and continuous. Thus L_θ has constant index on U so that G_θ is a semi-Riemannian bundle metric in H. Assume the Levi form L_θ to have signature (r, s). Let us consider the Hermitian form

$$\langle z, w \rangle_r = \sum_{j=1}^{r} z_j \overline{w}_j - \sum_{j=r+1}^{n} z_j \overline{w}_j$$

where $n = r + s$ and $z, w \in \mathbb{C}^n$. Let

$$\mathrm{U}(r,s) = \{A \in \mathrm{GL}(n,\mathbb{C}) : \langle Az, Aw\rangle_r = \langle z, w\rangle_r, \forall z, w \in \mathbb{C}^n\}.$$

Let $G \subset \mathrm{GL}(q,\mathbb{R})$ consist of all matrices of the form

$$\begin{bmatrix} g^\alpha_\beta & \Omega^\alpha_\beta & 0 \\ -\Omega^\alpha_\beta & g^\alpha_\beta & 0 \\ 0 & 0 & 1 \end{bmatrix}$$

with $[g^\alpha_\beta + i\Omega^\alpha_\beta] \in \mathrm{U}(r,s)$. Here $q = 2n+1$. Let $E_{\theta,x}$ consist of all $z \in E_x$ with $z(e_{2n+1}) = \xi(x)$ and $g_\theta(z(e_\alpha), z(e_\beta)) = \epsilon_\alpha \delta_{\alpha\beta}$, $g_\theta(z(e_\alpha), z(e_{\beta+n})) = 0$. Here ξ is the characteristic direction of $(\mathcal{F}, \theta)$. Also $\epsilon_\alpha = 1$ if $1 \leq \alpha \leq r$ and $\epsilon_\alpha = -1$ if $r+1 \leq \alpha \leq r+s$. Then E_θ is a transverse G-structure.

We end this section by proving Theorem 6.15. As N is orientable so is Q. Thus we may choose - as done above for arbitrary transversally oriented foliations - a nowhere vanishing globally defined section $\xi \in \Gamma^\infty(Q)$ (not satisfying (6.11) in general) such that $\theta(\xi) = 1$ and (6.12) holds. Next $(\mathcal{L}_X\theta)\xi = -\theta_T([X, Y_\xi])$ for some $Y_\xi \in \mathcal{X}(M)$ with $\Pi Y_\xi = \xi$ and

$$\theta_T([X, Y_\xi])_x = \lim_{t\to 0} \frac{1}{t}\{\theta_{T,x} Y_\xi(x) - \theta_{T,x}(d_{\varphi_t(x)}\varphi_{-t})Y_\xi(\varphi_t(x))\} =$$

$$= \lim_{t\to 0} \frac{1}{t}\{1 - ((\varphi_{-t})^*\theta_T)_{\varphi_t(x)} Y_\xi(\varphi_t(x))\} = 0$$

because (by (6.3))

$$\varphi_t^*\theta_T = \varphi_t^* f^*\theta_N = f^*\theta_N = \theta_T$$

for any $|t| < \epsilon$. Similarly $(\mathcal{L}_X\theta)s = 0$ for any $s \in \Gamma^\infty(H)$. Finally, if N is nondegenerate then one simply lifts the characteristic direction ξ_N of $(N, T_{1,0}(N), \theta_N)$ to get a section $\xi \in \Gamma^\infty_B(Q)$ satisfying (6.7). Details are left as an exercise to the reader.

6.4. Degenerate CR manifolds

Let $(M, T_{1,0}(M))$ and $(N, T_{1,0}(N))$ be two CR manifolds of hypersurface type of CR dimensions $N = n + k$ and n, respectively.

PROPOSITION 6.24. *Let $f : M \to N$ be a CR submersion, i.e. a C^∞ submersion and a CR map. Then $(M, T_{1,0}(M))$ is degenerate.*

Proof. We organize the proof in two steps, as follows.

Step 1. Let θ_M and θ_N be choices of pseudohermitian structures on M and N, respectively. Then $f^*\theta_N = \lambda\theta_M$ for some nowhere vanishing $\lambda \in \Omega^0(M)$.

If ω is a differential form on M, we adopt the notation

$$\mathrm{Sing}(\omega) = \{x \in M : \omega_x = 0\}.$$

Let $x \in \mathrm{Sing}(f^*\theta_N)$. Then $\theta_{N,f(x)} \circ (d_x f) = 0$. On the other hand $d_x f : T_x(M) \to T_{f(x)}(N)$ is onto. Thus $T_{f(x)}(N) \subseteq \mathrm{Ker}(\theta_{N,f(x)}) = H(N)_{f(x)}$, a contradiction. Thus

$$\mathrm{Sing}(f^*\theta_N) = \emptyset. \tag{6.13}$$

Let $X \in H(M)$. Then (as f is a CR map) $(df)X \in H(N)$. Hence

$$(f^*\theta_N)X = \theta_N((df)X) = 0$$

that is $X \in \text{Ker}(f^*\theta_N)$. Let $x \in M$ and $d = \dim_{\mathbb{R}} \text{Ker}(f^*\theta_N)_x$. Then $2N \leq d \leq 2N+1$. If $d = 2N+1$ then $x \in \text{Sing}(f^*\theta_N)$, a contradiction. It remains that $d = 2N$, that is

$$H(M) = \text{Ker}(f^*\theta_N). \tag{6.14}$$

By (6.13)-(6.14) it follows that $f^*\theta_N$ is a pseudohermitian structure on M (and Step 1 is proved).

Step 2. The Levi form G_M is degenerate on the vertical bundle $P = \text{Ker}(df)$.

Let $X \in P$. By Step 1 we have

$$\theta_M X = \frac{1}{\lambda}(f^*\theta_N)X = \frac{1}{\lambda}\theta_N(f_*X) = 0$$

so that $P \subseteq H(M)$. Then Step 2 follows from the calculation

$$G_M(X,Y) = (d\theta_M)(X, J_M Y) = d(\frac{1}{\lambda} f^*\theta_N)(X, J_M Y) =$$

$$= \left((d\frac{1}{\lambda}) \wedge f^*\theta_N + \frac{1}{\lambda} d(f^*\theta_N)\right)(X, J_M Y) =$$

$$= \frac{1}{\lambda}(d\theta_N)(f_*X, f_*J_M Y) = 0.$$

Proposition 6.24 is proved.

By our Theorem 6.15 the vertical bundle $P = \text{Ker}(df)$ of f is the tangent bundle of a CR foliation $\mathcal{F}$ of M whose transverse geometry is that of N. Also $\mathcal{F}$ is nondegenerate if so does $(N, T_{1,0}(N))$. Its transverse Levi distribution is given by $H = H(M)/P$. One may say loosely that by passing to the quotient $H(M)/P$ one *factors out* the degeneracy.

In general let $(M, T_{1,0}(M))$ be a CR manifold of hypersurface type of CR dimension $n+k$. Given a pseudohermitian structure θ_M let us set

$$P_{M,x} = \{v \in H(M)_x : (d\theta_M)_x(v,w) = 0\,,\, \forall w \in H(M)_x\}$$

for any $x \in M$. Then P_M is involutive and J_M-invariant (cf. e.g. [**111**]). Thus (by applying both the Frobenius and Newlander-Nirenberg theorems) if $\dim_{\mathbb{R}} P_{M,x} = 2k = const.$ then M carries a foliation $\mathcal{F}_L$ by complex k-manifolds (with $T(\mathcal{F}_L) = P_M$).

DEFINITION 6.25. $\mathcal{F}_L$ is called the *Levi foliation* of M. $\square$

Let us set $H = H(M)/P_M \subset Q$. If $X \in \Gamma^\infty(P_M)$ and $s \in \Gamma^\infty(H)$ then

$$0 = (d\theta_M)(X, Y_s) = -\frac{1}{2}\theta_M([X, Y_s])$$

(for some $Y_s \in \Gamma^\infty(H(M))$ with $\Pi Y_s = s$) yields $[X, Y_s] \in \Gamma^\infty(H(M))$ and thus

$$\overset{\circ}{\nabla}_X s = \Pi[X, Y_s] \in \Gamma^\infty(H)$$

i.e. H is parallel with respect to the Bott connection of $\mathcal{F}$. Let us define $J : H \to H$ by setting

$$Js = \Pi J_M Y_s \tag{6.15}$$

for any $s \in \Gamma^\infty(H)$. As J_M descends to a complex structure in P_M we may extend it (by $\mathbb{C}$-linearity) to $P_M \otimes \mathbb{C}$ and let $P_M^{1,0} = \text{Eigen}(i)$ be the eigenbundle corresponding to the eigenvalue i. Similarly, let us extend J to $H \otimes \mathbb{C}$ and let us set $\mathcal{H} = \text{Eigen}(i)$.

PROPOSITION 6.26. *One has*

$$\mathcal{H} = T_{1,0}(M)/P_M^{1,0} .$$

Proof. Let $\sigma \in \mathcal{H}$ and $Y_\sigma \in H(M) \otimes \mathbb{C}$ with $\Pi Y_\sigma = \sigma$. Then $J\sigma = i\sigma$ yields

$$J_M Y_\sigma - i Y_\sigma = Z \tag{6.16}$$

for some $Z \in P_M \otimes \mathbb{C}$. By applying J_M to (6.16) one gets $J_M = -iZ$, that is $Z \in P_M^{0,1}$ (we set $P_M^{0,1} = \overline{P_M^{1,0}}$). Let us define $W \in H(M) \otimes \mathbb{C}$ by setting

$$W = Y_\sigma + \frac{i}{2}(\overline{Z} - Z). \tag{6.17}$$

Then $W \in T_{1,0}(M)$ and (by (6.17)) $\Pi W = \sigma$ i.e. $\sigma \in T_{1,0}(M)/P_M^{1,0}$.

As $P_M \rfloor \theta_M = 0$ there is a unique $\theta \in \Gamma^\infty(Q^*)$ such that $\theta \circ \Pi = \theta_M$. We have

THEOREM 6.27. *Let us assume that*

$$\mathcal{L}_X J_M = 0 \ , \ \ \mathcal{L}_X \theta_M = 0$$

for any $X \in \Gamma^\infty(P_M)$. Then i) *$\mathcal{H}$ is a transverse almost CR structure, and* ii) *θ is a transverse pseudohermitian structure.*

REMARK 6.28. 1) Note that the Lie derivative $\mathcal{L}_X J_M$ is well defined because $[X, Y] \in \Gamma^\infty(H(M))$ for any $X \in \Gamma^\infty(P_M)$ and $Y \in \Gamma^\infty(H(M))$.
2) Due to

$$(\mathcal{L}_X \theta_M) Y = X(\theta_M Y) - \theta_M([X, Y]) = 0$$

for any $Y \in \Gamma^\infty(H(M))$, the hypothesis $\mathcal{L}_X \theta_M = 0$ in Theorem 6.27 may be weakened to

$$F \rfloor \mathcal{L}_X \theta_M = 0$$

for some complement F of $H(M)$ in $T(M)$.
3) Recall that any CR-straightenable complex foliation is a CR foliation. Thus (by a result of [**202**]) if M is realized in $\mathbb{C}^{N+1}$ then a sufficient condition for the integrability of $\mathcal{H}$ in Theorem 6.27 is that the Gauss map $x \in M \mapsto P_{M,x} \in G(k, N+1)$ be a CR map. Here $G(k, N+1)$ denotes the (complex) Grassmann manifold of all complex k-subspaces of $\mathbb{C}^{N+1}$. □

To prove Theorem 6.27 let $s \in \Gamma^\infty(H)$. Then

$$Y_{Js} - J_M Y_s = V$$

for some $V \in \Gamma^\infty(P_M)$. Thus $[X, V] \in \Gamma^\infty(P_M)$ and we may conduct the calculation

$$(\mathcal{L}_X J)s = \Pi\{[X, Y_{Js}] - J_M[X, Y_s]\} = \Pi(\mathcal{L}_X J_M)Y_s = 0.$$

Clearly $H = \mathrm{Ker}(\theta)$. Also $\mathcal{L}_X \theta = 0$ if and only if $\mathcal{L}_X \theta_M = 0$. Q.e.d.

REMARK 6.29. 1) Let $f : M \to N$ be a CR submersion and $P = \mathrm{Ker}(df)$. Then

$$P \subseteq P_M . \tag{6.18}$$

Then each leaf of the Levi foliation of M is a saturated set in M thought of as endowed with the CR foliation $\mathcal{F}$ tangent to P i.e. P is a subfoliation of P_M. In particular, any exotic characteristic class of the Levi foliation must be an exotic characteristic class of $\mathcal{F}$ (cf. Corollary 5.2 in [**71**], p. 236).

Note that $d_x f : H(M)_x \to H(N)_{f(x)}$ is onto. Indeed let $v \in H(N)_{f(x)}$. As f is a submersion there is $u \in T_x(M)$ such that $(d_x f)u = v$. Then

$$0 = \theta_{N,f(x)}(v) = \theta_{N,f(x)}(d_x f)u = (f^*\theta_N)_x(u) = \lambda(x)\theta_{M,x}(u)$$

yields $u \in H(M)_x$. Q.e.d. Let $X \in P_M$. Then $f_*X \in H(N)$ and

$$(d\theta_N)(f_*X, H(N)) = (d\theta_N)(f_*X, f_*H(M)) =$$

$$= (d\lambda \wedge \theta_M + \lambda d\theta_M)(X, H(M)) = 0.$$

Consequently

$$f_* P_M \subseteq P_N .$$

Assume $(N, T_{1,0}(N))$ to be nondegenerate. Then $P_N = \{0\}$. Hence $P_M \subseteq \mathrm{Ker}(df) = P$. Thus, if N is nondegenerate, then (by (6.18)) P is the Levi foliation of M.
2) In practice, the hypothesis of our Theorem 6.27 often hold good. For instance let $(N, T_{1,0}(N))$ be a nondegenerate CR manifold (of hypersurface type). The product manifold $M = N \times \mathbb{C}^k$ carries the complex foliation $\mathcal{F}$ whose leaves are $\{y\} \times \mathbb{C}^k$, $y \in N$. It is easy to see that M is a CR manifold and $\mathcal{F}$ its Levi foliation. Indeed if $x = (y, \zeta) \in M$ let $\phi_\zeta : N \to M$ and $\psi_y : \mathbb{C}^k \to M$ given by $\phi_\zeta(y) = \psi_y(\zeta) = x$. Then

$$(d_y\phi_\zeta)T_{1,0}(N)_y \oplus (d_\zeta\psi_y)T^{1,0}(\mathbb{C}^k)_\zeta , \quad (y, \zeta) \in M,$$

is a CR structure on M and ϕ_ζ is a CR immersion. Also the natural projection $f : M \to N$ is a CR submersion and $\mathcal{F}$ is tangent to $P_M = \mathrm{Ker}(df)$. Then (by Theorem 6.15) $\mathcal{F}$ is a CR foliation of M. In particular $(\mathcal{L}_X J_M)Y = (\mathcal{L}_X J)\Pi Y = 0$ for any $X \in P_M$ and $Y \in T(M)$. Next, let $\theta_M = f^*\theta_N$ (where θ_N is a pseudohermitian structure on N) and $\theta \in \Gamma^\infty(Q^*)$ given by $\theta \circ \Pi = \theta_M$. Then (by Theorem 4) $(\mathcal{L}_X\theta_M)Y = (\mathcal{L}_X\theta)\Pi Y = 0$. Clearly $\mathcal{F}$ is a CR-straightenable complex foliation of M. $\square$

Assume that the Levi form of M has a nontrivial kernel P_M of constant dimension and the transverse Levi form is positive definite so that one may factor out the degeneracy towards a strictly pseudoconvex transverse structure. The (local) embeddability problem for degenerate CR manifolds is open (in the C^∞ category). However, in the light of our Theorem 6.44 it is tempting to conjecture that 'foliated' versions of known embeddability results (e.g. any strictly pseudoconvex CR manifold M is locally realizable if M is compact (cf. [**55**]) or if M is noncompact yet of CR dimension ≥ 3 (cf. [**162**] and [**4**])) should hold for strictly pseudoconvex transverse CR structures occurring on degenerate CR manifolds.

6.5. The transverse Cauchy-Riemann complex

Let $(\mathcal{F}, \mathcal{H})$ be a CR foliation. We consider the differential operator

$$\overline{\partial}_Q : \Gamma^\infty_B(\Lambda^k\overline{\mathcal{H}}^*) \to \Gamma^\infty_B(\Lambda^{k+1}\overline{\mathcal{H}}^*)$$

defined by the following considerations. Let $\omega \in \Gamma^\infty_B(\Lambda^k\overline{\mathcal{H}}^*)$ and $\alpha_j \in \Gamma^\infty(\overline{\mathcal{H}})$, $1 \leq j \leq k+1$. Let $Y_j \in \Gamma^\infty(T(M) \otimes \mathbb{C})$ such that $\Pi Y_j = \alpha_j$, $1 \leq j \leq k+1$. Finally let us set

$$(\overline{\partial}_Q\omega)(\alpha_1, \cdots, \alpha_{k+1}) = \sum_{j=1}^{k+1}(-1)^{j+1}Y_j(\omega(\alpha_1, \cdots, \hat{\alpha}_j, \cdots, \alpha_{k+1}))+$$

$$+ \sum_{1\leq i<j\leq k+1} (-1)^{i+j}\omega(\Pi[Y_i,Y_j],\alpha_1,\cdots,\hat{\alpha}_i,\cdots,\hat{\alpha}_j,\cdots,\alpha_{k+1}). \tag{6.19}$$

The definition (6.19) does not depend upon the choice of representatives Y_j of α_j. Indeed let $Z_j \in \Gamma^\infty(T(M)\otimes\mathbb{C})$ such that $Y_j = Z_j + X_j$ for some $X_j \in \Gamma^\infty(P\otimes\mathbb{C})$. Then

$$\Pi[Y_i,Y_j] = \Pi[Z_i,Z_j] + \mathring{\nabla}_{X_i}\alpha_j - \mathring{\nabla}_{X_j}\alpha_i\,.$$

Next, the identity

$$\sum_{1\leq i<j\leq k+1} (-1)^{i+j}\omega\left(\mathring{\nabla}_{X_i}\alpha_j - \mathring{\nabla}_{X_j}\alpha_i,\alpha_1,\cdots,\hat{\alpha}_i,\cdots,\hat{\alpha}_j,\cdots,\alpha_{k+1}\right) =$$

$$= \sum_{i=2}^{k+1}\sum_{j=1}^{i-1}(-1)^i\omega\left(\alpha_1,\cdots,\mathring{\nabla}_{X_i}\alpha_j,\cdots,\hat{\alpha}_i,\cdots,\alpha_{k+1}\right) +$$

$$+\sum_{i=1}^{k}\sum_{j=i+1}^{k+1}(-1)^i\omega\left(\alpha_1,\cdots,\hat{\alpha}_i,\cdots,\alpha_{k+1}\right)$$

leads to

$$\sum_{i=1}^{k+1}(-1)^{i+1}X_i(\omega(\alpha_1,\cdots,\hat{\alpha}_i,\cdots,\alpha_{k+1}))+$$

$$+\sum_{1\leq i<j\leq k+1}(-1)^{i+j}\omega\left(\mathring{\nabla}_{X_i}\alpha_j -\right.$$

$$\left.-\mathring{\nabla}_{X_j}\alpha_i,\alpha_1,\cdots,\hat{\alpha}_i,\cdots,\hat{\alpha}_j,\cdots,\alpha_{k+1}\right) =$$

$$= \sum_{i=1}^{k+1}(-1)^{i+1}(\mathcal{L}_{X_i}\omega)(\alpha_1,\cdots,\hat{\alpha}_i,\cdots,\alpha_{k+1}) = 0.$$

THEOREM 6.30. *Let $(\mathcal{F},\mathcal{H})$ be a CR foliation of M. Then*

$$\Omega^0_B(\mathcal{F})\otimes\mathbb{C} \xrightarrow{\overline{\partial}_Q} \Gamma^\infty_B(\overline{\mathcal{H}}^*) \xrightarrow{\overline{\partial}_Q} \Gamma^\infty_B(\Lambda^2\overline{\mathcal{H}}^*) \xrightarrow{\overline{\partial}_Q} \cdots \tag{6.20}$$

is a cochain complex i.e.

$$\overline{\partial}_Q\circ\overline{\partial}_Q = 0. \tag{6.21}$$

This follows from (6.19) and the integrability property of $\mathcal{H}$. To illustrate our ideas we check (6.21) on basic functions $\lambda : M \to \mathbb{C}$. Let $\{\zeta_\alpha\}\subset\Gamma^\infty_B(Q\otimes\mathbb{C})$ be an admissible frame of $\mathcal{H}$. Let $Y_\alpha \in V(\mathcal{F})\otimes\mathbb{C}$ with $\Pi Y_\alpha = \zeta_{\overline{\alpha}}$. Then

$$(\overline{\partial}^2_Q\lambda)(\zeta_{\overline{\alpha}},\zeta_{\overline{\beta}}) = Y_\alpha\left((\overline{\partial}_Q\lambda)\zeta_{\overline{\beta}}\right) - Y_\beta\left((\overline{\partial}_Q\lambda)\zeta_{\overline{\alpha}}\right) - (\overline{\partial}_Q\lambda)[\zeta_{\overline{\alpha}},\zeta_{\overline{\beta}}] =$$

$$= Y_\alpha Y_\beta\lambda - Y_\beta Y_\alpha\lambda - [Y_\alpha,Y_\beta]\lambda = 0.$$

DEFINITION 6.31. We refer to (6.20) as the *transverse Cauchy-Riemann complex* of $(M,\mathcal{F},\mathcal{H})$. □

We may also adopt the following

DEFINITION 6.32. Let $\mathcal{F}$ be a CR foliation. A *basic CR function* is a function $f \in \Omega_B^0(\mathcal{F}) \otimes \mathbb{C}$ such that

$$\overline{\partial}_Q f = 0 \tag{6.22}$$

and (6.22) are the *transverse tangential Cauchy-Riemann equations.* □

When the transverse CR structure is embedded (cf. our Definition 6.42 later in this chapter) it is an open problem whether one may produce a normal form for a (transversally) CR foliation, in the spirit of Lemma 1 in [**50**], p. 103-104, or Theorem 1 in [**50**], p. 105-108. A foliated version of the Baouendi-Treves approximation theorem (cf. Theorem 1 in [**50**], p. 191) for basic CR functions is expected to depend upon the availability of such a normal form. We leave this as an open problem.

If a transverse pseudohermitian structure θ is fixed and ξ is the corresponding characteristic direction then each $\eta \in \Gamma_B^\infty(\Lambda^k Q^* \otimes \mathbb{C})$ with $\mathcal{H} \rfloor \eta = 0$ and $\xi \rfloor \eta = 0$ may be regarded as an element of $\Gamma_B^\infty(\Lambda^k \mathcal{H}^*)$, and conversely.

DEFINITION 6.33. η (respectively $\Phi_k \eta$) is referred to as a *transverse* $(0,k)$*-form* (respectively as a *basic* $(0,k)$*-form*) on M. □

In this pseudohermitian setting the complex (6.20) may be redefined by declaring $\overline{\partial}_Q \eta$ to be the unique transverse $(0,k+1)$-form which coincides with $d_Q \eta$ on $\overline{\mathcal{H}} \otimes \cdots \otimes \overline{\mathcal{H}}$ ($k+1$ factors). By taking into account Proposition 3.11 in [**155**], vol. I, p. 36, it follows that the two definitions of $\overline{\partial}_Q$ are equivalent. Our constructions may be summarized in the following diagram

$$\begin{array}{ccccc}
\Omega_B^0(\mathcal{F}) \otimes \mathbb{C} & & & & \Omega_B^0(\mathcal{F}) \otimes \mathbb{C} \\
\overline{\partial}_Q \downarrow & & & & \downarrow d \\
\Gamma_B^\infty(\overline{\mathcal{H}}^*) & \hookrightarrow & \Gamma_B^\infty(Q^* \otimes \mathbb{C}) & \xrightarrow{\Phi_1} & \Omega_B^1(\mathcal{F}) \otimes \mathbb{C} \\
\overline{\partial}_Q \downarrow & & d_Q \downarrow & & \downarrow d \\
\Gamma_B^\infty(\Lambda^2 \overline{\mathcal{H}}^*) & \hookrightarrow & \Gamma_B^\infty(\Lambda^2 Q^* \otimes \mathbb{C}) & \xrightarrow{\Phi_2} & \Omega_B^2(\mathcal{F}) \otimes \mathbb{C} \\
\overline{\partial}_Q \downarrow & & d_Q \downarrow & & \downarrow d \\
\vdots & & \vdots & & \vdots
\end{array}$$

DEFINITION 6.34. Let $H^k_{\overline{\partial}_Q}(\mathcal{F}) = H^k_{\overline{\partial}_Q}(M, \mathcal{F})$ be the cohomology groups of the complex (6.20). These are referred to as the *transverse Kohn-Rossi cohomology groups* of $(M, \mathcal{F}, \mathcal{H})$. □

THEOREM 6.35. *Let $\mathcal{F}$ be the simple CR foliation on M defined by a C^∞ submersion $f : M \to N$ onto a nondegenerate CR manifold N of hypersurface type. Then the transverse Kohn-Rossi cohomology groups of $(M, \mathcal{F})$ are isomorphic to the Kohn-Rossi cohomology groups of the base manifold N.*

Proof. Let θ_N be a fixed pseudohermitian structure on N and ξ_N the characteristic direction of (N, θ_N). Let α be a $(0,k)$-form on N i.e. $\alpha \in \Gamma^\infty(\Lambda^k T^*(N) \otimes$

$\mathbb{C})$, $T_{1,0}(N) \rfloor \alpha = 0$, and $\xi_N \rfloor \alpha = 0$. Pullbacks of forms on N via f are basic. Thus $f^*\alpha \in \Omega^k_B(\mathcal{F}) \otimes \mathbb{C}$. Let us set

$$\alpha_f = \Psi_k f^* \alpha.$$

If $\sigma \in \Gamma^\infty(\mathcal{H})$ then $f_* Y_\sigma \in \Gamma^\infty(T_{1,0}(N))$ (for some $Y_\sigma \in \mathcal{X}(M)$ with $\Pi Y_\sigma = \sigma$) and

$$\sigma \rfloor \alpha_f = \Psi_{k-1} f^* \left[(f_* Y_\sigma) \rfloor \alpha \right] = 0.$$

Similarly $f_* Y_\xi = \xi_N \circ f$ yields

$$\xi \rfloor \alpha_f = \Psi_{k-1} f^* \left[(f_* Y_\xi) \rfloor \alpha \right] = \Psi_{k-1} f^* (\xi_N \rfloor \alpha) = 0$$

where $\xi \in \Gamma^\infty_B(Q)$ is the characteristic direction of $(M, \mathcal{F}, \theta)$ and $\theta \circ \Pi = f^*\theta_N$. Thus $\alpha_f \in \Gamma^\infty_B(\Lambda^k \overline{\mathcal{H}}^*)$. Assume that $\overline{\partial}_N \alpha = 0$ where

$$\overline{\partial}_N : \Gamma^\infty(\Lambda^k T_{0,1}(N)^*) \to \Gamma^\infty(\Lambda^{k+1} T_{0,1}(N)^*)$$

is the tangential Cauchy-Riemann operator of $(N, T_{1,0}(N))$. Note that

$$d_Q \alpha_f = (d\alpha)_f \, .$$

Thus

$$(\overline{\partial}_Q \alpha_f)(\sigma_1, \cdots, \sigma_{k+1}) = (\overline{\partial}_N \alpha)(f_* Y_{\sigma_1}, \cdots, f_* Y_{\sigma_{k+1}}) = 0$$

for any $\sigma_j \in \Gamma^\infty(\overline{\mathcal{H}})$. Let $H^{0,k}(N)$ be the Kohn-Rossi cohomology groups of the tangential Cauchy-Riemann complex of N. Define

$$\phi : H^{0,k}(N) \to H^k_{\overline{\partial}_Q}(\mathcal{F})$$

by setting $\phi([\alpha]) = [\alpha_f]$. To see that $\phi([\alpha])$ is well defined let $\alpha' = \alpha + \overline{\partial}_N \beta$ for some $\beta \in \Gamma^\infty(\Lambda^{k-1} T_{0,1}(N)^*)$. Then

$$\alpha'_f = \alpha_f + \Psi_k f^* \overline{\partial}_N \beta = \alpha_f + \overline{\partial}_Q \beta_f \, .$$

Already $\alpha \mapsto \alpha_f$ is onto. Indeed $f^*\alpha = \Phi_k \omega$ with $\omega \in \Gamma^\infty_B(\Lambda^k \overline{\mathcal{H}}^*)$ may be solved for α as follows. Let us set

$$\alpha(Z_1, \cdots, Z_k) = \omega(\Pi Y_1, \cdots, \Pi Y_k)$$

for any $Z_j \in \Gamma^\infty(T_{0,1}(N))$ and some $Y_j \in \mathcal{X}(M) \otimes \mathbb{C}$ with $f_* Y_j = Z_j$. Then $\alpha(Z_1, \cdots, Z_k)$ is well defined because $X \rfloor \Phi_k \omega = 0$ for each $X \in \Gamma^\infty(P)$).

To check that ϕ is a monomorphism assume that $\alpha_f = \overline{\partial}_Q \eta$ for some $\eta \in \Gamma^\infty_B(\Lambda^{k-1} \overline{\mathcal{H}}^*)$. As

$$\Omega^{k-1}_B(\mathcal{F}) \approx \Gamma^\infty(\Lambda^{k-1} T^*(N))$$

there is a unique $\gamma \in \Gamma^\infty(\Lambda^{k-1} T^*(N) \otimes \mathbb{C})$ such that $f^*\gamma = \Phi_{k-1}\eta$ and therefore γ is a $(0, k-1)$-form on N and $\alpha = \overline{\partial}_N \gamma$. Q.e.d.

In analogy with the study of the basic cohomology of foliated manifolds, and encouraged by the substantial progress there (cf. A. El-Kacimi & G. Hector, [**146**], V. Sergiescu, [**218**]) one may raise several questions related to the cohomology of the complex (6.20) (e.g. existence of spectral sequences abutting on $H^*_{\overline{\partial}_Q}(\mathcal{F})$, finitude and vanishing theorems, etc.). However, we expect a lack of relationship between $H^*_{\overline{\partial}_Q}(\mathcal{F})$ and the basic cohomology of $(M, \mathcal{F})$ as a foliated counterpart of the - not sufficiently understood as yet - lack of relationship between the Kohn-Rossi and De Rham cohomologies of a CR manifold. A step towards explaining the relationship among the two might be a foliated analog of M. Rumin's result (cf. [**210**]) that given a compact strictly pseudoconvex CR manifold M, of CR dimension n, on which a contact form θ has been fixed, there is a differential operator

$$D : \Omega^n(M)/\mathcal{I}^n \to \mathcal{J}^{n+1}$$

such that

$$0 \to \mathbb{R} \to C^\infty(M) \stackrel{d_{H(M)}}{\longrightarrow} \Omega^1(M)/\mathcal{I}^1 \stackrel{d_{H(M)}}{\longrightarrow} \cdots \stackrel{d_{H(M)}}{\longrightarrow} \Omega^n(M)/\mathcal{I}^n \stackrel{D}{\to}$$

$$\stackrel{D}{\to} \mathcal{J}^{n+1} \stackrel{D}{\to} \mathcal{J}^{n+2} \stackrel{D}{\to} \cdots \stackrel{D}{\to} \mathcal{J}^{2n+1} \to 0$$

is a resolution of the constant sheaf $\mathbb{R}$, so that the corresponding cohomology coincides with the de Rham cohomology of M. Here $\mathcal{I}^\bullet$ and $\mathcal{J}^\bullet$ are the ideals of the de Rham algebra $\Omega^\bullet(M)$ given respectively by

$$\mathcal{I}^\bullet = \{\theta \wedge \beta + (d\theta) \wedge \gamma : \beta, \gamma \in \Omega^\bullet(M)\},$$

$$\mathcal{J}^\bullet = \{\alpha \in \Omega^\bullet(M) : \theta \wedge \alpha = 0, \ \ (d\theta) \wedge \alpha = 0\}.$$

Also $d_{H(M)}$ is naturally induced by the exterior differential operator d. We leave this as an open problem. We establish the following

THEOREM 6.36. *Let $(M, T_{1,0}(M))$ be a CR manifold whose Levi form has a nontrivial kernel P_M with $\dim_{\mathbb{R}} P_{M,x} = 2k$, $x \in M$, and let $\mathcal{F}$ be the foliation of M by complex k-manifolds tangent to P_M. Assume that $\mathcal{L}_X J_M = 0$, for any $X \in \Gamma^\infty(P_M)$, and that $\mathcal{H}$ is integrable. Then each $\lambda \in H^0_{\overline{\partial}_Q}(\mathcal{F})$ is a basic CR function on M. Also, there is a natural injection of $H^1_{\overline{\partial}_Q}(\mathcal{F})$ into the first Kohn-Rossi cohomology group of M.*

Proof. As $\mathcal{L}_X J_M = 0$ it follows (by Theorem 6.27) that $\mathcal{H}$ is a transverse almost CR structure. Next $\mathcal{H}$ is assumed to be integrable, so that $\mathcal{F}$ is a CR foliation. Let $\lambda \in \Omega^0_B(\mathcal{F}) \otimes \mathbb{C}$ with $\overline{\partial}_Q \lambda = 0$. As (by (6.19)) $\overline{\partial}_M \lambda = (\overline{\partial}_Q \lambda) \circ \Pi$ it follows that

$$H^0_{\overline{\partial}_Q}(\mathcal{F}) = \left[\Omega^0_B(\mathcal{F}) \otimes \mathbb{C}\right] \cap CR^\infty(M).$$

Next let η be a transverse $(0,1)$-form with $\overline{\partial}_Q \eta = 0$. Let $\tilde{\eta}$ be given by (in view of Proposition 6.26)

$$\tilde{\eta}(Z) = \eta(\Pi Z), \quad Z \in T_{0,1}(M).$$

Then $(\overline{\partial}_M \tilde{\eta})(Z, W) = (\overline{\partial}_Q \eta)(\Pi Z, \Pi W)$ for any $Z, W \in T_{0,1}(M)$. The map

$$H^1_{\overline{\partial}_Q}(\mathcal{F}) \to H^{0,1}_{KR}(M), \quad [\eta] \mapsto [\tilde{\eta}],$$

is one-to-one. Indeed if $[\tilde{\eta}] = 0$ there is $\lambda \in \Omega^0(M) \otimes \mathbb{C}$ such that $\tilde{\eta} = \overline{\partial}_M \lambda$. For any $X \in P_{0,1}$ we see that $0 = X \rfloor \tilde{\eta} = X(\lambda)$ so that λ is basic. Finally $\overline{\partial}_Q \lambda = \eta$. Q.e.d.

The Authors hope that a further development of the theory of CR foliations may lead to a better understanding of degenerate CR manifolds.

6.6. Canonical transverse connections

Let $(\mathcal{F}, \mathcal{H})$ be a nondegenerate CR foliation endowed with the transverse pseudohermitian structure θ. Let ξ be the characteristic direction of $(\mathcal{F}, \theta)$. Let H be the transverse Levi distribution. Let us extend its complex structure $J : H \to H$ to a bundle morphism $J : Q \to Q$ by requesting $J\xi = 0$. If g_θ is the transverse Webster metric let us set

$$g_{\theta,T}(Y, Z) = g_\theta(\Pi Y, \Pi Z), \quad Y, Z \in \mathcal{X}(M).$$

Let g be a semi-Riemannian (i.e. nondegenerate and of constant index) metric on M. Assume g is nondegenerate on P. Then

$$T(M) = P \oplus P^\perp$$

where $P^\perp$ is the orthogonal complement of P in $T(M)$ with respect to g. Let $\sigma_g : Q \to P^\perp$ be the natural isomorphism. Then $\Pi(\sigma_g s) = s$ for any $s \in Q$. Also g induces a bundle metric g_Q in Q given by

$$g_Q(s, r) = g(\sigma_g s, \sigma_g r), \quad s, r \in Q.$$

We recall that g is bundle-like if g_Q is holonmy invariant i.e. $\mathcal{L}_X g_Q = 0$ for any $X \in \Gamma^\infty(P)$. Also

DEFINITION 6.37.

$$g_T(Y, Z) = g_Q(\Pi Y, \Pi Z), \quad Y, Z \in \mathcal{X}(M),$$

is the *associated transverse metric* of g. □

By slightly generalizing Proposition 3.3 in [**179**], p. 80, we see that there is a bundle-like semi-Riemannian metric g on M whose associated transverse metric is precisely $g_{\theta,T}$. Indeed, let h be just any Riemannian metric on M and P_h the orthogonal complement of P in $T(M)$ with respect to h. If $Y \in \mathcal{X}(M)$ then Y_P and Y_{P_h} denote respectively its components with respect to the direct sum decomposition $T(M) = P \oplus P_h$. Then we define g by setting

$$g(Y, Z) = h(Y_P, Z_P) + g_{\theta,T}(Y_{P_h}, Z_{P_h}), \quad Y, Z \in \mathcal{X}(M). \tag{6.23}$$

If L_θ has signature (r, s), $r + s = n$, then g_θ has signature $(2r + 1, s)$. Hence g (given by (6.23)) has signature $(2r + p + 1, 2s)$ where $p = \dim_{\mathbb{R}} P_x$, $x \in M$.

Let ∇ be a connection in $Q \to M$ and T_∇ its torsion tensor field i.e. $T_\nabla(Y, Z) = \nabla_Y \Pi Z - \nabla_Z \Pi Y - \Pi[Y, Z]$ for any $Y, Z \in \mathcal{X}(M)$.

DEFINITION 6.38. ∇ is *adapted* if $\nabla_X = \mathring{\nabla}_X$ for any $X \in \Gamma^\infty(P)$. □

If ∇ is adapted then we define Tor and $\tau : Q \to Q$ by setting

$$Tor(s, t) = T_\nabla(Y_s, Y_t), \quad \tau(s) = Tor(\xi, s),$$

for any $s, t \in \Gamma^\infty(Q)$ and some $Y_s, Y_t \in \mathcal{X}(M)$ with $\Pi Y_s = s$, $\Pi Y_t = t$. It is easy to check that $Tor(s, t)$ is well defined. Indeed for any $X, X' \in \Gamma^\infty(P)$ we have $T_\nabla(X, X') = 0$ due to the integrability of P and $T_\nabla(Y_s, X) = 0$ because ∇ is adapted. We may state the following

THEOREM 6.39. (E. Barletta et al., [**21**])
Let $(\mathcal{F}, \mathcal{H})$ be a nondegenerate CR foliation and θ a fixed transverse pseudohermitian structure. Then there is a unique adapted connection ∇ in Q satisfying the following axioms

i) H *is parallel with respect to* ∇
ii) $\nabla J = 0\,,\ \nabla g_\theta = 0$
iii) $\tau J + J\tau = 0$
iv) $\forall \alpha, \beta \in \Gamma^\infty(\mathcal{H})\ :\ Tor(\alpha,\beta) = 0\,,\ Tor(\alpha,\overline{\beta}) = 2iL_\theta(\alpha,\overline{\beta})\xi$

Proof. The proof follows the steps in [**235**]. Let g be a bundle-like semi-Riemannian metric on M whose associated transverse metric is $g_{\theta,T}$. To establish uniqueness let ∇ be an adapted connection in Q obeying to i)-iv). By i) and $\nabla J = 0$ it follows that

$$\nabla_X \Gamma^\infty(\mathcal{H}) \subset \Gamma^\infty(\mathcal{H}), \quad \nabla_X \Gamma^\infty(\overline{\mathcal{H}}) \subset \Gamma^\infty(\overline{\mathcal{H}}),$$

for any $X \in \mathcal{X}(M)$. Let $\rho_+ : Q \otimes \mathbb{C} \to \mathcal{H}$ and $\rho_- : Q \otimes \mathbb{C} \to \overline{\mathcal{H}}$ be the natural projections associated with the direct sum decomposition

$$Q \otimes \mathbb{C} = \mathcal{H} \oplus \overline{\mathcal{H}} \oplus \mathbb{C}\xi.$$

By iv) we have

$$Tor(\overline{\alpha}, \beta) = -2iL_\theta(\beta, \overline{\alpha})\xi$$

or

$$\nabla_{\sigma_g \overline{\alpha}}\beta - \nabla_{\sigma_g \beta}\overline{\alpha} - \Pi[\sigma_g \overline{\alpha}, \sigma_g \beta] = -2iL_\theta(\beta, \overline{\alpha})\xi$$

which yields

$$\nabla_{\sigma_g \overline{\alpha}}\beta = \rho_+ \Pi[\sigma_g \overline{\alpha}, \sigma_g \beta] \tag{6.24}$$

for any $\alpha, \beta \in \Gamma^\infty(\mathcal{H})$. Let ω be given by

$$\omega = -d_Q \theta.$$

The axiom $\nabla g_\theta = 0$ may be written

$$X(g_\theta(s,r)) = g_\theta(\nabla_X s, r) + g_\theta(s, \nabla_X r)$$

for any $X \in T(M)$, $s, r \in Q$. In particular for $s = \xi$ one has

$$(\nabla_X \theta) r = g_\theta(\nabla_X \xi, r). \tag{6.25}$$

If $r \in H$ then (6.25) gives $g_\theta(\nabla_X \xi, r) = 0$ or $\Pi_H \nabla_X \xi = 0$ (where $\Pi_H : Q \to H$ is the natural projection associated with (6.12)). Similarly, if $r = \xi$ then (6.25) becomes $\theta(\nabla_X \xi) = 0$. Therefore

$$\nabla \xi = 0. \tag{6.26}$$

Note that

$$\nabla \omega = 0 \tag{6.27}$$

as a consequence of ii) and $\omega(\alpha, \overline{\beta}) = -ig_\theta(\alpha, \overline{\beta})$. Let us take the complex conjugate of (6.24) and use (6.27) so that to obtain

$$\omega(\nabla_{\sigma_g \alpha}\beta, \overline{\gamma}) = (\sigma_g \alpha)(\omega(\beta, \overline{\gamma})) - \omega(\beta, \rho_- \Pi[\sigma_g \alpha, \sigma_g \overline{\gamma}]) \tag{6.28}$$

for any $\alpha, \beta, \gamma \in \mathcal{H}$. Let us set

$$S(g) = \sigma_g(\xi).$$

Next let us define $T_{S(g)}$ by setting

$$T_{S(g)} = -\frac{1}{2} J \circ (\mathcal{L}_{S(g)} J).$$

Then (by (6.26))

$$\nabla_{S(g)} r = \mathcal{L}_{S(g)} r + \tau(r). \tag{6.29}$$

By ii) and (6.29) we have

$$\begin{aligned} 0 = (\nabla_{S(g)} J) r = \nabla_{S(g)} J r - J \nabla_{S(g)} r = \\ = \Pi[S(g), \sigma_g J r] + \tau(Jr) - J\Pi[S(g), \sigma_g r] - J\tau(r) = \\ = \mathcal{L}_{S(g)} J r - J \mathcal{L}_{S(g)} r + (\tau J - J\tau) r \end{aligned}$$

so that (by iii))

$$\tau = T_{S(g)}. \tag{6.30}$$

Summing up (by (6.24), (6.26) and (6.28)-(6.30)) we have

$$\begin{cases} \nabla_{\sigma_g \overline{\alpha}} \beta = \rho_+ \Pi[\sigma_g \overline{\alpha}, \sigma_g \beta] \\ \nabla_{\sigma_g \alpha} \beta = U_{\alpha\beta}(g) \\ \nabla_{S(g)} \beta = \mathcal{L}_{S(g)} \beta + T_{S(g)} \beta \\ \nabla \xi = 0 \end{cases} \tag{6.31}$$

where $U_{\alpha\beta}(g) \in \mathcal{H}$ is defined by

$$\omega(U_{\alpha\beta}(g), \overline{\gamma}) = (\sigma_g \alpha)(\omega(\beta, \overline{\gamma})) - \omega(\beta, \rho_- \Pi[\sigma_g \alpha, \sigma_g \overline{\gamma}]).$$

The uniqueness statement in Theorem 6.39 is completely proved. To establish existence let g be a bundle-like semi-Riemannian metric on M inducing the transverse Webster metric in Q. Let

$$\nabla : \Gamma^\infty(T(M) \otimes \mathbb{C}) \times \Gamma^\infty(Q \otimes \mathbb{C}) \to \Gamma^\infty(Q \otimes \mathbb{C})$$

be defined by (6.31) together with

$$\begin{cases} \nabla_{\sigma_g \alpha} \overline{\beta} = \overline{\nabla_{\sigma_g \overline{\alpha}} \beta} \\ \nabla_{\sigma_g \overline{\alpha}} \overline{\beta} = \overline{\nabla_{\sigma_g \alpha} \beta} \\ \nabla_{S(g)} \overline{\beta} = \overline{\nabla_{S(g)} \beta} \end{cases} \tag{6.32}$$

and

$$\nabla_X = \mathring{\nabla}_X$$

for any $\alpha, \beta \in \mathcal{H}$, $X \in P$. Before going any further note that the definition (6.31) does not depend upon the choice of g. Indeed if g' is another bundle-like semi-Riemannian metric inducing g_θ then there is a natural bundle morphism

$$\epsilon = \epsilon_{g,g'} : Q \to P$$

given by $\epsilon_{g,g'}(s) = \sigma_{g'}(s) - \sigma_g(s)$ for any $s \in Q$. Then

$$\begin{aligned} T_{S(g')} \beta = T_{S(g)} \beta + \\ + \frac{1}{2} J \{ \mathring{\nabla}_{\epsilon(J\beta)} \xi - \mathring{\nabla}_{\epsilon(\xi)} J\beta - J \mathring{\nabla}_{\epsilon(\beta)} \xi + J \mathring{\nabla}_{\epsilon(\xi)} \beta \}. \end{aligned}$$

Using $\mathring{\nabla} J = 0$ (as $\mathcal{H}$ is a transverse CR structure) and $\xi \in \Gamma_B^\infty(Q)$ we obtain

$$T_{S(g')} \beta = T_{S(g)} \beta \tag{6.33}$$

for any $\beta \in \mathcal{H}$. At this point we may use

$$[\epsilon(\overline{\alpha}), \epsilon(\beta)] \in P \otimes \mathbb{C}, \quad \mathring{\nabla}_{\epsilon(\beta)} \overline{\alpha} \in \overline{\mathcal{H}},$$

and the first identity in (6.31) to conduct the following calculation

$$\begin{aligned} \nabla_{\sigma_{g'} \overline{\alpha}} \beta = \nabla_{\sigma_g \overline{\alpha}} \beta + \mathring{\nabla}_{\epsilon(\overline{\alpha})} \beta = \\ = \rho_+ \Pi[\sigma_g \overline{\alpha}, \sigma_g \beta] + \mathring{\nabla}_{\epsilon(\overline{\alpha})} \beta = \rho_+ \Pi[\sigma_{g'} \overline{\alpha}, \sigma_{g'} \beta]. \end{aligned}$$

Next using $\overset{\circ}{\nabla}\omega = 0$ one may derive

$$U_{\alpha\beta}(g') = U_{\alpha\beta}(g) + \overset{\circ}{\nabla}_{\epsilon(\alpha)}\beta$$

so that (by the second identity in (6.31))

$$\nabla_{\sigma_{g'}\alpha}\beta = \nabla_{\sigma_g\alpha}\beta + \overset{\circ}{\nabla}_{\epsilon(\alpha)}\beta = U_{\alpha\beta}(g').$$

Finally

$$\nabla_{S(g')}\beta = \nabla_{S(g)}\beta + \overset{\circ}{\nabla}_{\epsilon(\xi)}\beta =$$
$$= \mathcal{L}_{S(g)}\beta + T_{S(g)}\beta + \overset{\circ}{\nabla}_{\epsilon(\xi)}\beta = \mathcal{L}_{S(g')}\beta + T_{S(g')}\beta.$$

Taking into account (6.33) we adopt the notation $T_\xi = T_{S(g)}$. Note that

$$J^2 = -I + \theta \otimes \xi. \tag{6.34}$$

Also

$$\mathcal{L}_{S(g)}\theta = 0 \tag{6.35}$$

as a consequence of

$$(\mathcal{L}_{S(g)}\theta)r = S(g)(\theta r) - \theta(\Pi[S(g), \sigma_g r]) = 2(d_Q\theta)(\xi, r) = 0.$$

It is straightforward that

$$J \circ (\mathcal{L}_{S(g)}J) + (\mathcal{L}_{S(g)}J) \circ J = 0, \tag{6.36}$$

$$\mathcal{L}_{S(g)}\xi = 0 \ , \ \ \theta(\mathcal{L}_{S(g)}r) = 0, \tag{6.37}$$

for any $r \in H$. Next (6.36) yields

$$JT_\xi + T_\xi J = 0. \tag{6.38}$$

Let $\beta \in \mathcal{H}$. We have (by (6.36) and (6.38))

$$J(\mathcal{L}_{S(g)}\beta + T_\xi\beta) = \mathcal{L}_{S(g)}J\beta - (\mathcal{L}_{S(g)}J)\beta + JT_\xi\beta =$$
$$= \mathcal{L}_{S(g)}J\beta + (\mathcal{L}_{S(g)}J)J^2\beta + JT_\xi\beta =$$
$$= \mathcal{L}_{S(g)}J\beta + JT_\xi\beta - J(\mathcal{L}_{S(g)}J)J\beta = i(\mathcal{L}_{S(g)}\beta + T_\xi\beta)$$

that is

$$\beta \in \mathcal{H} \Longrightarrow \mathcal{L}_{S(g)}\beta + T_\xi\beta \in \mathcal{H}. \tag{6.39}$$

We may conduct the following calculation

$$0 = (d_Q^2\theta)(\xi, \beta, \overline{\gamma}) = -(d_Q\omega)(\xi, \beta, \overline{\gamma}) =$$
$$= -\frac{1}{3}\{S(\omega(\beta, \overline{\gamma})) + (\sigma_g\beta)(\omega(\overline{\gamma}, \xi)) +$$
$$+(\sigma_g\overline{\gamma})(\omega(\xi, \beta)) - \omega(\Pi[S(g), \sigma_g\beta], \overline{\gamma}) -$$
$$-\omega(\Pi[\sigma_g\beta, \sigma_g\overline{\gamma}], \xi) - \omega(\Pi[\sigma\overline{\gamma}, S(g)], \beta)\} =$$
$$= -\frac{1}{3}\{S(g)(\omega(\beta, \overline{\gamma})) - \omega(\mathcal{L}_S\beta, \overline{\gamma}) + \omega(\mathcal{L}_{S(g)}\overline{\gamma}, \beta)\}$$

which yields

$$\mathcal{L}_{S(g)}\omega = 0. \tag{6.40}$$

Next the following calculation

$$\omega(T_\xi r, t) + \omega(r, T_\xi t) =$$
$$= \frac{1}{2}\{\omega((\mathcal{L}_{S(g)}J)r, Jt) + \omega(Jr, (\mathcal{L}_{S(g)}J)t)\} =$$

$$= \frac{1}{2}\{(\mathcal{L}_{S(g)}\omega)(r,t) - (\mathcal{L}_{S(g)}\omega)(Jr, Jt)\}$$

leads (by (6.40)) to

$$\omega(T_\xi r, t) + \omega(r, T_\xi t) = 0 \tag{6.41}$$

for any $r, t \in Q$. Finally

$$(d_Q\omega)(\alpha, \beta, \overline{\gamma}) = 0$$

yields

$$(\sigma_g\alpha)(\omega(\beta,\overline{\gamma})) + (\sigma_g\beta)(\omega(\overline{\gamma},\alpha)) + \omega(\overline{\gamma}, \Pi[\sigma_g\alpha, \sigma_g\beta]) + \tag{6.42}$$
$$+\omega(\alpha, \rho_-\Pi[\sigma_g\beta, \sigma_g\overline{\gamma}]) + \omega(\beta, \rho_-\Pi[\sigma_g\overline{\gamma}, \sigma_g\alpha]) = 0$$

for any $\alpha, \beta, \gamma \in \mathcal{H}$. At this point one may check the axioms. First (by (6.31)-(6.32))

$$Tor(\alpha, \overline{\beta}) = \nabla_{\sigma_g\alpha}\overline{\beta} - \nabla_{\sigma_g\overline{\beta}}\alpha - \Pi[\sigma_g\alpha, \sigma_g\overline{\beta}] =$$
$$= \rho_-\Pi[\sigma_g\alpha, \sigma_g\overline{\beta}] - \rho_+\Pi[\sigma_g\overline{\beta}, \sigma_g\alpha] - \Pi[\sigma_g\alpha, \sigma_g\overline{\beta}] =$$
$$-\theta(\Pi[\sigma_g\alpha, \sigma_g\overline{\beta}])\xi = 2(d_Q\theta)(\alpha, \overline{\beta})\xi = 2iL_\theta(\alpha, \overline{\beta})\xi.$$

LEMMA 6.40.

$$\alpha, \beta \in \mathcal{H} \Longrightarrow \Pi[\sigma_g\alpha, \sigma_g\beta] \in \mathcal{H}.$$

Proof. Let $\{\zeta_\mu\}$ be an admissible frame and write (locally) $\alpha = \alpha^\mu\zeta_\mu$, $\beta = \beta^\mu\zeta_\mu$. Let $Z_\mu \in L(\mathcal{F}) \otimes \mathbb{C}$ with $\Pi Z_\mu = \zeta_\mu$. Then $\sigma_g\zeta_\mu = Z_\mu + X_\mu$ for some $X_\mu \in P \otimes \mathbb{C}$ yields

$$\Pi[\sigma_g\zeta_\mu, \sigma_g\zeta_\nu] = [\zeta_\mu, \zeta_\nu]. \tag{6.43}$$

Finally (by (6.43))

$$\Pi[\sigma_g\alpha, \sigma_g\beta] = \alpha^\mu\beta^\nu[\zeta_\mu, \zeta_\nu] + \alpha^\mu(\sigma_g\zeta_\mu)(\beta^\nu)\zeta_\nu - \beta^\nu(\sigma_g\zeta_\nu)(\alpha^\mu)\zeta_\mu \in \mathcal{H}$$

and Lemma 6.40 is proved. As

$$Tor(\alpha, \beta) = U_{\alpha\beta}(g) - U_{\beta\alpha}(g) - \Pi[\sigma_g\alpha, \sigma_g\beta]$$

it follows (by Lemma 6.40) that $Tor(\alpha, \beta) \in \mathcal{H}$. Then (by (6.42)) we have $\omega(Tor(\alpha,\beta), \overline{\gamma}) = 0$ and axiom iv) is checked. Using (6.31) we have $Tor(\xi, \alpha) = \nabla_{S(g)}\alpha - \Pi[S(g), \sigma_g\alpha] = T_\xi\alpha$. Noting also that

$$T_\xi\xi = 0 \tag{6.44}$$

we may conclude that

$$Tor(\xi, r) = T_\xi r \tag{6.45}$$

for any $r \in Q$. Then (6.38) yields iii) in Theorem 6.39. Finally

$$(\nabla_{\sigma_g\alpha}\omega)(\beta, \overline{\gamma}) = (\sigma_g\alpha)(\omega(\beta, \overline{\gamma})) - \omega(\nabla_{\sigma_g\alpha}\beta, \overline{\gamma}) - \omega(\beta, \nabla_{\sigma_g\alpha}\overline{\gamma}) =$$
$$= (\sigma_g\alpha)(\omega(\beta, \overline{\gamma})) - \omega(U_{\alpha\beta}(g), \overline{\gamma}) - \omega(\beta, \rho_-\Pi[\sigma_g\alpha, \sigma_g\overline{\gamma}]) = 0$$

and

$$(\nabla_{S(g)}\omega)(\beta, \overline{\gamma}) = S(g)(\omega(\beta, \overline{\gamma})) - \omega(\nabla_{S(g)}\beta, \overline{\gamma}) - \omega(\beta, \nabla_{S(g)}\overline{\gamma}) =$$
$$= S(g)(\omega(\beta, \overline{\gamma})) - \omega(\mathcal{L}_{S(g)}\beta + T_\xi\beta, \overline{\gamma}) - \omega(\beta, \mathcal{L}_{S(g)}\overline{\gamma} + T_\xi\overline{\gamma}) =$$
$$= (\mathcal{L}_{S(g)}\omega)(\beta, \overline{\gamma}) - \{\omega(T_\xi\beta, \overline{\gamma}) + \omega(\beta, T_\xi\overline{\gamma})\} = 0$$

(by (6.40)-(6.41)) yield $\nabla\omega = 0$ (which together with $\nabla\xi = 0$ implies $\nabla g_\theta = 0$).

It is known that with the Tanaka-Webster connection of a nondegenerate CR manifold M (of CR dimension n) one may associate the pseudoconformal curvature tensor $C_\beta{}^\alpha{}_{\rho\overline{\sigma}}$ (cf. (3.8) in [**250**], p. 35) and $C_\beta{}^\alpha{}_{\rho\overline{\sigma}} = 0$ if and only if M is locally CR equivalent to the sphere $S^{2n+1} \subset \mathbb{C}^{n+1}$. In view of our Theorem 6.39 it is tempting to look for a foliated analogue of this result. We leave this as an open problem. Cf. [**164**], p. 388-389, and [**179**], p. 61-63, for the many notions of equivalence of foliations. Other interesting geometric problems are suggested by the existence and uniqueness of an adapted connection in the normal bundle of a nondegenerate transversally CR foliation, similar to the Tanaka-Webster connection of a nondegnerate CR manifold. For instance, is there a transverse pseudohermitian structure $\hat{\theta}$ such that the canonical connection $\hat{\nabla}$ furnished by Theorem 6.39 has constant scalar curvature (the foliated equivalent of the CR Yamabe problem, cf. D. Jerison & J.M. Lee, [**142**])?

6.7. The embedding problem

Let M be a m-dimensional C^∞ manifold, $m = 2n + k + p$, and $(\mathcal{F}, \mathcal{H})$ a CR foliation of M of codimension $q = 2n + k$ and transverse CR dimension n.

DEFINITION 6.41. An *embedding* of $(M, \mathcal{H})$ is a C^∞ immersion

$$\psi = (g_1, \cdots, g_{n+k}, f_1, \cdots, f_r) : M \to \mathbb{C}^N$$

with $N = n + k + r$, $r \geq p$, so that the following conditions are satisfied

i) $g_j \in \Omega^0_B(\mathcal{F}) \otimes \mathbb{C}$,

ii) $\overline{\partial}_Q g_j = 0$, $1 \leq j \leq n + k$.

Let us set $g = (g_1, \cdots, g_{n+k}) : M \to \mathbb{C}^{n+k}$. As each g_j is basic the differential $d_x g$ induces a map $G_x : Q_x \to T_{g(x)}(\mathbb{C}^{n+k})$. Finally, we request

iii) for any $x \in M$, G_x is one-to-one. □

DEFINITION 6.42. The embedding ψ is *generic* if $r = p$. A pair $(M, \mathcal{H})$ for which an embedding ψ exists is termed *embeddable.* □

Let $\psi : M \to \mathbb{C}^N$ be an embedding of $(M, \mathcal{H})$. Then

$$G_x(\mathcal{H}_x) \subset T^{1,0}(\mathbb{C}^{n+k})_{g(x)} \,. \tag{6.46}$$

Here $T^{1,0}(\mathbb{C}^{n+k})$ denotes the holomorphic tangent bundle of $\mathbb{C}^{n+k}$. To prove (6.46) let $\{\zeta_\alpha\}$ be an admissible frame and T_α some foliate complex vector fields with $\Pi T_\alpha = \zeta_\alpha$. Then (by ii) in the definition of ψ) we have $T_{\overline{\alpha}}(g_j) = 0$ where $T_{\overline{\alpha}} = \overline{T_\alpha}$. Let $(\zeta^1, \cdots, \zeta^{n+k})$ be the natural complex coordinates on $\mathbb{C}^{n+k}$. Finally

$$G_x \zeta_{\overline{\alpha}}(x) = (d_x g) T_{\overline{\alpha}}(x) =$$

$$= T_{\overline{\alpha}}(g_j) \frac{\partial}{\partial \zeta^j} + T_{\overline{\alpha}}(\overline{g_j}) \frac{\partial}{\partial \overline{\zeta}^j} \in T^{1,0}(\mathbb{C}^{n+k})_{g(x)} \,.$$

If additionally $q = 2n + 1$ then

$$G_x \mathcal{H}_x = T^{0,1}(\mathbb{C}^{n+1})_{g(x)} \cap G_x(Q_x \otimes \mathbb{C}) \tag{6.47}$$

for any $x \in M$. Indeed, let d be the complex dimension of the right hand term in (6.47). Then (by (6.46)) $n \leq d \leq n + 1$. If $d = n + 1$ then $T^{1,0}(\mathbb{C}^{n+1})_{g(x)} \subset G_x(Q_x \otimes \mathbb{C})$ and (by taking complex conjugates) one gets a contradiction. Thus $d = n$. Q.e.d.

DEFINITION 6.43. We call $(M, \mathcal{H})$ *locally embeddable* if for any $x \in M$ there is an open neighborhood U of x in M and an embedding $\psi : U \to \mathbb{C}^N$ of $(U, \mathcal{H}_U)$, where $\mathcal{H}_U$ denotes the portion of $\mathcal{H}$ over U. □

We may state the following

THEOREM 6.44. (E. Barletta et al., [**21**])
Let M be a m-dimensional real analytic manifold, $m = 2n+k+p$, and $(N, T_{1,0}(N))$ a real analytic CR manifold of type (n,k). Let $\mathcal{F}$ be a $\Gamma^\omega_{CR}(N)$-foliation of M of codimension $q = 2n+k$ and $\mathcal{H}$ its (real analytic) transverse CR structure. Then $(M, \mathcal{H})$ is locally embeddable.

Proof. Near $x_0 \in M$, $\mathcal{H}$ is generated by n real analytic sections $\zeta_\alpha \in \Gamma^\omega_B(\mathcal{H})$ so that $[\zeta_\alpha, \zeta_\beta] \in \Gamma^\omega(\mathcal{H})$. Using a real analytic foliated local coordinate system $(x^1, \cdots, x^p, y^1, \cdots, y^q)$ for M we may assume that M is an open subset of $\mathbb{R}^{q+p}$ containing the origin and that $\zeta_\alpha = \Pi L_\alpha$ for some real analytic foliate vector field L_α in $T(\mathbb{R}^{q+p}) \otimes \mathbb{C}$. We write

$$\overline{L}_\alpha = \sum_{j=1}^{q} a_{\alpha j}(y,x) \frac{\partial}{\partial y^j}, \quad 1 \le \alpha \le n,$$

for some C^ω functions $a_{\alpha j} : \mathbb{R}^{q+p} \to \mathbb{C}$. As both $\overline{L}_\alpha$ and $\partial/\partial y^j$ are foliate $a_{\alpha j}$ are basic i.e. $a_{\alpha j} = a_{\alpha j}(y)$. Since $\{\zeta_{\overline{\alpha}}\}_{1 \le \alpha \le n}$ are linearly independent and $\{\Pi(\partial/\partial y^j) : 1 \le j \le q\}$ is a local frame of Q the matrix $[a_{\alpha j}(0)]$ has complex rank n. By reordering the coordinates if needed we may assume the $n \times n$ block $A = [a_{\alpha\beta}]$ is nonsingular in a neighborhood U of $0 \in \mathbb{R}^{q+p}$. Set us set $y = (t,u)$, $t \in \mathbb{R}^n$, $u \in \mathbb{R}^{n+k}$. By multiplying the coefficients of $\{\zeta_{\overline{\alpha}}\}$ with A^{-1} we obtain another admissible frame of $\overline{\mathcal{H}}$ (over U) of the form $\zeta_{\overline{\alpha}} = \Pi \overline{L}_\alpha$ where

$$\overline{L}_\alpha = \frac{\partial}{\partial t^\alpha} + \sum_{j=1}^{n+k} \lambda_{\alpha j}(t,u) \frac{\partial}{\partial u^j}, \quad 1 \le \alpha \le n,$$

for some C^ω functions $\lambda_{\alpha j} : U \to \mathbb{C}$ (depending only on $y = (t,u)$). The Lie product $[\overline{L}_\alpha, \overline{L}_\beta]$ has no $(\partial/\partial t^\alpha)$-component. Also

$$[\zeta_{\overline{\alpha}}, \zeta_{\overline{\beta}}] = \sum_{\gamma=1}^{n} C^\gamma_{\alpha\beta} \zeta_{\overline{\gamma}}$$

(because $\{\zeta_\alpha\}$ is admissible) so that

$$[\zeta_{\overline{\alpha}}, \zeta_{\overline{\beta}}] = 0, \quad 1 \le \alpha, \beta \le n. \tag{6.48}$$

Let $\zeta \in \mathbb{C}^n$, $w \in \mathbb{C}^{n+k}$ and $z \in \mathbb{C}^p$ be the complexifications of $t \in \mathbb{R}^n$, $u \in \mathbb{R}^{n+k}$ and $x \in \mathbb{R}^p$, respectively (so that $t = \mathrm{Re}(\zeta)$, $u = \mathrm{Re}(w)$ and $x = \mathrm{Re}(z)$). By replacing t and u by ζ and w in the power series expansion of $\lambda_{\alpha j}$ (about 0) we get functions $\tilde{\lambda}_{\alpha j} : \mathbb{C}^{2n+k} \to \mathbb{C}$ which are holomorphic in a neighborhood $\tilde{U}$ of $0 \in \mathbb{C}^{2n+k}$ and $\tilde{\lambda}_{\alpha j}(t,u) = \lambda_{\alpha j}(t,u)$. Let us define

$$\tilde{L}_\alpha = \frac{\partial}{\partial \zeta^\alpha} + \sum_{j=1}^{n+k} \tilde{\lambda}_{\alpha j}(\zeta, w) \frac{\partial}{\partial w^j}, \quad 1 \le \alpha \le n.$$

Note that

$$\frac{\partial \tilde{\lambda}_{\alpha j}}{\partial \zeta^\alpha}(t,u) = \frac{\partial \lambda_{\alpha j}}{\partial t^\alpha}(t,u), \quad \frac{\partial \tilde{\lambda}_{\alpha j}}{\partial w^\ell}(t,u) = \frac{\partial \lambda_{\alpha j}}{\partial u^\ell}(t,u), \tag{6.49}$$

(because $\tilde{\lambda}_{\alpha j}(\zeta, w)$ are holomorphic in ζ and w). At this point (6.48)-(6.49) and the identity theorem for holomorphic functions yield

$$[\tilde{L}_\alpha, \tilde{L}_\beta] = 0$$

on $\tilde{U} \subset \mathbb{C}^{2n+k}$. Therefore we may apply Lemma 1 in A. Boggess, [**50**], p. 56, to conclude that there is a holomorphic map

$$(W_1, \cdots, W_{n+k}) : \mathbb{C}^n \times \mathbb{C}^{n+k} \to \mathbb{C}^{n+k}$$

(defined on a possibly smaller neighborhood $\tilde{U}$ of $0 \in \mathbb{C}^{2n+k}$) such that

$$\tilde{L}_\alpha W_j = 0,$$

$$W_j(0, w) = w_j \,, \quad (0, w) \in \tilde{U}.$$

Similarly i.e. again by Lemma 1 in [**50**], p. 56, for the operators

$$\tilde{L}_\alpha + \sum_{s=1}^{p} \mu_{\alpha s} \frac{\partial}{\partial z^s}$$

with $\mu_{\alpha s} = 0$, there is a holomorphic function

$$(V_1, \cdots, V_{n+k}, \tilde{f}_1, \cdots, \tilde{f}_p) : \mathbb{C}^n \times \mathbb{C}^{n+k} \times \mathbb{C}^p \to \mathbb{C}^{n+k} \times \mathbb{C}^p$$

such that

$$\tilde{L}_\alpha V_j = 0, \quad \tilde{L}_\alpha \tilde{f}_s = 0,$$

$$V_j(0, w, z) = w_j \,, \quad \tilde{f}_s(0, w, z) = z_s \,.$$

Let ρ be the projection $(\zeta, w, z) \mapsto (\zeta, w)$ and let us set $\tilde{g}_j = W_j \circ \rho$ for $1 \le j \le n+k$. Next let us consider $\tilde{\psi} = (\tilde{g}_1, \cdots, \tilde{g}_{n+k}, \tilde{f}_1, \cdots, \tilde{f}_p)$ and let us define the C^ω map $\psi = (g_1, \cdots, g_{n+k}, f_1, \cdots, f_p)$ by setting

$$\psi(t, u, x) = \tilde{\psi}(t, u, x)$$

for $(t, u, x) \in \mathbb{R}^n \times \mathbb{R}^{n+k} \times \mathbb{R}^p$. Note that

$$\overline{L}_\alpha g_j = \tilde{L}_\alpha \tilde{g}_j = 0 \quad \text{on} \quad \tilde{U} \cap (\mathbb{R}^n \times \mathbb{R}^{n+k}).$$

Also g_j are basic. Moreover

$$\psi(0, u, x) = \tilde{\psi}(0, u, x) = (u, x).$$

Let us show that ψ is a generic embedding. As $\overline{L}_\alpha g_j = 0$ yields $\overline{\partial}_Q g_j = 0$ one should only check that $d_0\psi$ has real rank $2n + k + p$. Let us et $\psi = X + iY$. Note that

$$(u, x) = \psi(0, u, x) = X(0, u, x) + iY(0, u, x)$$

yields

$$d_0\psi = \begin{pmatrix} (\partial X/\partial t)(0) & I_{n+k+p} \\ (\partial Y/\partial t)(0) & 0 \end{pmatrix}.$$

Also the imaginary parts of $\overline{L}_\alpha g_j = 0$ and $\overline{L}_\alpha f_s = 0$ may be written in the following matrix form

$$\frac{\partial Y}{\partial t}(0) = -\begin{pmatrix} I_{n+k} \\ 0 \end{pmatrix} \cdot (\operatorname{Im} \lambda)^t(0)$$

where $\operatorname{Im}\lambda = [Im(\lambda_{\alpha j})]$. Next (by $\mathcal{H}\cap\overline{\mathcal{H}} = \{0\}$) $\{\zeta_1 - \zeta_{\overline{1}}, \cdots, \zeta_n - \zeta_{\overline{n}}\}$ are linearly independent (over $\mathbb{C}$) and therefore

$$\frac{1}{2i}(\zeta_{\overline{\alpha}} - \zeta_\alpha) = \sum_{j=1}^{n+k} \operatorname{Im}(\lambda_{\alpha j})\Pi\frac{\partial}{\partial u^j}$$

shows that $(\operatorname{Im}\lambda)(0)$ has rank n. Let us set $g = (g_1, \cdots, g_{n+k})$. Similarly, to show that $\operatorname{rank}(d_0 g) = 2n + k$ set $g = U + iV$. Then $g(0, u, x) = u$ yields

$$d_0 g = \begin{pmatrix} (\partial U/\partial t)(0) & I_{n+k} & 0 \\ (\partial V/\partial t)(0) & 0 & 0 \end{pmatrix}$$

and the imaginary part of $\overline{L}_\alpha g_j = 0$ may be written

$$\frac{\partial V}{\partial t} = -(\operatorname{Im}\lambda)^t(0)$$

so that $\operatorname{rank}(\partial V/\partial t)(0) = n$. Q.e.d.

Let us give an example of embedded transverse CR structure.

EXAMPLE 6.45. Let $N \subset \mathbb{C}^{n+1}$ be a nondegenerate real hypersurface and $M = N \times \mathbb{C}^{k+1}$ with the natural complex foliation $\mathcal{F}$. Let $\mathcal{H}$ be the transverse CR structure of $\mathcal{F}$. Then

$$\psi = (g_1, \cdots, g_{n+1}, f_1, \cdots, f_{2k}) : M \to \mathbb{C}^{n+1+2k}$$

given by $g_j(z,\zeta) = z^j$ and $f_a(z,\zeta) = \zeta^a$, $f_{k+a}(z,\zeta) = 0$ for $1 \leq j \leq n+1$ and $1 \leq a \leq k$, is a generic embedding of $(M, \mathcal{H})$. Indeed let $\{T_\alpha\}$ be a (local) frame of $T_{1,0}(N)$. Thus $\{T_\alpha, \partial/\partial\zeta^a\}$ is a (local) frame of $T_{1,0}(M)$. The coordinate functions z^j are holomorphic so that $z^j_{|N} \in CR(N)$ and hence g_j are CR functions on M. For each leaf $S = \{z\} \times \mathbb{C}^k$ of $\mathcal{F}$ we have $(g_j)_{|S} = const.$ so that $g_j \in \Omega^0_B(\mathcal{F})$. Then (by Theorem 9) $\overline{\partial}_Q g_j = 0$. If E is a vector bundle over $\mathbb{C}^k$ let g^*E be the pullback of E by $g : M \to \mathbb{C}^{n+1}$, $g(z,\zeta) = z$. Finally we need to check that $G : Q \to g^*T(\mathbb{C}^{n+1})$ is a bundle monomorphism. To this end let $G_x s = 0$. There is $Y = (V, W)$ such that $\Pi_x Y = s$ and $V \in T_z(N)$, $W \in T_\zeta(\mathbb{C}^k)$. Then $V = (d_x g)Y = Gs = 0$ so that $Y = (0, W) \in P_{M,x}$. Q.e.d. □

In the spirit of Nirenberg's nonimbeddable example (cf. e.g. A. Boggess, [**50**], p. 172-178) it is a natural question whether a deformation of our Example 1.18 may be produced such that to obtain a nonimbeddable (transversally) CR foliation. We leave this as an open problem.

CHAPTER 7

$\mathcal{G}$-Lie foliations

A theory of CR structures on real Lie algebras has been developed in a series of recent papers by G. Gigante & G. Tomassini, [**116**], and S. Donnini & G. Gigante, [**82**].

DEFINITION 7.1. If $\mathcal{G}$ is a real q-dimensional Lie algebra, a complex subalgebra $\mathbf{a} \subset \mathcal{G} \otimes_{\mathbb{R}} \mathbb{C}$ is a *CR structure* on $\mathcal{G}$ if $\mathbf{a} \cap \overline{\mathbf{a}} = (0)$. □

DEFINITION 7.2. A $\mathcal{G}$-valued 1-form $\omega \in \Omega^1(M, \mathcal{G})$ on M is called a *Maurer-Cartan* form if it satisfies the Maurer-Cartan equation

$$d\omega + \frac{1}{2}[\omega, \omega] = 0.$$

□

Given a Maurer-Cartan form ω on M let us set $P_x = \text{Ker}(\omega_x)$ for any $x \in M$. If $\omega_x : T_x(M) \to \mathcal{G}$ is surjective for any $x \in M$ then P is a smooth involutive distribution on M and therefore gives rise to a foliation $\mathcal{F}$ of codimension q of M.

DEFINITION 7.3. $\mathcal{F}$ is called a *$\mathcal{G}$-Lie foliation.* □

The terminology is due to E. Fedida, [**103**]. The aim of this chapter is to investigate the interplay between CR structures on real Lie algebras and transverse CR structures on $\mathcal{G}$-Lie foliations. Given a $\mathcal{G}$-Lie foliation $\mathcal{F}$ of M, any CR structure on $\mathcal{G}$ is observed to give rise naturally to a transverse CR structure on $(M, \mathcal{F})$. If $\mathcal{F}$ has dense leaves the converse is shown to hold as well. Moreover, if $\mathcal{F}$ is a complete $\mathcal{G}$-Lie foliation with dense leaves carrying the transverse CR structure $\mathcal{H}$ arising from a CR structure $\mathbf{a}$ on $\mathcal{G}$ then we show that the transverse Kohn-Rossi cohomology groups of $(M, \mathcal{F}, \mathcal{H})$ are isomorphic with the Kohn-Rossi cohomology groups of $(\mathcal{G}, \mathbf{a})$ (cf. Theorem 7.12).

With any nondegenerate transverse CR structure on a foliation $\mathcal{F}$ one may associate a natural *f-structure* in the normal bundle of $\mathcal{F}$. We give a homotopy classification of f-structures in the normal bundle of a $\mathcal{G}$-Lie foliation (cf. Theorem 7.15).

7.1. $\mathcal{G}$-Lie foliations and transverse CR structures

Let $\mathcal{F}$ be a codimension q foliation on the C^∞ manifold M. Let $P = T(\mathcal{F})$ and $Q = \nu(\mathcal{F})$ be respectively the tangent and normal bundles of the foliation and let $\Pi : T(M) \to Q$ be the natural bundle morphism. Let $V(M, \mathcal{F})$ be the Lie algebra of foliated vector fields on M.

DEFINITION 7.4. A foliation $\mathcal{F}$ of M is *transversally parallelizable* if there are q globally defined foliated vector fields $Y_1, \cdots, Y_q \in V(M, \mathcal{F})$ such that the associated transverse vector fields $\Pi Y_1, \cdots, \Pi Y_q$ are linearly independent at any point $x \in M$. □

Any $\mathcal{G}$-Lie foliation is known to be transversally parallelizable. Indeed let $\mathcal{F}$ be a $\mathcal{G}$-Lie foliation of M and $\{E_1, \cdots, E_q\}$ a basis of $\mathcal{G}$. The map ω_x induces a $\mathbb{R}$-linear isomorphism

$$\hat{\omega}_x : Q_x \to \mathcal{G}, \quad x \in M.$$

Let $s_j \in \Gamma^\infty(Q)$ such that $\hat{\omega}_x(s_j(x)) = E_j$ and $Y_j \in \mathcal{X}(M)$ such that $\Pi Y_j = s_j$, $1 \leq j \leq q$. Then $Y_j \in L(M, \mathcal{F})$ (so that $\mathcal{F}$ is transversally parallelizable) and

$$\omega([Y_i, Y_j]) = [E_i, E_j], \quad 1 \leq i, \ j \leq q. \tag{7.1}$$

Cf. e.g. Lemma 11.1 in P. Tondeur, [**243**], p. 145.

DEFINITION 7.5. If the foliated vector fields $Y_1, \cdots, Y_q$ can be chosen to be complete, i.e. such that each Y_i induces a global 1-parameter group of global transformations of M, then $\mathcal{F}$ is a *complete* $\mathcal{G}$-Lie foliation. □

Let $\mathcal{G}$ be a real Lie algebra and $\mathbf{a} \subset \mathcal{G} \otimes_{\mathbb{R}} \mathbb{C}$ a CR structure on $\mathcal{G}$. Set $A = Re\{\mathbf{a} \oplus \overline{\mathbf{a}}\}$. The integer $k = \dim_{\mathbb{R}} \mathcal{G}/A$ is the *codimension* of $\mathbf{a}$. Note that A carries the complex structure $J : A \to A$ given by $J(Z + \overline{Z}) = i(Z - \overline{Z})$ for any $Z \in \mathbf{a}$. Here $i = \sqrt{-1}$.

Let $\mathcal{G}$ be a real Lie algebra and $\mathbf{a} \subset \mathcal{G} \otimes_{\mathbb{R}} \mathbb{C}$ a CR structure. By a classical result in Lie group theory, there is a unique connected and simply connected Lie group such that its Lie algebra (of left invariant vector fields) is $\mathcal{G}$. Then G is a CR Lie group. Indeed let us set

$$T_{1,0}(G)_g = (d_e L_g) \, \mathrm{ev}_e \, \mathbf{a}$$

for any $g \in G$. Here ev_e is the ($\mathbb{C}$-linear extension to $\mathcal{G} \otimes_{\mathbb{R}} \mathbb{C}$ of the) $\mathbb{R}$-linear isomorphism $\mathcal{G} \approx T_e(G)$ given by the evaluation of invariant vector fields at e (and e is the identity in G). Then $T_{1,0}(G)$ is a left invariant CR structure on G.

We are mainly interested in CR structures (on real Lie algebras) of codimension $k = 1$. If this is the case one may recover the tools of pseudohermitian geometry. Precisely let $\mathcal{G}$ be a real Lie algebra and

$$d_{\mathcal{G}} : \Lambda^s \mathcal{G}^* \to \Lambda^{s+1} \mathcal{G}^*, \quad s \geq 0,$$

the Chevalley-Eilenberg complex of $\mathcal{G}$. Let $\mathbf{a}$ be a CR structure on $\mathcal{G}$.

DEFINITION 7.6. A form $\theta \in \mathcal{G}^*$ is a *pseudohermitian structure* on $(\mathcal{G}, \mathbf{a})$ if $\mathrm{Ker}(\theta) = A$. □

If θ, θ' are two pseudohermitian structures on $(\mathcal{G}, \mathbf{a})$ then $\theta' = \lambda\theta$ for some $\lambda \in \mathbb{R}$, $\lambda \neq 0$.

DEFINITION 7.7. The *Levi form* of $(\mathcal{G}, \mathbf{a})$ is given by

$$G_\theta(X, Y) = (d_{\mathcal{G}}\theta)(X, JY)$$

for any $X, Y \in A$. □

Clearly $G_{\lambda\theta} = \lambda G_\theta$. Next

DEFINITION 7.8. $(\mathcal{G}, \mathbf{a})$ is *nondegenerate* if G_θ is nondegenerate for some pseudohermitian structure θ on $(\mathcal{G}, \mathbf{a})$. □

If $(\mathcal{G}, \mathbf{a})$ is nondegenerate and a pseudohermitian structure θ has been fixed then there is a unique $T \in \mathcal{G}$, $T \neq 0$, such that

$$\theta(T) = 1, \quad T \rfloor d_{\mathcal{G}}\theta = 0.$$

DEFINITION 7.9. T is the *characteristic direction* of $d_{\mathcal{G}}\theta$. □

Let $(\mathcal{G}, \mathbf{a})$ be a real Lie algebra carrying a CR structure and let us consider the corresponding CR Lie group $(G, T_{1,0}(G))$. As $(G, T_{1,0}(G))$ is a CR manifold, we may consider its tangential Cauchy-Riemann complex

$$\overline{\partial}_G : \Gamma^\infty(\Lambda^s T_{0,1}(G)^*) \to \Gamma^\infty(\Lambda^{s+1} T_{0,1}(G)^*), \quad s \geq 0.$$

DEFINITION 7.10. An element $\alpha \in \Gamma^\infty(\Lambda^s T_{0,1}(G)^*)$ is *left invariant* if

$$\alpha_{ag}((d_g L_a)V_1, \cdots, (d_g L_a)V_s) = \alpha_g(V_1, \cdots, V_s) \tag{7.2}$$

for any $V_1, \cdots, V_s \in T_{0,1}(G)_g$, $g \in G$, $a \in G$. □

The left hand side of (7.2) makes sense because L_a is a CR map. Let $\Gamma^\infty_{inv}(\Lambda^s T_{0,1}(G)^*)$ be the space of all left invariant C^∞ sections α in $\Lambda^s T_{0,1}(G)^*$. The tangential Cauchy-Riemann operator $\overline{\partial}_G$ descends (because it commutes with the pullback of forms by left translations) to a differential operator

$$\overline{\partial}_G : \Gamma^\infty_{inv}(\Lambda^s T_{0,1}(G)^*) \to \Gamma^\infty_{inv}(\Lambda^{s+1} T_{0,1}(G)^*).$$

There is a natural $\mathbb{C}$-linear isomorphism

$$I_s : \Lambda^s \overline{\mathbf{a}}^* \to \Gamma^\infty_{inv}(\Lambda^s T_{0,1}(G)^*), \quad s \geq 0.$$

Let us set

$$\overline{\partial}_{\mathcal{G}} = I_{s+1}^{-1} \circ \overline{\partial}_G \circ I_s \, .$$

We obtain a complex

$$\overline{\partial}_{\mathcal{G}} : \Lambda^s \overline{\mathbf{a}}^* \to \Lambda^{s+1} \overline{\mathbf{a}}^*, \quad s \geq 0. \tag{7.3}$$

DEFINITION 7.11. (7.3) is the *Cauchy-Riemann complex* of $(\mathcal{G}, \mathbf{a})$ and its cohomology

$$H^{0,s}(\mathcal{G}, \mathbf{a}) = H^s(\Lambda^\bullet \overline{\mathbf{a}}^*, \overline{\partial}_{\mathcal{G}})$$

is the *Kohn-Rossi cohomology* of $(\mathcal{G}, \mathbf{a})$. □

We may state the following

THEOREM 7.12. (E. Barletta et al., [**19**])
Let $\mathcal{F}$ be a $\mathcal{G}$-Lie foliation of M determined by the Maurer-Cartan form $\omega \in \Omega^1(M, \mathcal{G})$. Then

1) *If $\mathbf{a}$ is a CR structure on $\mathcal{G}$ (of codimension k) then*

$$\mathcal{H}_x = \hat{\omega}_x^{-1}(\mathbf{a}), \quad x \in M,$$

is a transverse CR structure on $(M, \mathcal{F})$ (of transverse CR codimenssion k). If additionally $\mathcal{F}$ has at least a dense leaf then any transverse CR structure $\mathcal{H}$ on $(M, \mathcal{F})$ determines a unique CR structure $\mathbf{a}$ on $\mathcal{G}$.

2) *Let $\mathcal{F}$ be complete and let $\mathbf{a}$ be a CR structure on $\mathcal{G}$. If $(\mathcal{G}, \mathbf{a})$ is nondegenerate and $\mathcal{F}$ has dense leaves then*

$$H^s_{\overline{\partial}_Q}(M, \mathcal{F}) \approx H^{0,s}(\mathcal{G}, \mathbf{a}), \quad s \geq 0,$$

that is the transverse Kohn-Rossi cohomology of $(\mathcal{F}, \mathcal{H})$ is isomorphic to the Kohn-Rossi cohomology of $(\mathcal{G}, \mathbf{a})$.

We shall prove Theorem 7.12 in section later on. The complex (7.3) admits a simple description when $(\mathcal{G}, \mathbf{a})$ is nondegenerate. Indeed if this is the case then let $T \in \mathcal{G}$, $T \neq 0$, such that $\theta(T) = 1$ and $T \rfloor d_{\mathcal{G}}\theta = 0$.

DEFINITION 7.13. A s-form $\alpha \in \Lambda^s \mathcal{G}^* \otimes \mathbb{C}$ is a $(0,s)$-*form* (or a form of *type* $(0,s)$) if $\mathbf{a} \rfloor \alpha = 0$ and $T \rfloor \alpha = 0$. □

There is a natural identification of $\Lambda^s \overline{\mathbf{a}}^*$ with the space of all $(0,s)$-forms on $\mathcal{G}$. Then one may redefine $\overline{\partial}_{\mathcal{G}}$ as follows. Let α be a $(0,s)$-form on $\mathcal{G}$. Then $\overline{\partial}_{\mathcal{G}}\alpha$ is the unique $(0,s+1)$-form on $\mathcal{G}$ such that $\overline{\partial}_{\mathcal{G}}\alpha$ and $d_{\mathcal{G}}\alpha$ coincide when both are restricted to $\overline{\mathbf{a}} \otimes \cdots \otimes \overline{\mathbf{a}}$ ($s+1$ terms).

7.2. Transverse f-structures

Let $\mathcal{F}$ be a codimension $q = 2n+1$ foliation of M and $\mathcal{H}$ a nondegenerate transverse CR structure of transverse CR dimension n on $(M, \mathcal{F})$. Fix a transverse pseudohermitian structure θ and let ξ be the characteristic direction of $d_Q\theta$. We may prolongate the complex structure J_Q of the transverse Levi distribution H to a (holonomy invariant) endomorphism of Q by requesting that $J_Q\xi = 0$. Then $J_Q^3 + J_Q = 0$.

DEFINITION 7.14. A f-*structure* in Q is a bundle endomorphism $J : Q \to Q$ such that $J^3 + J = 0$ and $rank(J) = 2n$. □

Then $J_Q : Q \to Q$ is a (holonomy invariant) f-structure in Q (induced by $(\mathcal{H}, \theta)$). Let us set

$$G = \{g \in GL(2n+1, \mathbb{R}) : g\, J_0 = J_0\, g\}$$

where

$$J_0 = \begin{pmatrix} 0 & 0 & 0 \\ 0 & 0 & -I_n \\ 0 & I_n & 0 \end{pmatrix}.$$

Let $p_T^1 : B_T^1(M, \mathcal{F}) \to M$ be the principal $\mathrm{GL}(2n+1, \mathbb{R})$-bundle of all transverse frames and $Y(M, \mathcal{F})$ the associated bundle with standard fibre the homogeneous space $\mathrm{GL}(2n+1, \mathbb{R})/G$. Any f-structure in Q is a cross-section in $Y(M, \mathcal{F})$. We may state the following

THEOREM 7.15. (E. Barletta et al., [**19**])
Let $\mathcal{F}$ be a $\mathcal{G}$-Lie foliation of M of codimension $2n+1$. Then the set of homotopy classes of f-structures in $\nu(\mathcal{F})$ is in a one-to-one and on-to correspondence with the set of homotopy classes of continuous maps from M to $\mathrm{GL}^+(2n+1, \mathbb{R})/\mathrm{GL}_1(n, \mathbb{C})$ where $\mathrm{GL}_1(n, \mathbb{C}) = \mathrm{GL}(n, \mathbb{C}) \cap \mathrm{SL}(2n, \mathbb{R})$.

Let $\mathcal{F}$ be a codimension $q = 2n+k$ foliation of M and $\overset{\circ}{\nabla}$ its Bott connection. Let $\mathcal{H} \subset Q \otimes \mathbb{C}$ be a complex subbundle of complex rank n. Let us set $H = \mathrm{Re}\{\mathcal{H} \oplus \overline{\mathcal{H}}\}$. Then H carries the complex structure $J_Q : H \to H$ given by $J_Q(\alpha + \overline{\alpha}) = i(\alpha - \overline{\alpha})$ for any $\alpha \in \Gamma^\infty(\mathcal{H})$. We recall that $\mathcal{H}$ is a transverse almost CR structure (of transverse CR dimension n and transverse CR codimension k) if 1) $\mathcal{H} \cap \overline{\mathcal{H}} = (0)$, 2) H is parallel with respect to the Bott connection of $\mathcal{F}$, and 3) $\mathcal{L}_X J_Q = 0$ for any $X \in \Gamma^\infty(P)$. Let

$$d_B : \Omega_B^s(M, \mathcal{F}) \to \Omega_B^{s+1}(M, \mathcal{F}), \quad s \geq 0,$$

be the basic complex of the foliated manifold $(M, \mathcal{F})$. As shown previously there exist natural isomorphisms

$$\Phi_s : \Gamma_B^\infty(\Lambda^s Q^*) \to \Omega_B^s(M, \mathcal{F}), \quad s \geq 0,$$

and therefore an induced complex

$$d_Q : \Gamma_B^\infty(\Lambda^s Q^*) \to \Gamma_B^\infty(\Lambda^{s+1} Q^*), \quad s \geq 0.$$

As to the geometric meaning of the requirements 1)-3) in the definition of the notion of a transverse almost CR structure, let us mention that given a leaf L of $\mathcal{F}$ and $\gamma : [0,1] \to L$ a smooth curve in L then

$$\tau_\gamma \mathcal{H}_{\gamma(0)} = \mathcal{H}_{\gamma(1)} \tag{7.4}$$

where $\tau_\gamma : Q_{\gamma(0)} \to Q_{\gamma(1)}$ is the holonomy map. Indeed, let s be a solution of the ODE

$$\left(\mathring{\nabla}_{d\gamma/dt} s\right)_{\gamma(t)} = 0 \tag{7.5}$$

of initial data $s(\gamma(0)) \in \mathcal{H}_{\gamma(0)}$. Then

$$\frac{d}{dt}\{\theta(s(\gamma(t)))\} = \{(\mathcal{L}_{d\gamma/dt}\theta)s\}_{\gamma(t)} = 0$$

hence $\theta(s) \circ \gamma =$ constant on $[0,1]$. Since $s(\gamma(0)) \in \mathcal{H}_{\gamma(0)}$ then $0 = \theta(s)_{\gamma(0)} = \theta(s)_{\gamma(1)}$ that is $s(\gamma(1)) \in H_{\gamma(1)} \otimes_{\mathbb{R}} \mathbb{C}$. In a similar way, we may show (as $\mathcal{L}_X J_Q = 0$) that

$$J_Q \circ \tau_\gamma = \tau_\gamma \circ J_Q \,.$$

Then $J_{Q,\gamma(1)} s(\gamma(1)) = i\, s(\gamma(1))$ hence $s(\gamma(1)) \in \mathcal{H}_{\gamma(1)}$.

Let $(N, T_{1,0}(N))$ be a CR manifold and

$$\overline{\partial}_N : \Gamma^\infty(\Lambda^s T_{0,1}(N)^*) \to \Gamma^\infty(\Lambda^{s+1} T_{0,1}(N)^*), \quad s \geq 0,$$

its tangential Cauchy-Riemann complex. Assume that $(N, T_{0,1}(N))$ is nondegenerate of hypersurface type. Let θ_N be a fixed pseudohermitian structure on N and T_N the global nowhere zero tangent vector field on N such that $\theta_N T_N = 1$ and $T \rfloor d\theta_N = 0$.

DEFINITION 7.16. A CR map $f : N \to N$ is *pseudohermitian* if $f^*\theta_N = \theta_N$. □

If $f : N \to N$ is pseudohermitian then $(d_x f) T_{N,x} = T_{N,f(x)}$ for any $x \in N$ and $f^* \circ \overline{\partial}_N = \overline{\partial}_N \circ f^*$.

Let G be a CR Lie group and $\mathcal{G}$ its Lie algebra. Let $\mathbf{a}$ be the CR structure of $\mathcal{G}$ (associated with the left invariant CR structure of G). Let $\theta_0 \in \mathcal{G}^*$ be a pseudohermitian structure on $(\mathcal{G}, \mathbf{a})$. Then $\theta = I_1\theta_0$ is a left invariant pseudohermitian structure on G. Consequently any left translation L_a is a pseudohermitian map of (G, θ) into itself (and $L_a^* \overline{\partial}_G = \overline{\partial}_G L_a^*$).

Proof of Theorem 7.12. Let $\mathbf{a} \subset \mathcal{G} \otimes_{\mathbb{R}} \mathbb{C}$ be a CR structure on $\mathcal{G}$ and let us set

$$\mathcal{H}_x = \hat{\omega}_x^{-1}(\mathbf{a}) \subset Q_x \otimes_{\mathbb{R}} \mathbb{C}, \quad x \in M.$$

As $\hat{\omega}_x$ is a real operator it commutes with complex conjugation. Thus $\mathcal{H}_x \cap \overline{\mathcal{H}}_x = (0)$. We need to check that H and J_Q are parallel with respect to the Bott connection of $(M, \mathcal{F})$. Let us assume the basis $\{E_1, \cdots, E_{2n+k}\}$ of $\mathcal{G}$ is chosen such that $\{E_1, \cdots, E_{2n}\} \subset A$ and $E_{\alpha+n} = JE_\alpha$. Let $X \in \Gamma^\infty(P)$ and $s \in \Gamma^\infty(H)$. There exist functions $f^j \in \Omega^0(M)$, $1 \leq j \leq 2n$, such that $s = f^j s_j$. Let $Y_s \in \mathcal{X}(M)$ such that $\Pi Y_s = s$. Then

$$Y_s = f^j Y_j + X_s$$

for some $X_s \in \Gamma^\infty(P)$. Since $Y_j \in V(M, \mathcal{F})$ we have $\Pi[X, Y_j] = 0$ so that

$$\mathring{\nabla}_X s = X(f^j) s_j \in \Gamma^\infty(H).$$

Note that

$$(J_Q)_x = \hat{\omega}_x^{-1} \circ J \circ \hat{\omega}_x \,, \quad x \in M.$$

Then $J_Q s_\alpha = s_{\alpha+n}$ and $J_Q s_{\alpha+n} = -s_\alpha$. Finally $\mathcal{L}_X s_j = 0$ yields $(\mathcal{L}_X J_Q) s_j = 0$. Let us check that $\mathcal{H}$ is integrable. Let $\zeta_\alpha \in \ell(M, \mathcal{F}) \otimes \mathbb{C}$ defined by $\zeta_\alpha(x) = \hat{\omega}_x^{-1}(E_\alpha - iE_{\alpha+n})$ for any $x \in M$, $1 \leq \alpha \leq n$. Then $\{\zeta_\alpha\}$ is a global admissible frame of $\mathcal{H}$. Indeed (by (7.1)) we have

$$\hat{\omega}([\zeta_\alpha, \zeta_\beta])_x = [E_\alpha - iE_{\alpha+n}, E_\beta - iE_{\beta+n}] \in \mathbf{a}$$

as $\mathbf{a}$ is an algebra. Therefore $\mathcal{H}$ is a transverse CR structure. Viceversa let $\mathcal{H}$ be a transverse CR structure on $(M, \mathcal{F})$. Let $\{E_1, \cdots, E_{2n+k}\}$ be a basis of $\mathcal{G}$ and $s_j \in \Gamma^\infty(Q)$ such that $\hat{\omega}(s_j) = E_j$. Let $x \in M$ and let $U \subseteq M$ be an open neighborhood of x in M. Let $\{\zeta_1, \cdots, \zeta_n\}$ an admissible frame of $\mathcal{H}$ on U. Let us set

$$\mathbf{a}_x = \sum_{\alpha=1}^{n} \mathbb{C}\, \hat{\omega}(\zeta_\alpha)_x \subset \mathcal{G} \otimes_{\mathbb{R}} \mathbb{C}.$$

The definition of $\mathbf{a}_x$ doesn't depend upon the choice of admissible frame $\{\zeta_1, \cdots, \zeta_n\}$ on U. The resulting map $x \mapsto \mathbf{a}_x$ is locally constant. Indeed, there exist C^∞ functions $\lambda_\alpha^j : U \to \mathbb{C}$ such that $\zeta_\alpha = \lambda_\alpha^j s_j$. As $s_j \in \Gamma_B^\infty(Q)$ then λ_α^j are basic functions. Since at least one leaf of $\mathcal{F}$ is dense, each basic function is a constant. Thus

$$\mathbf{a}_x = \sum_{\alpha=1}^{n} \mathbb{C} \lambda_\alpha^j E_j = \text{ constant}$$

on U. Yet M is connected so that $x \mapsto \mathbf{a}_x$ is a constant map. Let us set $\mathbf{a} = \mathbf{a}_x$, $x \in M$. Then $\mathbf{a}$ is a CR structure on $\mathcal{G}$.

To prove the second statement in Theorem 7.12 we need to recall a few facts on the structure of complete $\mathcal{G}$-Lie foliations (cf. e.g. [**179**], p. 112-117). Let $\mathcal{F}$ be a complete $\mathcal{G}$-Lie foliation of M. Let G be the unique connected and simply connected Lie group whose Lie algebra is $\mathcal{G}$. Let $M \times G \to M$ be the trivial principal G-bundle (whose right translations R_h are given by $R_h(x, g) = (x, hg)$, for any $x \in M$, $g, h \in G$). Let $\mathcal{G}_\omega$ be the real q-dimensional Lie algebra spanned (over $\mathbb{R}$) by $\{s_1, \cdots, s_q\} \subset \Gamma_B^\infty(Q)$. Then $\mathcal{G}_\omega$ is a subalgebra of $\ell(M, \mathcal{F})$ (the inclusion $\mathcal{G}_\omega \subset \ell(M, \mathcal{F})$ is strict, in general) isomorphic to $\mathcal{G}$. Let L_ω be the Lie subalgebra of $V(M, \mathcal{F})$ consisting of all foliated vector fields whose associated transverse vector fields are elements of $\mathcal{G}_\omega$.

DEFINITION 7.17. The *lift* $\tilde{Y} \in \mathcal{X}(M \times G)$ of $Y \in L_\omega$ is given by

$$\tilde{Y}_{(x,g)} = (d_x \psi^g) Y_x + (d_g \psi_x)(\omega_x Y_x)_g \tag{7.6}$$

for any $(x, g) \in M \times G$. Here $\psi^g(x) = \psi_x(g) = (x, g)$. □

Let us set

$$\Gamma_{(x,g)} = \{\tilde{Y}_{(x,g)} \in T_{(x,g)}(M \times G) : Y \in L_\omega\}.$$

Then Γ is a connection in the principal G-bundle $M \times G$ over M. By (4.3) in [**179**], p. 113, Γ is flat and the leaves of the arising foliation are the holonomy bundles of Γ. Let $\tilde{M}$ be a leaf of the foliation determined by Γ. Let $p_1 : M \times G \to M$ and $p_2 : M \times G \to G$ be the natural projections and let us consider the maps $p : \tilde{M} \to M$ and $f_\omega : \tilde{M} \to G$ got respectively as restrictions of p_i, $i = 1, 2$, to the leaf $\tilde{M}$.

DEFINITION 7.18. $f_\omega : \tilde{M} \to G$ is called the *developing map* of the complete $\mathcal{G}$-Lie foliation $\mathcal{F}$. □

Then the central result of [**103**] states that $p : \tilde{M} \to M$ is a covering map while $f_\omega : \tilde{M} \to G$ is a locally trivial bundle. Moreover the pullback $p^*\mathcal{F}$ of $\mathcal{F}$ via $p : \tilde{M} \to M$ and the simple foliation defined by the submersion $f_\omega : \tilde{M} \to G$ actually coincide.

Let $\mathcal{G}$ be a real $(2n+1)$-dimensional Lie algebra. Let $\mathbf{a} \subset \mathcal{G} \otimes_{\mathbb{R}} \mathbb{C}$ be a nondegenerate CR structure on $\mathcal{G}$ and $\theta_0 \in \mathcal{G}^*$ a pseudohermitian structure on $(\mathcal{G}, \mathbf{a})$. Let $T \in \mathcal{G}$, $T \neq 0$, be the characteristic direction of $d_{\mathcal{G}}\theta_0$. Let $\mathcal{F}$ be a complete $\mathcal{G}$-Lie foliation of M. Let $\{E_1, \cdots, E_{2n+1}\}$ be a basis of $\mathcal{G}$ such that $E_{2n+1} = T$. Let us set

$$\xi_x = \hat{\omega}_x^{-1}(T), \quad x \in M. \tag{7.7}$$

Then $\xi \in \Gamma_B^\infty(Q)$. Moreover, let us set

$$(\theta s)_x = \theta_0\, \hat{\omega}(s)_x\,, \quad x \in M,$$

for any $s \in \Gamma^\infty(Q)$. It is then straightforward that $\theta \in \Gamma_B^\infty(Q^*)$ and

$$\theta(\xi) = 1, \quad \xi \rfloor d_Q\theta = 0.$$

That is, as $(\mathcal{G}, \mathbf{a})$ is nondegenerate $(\mathcal{F}, \mathcal{H})$ is nondegenerate as well, and ξ is the characteristic direction of $d_Q\theta$. Let $\alpha \in \Gamma_B^\infty(\Lambda^s \overline{\mathcal{H}}^*)$. Let us set

$$\tilde{\alpha} = p^* \Phi_s \alpha.$$

As $\Phi_s\alpha \in \Omega_B^s(M, \mathcal{F}) \otimes \mathbb{C}$ we have $\tilde{\alpha} \in \Omega_B^s(\tilde{M}, p^*\mathcal{F}) \otimes \mathbb{C}$ (cf. also [**243**], p. 148). Let $g \in G$ and $X_1, \cdots, X_s \in T_g(G)$. Consider $\tilde{x} \in f_\omega^{-1}(g)$ and $V_1, \cdots, V_s \in T_{\tilde{x}}(\tilde{M})$ such that $(d_{\tilde{x}} f_\omega)V_j = X_j$, $1 \le j \le s$. We define a s-form $f_\omega\tilde{\alpha}$ on G by setting

$$(f_\omega\tilde{\alpha})_g(X_1, \cdots, X_s) = \tilde{\alpha}_{\tilde{x}}(V_1, \cdots, V_s).$$

Step 1. *The definition of $(f_\omega\tilde{\alpha})_g(X_1, \cdots, X_s)$ doesn't depend upon the choice of $\tilde{x} \in f_\omega^{-1}(g)$ and $V_1, \cdots, V_s \in T_{\tilde{x}}(\tilde{M})$ so that $(d_{\tilde{x}} f_\omega)V_j = X_j$, $1 \le j \le s$.*

For the sake of simplicity we check this statement for $s = 1$ only. Let $\tilde{x}, \tilde{x}' \in f_\omega^{-1}(g)$ and $V \in T_{\tilde{x}}(\tilde{M})$, $V' \in T_{\tilde{x}'}(\tilde{M})$ such that

$$(d_{\tilde{x}} f_\omega)V = X, \quad (d_{\tilde{x}'} f_\omega)V' = X.$$

There are $x, x' \in M$ such that $\tilde{x} = (x, g)$ and $\tilde{x}' = (x', g)$. We distinguish two cases as I) there is a connected component $\tilde{L}$ of $f_\omega^{-1}(g)$ such that $\tilde{x}, \tilde{x}' \in \tilde{L}$, or II) $\tilde{x}$ and $\tilde{x}'$ lie in two distinct connected components of $f_\omega^{-1}(g)$. If case I) occurs then $\tilde{L}$ is a leaf of $p^*\mathcal{F}$. Also $L = p(\tilde{L})$ is a leaf of $\mathcal{F}$ and $p : \tilde{L} \to L$ is a Galois covering. As $(d_{(x,g)}p)T(p^*\mathcal{F})_{(x,g)} = P_x$ the map $d_{(x,g)}p$ induces a $\mathbb{R}$-linear isomorphism

$$[d_{(x,g)}p] : \nu(p^*\mathcal{F})_{(x,g)} \to Q_x .$$

It commutes with the holonomy maps. Indeed, let $\tilde{\gamma} : [0,1] \to \tilde{L}$ be a smooth curve such that $\tilde{\gamma}(0) = \tilde{x}$ and $\tilde{\gamma}(1) = \tilde{x}'$. Let us set $\gamma = p \circ \tilde{\gamma}$. Then γ is a smooth curve in the leaf L (connecting x and x'). Let $\tau_\gamma : Q_x \to Q_{x'}$ and $\tau_{\tilde{\gamma}} : \nu(p^*\mathcal{F})_{(x,g)} \to \nu(p^*\mathcal{F})_{(x',g)}$ be the corresponding holonomy maps. To show that

$$\tau_\gamma \circ [d_{(x,g)}p] = [d_{(x',g)}p] \circ \tau_{\tilde{\gamma}}$$

consider the solution $\tilde{s}$ of the ODE

$$\left(\mathring{\nabla}_{d\tilde{\gamma}/dt}\tilde{s}\right)_{\tilde{\gamma}(t)} = 0$$

with initial data $\tilde{s}((x,g)) \in \nu(p^*\mathcal{F})_{(x,g)}$ (the same symbol $\mathring{\nabla}$ denotes the Bott connection of $p^*\mathcal{F}$, as well). It suffices to show that

$$s(\gamma(t)) = [d_{\tilde{\gamma}(t)}p]\tilde{s}(\tilde{\gamma}(t))$$

satisfies the ODE (7.5). Let us set

$$Y_s(\gamma(t)) = (d_{\tilde{\gamma}(t)}p)Y_{\tilde{s}}(\tilde{\gamma}(t))$$

where $Y_{\tilde{s}} \in \mathcal{X}(\tilde{M})$ is chosen such that $\tilde{\Pi}Y_{\tilde{s}} = \tilde{s}$ (and $\tilde{\Pi} : T(\tilde{M}) \to \nu(p^*\mathcal{F})$ is the natural bundle morphism). Then $\Pi Y_s = s$ and we may conduct the following computation

$$0 = [d_{\tilde{\gamma}(t)}p]\left(\mathring{\nabla}_{d\tilde{\gamma}/dt}\tilde{s}\right)_{\tilde{\gamma}(t)} = [d_{\tilde{\gamma}(t)}p]\tilde{\Pi}_{\tilde{\gamma}(t)}\left[\frac{d\tilde{\gamma}}{dt}, Y_{\tilde{s}}\right]_{\tilde{\gamma}(t)} =$$

$$= \Pi_{\gamma(t)}(d_{\tilde{\gamma}(t)}p)\left[\frac{d\tilde{\gamma}}{dt}, Y_{\tilde{s}}\right]_{\tilde{\gamma}(t)} = \Pi_{\gamma(t)}\left[\frac{d\gamma}{dt}, Y_s\right]_{\gamma(t)} = (\mathring{\nabla}_{d\gamma/dt}s)_{\gamma(t)} .$$

To show that

$$\tilde{\alpha}_{(x,g)}V = \tilde{\alpha}_{(x',g)}V' \tag{7.8}$$

we need two facts. First let $[d_{(x,g)}f_\omega] : \nu(p^*\mathcal{F})_{(x,g)} \to T_g(G)$ be the $\mathbb{R}$-linear isomorphism induced by $d_{(x,g)}f_\omega$ (as $\mathrm{Ker}(d_{(x,g)}f_\omega) = T(p^*\mathcal{F})_{(x,g)}$). Then (cf. [**179**], p. 24) we have

$$\tau_{\tilde{\gamma}} = [d_{(x',g)}f_\omega]^{-1} \circ [d_{(x,g)}f_\omega]. \tag{7.9}$$

Next

$$\alpha_x = \alpha_{x'} \circ \tau_\gamma . \tag{7.10}$$

Indeed let $s_0 \in \overline{\mathcal{H}}_x$ and let $s(\gamma(t))$ be the solution of the ODE (7.5) with $s(\gamma(0)) = s_0$. Then $\tau_\gamma s_0 \in \overline{\mathcal{H}}_{x'}$ (by (7.4)). Moreover as $\alpha \in \Gamma^\infty_B(\overline{\mathcal{H}}^*)$ we have $\mathcal{L}_{d\gamma/dt}\alpha = 0$ and therefore

$$\frac{d}{dt}\left\{\alpha(s)_{\gamma(t)}\right\} = 0$$

i.e. $\alpha(s)_{\gamma(t)} =$ constant, etc. Using (7.10) we may conduct the following computation

$$\tilde{\alpha}_{(x',g)}V' = (p^*\Phi_1\alpha)_{(x',g)}V' = (\Phi_1\alpha)_{x'}(d_{(x',g)}p)V' =$$

$$= \alpha_{x'}[d_{(x',g)}p]\tilde{\Pi}_{(x',g)}V' = \alpha_x\tau_\gamma^{-1}[d_{(x',g)}p]\tilde{\Pi}_{(x',g)}V' =$$

$$= \alpha_x[d_{(x,g)}p]\tau_{\tilde{\gamma}}^{-1}\tilde{\Pi}_{(x',g)}V'.$$

Moreover

$$(d_{(x,g)}f_\omega)V = (d_{(x',g)}f_\omega)V'$$

so that (by (7.9))

$$[d_{(x,g)}f_\omega]\tilde{\Pi}_{(x,g)}V = [d_{(x',g)}f_\omega]\tilde{\Pi}_{(x',g)}V' = [d_{(x,g)}f_\omega]\tau_{\tilde{\gamma}}^{-1}\tilde{\Pi}_{(x',g)}V'$$

that is

$$\tau_{\tilde{\gamma}}(\tilde{\Pi}_{(x,g)}V) = \tilde{\Pi}_{(x',g)}V'.$$

Therefore we may conclude with the following computation

$$\tilde{\alpha}_{(x',g)}V' = \alpha_x[d_{(x,g)}p]\tau_{\tilde{\gamma}}^{-1}\tilde{\Pi}_{(x',g)}V' = \alpha_x[d_{(x,g)}p]\tilde{\Pi}_{(x,g)}V =$$
$$= \alpha_x\Pi_x(d_{(x,g)}p)V = (\Phi_1\alpha)_x(d_{(x,g)}p)V = (p^*\Phi_1\alpha)_{(x,g)}V = \tilde{\alpha}_{(x,g)}V$$

and (7.8) is completely proved.

If case II) occurs, let $\tilde{L}$ be the connected component of $\tilde{x}$ in $f_\omega^{-1}(g)$ (so that $\tilde{L}$ is a leaf of $p^*\mathcal{F}$) and let $L = p(\tilde{L})$ be the corresponding leaf of $\mathcal{F}$. Since $\mathcal{F}$ has at least one dense leaf one has $\Omega^0_B(M,\mathcal{F}) = \mathbb{R}$. Yet $\mathcal{F}$ is complete so that (by Prop. 4.2 in [**179**]) all leaves of $\mathcal{F}$ are dense in M. As L is dense then there is a sequence $(x_j)_{j\in\mathbf{N}}$ in L which tends to x' as $j\to\infty$. Let $\tilde{x}_j \in \tilde{L}$ such that $p(\tilde{x}_j) = x_j$, $j \in \mathbf{N}$. By the arguments in case I) we obtain

$$\tilde{\alpha}_{\tilde{x}}V = \tilde{\alpha}_{\tilde{x}_j}V_j \tag{7.11}$$

where $V_j \in T_{\tilde{x}_j}(\tilde{M})$ are chosen such that $(d_{\tilde{x}_j}f_\omega)V_j = X$, $j \in \mathbf{N}$. As p is a covering map we may choose open neighborhoods $\tilde{U} \subseteq \tilde{M}$ and $U \subseteq M$ of $\tilde{x}'$ and x' respectively such that $p : \tilde{U} \to U$ is a diffeomorphism. Then $\tilde{x}_j \in \tilde{U}$ for any $j \geq j_0$ and some $j_0 \geq 1$ (and thus $\lim_{j\to\infty}\tilde{x}_j = \tilde{x}'$). However this remark and (7.11) do not yield (7.8) directly (since there is no natural candidate for V' there). Indeed (7.11) doesn't necessarily imply that $(V_j)_{j\in\mathbf{N}}$ is convergent in $T(\tilde{M})$. We circumvent these difficulties as follows. Since $V_j \in T_{\tilde{x}_j}(\tilde{M}) = \Gamma_{\tilde{x}_j}$ (and Γ is determined by the Lie algebra L_ω) then there is $X_j \in T_{x_j}(M)$ such that

$$V_j = (d_{x_j}\Psi^g)X_j + (d_g\Psi_{x_j})(\omega_{x_j}X_j)_g\,.$$

Let $\mathrm{ev}_g : \mathcal{G} \to T_g(G)$ be the evaluation of (invariant) fields at g (an isomorphism). Then

$$(d_{\tilde{x}_j}p)V_j = X_j\,,\quad \Pi_{x_j}X_j = \hat{\omega}_{x_j}^{-1}(\mathrm{ev}_g^{-1}X)$$

as $p\circ\Psi^g = 1$ and $p\circ\Psi_x = $ constant, respectively $f_\omega\circ\Psi^g = $ constant and $f_\omega\circ\Psi_x = 1$. We may conduct the computation

$$\tilde{\alpha}_{\tilde{x}_j}V_j = (p^*\Phi_1\alpha)_{\tilde{x}_j}V_j = (\Phi_1\alpha)_{x_j}(d_{\tilde{x}_j}p)V_j =$$
$$= \alpha_{x_j}\Pi_{x_j}X_j = \alpha_{x_j}\hat{\omega}_{x_j}^{-1}(\mathrm{ev}_g^{-1}X).$$

Yet $x \mapsto \alpha_x\hat{\omega}_x^{-1}(\mathrm{ev}_g^{-1}X)$ is an element of $\Omega^0(M)\otimes\mathbb{C}$ and therefore continuous. Thus

$$\lim_{j\to\infty}\tilde{\alpha}_{\tilde{x}_j}V_j = \alpha_{x'}\hat{\omega}_{x'}^{-1}(\mathrm{ev}_g^{-1}X).$$

Let $s \in \mathcal{G}_\omega$ be defined by

$$s(y) = \hat{\omega}_y^{-1}(\mathrm{ev}_g^{-1}X)$$

for any $y \in M$. Let us choose $Y \in L_\omega$ such that $\Pi Y = s$ and let us set $V'' = \tilde{Y}_{\tilde{x}'}$ where $\tilde{Y}$ is the lift of Y (given by (7.6)). Then

$$\tilde{\alpha}_{\tilde{x}'}V'' = \alpha_{x'}\Pi_{x'}Y_{x'} = \alpha_{x'}s(x') = \alpha_{x'}\hat{\omega}_{x'}^{-1}(\Phi_{x'}^{-1}(\mathrm{ev}_g^{-1}X))$$

such that

$$\lim_{j\to\infty} \tilde{\alpha}_{\tilde{x}_j} V_j = \tilde{\alpha}_{\tilde{x}'} V''.$$

Let $j \to \infty$ in (7.11). We obtain

$$\tilde{\alpha}_{\tilde{x}} V = \tilde{\alpha}_{\tilde{x}'} V''. \tag{7.12}$$

Note that $V'' - V' \in \operatorname{Ker}(d_{\tilde{x}'} f_\omega) = T(p^*\mathcal{F})_{\tilde{x}'}$. Yet $p_* T(p^*\mathcal{F}) = P$ so that $p_* V'' = p_* V' + Y$ for some $Y \in P_{x'}$. Finally $\tilde{\alpha} V'' = \tilde{\alpha} V' + Y \rfloor \Phi_1 \alpha$ (and $\Phi_1\alpha$ is a basic form on $(M, \mathcal{F})$) so that (7.12) may be written in the form (7.8). This ends the proof of Step 1.

Step 2. *$f_\omega \tilde{\alpha}$ is a left invariant form on G.*

Let $x \in M$ and $\tilde{x} \in p^{-1}(x)$. Let us set

$$H = \{g \in G : R_g(\tilde{x}) \in \tilde{M}\}.$$

Then H is a subgroup of G. Moreover the definition of H does not depend upon the choice of $x \in M$ and $\tilde{x} \in p^{-1}(x)$ (cf. e.g. [**179**], p. 115). Let $a \in H, g \in G$ and $\tilde{x} \in f_\omega^{-1}(g)$. Let $X \in T_g(G) \otimes \mathbb{C}$. We wish to compute $(L_a^* f_\omega \tilde{\alpha})_g X$. As $f_\omega \circ R_a = L_a \circ f_\omega$ we observe that $R_a(\tilde{x}) \in f_\omega^{-1}(ag)$. Let us set $X' = (d_g L_a) X$ and $V' = (d_{\tilde{x}} R_a) V$ where $V \in T_{\tilde{x}}(\tilde{M}) \otimes \mathbb{C}$ is chosen such that $(d_{\tilde{x}} f_\omega) V = X$. Then

$$(d_{R_a(\tilde{x})} f_\omega) V' = X'$$

so that (by $p \circ R_a = p$)

$$L_a^* f_\omega \tilde{\alpha} = f_\omega \tilde{\alpha} \tag{7.13}$$

for any $a \in H$. Nevertheless, as $\mathcal{F}$ has dense leaves H is dense in G (cf. e.g. [**243**], p. 148) so that (7.13) holds at any $a \in G$. It follows that $f_\omega \tilde{\alpha}$ is a left invariant form. Step 2 is completely proved.

Step 3. *If $\alpha_0 = I_s^{-1}(f_\omega \tilde{\alpha})$ then $\alpha_0 \in \Lambda^s \mathbf{a}^*$.*

Again, we prove Step 3 for $s = 1$ only. Indeed as $\alpha \in \Gamma_B^\infty(\overline{\mathcal{H}}^*)$ we have $\xi \rfloor \alpha = 0$ and $\mathcal{H} \rfloor \alpha = 0$, where ξ is given by (7.7). Let $T \in \mathcal{G}$ be the characteristic direction of $d_{\mathcal{G}} \theta_0$. Let $\tilde{x} \in f_\omega^{-1}(e)$ and $V \in T_{\tilde{x}}(\tilde{M})$ such that $(d_{\tilde{x}} f_\omega) V = T_e$. Since $T_{\tilde{x}}(\tilde{M}) = \Gamma_{(x,e)}$, $x = p(\tilde{x})$, there is $Y \in L_\omega$ such that $V = \tilde{Y}_{(x,e)}$ where $\tilde{Y}$ is the lift of Y. Then

$$(d_{\tilde{x}} p) V = Y,$$
$$\Pi_x Y_x = \hat{\omega}_x^{-1}(T) = \xi_x\,,$$

so that we may conduct the following computation

$$\alpha_0(T) = (I_1^{-1} f_\omega \tilde{\alpha}) T = (f_\omega \tilde{\alpha})_e T_e = \tilde{\alpha}_{\tilde{x}} V = (p^* \Phi_1 \alpha)_{\tilde{x}} V =$$
$$= (\Phi_1 \alpha)_x (d_{\tilde{x}} p) V = \alpha_x \Pi_x Y_x = \alpha(\xi)_x = 0.$$

If $Z \in \mathbf{a}$ then it may be shown in a similar way that

$$\alpha_0(Z) = \alpha_x \hat{\omega}_x^{-1}(Z) = 0$$

(as $\hat{\omega}_x^{-1}(Z) \in \mathcal{H}_x$). Step 3 is completely proved. To end the proof of Theorem 7.12 we need to establish the following

Step 4. *The map*

$$\Gamma_B^\infty(\Lambda^s \overline{\mathcal{H}}^*) \to \Lambda^s \overline{\mathbf{a}}^* \quad , \quad \alpha \mapsto \alpha_0\,, \tag{7.14}$$

induces an isomorphism

$$H^s_{\overline{\partial}_Q}(M,\mathcal{F}) \to H^{0,s}(\mathcal{G},\mathbf{a}) \ , \ \ [\alpha] \mapsto [\alpha_0].$$

Here brackets indicate cohomology classes. We need the transverse Cauchy-Riemann complex of a CR foliation. Let $(\mathcal{F},\mathcal{H})$ be a CR foliation. There is a complex

$$\overline{\partial}_Q : \Gamma^\infty_B(\Lambda^s\overline{\mathcal{H}}^*) \to \Gamma^\infty_B(\Lambda^{s+1}\overline{\mathcal{H}}^*), \quad s \geq 0, \tag{7.15}$$

which is most easily described when $(\mathcal{F},\mathcal{H})$ is nondegenerate. We recall that elements in $\Gamma^\infty_B(\Lambda^s\overline{\mathcal{H}}^*)$ are transverse $(0,s)$-forms invariant by holonomy i.e. those $\alpha \in \Gamma^\infty_B(\Lambda^s Q^* \otimes \mathbb{C})$ such that $\xi \rfloor \alpha = 0$ and $\mathcal{H} \rfloor \alpha = 0$. Next $\overline{\partial}_Q\alpha$ is the unique transverse $(0,s+1)$-form which coincides with $d_Q\alpha$ when both are restricted to $\overline{\mathcal{H}} \otimes \cdots \otimes \overline{\mathcal{H}}$ ($s+1$ terms). Finally, the cohomology

$$H^s_{\overline{\partial}_Q}(M,\mathcal{F}) = H^s(\Gamma^\infty_B(\Lambda^\bullet\overline{\mathcal{H}}^*),\overline{\partial}_Q), \quad s \geq 0,$$

of (7.15) is the transverse Kohn-Rossi cohomology of $(\mathcal{F},\mathcal{H})$.

As (7.14) is already an isomorphism, to prove Step 4 we only need to check that $[\alpha] \mapsto [\alpha_0]$ is well defined. This amounts to checking that $(\overline{\partial}_Q\beta)_0$ is a coboundary for any $\beta \in \Gamma^\infty_B(\Lambda^{s-1}\overline{\mathcal{H}}^*)$. Note first that

$$d\tilde{\alpha} = (d_Q\alpha)^\sim. \tag{7.16}$$

Indeed

$$d\tilde{\alpha} = dp^*\Phi_s\alpha = p^*d\Phi_s\alpha = p^*\Phi_{s+1}d_Q\alpha = (d_Q\alpha)^\sim.$$

By (7.16) we are entitled to consider $f_\omega d\tilde{\alpha}$. Moreover we have

$$f_\omega d\tilde{\alpha} = df_\omega\tilde{\alpha} \tag{7.17}$$

for any $\alpha \in \Gamma^\infty_B(\Lambda^s\overline{\mathcal{H}}^*)$. The identity (7.17) follows from Prop. 3.11 in [**155**], vol. I, p. 36. Finally a computation based on (7.17) leads to

$$\overline{\partial}_{\mathcal{G}}\beta_0 = (\overline{\partial}_Q\beta)_0$$

and Step 4 is completely proved.

Proof of Theorem 7.15. Let

$$\varphi : B^1_T(M,\mathcal{F}) \to \mathrm{End}(Q)$$

be the bundle morphism $x \mapsto \varphi_x$ given by $\varphi_x(z) : Q_x \to Q_x$, $x = p^1_T(z)$, where $\varphi_x(z)$ is the linear map whose matrix with respect to the basis $\{z(e_1),\cdots,z(e_q)\}$ is J_0 and $\{e_1,\cdots,e_q\}$ is the canonical basis in $\mathbb{R}^q$. We need the following

LEMMA 7.19. *Let $z \in B^1_T(M,\mathcal{F})$ with $x = p^1_T(z)$ and $g \in \mathrm{GL}(q,\mathbb{R})$. Then $\varphi_x(z) = \varphi_x(zg)$ if and only if $g \in G$.*

The proof is straightforward. By Lemma 7.19 we have

$$\mathrm{Im}(\varphi) \approx B^1_T(M,\mathcal{F})/G.$$

On the other hand (cf. [**155**], vol. I, p. 57)

$$Y(M,\mathcal{F}) = \frac{B^1_T(M,\mathcal{F}) \times (\mathrm{GL}(q,\mathbb{R})/G)}{\mathrm{GL}(q,\mathbb{R})} \approx B^1_T(M,\mathcal{F})/G.$$

Let $J \in \Gamma^\infty(\text{End}(Q))$ be a f-structure in Q. Then $J \in \Gamma^\infty(\text{Im}(\varphi))$ that is any f-structure in Q may be thought of (via $\text{Im}(\varphi) \approx B^1_T(M,\mathcal{F})/G \approx Y(M,\mathcal{F})$) as a section in $Y(M,\mathcal{F})$. Let $\mathcal{Y}(M,\mathcal{F})$ be the set of all homotopy classes of C^∞ sections in $Y(M,\mathcal{F})$. As $\mathcal{F}$ is a $\mathcal{G}$-Lie foliation it is transversally parallelizable hence

$$B^1_T(M,\mathcal{F}) \approx M \times \text{GL}(2n+1,\mathbb{R})$$

and consequently the associated bundle $Y(M,\mathcal{F})$ is trivial as well

$$Y(M,\mathcal{F}) \approx M \times (\text{GL}(2n+1,\mathbb{R})/G)$$

Thus (cf. [**230**], section 5.7) $\mathcal{Y}(M,\mathcal{F})$ is in a one-to-one and on-to correspondence with the set of homotopy classes of continuous maps from M to $\text{GL}(2n+1,\mathbb{R})/G$. Note that

$$\text{GL}(2n+1,\mathbb{R})/G \approx \text{GL}^+(2n+1,\mathbb{R})/G^+$$

where $\text{GL}^+(2n+1,\mathbb{R}) = \{g \in \text{GL}(2n+1,\mathbb{R}) : \det(g) > 0\}$ and $G^+ = G \cap \text{GL}^+(2n+1,\mathbb{R})$. Define $\text{GL}_1(n,\mathbb{C}) = \{g \in \text{GL}(n,\mathbb{C}) : |\det(g)| = 1\}$ (e.g. $\text{SL}(n,\mathbb{C}) \subset \text{GL}_1(n,\mathbb{C})$ yet inclusion is strict). Then $\text{GL}(n,\mathbb{C}) \to \text{GL}^+(2n+1,\mathbb{R})$ induces a group monomorphism $\text{GL}_1(n,\mathbb{C}) \to \text{GL}^+(2n+1,\mathbb{R})$. We need the following

LEMMA 7.20. *Let $\mathbb{R}_+ = (0,\infty)$ be the multiplicative positive reals. Then*

$$\frac{\text{GL}^+(2n+1,\mathbb{R})}{\text{GL}_1(n,\mathbb{C})} \to \frac{\text{GL}^+(2n+1,\mathbb{R})}{G^+}$$

is a principal $\mathbb{R}^2_+$-bundle.

Proof. The following short sequence of groups and group homomorphisms

$$1 \to \text{GL}_1(n,\mathbb{C}) \longrightarrow G^+ \xrightarrow{\rho} \mathbb{R}_+ \times \mathbb{R}_+ \to 1$$

where

$$\rho : \begin{pmatrix} a & 0 & 0 \\ 0 & A & -B \\ 0 & B & A \end{pmatrix} \mapsto (a, |\det(A+iB)|)$$

is exact. Then Lemma 7.20 is got from the following computation

$$\left(\text{GL}^+(2n+1,\mathbb{R})/\text{GL}_1(n,\mathbb{C})\right)/\mathbb{R}^2_+ \approx \frac{\text{GL}^+(2n+1,\mathbb{R})/\text{GL}_1(n,\mathbb{C})}{G^+/\text{Ker}(\rho)} \approx$$

$$\approx \frac{\text{GL}^+(2n+1,\mathbb{R})/\text{GL}_1(n,\mathbb{C})}{G^+/\text{GL}_1(n,\mathbb{C})} \approx \text{GL}^+(2n+1,\mathbb{R})/G^+.$$

Cf. Theorem 5.7 in [**155**], vol. I, each bundle whose standard fibre diffeomorphic to $\mathbb{R}^m$ (for some m) admits global sections (and is therefore trivial). Thus (by Lemma 7.20)

$$\text{GL}^+(2n+1,\mathbb{R})/\text{GL}_1(n,\mathbb{C}) \approx (\text{GL}(2n+1,\mathbb{R})/G) \times \mathbb{R}^2_+ .$$

Yet $\mathbb{R}^2_+$ is nullhomotopic so that $\text{GL}(2n+1,\mathbb{R})/G$ is homotopically equivalent to $\text{GL}^+(2n+1,\mathbb{R})/\text{GL}_1(n,\mathbb{C})$, and Theorem 7.15 is completely proved.

CHAPTER 8

Transverse Beltrami equations

Let M be a CR manifold of hypersurface type of CR dimension N. Let μ be a pointwise $\mathbb{C}$-anti-linear endomorphism of $T_{1,0}(M)$. Let $\{T_1, \cdots, T_N\}$ be a (local) frame of $T_{1,0}(M)$ and consider the first order PDE (with variable coefficients)

$$T_{\bar{j}}(f) = \mu^{\ell}_{\bar{j}} T_{\ell}(f) \tag{8.1}$$

where $\mu T_j = \mu^{\ell}_{\bar{j}} T_{\ell}$.

DEFINITION 8.1. (8.1) is called the *tangential Beltrami equation.* □

The terminology is essentially due to A. Korányi & H.M. Reimann, [**159**]-[**160**]. The tangential Beltrami equation is a CR analogue of the Beltrami equation $\partial_{\bar{z}} f = \mu \, \partial_z f$ in one complex variable, cf. e.g. J.A. Cima & W.R. Derrick, [**69**]. It is known that one of the ways to produce quasiconformal homeomorphisms of domains in the plane $\mathbb{C}$ is to solve the Beltrami equation, cf. e.g. L.V. Ahlfors, [**3**]. As to the CR case, A. Korányi & H.M. Reimann, [**159**], have demonstrated the connection between (8.1) and the K-quasiconformal automorphisms of the given CR manifold M. Also, they indicate a class of nonconstant solutions of (8.1) when $M = \mathbb{H}_1$ (the lowest dimensional Heisenberg group). Nevertheless, A. Korányi & H.M. Reimann's theory is confined to the case of strictly pseudoconvex CR manifolds, and breaks down if, for instance, one allows some degeneracy for the Levi form of the manifold.

Let M be a CR manifold and θ a pseudohermitian structure on M. Let $P = \{X \in H(M) : (d\theta)(X, \cdot) = 0 \text{ on } H(M)\}$ where $H(M)$ is the Levi distribution of M. If $\dim_{\mathbb{R}} P_x = 2k$, $x \in M$, then M is foliated by complex k-manifolds. If the complex structure induced by J in $H = H(M)/P$ is invariant under sliding along the leaves of the foliation $\mathcal{F}$ determined by P then $\mathcal{F}$ is a (transversally) CR foliation of M. While A. Korányi & H.M. Reimann's results do not apply (as M is degenerate) we may use foliation theory to deal with the following class of Beltrami equations. Let $Q = T(M)/P$ be the normal bundle of $\mathcal{F}$ and $\Pi : T(M) \to Q$ the projection. Let $N = n + k$. Assume the local frame $\{T_j\}$ is chosen in such a way that $\{T_{n+1}, \cdots, T_N\}$ is a frame of $P^{1,0}$ and $\zeta_\alpha = \Pi T_\alpha$, $1 \le \alpha \le n$, is an admissible frame of the transverse CR structure $\mathcal{H} = T_{1,0}(M)/P^{1,0}$ of $(M, \mathcal{F})$. Next let us assume that $P \subseteq \text{Ker}(\mu)$ so that μ descends to an endomorphism of $\mathcal{H}$. Finally if we restrict ourselves to *basic* unknown functions $f \in \Omega^0_B(\mathcal{F})$ then (8.1) may be written as

$$\zeta_{\bar{\alpha}}(f) = \mu^{\beta}_{\bar{\alpha}} \zeta_{\beta}(f). \tag{8.2}$$

This makes sense for an arbitrary CR foliation $\mathcal{F}$ of a C^∞ manifold M endowed with a $\mathbb{C}$-anti-linear endomorphism μ of its transverse CR structure and is invariant under a change of admissible frame.

DEFINITION 8.2. We refer to (8.2) as the (*transverse*) *Beltrami equation* of $(M, \mathcal{F})$. $\square$

We use the theory of CR foliations as developed in the previous sections to show that the components of an automorphism ϕ preserving the transverse contact structure of a given embedded strictly pseudoconvex CR foliation satisfy (8.2) where μ is the complex dilatation of ϕ (cf. Theorem 8.9) and conversely (cf. Theorem 8.10). For transversally Heisenberg CR foliations we characterize K-quasiconformality of a foliation automorphism (cf. Theorem 8.14).

8.1. Automorphisms of the transverse contact structure

Let $\mathcal{F}$ be a CR foliation of M.

DEFINITION 8.3. A C^∞ diffeomorphism $\phi : M \to M$ is an *automorphism* of $(M, \mathcal{F})$ if $(d_x\phi)P_x = P_{\phi(x)}$ for any $x \in M$. $\square$

Let $\mathrm{Aut}(\mathcal{F})$ be the group of all global automorphisms of $(M, \mathcal{F})$. The differential $d_x\phi$ of $\phi \in \mathrm{Aut}(\mathcal{F})$ at a point $x \in M$ induces a $\mathbb{R}$-linear isomorphism $[d_x\phi] : Q_x \to Q_{\phi(x)}$. Next

DEFINITION 8.4. $\phi \in \mathrm{Aut}(\mathcal{F})$ is an *automorphism of the transverse contact structure* if it preserves the transverse Levi distribution i.e.

$$[d_x\phi]H_x = H_{\phi(x)}\,, \quad x \in M.$$

If additionally

$$[d_x\phi] \circ J_x = J_{\phi(x)} \circ [d_x\phi], \quad x \in M,$$

then ϕ is an *automorphism of the transverse CR structure.* $\square$

Let $\mathrm{Aut}_H(\mathcal{F})$ (respectively $\mathrm{Aut}_{CR}(\mathcal{F})$) be the group of all automorphisms of the transverse contact structure (respectively of the transverse CR structure). Then

$$\mathrm{Aut}_{CR}(\mathcal{F}) \subset \mathrm{Aut}_H(\mathcal{F}) \subset \mathrm{Aut}(\mathcal{F}).$$

Inclusions are strict, in general. Let $\phi \in \mathrm{Aut}(\mathcal{F})$ and let ω be a C^∞ section in $\Lambda^k Q^*$. Define ω_T and ω_ϕ by setting

$$\omega_T(Y_1, \cdots, Y_k) = \omega(\Pi Y_1, \cdots, \Pi Y_k)$$

for any $Y_j \in T(M)$, and

$$\omega_{\phi,x}(z_1, \cdots, z_k) = \omega_{\phi(x)}\left([d_x\phi]z_1, \cdots, [d_x\phi]z_k\right)$$

for any $z_j \in Q_x$, $x \in M$. Then

$$(\omega_\phi)_T = \phi^*\omega_T\,.$$

The restriction of $\omega \mapsto \omega_T$ to $\Omega^k_B(\mathcal{F})$ gives an isomorphism onto the $\Omega^0_B(\mathcal{F})$-module of all basic k-forms on M. For any $\omega \in \Gamma^\infty_B(\Lambda^k Q^*)$ the pullback $\phi^*\omega_T$ is a basic k-form hence $\omega_\phi \in \Gamma^\infty_B(\Lambda^k Q^*)$. Also

$$d_Q\omega_\phi = (d_Q\omega)_\phi\,. \tag{8.3}$$

Let $\mathcal{F}$ be a CR foliation and $\theta \in \Gamma^\infty_B(Q^*)$ a transverse pseudohermitian structure. Let $\phi \in \mathrm{Aut}(\mathcal{F})$ and $\theta_\phi \in \Gamma^\infty_B(Q^*)$. It is easily seen that θ_ϕ is a transverse pseudohermitian structure provided that $\phi \in \mathrm{Aut}_H(\mathcal{F})$. We shall need the following

LEMMA 8.5. *Let $(\mathcal{F}, \mathcal{H})$ be a strictly pseudoconvex CR foliation. Let $\phi \in \mathrm{Aut}_H(\mathcal{F})$ be orientation preserving. Then*

$$[d_x\phi]\overline{\zeta} \notin \mathcal{H}_{\phi(x)}$$

for any $\zeta \in \mathcal{H}_x$, $\zeta \neq 0$, and any $x \in M$.

The proof is by contradiction. Assume that $\eta = [d_x\phi]\overline{\zeta} \in \mathcal{H}_{\phi(x)}$ for some $\zeta \in \mathcal{H}_x$, $\zeta \neq 0$, and some $x \in M$. By hypothesis, there is a transverse pseudohermitian structure $\theta \in \Gamma_B^\infty(Q^*)$ such that L_θ is positive-definite. Thus $\theta_\phi = \lambda\theta$ for some $\lambda \in \Gamma_B^\infty(Q^*)$, and $\lambda > 0$ everywhere (because ϕ is orientation preserving). Then (by (8.3))

$$\begin{aligned} 0 < L_{\theta,\phi(x)}(\eta, \overline{\eta}) &= -i(d_Q\theta)_{\phi,x}(\overline{\zeta}, \zeta) = \\ &= i(d_Q\theta_\phi)_x(\zeta, \overline{\zeta}) = -\lambda(x)L_{\theta,x}(\zeta, \overline{\zeta}) < 0 \end{aligned}$$

a contradiction.

If $\phi \in \mathrm{Aut}_{CR}(\mathcal{F})$ then $[d_x\phi]\mathcal{H}_x = \mathcal{H}_{\phi(x)}$ for any $x \in M$. On the other hand, this is not necessarily true for $\phi \in \mathrm{Aut}_H(\mathcal{F})$. If this is the case let

$$\mathcal{H}_{*,x} = \left\{\zeta \in H_x \otimes_{\mathbb{R}} \mathbb{C} : [d_x\phi]\zeta \in \mathcal{H}_{\phi(x)}\right\}.$$

We establish

PROPOSITION 8.6. *Let $(\mathcal{F}, \mathcal{H})$ be a strictly pseudoconvex CR foliation of M and $\phi \in \mathrm{Aut}_H(\mathcal{F})$, orientation preserving. Then there is a unique holonomy invariant, fibrewise $\mathbb{C}$-anti-linear bundle morphism $\mu : \mathcal{H} \to \mathcal{H}$ such that*

$$\mathcal{H}_{*,x} = \left\{\zeta - \overline{\mu_x\zeta} : \zeta \in \mathcal{H}_x\right\}$$

for any $x \in M$.

Proof. Let $\{\zeta_\alpha\}$ be an admissible local frame of $\mathcal{H}$ i.e. $\zeta_\alpha \in \ell(\mathcal{F}) \otimes \mathbb{C}$ and $[\zeta_\alpha, \zeta_\beta] \in \mathcal{H}$, $1 \leq \alpha, \beta \leq n$. We look for μ in the form

$$\mu\zeta_\alpha = \mu_{\overline{\alpha}}^{\beta}\zeta_\beta .$$

Let $\pi^{0,1} : H \otimes \mathbb{C} \to \overline{\mathcal{H}}$ be the natural projection associated with the direct sum decomposition $H \otimes \mathbb{C} = \mathcal{H} \oplus \overline{\mathcal{H}}$. Note that

$$\mathcal{H}_{*,x} = \mathrm{Ker}\left(\pi^{0,1}_{\phi(x)} \circ [d_x\phi]\right).$$

Let us set

$$[d_x\phi]\zeta_\alpha = \phi_\alpha^\beta\zeta_\beta + \phi_\alpha^{\overline{\beta}}\zeta_{\overline{\beta}}.$$

Then $\zeta_\alpha - \overline{\mu\zeta_\alpha} \in \mathcal{H}_*$ if and only if

$$\phi_{\overline{\alpha}}^{\beta} = \mu_{\overline{\alpha}}^{\gamma}\phi_\gamma^\beta \tag{8.4}$$

where $\phi_{\overline{\alpha}}^{\beta} = \overline{\phi_\alpha^{\overline{\beta}}}$. To solve (8.4) for the unknown functions $\mu_{\overline{\alpha}}^{\beta}$ it suffices to show that $[\phi_\alpha^\beta]$ is nonsingular. Indeed if $\det[\phi_\alpha^\beta(x)] = 0$ for some $x \in M$ then there is $(Z^1, \cdots, Z^n) \in \mathbb{C}^n \setminus \{0\}$ such that $\phi_\alpha^\beta(x)Z^\alpha = 0$. Let $\zeta = Z^\beta\zeta_\beta \in \mathcal{H}_x \setminus \{0\}$. Then

$$[d_x\phi]\overline{\zeta} = \overline{Z^\beta}\phi_{\overline{\beta}}^{\alpha}(x)\zeta_{\alpha,\phi(x)} \in \mathcal{H}_{\phi(x)},$$

a contradiction (by Lemma 8.5). Finally $\mathcal{L}_X\mu = 0$ for any $X \in P$ because ϕ_α^β, $\phi_\alpha^{\overline{\beta}}$ are basic functions.

DEFINITION 8.7. Given $\phi \in \mathrm{Aut}_H(\mathcal{F})$ we call $\mu : \mathcal{H} \to \mathcal{H}$ (furnished by Proposition 8.6) the *complex dilatation* of ϕ. □

Then

PROPOSITION 8.8. *Let $(\mathcal{F},\mathcal{H})$ be a strictly pseudoconvex CR foliation and $\phi \in \mathrm{Aut}_H(\mathcal{F})$, orientation preserving. Let $\mu : \mathcal{H} \to \mathcal{H}$ be the complex dilatation of ϕ. Then*

i) $(d_Q\theta)(\zeta, \overline{\mu\eta}) + (d_Q\theta)(\overline{\mu\zeta}, \eta) = 0$ *for any* $\zeta,\, \eta \in \mathcal{H}$.

ii) $\mu = 0$ *if and only if* $\phi \in \mathrm{Aut}_{CR}(\mathcal{F})$.

The proof is straightforward. Let $(\mathcal{F},\mathcal{H})$ be a CR foliation of M. We recall that an embedding of $(M,\mathcal{H})$ is a C^∞ immersion

$$\psi = (g^1, \cdots, g^{n+1}, h^1, \cdots, h^r) : M \to \mathbb{C}^N\,, \quad N = n+1+r,$$

where $r \geq p = \dim_{\mathbb{R}} P_x$, $x \in M$, such that the following properties hold i) $g^j \in \Omega^0_B(\mathcal{F}) \otimes \mathbb{C}$, and ii) $\overline{\partial}_Q g^j = 0$ for any $1 \leq j \leq n+1$. To recall the last requirement let

$$g = (g^1, \cdots, g^{n+1}) : M \to \mathbb{C}^{n+1}.$$

As the components of g are basic functions its differential $d_x g$ induces a $\mathbb{R}$-linear map $G_x = [d_x g] : Q_x \to T_{g(x)}(\mathbb{C}^{n+1})$. We request that iii) $G : Q \to g^*T(\mathbb{C}^{n+1})$ is a bundle monomorphism. Here $g^*T(\mathbb{C}^{n+1})$ is the pullback of $T(\mathbb{C}^{n+1})$ via g. Also $\overline{\partial}_Q : \Omega^0_B(\mathcal{F}) \otimes \mathbb{C} \to \Omega^{0,1}_B(\mathcal{F})$ is the transverse Cauchy-Riemann operator. An embedding ψ of $(M,\mathcal{H})$ is generic if $r = p$. For any embedding $\psi : M \to \mathbb{C}^N$ (of the transverse CR structure $\mathcal{H}$)

$$G_x(\mathcal{H}_x) \subset T^{1,0}(\mathbb{C}^{n+1})_{g(x)}\,.$$

As previously shown, any real analytic transverse CR structure (in the normal bundle of a C^ω foliation $\mathcal{F}$ on a C^ω manifold M) locally embedds in the above sense. We establish the following

THEOREM 8.9. (E. Barletta, [**16**])
Let $(\mathcal{F},\mathcal{H})$ be a strictly pseudoconvex CR foliation of M and $\phi \in \mathrm{Aut}_H(\mathcal{F})$, orientation preserving. Let

$$\psi = (g^1, \cdots, g^{n+1}, h^1, \cdots, h^r) : M \to \mathbb{C}^N, \quad N = n+1+r,$$

be an embedding of $(M,\mathcal{H})$. Then

i) $\phi^j = g^j \circ \phi \in \Omega^0_B(\mathcal{F})$ *for any* $1 \leq j \leq n+1$.

ii) *The complex dilatation μ of ϕ obeys to*

$$\zeta_{\overline{\alpha}}(\phi^j) - \mu^{\beta}_{\overline{\alpha}}\zeta_\beta(\phi^j) = 0 \tag{8.5}$$

for any admissible frame $\{\zeta_\alpha\}$ of $\mathcal{H}$.

Proof. First note that

$$G_{\phi(x)}[d_x\phi]\mathcal{H}_{*,x} \subset T^{1,0}(\mathbb{C}^{n+1})_{g(\phi(x))}. \tag{8.6}$$

for any $x \in M$. Moreover if $\{z^j\}$ are the natural complex coordinates in $\mathbb{C}^{n+1}$ then

$$d(g \circ \phi) = (d\phi^j) \otimes \frac{\partial}{\partial z^j} + (d\phi^{\overline{j}}) \otimes \frac{\partial}{\partial \overline{z}^j} \tag{8.7}$$

where $\phi^j = g^j \circ \phi$ and $\phi^{\overline{j}} = \overline{\phi^j}$. Let $\{\zeta_\alpha\}$ be an admissible local frame of $\mathcal{H}$. Let $X \in P_x$. Then $(d_x\phi)X \in P_{\phi(x)}$ so that

$$X\left(\phi^j\right) = \left((d_x\phi)\,X\right)\left(g^j\right) = 0,$$

that is each ϕ^j is a basic function. Let $\Pi^{0,1} : T(\mathbb{C}^{n+1}) \otimes \mathbb{C} \to T^{0,1}(\mathbb{C}^{n+1})$ be the natural bundle map. Using (8.6)-(8.7) we may conduct the calculation

$$0 = \Pi^{0,1}_{g(\phi(x))} G_{\phi(x)}[d_x\phi](\zeta_\alpha - \mu^{\overline{\beta}}_{\alpha}(x)\zeta_{\overline{\beta}}) =$$

$$= \left(\zeta_\alpha(\phi^{\overline{j}}) - \mu^{\overline{\beta}}_{\alpha}(x)\zeta_{\overline{\beta}}(\phi^{\overline{j}})\right) \frac{\partial}{\partial \overline{z}^j}$$

which leads (by complex conjugation) to (8.5). Q.e.d.

THEOREM 8.10. (E. Barletta, [**16**])
Let $(\mathcal{F}, \mathcal{H})$ be a strictly pseudoconvex CR foliation and $\phi \in \mathrm{Aut}(\mathcal{F})$. Let $\psi = (g^1, \cdots, g^{n+1}, h^1, \cdots, h^r) : M \to \mathbb{C}^N$ be an embedding of $(M, \mathcal{H})$. Let us set $\phi^j = g^j \circ \phi$. Let $\mu : \mathcal{H} \to \mathcal{H}$ be a fibrewise $\mathbb{C}$-anti-linear bundle morphism such that (8.5) holds for some admissible frame $\{\zeta_\alpha\}$ of $\mathcal{H}$ (and $\mu^{\beta}_{\overline{\alpha}}$ are given by $\mu\zeta_\alpha = \mu^{\beta}_{\overline{\alpha}}\zeta_\beta$). Then

i) Let $\theta \in \Gamma^\infty_B(Q^*)$ be a transverse pseudohermitian structure such that L_θ is positive-definite. Then there is a transverse pseudohermitian structure $\theta' \in \Gamma^\infty_B(Q^*)$ such that $\theta'_\phi = \theta$.

ii) ϕ is an orientation preserving automorphism of the transverse contact structure whose complex dilatation is μ.

Proof. Let $\{\zeta_\alpha\}$ be an admissible local frame of $\mathcal{H}$. Let us set

$$v_\alpha = \zeta_\alpha(\phi^j)\frac{\partial}{\partial z^j} \quad , \quad w_\alpha = \zeta_{\overline{\alpha}}(\phi^j)\frac{\partial}{\partial z^j} \, .$$

Then (by the transverse Beltrami equation (8.5))

$$w_\alpha = \mu^{\beta}_{\overline{\alpha}} v_\beta \, . \tag{8.8}$$

Moreover, note that

$$T^{1,0}(\mathbb{C}^{n+1})_{g(x)} \cap G_x(H_x \otimes \mathbb{C}) = G_x(\mathcal{H}_x), \quad x \in M. \tag{8.9}$$

Using (8.8)-(8.9) we have

$$G_{\phi(x)} \circ [d_x\phi](\zeta_\alpha - \overline{\mu\zeta_\alpha})_x = v_\alpha - \mu^{\overline{\beta}}_{\alpha}(x) w_\beta \in G_{\phi(x)}(\mathcal{H}_{\phi(x)})$$

hence (as $G_{\phi(x)}$ is one-to-one)

$$[d_x\phi](\zeta_\alpha - \overline{\mu\zeta_\alpha})_x \in \mathcal{H}_{\phi(x)} \, . \tag{8.10}$$

Let us set $t_\alpha = \zeta_\alpha - \overline{\mu\zeta_\alpha}$ for simplicity. Let $\theta_\phi \in \Gamma^\infty_B(Q^*)$. As, in general, ϕ may not preserve H one cannot directly conclude that θ_ϕ is a transverse pseudohermitian structure. However, note that $\{t_\alpha\}$ are independent (over $\mathbb{C}$) because of $\mathcal{H} \cap \overline{\mathcal{H}} = \{0\}$. Let L_x be the span of $t_\alpha(x)$, $1 \le \alpha \le n$. Then

$$\dim_{\mathbb{R}} \mathrm{Re}\{L_x \oplus \overline{L}_x\} = 2n.$$

On the other hand if $z \in \mathrm{Re}\{L_x \oplus \overline{L}_x\}$ then

$$z = \lambda^\alpha t_\alpha + \lambda^{\overline{\alpha}} t_{\overline{\alpha}} = (\lambda^\beta - \lambda^{\overline{\alpha}}\mu^{\beta}_{\overline{\alpha}})\zeta_\beta + (\lambda^{\overline{\beta}} - \lambda^\alpha \mu^{\overline{\beta}}_{\alpha})\zeta_{\overline{\beta}} \in H_x$$

so that (by looking at dimensions)

$$H = \mathrm{Re}\{L \oplus \overline{L}\}. \tag{8.11}$$

Finally $\theta_\phi(t_\alpha) = 0$ because of (8.10) so that (by (8.11)) θ_ϕ is a transverse pseudohermitian structure. Then $\theta_\phi = \lambda\theta$ for some nowhere vanishing basic function λ on M. Then

$$\theta' = \frac{1}{\lambda \circ \phi^{-1}}\,\theta$$

is the transverse pseudohermitian structure we are looking for (i.e. with $\theta'_\phi = \theta$). To prove the second statement in Theorem 8.10 note first that $\theta'_\phi = \theta$ suffices for concluding that $\phi \in \mathrm{Aut}_H(\mathcal{F})$. Let then $\mu_\phi : \mathcal{H} \to \mathcal{H}$ be the complex dilatation of ϕ. It remains to be shown that $\mu_\phi = \mu$. Let us set

$$\mathcal{H}_\mu = \{\zeta - \overline{\mu\zeta} \,:\, \zeta \in \mathcal{H}\}.$$

By the uniqueness statement in Proposition 8.6 it suffices to show that

$$\mathcal{H}_\mu = \mathcal{H}_* \,.$$

First note that (because μ is $\mathbb{C}$-anti-linear) $\mathcal{H}_\mu$ is the span of $\{t_\alpha\}$. Then (8.10) yields $\mathcal{H}_\mu \subseteq \mathcal{H}_*$. The converse follows from (8.11) and definitions (as $\mathcal{H}_* \subset H \otimes \mathbb{C}$). Q.e.d.

Let $(\mathcal{F}, \mathcal{H})$ be a strictly pseudoconvex CR foliation and θ a transverse pseudohermitian structure such that L_θ is positive definite. Let $\mu : \mathcal{H} \to \mathcal{H}$ be a fibrewise $\mathbb{C}$-anti-linear bundle morphism. Let us define

$$\|\mu\|^2_\theta = \sup_{\zeta \in \mathcal{H}} \frac{L_\theta(\mu\zeta, \overline{\mu\zeta})}{L_\theta(\zeta, \overline{\zeta})}.$$

We establish the following

PROPOSITION 8.11. *Let $(\mathcal{F}, \mathcal{H})$ be a strictly pseudoconvex CR foliation of M and $\psi : M \to \mathbb{C}^N$ an embedding of $(M, \mathcal{H})$. Let $\phi \in \mathrm{Aut}(\mathcal{F})$. Let $\mu : \mathcal{H} \to \mathcal{H}$ be a fibrewise $\mathbb{C}$-anti-linear bundle morphism such that* (8.5) *holds for some admissible frame $\{\zeta_\alpha\}$ of $\mathcal{H}$. Let $\theta \in \Omega^1_B(\mathcal{F})$ be a transverse pseudohermitian structure such that L_θ is positive-definite and $\theta' \in \Gamma^\infty_B(Q^*)$ the transverse pseudohermitian structure furnished by Theorem* 8.10 *i.e. such that $\theta'_\phi = \theta$. If $\|\mu\|_{\theta'} \leq 1$ then $L_{\theta'}$ is positive definite.*

Proof. By the proof of Theorem 8.9, $\mathcal{H}_{\phi(x)}$ is the span of

$$[d_x\phi]t_\alpha(x), \quad 1 \leq \alpha \leq n.$$

Thus for any $\zeta \in \mathcal{H}_{\phi(x)}$ there is $\eta \in \mathcal{H}_x$ such that $\zeta = [d_x\phi](\eta - \overline{\mu_x\eta})$. Then

$$L_{\theta', \phi(x)}(\zeta, \overline{\zeta}) = L_{\theta,x}(\eta, \overline{\eta})\left(1 - \frac{L_{\theta,x}(\mu\eta, \overline{\mu\eta})}{L_{\theta,x}(\eta, \overline{\eta})}\right) \geq$$

$$\geq \left(1 - \|\mu\|^2_{\theta',x}\right) L_{\theta,x}(\eta, \overline{\eta}) > 0$$

so that $L_{\theta'}$ is positive definite. Q.e.d.

8.2. K-quasiconformal automorphisms

Let $\mathcal{F}$ be a CR foliation of M and $\theta \in \Gamma^\infty_B(Q^*)$ a transverse pseudohermitian structure. Define G_θ by setting

$$G_\theta(s, r) = (d_Q\theta)(s, Jr)$$

for any s, $r \in H$. Let $\phi \in \mathrm{Aut}_H(\mathcal{F})$ be orientation preserving. Let $K \in (0, +\infty)$.

DEFINITION 8.12. ϕ is a *K-quasiconformal automorphism* if

$$\lambda(x)K^{-1}G_\theta(r,r)_x \leq (G_\theta)_\phi(r,r)_x \leq \lambda(x)KG_\theta(r,r)_x \tag{8.12}$$

for any $r \in \Gamma^\infty(H)$ and any $x \in M$. □

Here $(G_\theta)_\phi$ is given by

$$(G_\theta)_\phi(r,s)_x = G_{\theta,\phi(x)}\left([d_x\phi]r_x, [d_x\phi]s_x\right).$$

Also $\lambda \in \Omega^0_B(\mathcal{F})$ is the unique basic function such that $\theta_\phi = \lambda\theta$ and $\lambda > 0$.

DEFINITION 8.13. A *(transversally) Heisenberg* foliation $\mathcal{F}$ of M is a $\Gamma^\infty_{CR}(\mathbb{H}_n)$-foliation of M, where $\mathbb{H}_n$ is the Heisenberg group. □

That is a Heisenberg foliation is a CR foliation whose transverse geometry is modelled on $\mathbb{H}_n$. We establish

THEOREM 8.14. (E. Barletta, [**16**])
Let $\mathcal{F}$ be a Heisenberg foliation of M. Let $\phi \in \mathrm{Aut}_H(\mathcal{F})$ be orientation preserving. Let μ be its complex dilatation. The following statements are equivalent
i) *ϕ is K-quasiconformal*
ii) *$\|\mu\| \leq (K-1)/(K+1)$ everywhere on M.*

Proof. Let $(z^1, \cdots, z^n, t)$ be natural coordinates on $\mathbb{H}_n = \mathbb{C}^n \times \mathbb{R}$. Let

$$Z_\alpha = \frac{\partial}{\partial z^\alpha} + i\overline{z}^\alpha\frac{\partial}{\partial t}, \quad \theta_0 = dt + i\sum_{\alpha=1}^n\{z^\alpha d\overline{z}^\alpha - \overline{z}^\alpha dz^\alpha\}.$$

We recall that $\{Z_\alpha\}$ span $T_{1,0}(\mathbb{H}_n)$ while θ_0 is a pseudohermitian structure on $\mathbb{H}_n$. Let $\{U_j, f_j, \gamma_{ij}\}$ be the local data defining a $\Gamma^\infty_{CR}(\mathbb{H}_n)$-foliation $\mathcal{F}$ of M i.e. $f_j : U_j \to \mathbb{H}_n$ are C^∞ submersions and $\gamma_{ij} \in \Gamma^\infty_{CR}(\mathbb{H}_n)$ are such that $f_i = \gamma_{ij}f_j$ on $U_i \cap U_j \neq \emptyset$. Let $\theta_j \in \Gamma^\infty_B(Q^*)$ be defined by $\theta_{j,T} = f_j^*\theta_0$. Next if θ is a transverse pseudohermitian structure on $(M, \mathcal{F})$ then $\theta = \lambda_j\theta_j$ for some C^∞ basic function $\lambda_j : U_j \to \mathbb{R}^*$. Hence $\mathcal{F}$ is transversally nondegenerate and either θ or $-\theta$ yield the corresponding transverse Levi form positive-definite. Finally

$$\zeta_{\alpha,x} = [d_xf_j]^{-1}Z_{\alpha,f_j(x)}, \quad x \in U_j,$$

is an admissible frame of $\mathcal{H}$ on U_j and

$$d_Q\theta = 2i\delta_{\alpha\beta}\lambda_j\,\theta^\alpha \wedge \theta^{\overline{\beta}}. \tag{8.13}$$

To prove Theorem 8.10 we shall also use the (real) frame

$$\{e_A\} = \{e_\alpha, e_{\alpha+n}\}$$

of H given by

$$e_\alpha = \zeta_\alpha + \zeta_{\overline{\alpha}}, \quad e_{\alpha+n} = i(\zeta_\alpha - \zeta_{\overline{\alpha}}).$$

Let us set

$$[d_x\phi]e_{A,x} = \Phi^B_A(x)e_{B,\phi(x)}, \quad g = [\lambda^{-1/2}\Phi^B_A]_{1\leq A,B\leq 2n}.$$

A calculation based on (8.13) shows that $g(x) \in \mathrm{Sp}(2n,\mathbb{R})$ for any $x \in U_j$. Also each entry of g is a real valued basic function on U_j. Dropping the point x in the notation, for simplicity, we may represent g as $g = k\,a\,k'$ for some k, $k' \in \mathrm{GL}(2n,\mathbb{R})$ of the type

$$\begin{pmatrix} A & -B \\ B & A \end{pmatrix}, \quad A,\, B \in \mathrm{GL}(n,\mathbb{R}),$$

$$A\overline{A}^t + B\overline{B}^t = I_n\,, \quad A\overline{B}^t - B\overline{A}^t = 0,$$

and

$$a = \operatorname{diag}(e^{t_1}, \cdots, e^{t_n}, e^{-t_1}, \cdots, e^{-t_n}),$$
$$t_1 \geq t_2 \geq \cdots \geq t_n \geq 0, \quad t_j = t_j(x), \quad 1 \leq j \leq n.$$

Then the complex coefficients of $[d_x\phi]$ i.e.

$$[d_x\phi]\zeta_{\alpha,x} = \phi^\beta_\alpha(x)\zeta_{\beta,\phi(x)} + \phi^{\overline{\beta}}_\alpha(x)\zeta_{\overline{\beta},\phi(x)}$$

where

$$\phi^{\overline{\beta}}_{\alpha} = \mu^{\overline{\beta}}_{\alpha}\,\phi^{\overline{\beta}}_{\overline{\sigma}} \tag{8.14}$$

satisfy

$$\phi^\alpha_\beta = \lambda^{1/2}(k^\alpha_\sigma - ik^\alpha_{\sigma+n})(k'^\sigma_\beta - k'^\sigma_{\beta+n}) \cosh t_\sigma\,,$$
$$\phi^\alpha_{\overline{\beta}} = \lambda^{1/2}(k^\alpha_\sigma - ik^\alpha_{\sigma+n})(k'^\sigma_\beta + ik'^\sigma_{\beta+n}) \sinh t_\sigma\,.$$

Let us set

$$v^\alpha_\sigma = k^\alpha_\sigma - ik^\alpha_{\sigma+n}\quad,\quad v'^\alpha_\sigma = k'^\alpha_\sigma - ik'^\alpha_{\sigma+n}$$

(so that $v = (v^\alpha_\sigma)$, $v' = (v'^\alpha_\sigma)$ are U(n)-valued). Then (by (8.14)) $\mu = (\mu^\alpha_{\overline{\beta}})$ may be written as

$$\mu = v'^{-1}\operatorname{diag}(\tanh t_1, \cdots, \tanh t_n)\, v'. \tag{8.15}$$

Next (by (8.13))

$$L_\theta(\zeta, \overline{\zeta}) = \lambda_j \sum_{\alpha=1}^{n} |a^\alpha|^2$$

for any $\zeta = a^\alpha\zeta_\alpha \in \mathcal{H}_x$. Hence

$$\|\mu_x\|^2_\theta = \sup_{a\in\mathbb{C}^n} \frac{\sum_{\alpha,\beta,\gamma} a^\alpha\overline{a^\gamma}\mu^\beta_{\overline{\alpha}}\,\mu^{\overline{\beta}}_{\gamma}}{\sum_\alpha |a^\alpha|^2}.$$

On the other hand (using (8.15))

$$\sum_\beta \mu^\beta_{\overline{\alpha}}\,\mu^{\overline{\beta}}_\gamma = \sum_\beta (\tanh t_\beta)^2\, v'^{\overline{\beta}}_{\overline{\alpha}} v'^\beta_\gamma \leq \left(\max_{1\leq\sigma\leq n} (\tanh t_\sigma)^2\right) \sum_\beta v'^{\overline{\beta}}_{\overline{\alpha}} v'^\beta_\sigma$$

hence

$$\sum_\beta \mu^\beta_{\overline{\alpha}}(x)\,\overline{\mu^\beta_{\overline{\gamma}}(x)} \leq (\tanh t_1)^2\,\delta_{\alpha\gamma}\,.$$

Consequently

$$\|\mu_x\|^2_\theta \leq (\tanh t_1(x))^2 \sup_{\|\alpha\|=1} \sum_{\alpha,\gamma} a^\alpha\overline{a^\gamma}\delta_{\alpha\gamma}$$

or

$$\|\mu_x\|_\theta \leq \tanh t_1(x). \tag{8.16}$$

Let $\zeta = a^\alpha\zeta_{\alpha,x} \in \mathcal{H}_x$ where a^α are determined by

$$\sum_\alpha a^\alpha v'^{\overline{\sigma}}_{\overline{\alpha}} = \delta_{1\sigma}$$

or $a^\beta = v'^1_\beta$. In particular $\sum_\alpha a^\alpha\overline{a^\alpha} = 1$. With this choice of $\zeta \in \mathcal{H}_x$ we have $L_\theta(\zeta, \overline{\zeta}) = \lambda_j(x)$ and (again by (8.15))

$$L_\theta(\mu\,\zeta, \overline{\mu\,\zeta}) = \lambda_j(x)\,(\tanh t_1)^2$$

Consequently in (8.16) actually equality holds

$$\|\mu\|_\theta = \tanh t_1 \, .$$

Next

$$1 + \|\mu\|_\theta = \frac{2e^{t_1}}{e^{t_1} + e^{-t_1}} \; , \; 1 - \|\mu\|_\theta = \frac{2e^{-t_1}}{e^{t_1} + e^{-t_1}} \, ,$$

hence

$$\frac{1 + \|\mu\|_\theta}{1 - \|\mu\|_\theta} = \frac{e^{t_1}}{e^{-t_1}} = \frac{\max_{\|a\|=1} \|g\, a\|}{\min_{\|a\|=1} \|g\, a\|} \, .$$

Let $r \in H_x$, $r = \zeta + \overline{\zeta}$, where $\zeta = a^\alpha \zeta_{\alpha,x} \in \mathcal{H}_x$. Then

$$G_\theta(r, r) = 2\lambda_j(x)\|a\|^2.$$

Moreover (by (8.12))

$$K^{-1}\|a\|^2 \leq \|g\, a\|^2 \leq K\|a\|^2. \tag{8.17}$$

Here we chose θ such that $\lambda_j > 0$. Then

$$\frac{\max_{\|a\|=1} \|g\, a\|^2}{\min_{\|a\|=1} \|g\, a\|^2} \leq K^2$$

yields

$$\frac{1 + \|\mu\|_\theta}{1 - \|\mu\|_\theta} \leq K$$

and i) $\Longrightarrow$ ii) is proved. Viceversa, if

$$\|\mu\|_\theta \leq \frac{K - 1}{K + 1}$$

then (8.17) holds for any $a \in \mathbb{C}^n$ hence ϕ is K-quasiconformal. The results in this section are due to E. Barletta, [**16**].

CHAPTER 9

Review of orbifold theory

The exposition in the present chapter follows the presentation by W.L. Baily, Jr., [**8**]-[**10**], I. Satake, [**213**]-[**216**], and J. Girbau & M. Nicolau, [**120**]. A section devoted to orbifolds may also be found in the recent work by I. Moerdijk & J. Mrčun, [**176**].

9.1. Defining families

Let B be a Hausdorff space and U an open subset of B.

DEFINITION 9.1. A *local uniformizing system* (l.u.s.) of dimension n of B over U is a synthetic object $\{\Omega, G, \varphi\}$ consisting of 1) a connected open subset Ω of $\mathbb{R}^n$, 2) a finite group G of C^∞ diffeomorphisms of Ω in itself, and 3) a continuous map $\varphi : \Omega \to U$ such that $\varphi \circ \sigma = \varphi$ for any $\sigma \in G$ and such that the map $\Omega/G \to U$ induced by φ is a homeomorphism. U is the *support* of the l.u.s. $\{\Omega, G, \varphi\}$ and φ is its *projection* onto U. $\square$

Let $\{\Omega, G, \varphi\}$ and $\{\Omega', G', \varphi'\}$ be two l.u.s.'s on B (*a posteriori* of the same dimension n) of supports U and U' respectively. Assume that $U \subseteq U'$.

DEFINITION 9.2. An *injection* of $\{\Omega, G, \varphi\}$ into $\{\Omega', G', \varphi'\}$, commonly denoted by

$$\lambda : \{\Omega, G, \varphi\} \to \{\Omega', G', \varphi'\},$$

is a C^∞ map $\lambda : \Omega \to \Omega'$ such that 1) λ is a C^∞ diffeomorphism of Ω onto some open subset of Ω', and 2) $\varphi' \circ \lambda = \varphi$. $\square$

For instance each $\sigma \in G$ is an injection of $\{\Omega, G, \varphi\}$ in itself. Let $\mathcal{A}$ be a family of l.u.s.'s of dimension n on B and denote by $\mathcal{H}$ the family consisting of all supports $U \subseteq B$ of l.u.s.'s belonging to $\mathcal{A}$.

DEFINITION 9.3. $\mathcal{A}$ is a *defining family* for B if 1) for any $\{\Omega, G, \varphi\}$, $\{\Omega', G', \varphi'\} \in \mathcal{A}$ of supports U, U' if $U \subseteq U'$ then there is an injection λ of $\{\Omega, G, \varphi\}$ into $\{\Omega', G', \varphi'\}$, and 2) $\mathcal{H}$ is an (open) cover of B and for any $U_1, U_2 \in \mathcal{H}$ and any $x \in U_1 \cap U_2$ there is $U \in \mathcal{H}$ such that $x \in U \subseteq U_1 \cap U_2$. $\square$

It is customary (cf. e.g. W.L. Baily, [**8**]) to request $\mathcal{H}$ to be a basis of open sets for the topology of B, yet the weaker requirement 2) above is known to suffice (cf. J. Girbau & M. Nicolau, [**120**]) for establishing most basic properties of orbifolds.

DEFINITION 9.4. Two defining families $\mathcal{A}, \mathcal{A}'$ are *directly equivalent* if there is a third defining family $\mathcal{A}''$ so that $\mathcal{A} \cup \mathcal{A}' \subseteq \mathcal{A}''$. Also $\mathcal{A}, \mathcal{A}'$ are *equivalent* if there is a chain $\{\mathcal{A}_1, \cdots, \mathcal{A}_r\}$ of defining families so that $\mathcal{A} = \mathcal{A}_1, \mathcal{A}' = \mathcal{A}_r$, and $\mathcal{A}_i$ is directly equivalent to $\mathcal{A}_{i+1}$, for any $1 \leq i \leq r-1$. A pair $(B, [\mathcal{A}])$ consisting of a connected paracompact Hausdorff space B and a class $[\mathcal{A}]$ of equivalent defining families for B, with $\mathcal{A}$ consisting of l.u.s.'s of dimension n on B, is a n-dimensional *orbifold* of class C^∞. $\square$

Often a pair $(B, \mathcal{A})$ is referred to as an orbifold, as well. Given an orbifold $(B, [\mathcal{A}])$ let $\{\Omega, G, \varphi\} \in \mathcal{A}$ and $x \in \Omega$.

DEFINITION 9.5. The *isotropy group* G_x of x is given by $G_x = \{\sigma \in G : \sigma(x) = x\}$. $\square$

Next let $\tilde{S}$ be given by

$$\tilde{S} = \{x \in \Omega : G_x \neq \{e\}\} \tag{9.1}$$

where e is the identity in G. As to the local structure of $\tilde{S}$ we may state

THEOREM 9.6. *Let $(B, [\mathcal{A}])$ be a n-dimensional orbifold of class C^∞. For any $x \in \tilde{S}$ there is an open set $\tilde{V} \subseteq \Omega$ with $x \in \tilde{V}$ and so that $\tilde{V} \cap \tilde{S}$ is a finite union of submanifolds of Ω of dimension $< n$. Moreover, for any $\tilde{q} \in \tilde{V}$ the isotropy group of $\tilde{q}$ is a subgroup of G_x.*

The proof of Theorem 9.6 is given in the next section. Before going any further we wish to give the following

EXAMPLE 9.7. Let $D = \{x \in \mathbb{R}^2 : |x| < 1\}$ be the unit disc and let G be the group of rotations of D of angles 0, $\pi/2$, π and $3\pi/2$. Let $B = D/G$ be the quotient space, carrying the quotient topology. Let $I = \{x = (x_1, x_2) \in D : x_1 \geq 0, x_2 \geq 0\}$ and $\sim$ the equivalence relation on I given by $(0, a) \sim (a, 0)$ for any $0 \leq a < 1$. Clearly, there is an identification of B and the quotient set $I/\sim$. Let $\pi : D \to B$ be the natural map. Let $\Omega = D(O, r)$ be the disc of center the origin and radius $r < 1$ and set $U = \pi(\Omega)$. Clearly $\pi^{-1}(U) = \Omega$ (as G is a group of rotations of center O) hence U is open in B. Next, set $\varphi = \pi|_\Omega$. Then $\{\Omega, G, \varphi\}$ is a l.u.s. of dimension 2 of B, of support U (each element of G is thought of as restricted to Ω) and

$$\mathcal{A} = \{\{\Omega, G, \varphi\} : \Omega = D(O, r), 0 < r < 1\}$$

is a defining family for B. Hence $(B, \mathcal{A})$ is a 2-dimensional orbifold of class C^∞. It is noteworthy that $\mathcal{A}$ is not a maximal defining family. Indeed, let $\tilde{V}$ be an open disc contained in the interior of I and $V = \pi(\tilde{V})$. Then $\pi^{-1}(V) = \tilde{V}$, hence V is open in B. Set $\psi = \pi|_{\tilde{V}}$. Then $\{\tilde{V}, \{e\}, \psi\}$ is a l.u.s. of dimension 2 of B. Here e is the rotation of center O and angle 0. Let $\{\Omega, G, \varphi\} \in \mathcal{A}$ and $\tilde{S}$ given by (9.1). Then $\tilde{S} = \{O\}$ (as $G_O = G$ and $x \neq O \Longrightarrow G_x = \{e\}$). Finally, if $\{\tilde{V}, \{e\}, \psi\}$ is the l.u.s. corresponding to a small disc $\tilde{V}$ in I as above, then the corresponding $\tilde{S}$ is empty (each point in $\tilde{V}$ has a trivial isotropy group). $\square$

9.2. The local structure of $\tilde{S}$

We proceed by proving Theorem 9.6. To this end let $x \in \tilde{S}$ and let $(u^1, \cdots, u^n)$ be the Cartesian coordinates on Ω. Then let us define a set of n^2 functions $a^i_j : G_x \to \mathbb{R}$ by setting

$$a^i_j(\sigma) = \frac{\partial(u^i \circ \sigma)}{\partial u^j}(x)$$

for any $\sigma \in G_x$. Let $n(x)$ be the order of G_x. Define $v^i : \Omega \to \mathbb{R}$ by setting

$$v^i = \frac{1}{n(x)} \sum_{\sigma \in G_x} a^i_j(\sigma^{-1}) u^j \circ \sigma \tag{9.2}$$

Then

$$\frac{\partial v^i}{\partial u^k}(x) = \frac{1}{n(x)} \sum_{\sigma \in G_x} a^i_j(\sigma^{-1}) a^j_k(\sigma) = \delta^i_k$$

hence the map $(v^1, \cdots, v^n) : \Omega \to \mathbb{R}^n$ is a local C^∞ diffeomorphism at x i.e. there is an open set $\tilde{V}_1$ such that $x \in \tilde{V}_1 \subseteq \Omega$ and such that $(v^1, \cdots, v^n) : \tilde{V}_1 \to \mathbb{R}^n$ is a diffeomorphism of $\tilde{V}_1$ onto its image. At this point we may build an open set $\tilde{V}$ such that $x \in \tilde{V} \subseteq \tilde{V}_1$ and satisfying the following additional requirements i) $\sigma(\tilde{V}) \subseteq \tilde{V}$ for any $\sigma \in G_x$, and ii) $\sigma(\tilde{V}) \cap \tilde{V} = \emptyset$ for any $\sigma \in G \setminus G_x$.

Indeed if $\sigma \in G \setminus G_x$ then, as $\sigma(x) \neq x$, there is an open neighborhood $\Omega_\sigma \subseteq \tilde{V}_1$ of x such that $\sigma(\Omega_\sigma) \cap \Omega_\sigma = \emptyset$. Let us set

$$\tilde{V}_0 = \bigcap_{\sigma \in G \setminus G_x} \Omega_\sigma .$$

As G is finite $\tilde{V}_0$ is open. Finally, let us set

$$\tilde{V} = \bigcap_{\sigma \in G_x} \sigma(\tilde{V}_0).$$

We leave it to the reader to check that $\tilde{V}$ satisfies the requirements i)-ii) above, as claimed. Under the coordinate transformation (9.2) the isotropy group G_x of x becomes a finite group of linear transformations. Indeed let $\tau \in G_x$. By the construction of $\tilde{V}$ the restriction of τ to $\tilde{V}$ corestricts to $\tilde{V}_1$ hence $v^i \circ \tau$ makes sense. It may be calculated as follows

$$v^i \circ \tau = \frac{1}{n(x)} \sum_{\sigma \in G_x} a^i_j(\sigma^{-1})(u^j \circ \sigma) \circ \tau =$$

$$= \frac{1}{n(x)} \sum_{\alpha \in G_x} a^i_j(\tau \circ \alpha^{-1}) u^j \circ \alpha = \frac{1}{n(x)} \sum_{\alpha \in G_x} a^i_k(\tau) a^k_j(\alpha^{-1}) u^j \circ \alpha$$

that is

$$v^i \circ \tau = a^i_j(\tau) v^j . \tag{9.3}$$

Let $\tau \in G_x$, $\tau \neq e$ (here $G_x \neq \{e\}$ as $x \in \tilde{S}$). Let F_τ consist of all $\tilde{q} \in \tilde{V}$ such that $\tau(\tilde{q}) = \tilde{q}$. Let us set $\phi = (v^1, \cdots, v^n)$. Then $\phi(F_\tau)$ is the eigenspace of $[a^i_j(\tau)]$ corresponding to the eigenvalue 1 (and clearly $\dim_{\mathbb{R}} \phi(F_\tau) < n$ as $\tau \neq e$) hence F_τ is a C^∞ manifold whose dimension equals the algebraic dimension of $\phi(F_\tau)$. Let us set

$$\tilde{X} = \bigcup_{\tau \in G_x - \{e\}} F_\tau \subset \tilde{V}.$$

At this point we may check that $G_{\tilde{q}}$ is a subgroup of G_x, for any $\tilde{q} \in \tilde{V}$. This follows from

$$G_{\tilde{q}} = \{\alpha \in G_x : \alpha(\tilde{q}) = \tilde{q}\}.$$

The inclusion $\supseteq$ is obvious. As to the opposite inclusion let $\alpha \in G_{\tilde{q}}$. Then $\alpha(\tilde{V}) \cap \tilde{V} \neq \emptyset$ and, by the very construction of $\tilde{V}$, it follows that $\alpha \in G_x$. To end the proof of Theorem 9.6 it suffices to show that

$$\tilde{S} \cap \tilde{V} = \tilde{X}.$$

To prove the inclusion $\subseteq$ let $\tilde{q} \in \tilde{S} \cap \tilde{V}$. Then $G_{\tilde{q}} \neq \{e\}$. Let $\tau \neq e$ with $\tau(\tilde{q}) = \tilde{q}$. Then $\tau \in F_\tau \subset \tilde{X}$. Q.e.d. To establish the inclusion $\supseteq$ let $\tilde{q} \in \tilde{X} \subset \tilde{V}$ and assume that $\tilde{q} \notin \tilde{S}$ i.e. $G_{\tilde{q}} = \{e\}$. If this is the case then $\tilde{q} \notin F_\tau$ for any $\tau \neq e$ i.e. $\tilde{q} \notin \tilde{X}$, a contradiction.

9.3. The monomorphism η

Let $(B, \mathcal{A})$ be a n-dimensional orbifold of class C^∞. Let

$$\{\Omega, G, \varphi\}, \{\Omega', G', \varphi'\} \in \mathcal{A}$$

of supports U, U' respectively. Assume that $U \subseteq U'$. The purpose of the present section is to establish the rather difficult result that

PROPOSITION 9.8. *Given two injections*

$$\lambda, \mu : \{\Omega, G, \varphi\} \to \{\Omega', G', \varphi'\}$$

there is a unique element $\sigma'_1 \in G'$ such that

$$\mu = \sigma'_1 \circ \lambda. \tag{9.4}$$

In particular

COROLLARY 9.9. *With any injection λ of $\{\Omega, G, \varphi\}$ into $\{\Omega', G', \varphi'\}$ one may associate a group monomorphism $\eta : G \to G'$ such that*

$$\lambda \circ \sigma = \eta(\sigma) \circ \lambda \tag{9.5}$$

for any $\sigma \in G$.

Note that the existence of $\eta(\sigma)$ is *postulated* in [**66**], p. 317 (and uniqueness is then immediate). Due to the work by J. Girbau & M. Nicolau, [**120**], it is now known that the existence of $\eta(\sigma)$ can be proved. Indeed, by applying (9.4) with $\mu = \lambda \circ \sigma$ for some fixed (otherwise arbitrary) $\sigma \in G$, there is a unique $\sigma'_1 \in G'$ such that $\lambda \circ \sigma = \sigma'_1 \circ \lambda$. We leave it to the reader to check that η given by $\eta(\sigma) = \sigma'_1$ is indeed a group homomorphism with $\text{Ker}(\eta) = \{e\}$.

To prove the existence of an element $\sigma'_1 \in G'$ satisfying (9.4) we need the following

LEMMA 9.10. *Let G be a finite group of linear automorphisms of $\mathbb{R}^n$ such that $G \neq \{e\}$. Assume that there is a $(n-1)$-dimensional subspace $V \subset \mathbb{R}^n$ consisting of fixed points of G (i.e. $G_x = G$ for any $x \in V$). Then G consists of precisely two elements, the identity and a symmetry with respect to V.*

Proof. Let $\{e_1, \cdots, e_n\}$ be a basis of $\mathbb{R}^n$ such that $\{e_1, \cdots, e_{n-1}\}$ is a basis of V. Then for any $\sigma \in G$ one has $\sigma(e_\alpha) = e_\alpha$, $1 \leq \alpha \leq n-1$, and $\sigma(e_n) = a^i e_i$ for some $a = (a^1, \cdots, a^n) \in \mathbb{R}^n$. Hence there is a representation $\rho : G \to \text{GL}(n, \mathbb{R})$ given by $\rho(\sigma) = g(a)$ where

$$g(a) = \begin{bmatrix} 1 & \cdots & 0 & a^1 \\ \vdots & & \vdots & \vdots \\ 0 & \cdots & 1 & a^{n-1} \\ 0 & \cdots & 0 & a^n \end{bmatrix}$$

Note that

$$g(a)g(b) = g(b^1 + a^1 b^n, \cdots, b^{n-1} + a^{n-1} b^n, a^n b^n)$$

hence

$$g(a)^2 = g(a^1(1 + a^n), \cdots, a^{n-1}(1 + a^n), (a^n)^2)$$

and in general $g(a)^r = g(\cdots, (a^n)^r) \in \rho(G)$ hence $a^n \in \{\pm 1\}$, as G has finite order. Let I_n be the $n \times n$ unit matrix. We may show that

$$g(a', 1) = I_n .$$

The proof is by contradiction. Assume that $a' \neq 0$ i.e. $a^i \neq 0$ for some $1 \leq i \leq n-1$. Then $\{ra^i\}_{r \geq 1}$ is an infinite sequence of real numbers and due to the identity

$$g(a', 1)^r = g(ra', 1) \in \rho(G)$$

it follows that G is infinite, a contradiction. Hence for any $\sigma \in G \setminus \{e\}$ there is $a' \in \mathbb{R}^{n-1}$ such that $\rho(\sigma) = g(a', -1)$. Finally, we may show that G contains precisely one element of this form. Indeed

$$g(a', -1)g(b', -1) = g(b' - a', 1) = I_n$$

hence $b' = a'$. We may conclude that

$$\rho(G) = \{I_n \ , \ \ g(a', -1)\}$$

for some uniquely determined $a' \in \mathbb{R}^{n-1}$ (depending on (G, V)). The proof of Lemma 9.10 is complete.

As $\lambda(\Omega)$ is an open subset of Ω' there is $x_1 \in \Omega$ such that

$$G'_{\lambda(x_1)} = \{e'\}.$$

Indeed if $G'_{\lambda(x)} \neq \{e'\}$ for any $x \in \Omega$ then $\lambda(\Omega) \subseteq \tilde{S}'$ (here $\tilde{S}'$ is the set of all points in Ω' having nontrivial isotropy groups). Let $x' \in \lambda(\Omega)$. By Theorem 9.6 there is an open set $\tilde{V}'$ with $x' \in \tilde{V}' \subseteq \Omega'$ such that $\tilde{V}' \cap \tilde{S}'$ is a finite union of submanifolds of Ω' of dimension $< n$. Yet $\tilde{V}' \cap S'$ contains $\tilde{V}' \cap \lambda(\Omega)$ which is open and consequently a n-dimensional manifold, a contradiction. By the very definition of an injection we have

$$\varphi'(\mu(x_1)) = \varphi(x_1) = \varphi'(\lambda(x_1))$$

hence (by $\Omega/G' \approx U'$) there is $\sigma'_1 \in G'$ with

$$\mu(x_1) = \sigma'_1(\lambda(x_1)) \tag{9.6}$$

i.e. the identity (9.4) holds at x_1. Let S consist of all $x \in \Omega$ such that $G'_{\lambda(x)} \neq \{e'\}$ and let us set

$$C = \{x \in \Omega \setminus S : \mu(x) = \sigma'_1(\lambda(x))\}.$$

Note that $x_1 \in C$ hence $C \neq \emptyset$. Next we shall show that C is both open and closed. To see that C is open let $x \in C$. Then $G'_{\lambda(x)} = \{e\}$ as $x \notin S$. Hence $\tau'(\lambda(x)) \neq \sigma'(\lambda(x))$ for any $\tau', \sigma' \in G'$ with $\tau' \neq \sigma'$. Let $V'_{\sigma'}$ be an open neighborhood of $\sigma'(\lambda(x))$ in Ω' (one for each $\sigma' \in G'$) such that 1) $\tau' \neq \sigma' \Longrightarrow V'_{\tau'} \cap V'_{\sigma'} = \emptyset$, and 2) $\sigma'(V'_{e'}) \subseteq V'_{\sigma'}$. As $\lambda : \Omega \to \Omega'$ is a diffeomorphism of Ω onto $\lambda(\Omega)$ and $\lambda(x) \in V'_{e'} \subseteq \Omega'$ there exists an open neighborhood V^{λ}_x of x in Ω such that $\lambda(V^{\lambda}_x) \subseteq V'_{e'}$. It is easily seen that actually $V^{\lambda}_x \subseteq \Omega \setminus S$ (indeed, if there is $\tilde{q} \in V^{\lambda}_x \cap S$ then $\lambda(\tilde{q}) \in V'_{e'}$ and $G'_{\lambda(\tilde{q})} \neq \{e'\}$ hence there is $\sigma' \in G', \sigma' \neq e'$, with $\lambda(\tilde{q}) = \sigma'(\lambda(\tilde{q})) \in V'_{\sigma'}$, i.e. $\lambda(\tilde{q}) \in V'_{e'} \cap V'_{\sigma'}$, a contradiction (as $\sigma' \neq e'$)). Summing up, until now we built an open neighborhood V^{λ}_x of x in Ω such that

$$x \in V^{\lambda}_x \subseteq \Omega \setminus S, \quad \lambda(V^{\lambda}_x) \subseteq V'_{e'}.$$

Similarly, there is an open neighborhood V^{μ}_x of x in Ω such that

$$\mu(V^{\mu}_x) \subseteq V'_{\sigma'_1}$$

(because of $V'_{\sigma'_1} \ni \sigma'_1(\lambda(x)) = \mu(x)$). Let us set

$$V_x = V_x^\lambda \cap V_x^\mu .$$

Then

$$x \in V_x \subseteq \Omega \setminus S, \quad \lambda(V_x) \subseteq V'_{e'} , \quad \mu(V_x) \subseteq V'_{\sigma'_1} .$$

At this point we may show that $V_x \subseteq C$ (hence C is a neighborhood of each of its points). To this end let $\tilde{q} \in V_x$. As

$$\varphi'(\mu(\tilde{q})) = \varphi(\tilde{q}) = \varphi'(\lambda(\tilde{q}))$$

there is $\sigma'_2 \in G'$ with

$$\mu(\tilde{q}) = \sigma'_2(\lambda(\tilde{q})).$$

Let us show that $\sigma'_2 = \sigma'_1$ (hence $\tilde{q} \in C$). Indeed

$$\mu(\tilde{q}) = \sigma'_2(\lambda(\tilde{q})) \in \sigma'_2(\lambda(V_x)) \subseteq \sigma'_2(V'_{e'}) \subseteq V'_{\sigma'_2}$$

i.e.

$$\mu(\tilde{q}) \in V'_{\sigma'_2} . \tag{9.7}$$

On the other hand

$$\mu(\tilde{q}) \in \mu(V_x) \subseteq V'_{\sigma'_1}$$

which together with (9.7) yields $\mu(\tilde{q}) \in V'_{\sigma'_1} \cap V'_{\sigma'_2}$ i.e. $V'_{\sigma'_1} \cap V'_{\sigma'_2} \neq \emptyset$ and then $\sigma'_1 = \sigma'_2$.

The fact that C is closed follows from the continuity of the maps μ, σ'_1 and λ.

The hard step is to show that $C = \Omega \setminus S$. If $\Omega \setminus S$ is connected this readily follows from the fact that C is nonempty and both open and closed. Assume that $\Omega \setminus S$ is not connected. Then, as an open and closed subset, C must be a union of connected components of $\Omega \setminus S$. Let A_1 be the connected component of $\Omega \setminus S$ that contains x_1. As $A_1 \subseteq C$ we have $\mu(\tilde{q}) = \sigma'_1(\lambda(\tilde{q}))$ for any $\tilde{q} \in A_1$. As Ω is connected, the connected components of $\Omega \setminus S$ are separated by S. Also, as $\Omega \setminus S$ is (by assumption) not connected, we know that $A_1 \neq \Omega \setminus S$.

Let A_0 be another connected component of $\Omega \setminus S$ such that $\overline{A}_0 \cap \overline{A}_1 \neq \emptyset$, where the closures are taken in Ω. Let $x_0 \in \overline{A}_0 \cap \overline{A}_1$. Let $\tilde{S}'$ consist of all $x' \in \Omega'$ such that $G'_{x'} \neq \{e'\}$. Then

$$\lambda(S) = \tilde{S}' \cap \lambda(\Omega). \tag{9.8}$$

The inclusion $\supseteq$ is obvious. Viceversa let $x \in \tilde{S}$. Then $x \in \Omega$ and $G'_{\lambda(x)} \neq \{e'\}$ hence $\lambda(x) \in \tilde{S}'$. As $\lambda(\Omega)$ is an open subset of Ω' it follows (by (9.8) and Theorem 9.6) that, in some neighborhood of $\lambda(x_0)$, $\lambda(S)$ is a finite union of submanifolds of Ω', of dimension $< n$. Next as λ is a diffeomorphism of Ω onto $\lambda(\Omega)$ it follows that S itself is a union of submanifolds of Ω, of dimension $< n$, in some neighborhood of x_0. Then x_0 has to lie on one of these submanifolds of dimension $n-1$, say S_{n-1}, and S_{n-1} may be chosen such that $S_{n-1} \cap \overline{A}_1$ itself contains a $(n-1)$-dimensional[1] submanifold. If this is the case, we may choose a point $x \in S_{n-1} \cap \overline{A}_1$ such that there is a neighborhood W_x of x with $W_x \cap S = W_x \cap S_{n-1}$ (eventually $x \neq x_0$).

[1]Submanifolds of dimension $\leq n-2$ do not locally separate two distinct connected components.

Then $W_x - \overline{A}_1$ is contained in a connected component of $\Omega \setminus S$, say A_2 (eventually $A_2 \neq A_0$). Let $x_2 \in A_2$ with $G'_{\lambda(x_2)} = \{e'\}$. As

$$\varphi'(\mu(x_2)) = \varphi(x_2) = \varphi'(\lambda(x_2))$$

there is $\sigma'_2 \in G'$ such that

$$\mu(x_2) = \sigma'_2(\lambda(x_2))$$

An argument analogous to the above (applied to x_2 and σ'_2 instead of x_1 and σ'_1) leads to

$$\mu(\tilde{q}) = \sigma'_2(\lambda(\tilde{q}))$$

for any $\tilde{q} \in A_2$. At this point we may show that actually $\sigma'_1 = \sigma'_2$. By continuity, $\mu(x) = \sigma'_1(\lambda(x))$ and $\mu(x) = \sigma'_2(\lambda(x))$ hence

$$\sigma_1'^{-1} \circ \sigma'_2 \in G'_{\lambda(x)}. \tag{9.9}$$

Let us choose a neighborhood $V_x \subset W_x$ of x in Ω such that there is a coordinate system on $V'_{\lambda(x)} = \lambda(V_x)$ centered at $\lambda(x)$ with respect to which $S'_{n-1} = \lambda(S_{n-1} \cap V_x)$ is a linear submanifold. Set $V''_{\mu(x)} = \mu(V_x)$. By the very choice of x and V_x, S'_{n-1} is precisely the set of all points of $\tilde{V}_x$ fixed by $G'_{\lambda(x)}$. Hence (by (9.9)) either $\sigma_1'^{-1} \circ \sigma'_2 = e'$ or (by Lemma 9.10) there is a symmetry s with respect to S'_{n-1} (in the local coordinates considered on $V'_{\lambda(x)}$) such that $\sigma'_2 = \sigma'_1 \circ s$.

Summing up, we know that $\mu(\tilde{q}) = \sigma'_i(\lambda(\tilde{q}))$ for any $\tilde{q} \in A_i, i = 1, 2$. Consequently σ'_i and $\mu \circ \lambda^{-1}$ coincide on $\lambda(A_i), i = 1, 2$. Therefore σ'_2 maps $\lambda(A_2)$ into $\mu(A_2)$ and $\lambda(A_2) \cap V'_{\lambda(x)}$ into $\mu(A_1) \cap V''_{\mu(x)}$, where from one may conclude that $\sigma'_2 = \sigma'_1 \circ s$ maps $\lambda(A_2) \cap V'_{\lambda(x)}$ into $\mu(A_1) \cap V''_{\mu(x)}$. This is not possible unless $\sigma'_1 = \sigma'_2$, hence $A_2 \subset C$.

Now, if $A_1 \cup A_2 = \Omega \setminus S$ then $C = \Omega \setminus S$ and we are done. Otherwise, there will exist another connected component A_3 such that $\overline{A}_3 \cap S$ contains a submanifold of dimension $n - 1$ which is contained in $\overline{A}_1 \cup S$, or in $\overline{A}_2 \cup S$, too. By repeating the argument above with A_3 instead of A_2 (and eventually with A_2 instead of A_1) one shows that $A_3 \subset C$. In the end, an iteration of this argument leads to $C = \Omega \setminus S$. Therefore (9.4) holds on $\Omega \setminus S$. By continuity (9.4) must hold on the whole of Ω. This proves the existence of an element $\sigma'_1 \in G'$ satisfying (9.4). To establish uniqueness it suffices to choose $\tilde{q}_0$ such that $G'_{\lambda(\tilde{q}_0)} = \{e'\}$. Then there is a unique $\sigma'_1 \in G'$ such that (9.4) holds at $\tilde{q}_0$.

9.4. The singular locus

We start be establishing the following

THEOREM 9.11. *Let $(B, \mathcal{A})$ be an orbifold and $\{\Omega, G, \varphi\}, \{\Omega', G', \varphi'\} \in \mathcal{A}$ two l.u.s.'s on B of supports U, U'. Assume that $U \subseteq U'$ and let λ be an injection of $\{\Omega, G, \varphi\}$ into $\{\Omega', G', \varphi'\}$ and $\eta : G \to G'$ the corresponding group monomorphism. Let $\sigma' \in G'$. If $\sigma'(\lambda(\Omega)) \cap \lambda(\Omega) \neq \emptyset$ then $\sigma'(\lambda(\Omega)) = \lambda(\Omega)$ and $\sigma' \in \eta(G)$.*

Proof. Let $\sigma' \in G'$ and assume that $\sigma'(\lambda(\Omega)) \cap \lambda(\Omega) \neq \emptyset$ i.e. there are $x, \tilde{q} \in \Omega$ such that

$$\sigma'(\lambda(x)) = \lambda(\tilde{q}). \tag{9.10}$$

We have (by (9.10))

$$\varphi(x) = \varphi'(\lambda(x)) = \varphi'(\sigma'(\lambda(x))) = \varphi'(\lambda(\tilde{q})) = \varphi(\tilde{q})$$

hence there is $\tau \in G$ such that

$$\tilde{q} = \tau(x). \tag{9.11}$$

Let us set $\tau' = \eta(\tau) \in G'$. As $\sigma'(\lambda(\Omega)) \cap \lambda(\Omega)$ is an open set

$$\tilde{V} = \lambda^{-1}(\sigma'(\lambda(\Omega)) \cap \lambda(\Omega))$$

is an open subset of Ω hence (by the local structure of $S = \{x \in U : G'_{\lambda(x)} \neq \{e'\}\}$) there is $x \in \tilde{V}$ with $G'_{\lambda(x)} = \{e'\}$. Then $\lambda \circ \tau = \eta(\tau) \circ \lambda$ (cf. (9.4)) may be written as $\lambda \circ \tau = \tau' \circ \lambda$, hence $\lambda(\tau(x)) = \tau'(\lambda(x))$ or (by (9.11))

$$\tau'(\lambda(x)) = \lambda(\tilde{q}) \tag{9.12}$$

The identities (9.10) and (9.12) yield $\sigma'(\lambda(x)) = \tau'(\lambda(x))$ or $\sigma' \circ \tau'^{-1} \in G'_{\lambda(x)} = \{e'\}$, hence $\sigma' = \tau'$ i.e. $\sigma' \in \eta(G)$. Finally

$$\sigma'(\lambda(\Omega)) = \tau'(\lambda(\Omega)) = \eta(\tau)\lambda(\Omega) = \lambda(\tau(\Omega)) = \lambda(\Omega).$$

Q.e.d.

THEOREM 9.12. *Let $\lambda : \{\Omega, G, \varphi\} \to \{\Omega', G', \varphi'\}$ be an injection and let $\eta : G \to G'$ be the corresponding group monomorphism. If $U = U'$ i.e. the l.u.s.'s $\{\Omega, G, \varphi\}$ and $\{\Omega', G', \varphi'\}$ have the same support, then $\Omega' = \lambda(\Omega)$ and $G' = \eta(G)$. In particular $\Omega \approx \Omega'$ (a diffeomorphism) and $G \approx G'$ (a group isomorphism).*

Proof. Note that

$$\Omega' = \bigcup_{\sigma' \in G'} \sigma'(\lambda(\Omega)). \tag{9.13}$$

The inclusion $\supseteq$ is obvious. To establish $\subseteq$ let $x' \in \Omega'$. Then $\varphi'(x') \in \varphi(\Omega)$ (as $\varphi(\Omega) = \varphi'(\Omega')$) i.e. there is $x \in \Omega$ with $\varphi'(x') = \varphi(x) = \varphi'(\lambda(x))$. Hence there is $\sigma' \in G'$ with $x' = \sigma'(\lambda(x))$. Q.e.d. Next note that

$$\sigma'(\lambda(\Omega)) = \lambda(\Omega) \tag{9.14}$$

for any $\sigma' \in G'$. Indeed one may distinguish two cases. Either the sets $\sigma'(\lambda(\Omega))$ and $\lambda(\Omega)$ intersect, and then they coincide (by Theorem 9.11) or $\sigma'(\lambda(\Omega)) \cap \lambda(\Omega) = \emptyset$. If this is the case we may exploit the fact that Ω' is connected and that each $\sigma'(\lambda(\Omega))$ is open as follows (by (9.13))

$$\Omega' = \bigcup_{\sigma' \in G'} \sigma'(\lambda(\Omega)) =$$

$$= \left[\bigcup_{\sigma'(\lambda(\Omega)) \cap \lambda(\Omega) \neq \emptyset} \sigma'(\lambda(\Omega)) \right] \cup \left[\bigcup_{\sigma'(\lambda(\Omega)) \cap \lambda(\Omega) = \emptyset} \sigma'(\lambda(\Omega)) \right] =$$

$$= \lambda(\Omega) \cup \left[\bigcup_{\sigma'(\lambda(\Omega)) \cap \lambda(\Omega) = \emptyset} \sigma'(\lambda(\Omega)) \right].$$

As Ω' is connected and $\lambda(\Omega) \neq \emptyset$ it must be that

$$\bigcup_{\sigma'(\lambda(\Omega)) \cap \lambda(\Omega) = \emptyset} \sigma'(\lambda(\Omega)) = \emptyset$$

i.e. the index set is empty

$$\{\sigma' \in G' : \sigma'(\lambda(\Omega)) \cap \lambda(\Omega) = \emptyset\} = \emptyset.$$

Thus (9.14) holds for any $\sigma' \in G'$. In particular (by the calculation above) we have $\Omega' = \lambda(\Omega)$. Finally (by (9.14) and Theorem 9.11) $G' \subseteq \eta(G)$ i.e. η is a group epimorphism. Q.e.d.

COROLLARY 9.13. *Let $(B, \mathcal{A})$ be an orbifold and $\{\Omega, G, \varphi\}$, $\{\Omega', G', \varphi'\} \in \mathcal{A}$ two l.u.s.'s of B with supports $U \subseteq U'$. Let λ be an injection of $\{\Omega, G, \varphi\}$ into $\{\Omega', G', \varphi'\}$ and let $\eta : G \to G'$ be the corresponding group monomorphism. Then for an arbitrary fixed pair of elements $\sigma', \tau' \in G'$, either $\sigma'(\lambda(\Omega)) \cap \tau'(\lambda(\Omega)) = \emptyset$, and then ${\tau'}^{-1} \circ \sigma' \in \eta(G)$, or $\sigma'(\lambda(\Omega)) = \tau'(\lambda(\Omega))$, and then ${\tau'}^{-1} \circ \sigma' \in \eta(G)$.*

The proof follows easily from Theorem 9.11.

THEOREM 9.14. *Let $\lambda : \{\Omega, G, \varphi\} \to \{\Omega', G', \varphi'\}$ be an injection and $\eta : G \to G'$ the corresponding group monomorphism. Then*

$$G'/\eta(G) \approx \pi_0(V_\lambda) \tag{9.15}$$

(a bijection) where

$$V_\lambda = \bigcup_{\sigma' \in G'} \sigma'(\lambda(\Omega)) \tag{9.16}$$

i.e. the quotient space $G'/\eta(G)$ may be identified with the set $\pi_0(V_\lambda)$ of all connected components of V_λ.

Proof. By Corollary 9.13, given two arbitrary elements $\sigma', \tau' \in G'$ the sets $\sigma'(\lambda(\Omega))$ and $\tau'(\lambda(\Omega))$ are either disjoint, or they coincide. It follows that we may fix elements $\sigma'_1, \cdots, \sigma'_s \in G'$ such that

$$V_\lambda = \bigcup_{i=1}^{s} \sigma'_i(\lambda(\Omega)),$$

$$i \neq j \Longrightarrow \sigma'_i(\lambda(\Omega)) \cap \sigma'_j(\lambda(\Omega)) = \emptyset.$$

Given $\sigma' \in G'$ let $H_{\sigma'}$ consist of all $\tau' \in G'$ with $\tau'(\lambda(\Omega)) = \sigma'(\lambda(\Omega))$. Clearly for any $\sigma' \in G'$ there is $i \in \{1, \cdots, s\}$ such that $\sigma' \in H_{\sigma'_i}$ hence (by Corollary 9.13) ${\sigma'}^{-1} \circ \sigma'_i \in \eta(G)$ i.e. $\sigma' \equiv \sigma'_i, \bmod \eta(G)$. Then we may represent the quotient space $G'/\eta(G)$ as

$$G'/\eta(G) = \{\sigma'_i \eta(G) : 1 \leq i \leq s\}$$

and define a map

$$G'/\eta(G) \to \pi_0(V_\lambda)\,,\ \sigma'_i \eta(G) \mapsto \sigma'_i(\lambda(\Omega)).$$

We leave it to the reader to check that this map is bijective.

THEOREM 9.15. *The open set V_λ given by (9.16) does not depend upon the choice of injection λ of $\{\Omega, G, \varphi\}$ into $\{\Omega', G', \varphi'\}$ (i.e. if μ is another injection of $\{\Omega, G, \varphi\}$ into $\{\Omega', G', \varphi'\}$ then $V_\lambda = V_\mu$).*

We stress the conclusion of Theorem 9.15 by giving to V_λ the new name $V(\Omega, \Omega')$ (hence (9.15) reads $G'/\eta(G) \approx \pi_0(V(\Omega, \Omega'))$). To prove Theorem 9.15 let μ be another injection of $\{\Omega, G, \varphi\}$ into $\{\Omega', G', \varphi'\}$ (besides from λ). Then there is a unique element $\sigma' \in G'$ such that $\mu = \sigma' \circ \lambda$. We distinguish two cases as I) $\sigma'_1 \in \eta(G)$ or II) $\sigma' \notin \eta(G)$. For the first case $\sigma' = \eta(\sigma)$ for some $\sigma \in G$, hence $\mu = \sigma' \circ \lambda = \eta(\sigma) \circ \lambda = \lambda \circ \sigma$ so that $\mu(\Omega) = \lambda(\Omega)$. As to the second case, as $\sigma' \in G'$

there is $i \in \{1, \cdots, s\}$ such that $\sigma' \equiv \sigma'_i$, mod $\eta(G)$ i.e. ${\sigma'}^{-1} \circ \sigma'_i = \eta(\tau)$ for some $\tau \in G$. Then

$$({\sigma'}^{-1} \circ \sigma'_i)\lambda(\Omega) = \eta(\eta)\lambda(\Omega) = \lambda(\tau(\Omega)) = \lambda(\Omega)$$

i.e.

$$\sigma'_i(\lambda(\Omega)) = \sigma'(\lambda(\Omega)) = \mu(\Omega).$$

We may conclude that in the second case $\mu(\Omega) \in \pi_0(V_\lambda)$ and $\mu(\Omega) \neq \lambda(\Omega)$ (otherwise $\sigma'(\lambda(\Omega)) = \lambda(\Omega)$ yields (by Theorem 9.11) $\sigma' \in \eta(G)$, a contradiction). At this point we may show that $V_\lambda = V_\mu$. To this end, let $x' \in V_\lambda$. Then there is $\tau' \in G'$ such that $x' \in \tau'(\lambda(\Omega))$. If case I occurs then $\lambda(\Omega) = \mu(\Omega)$ yields $x' \in V_\mu$. If case II occurs then

$$x' = \tau'(\lambda(x)) = \tau'{\sigma'}^{-1}\sigma'\lambda(x) = \tau'{\sigma'}^{-1}\mu(x)$$

for some $x \in \Omega$ i.e. $x' \in \tau'{\sigma'}^{-1}(\mu(\Omega)) \subseteq V_\mu$.

THEOREM 9.16. *Let $(B, \mathcal{A})$ be an orbifold and $\{\Omega, G, \varphi\}, \{\Omega', G', \varphi'\} \in \mathcal{A}$ two l.u.s.'s of B of supports U, U' such that $U \cap U' \neq \emptyset$. Let $p \in U \cap U'$ and $x \in \Omega$, $x' \in \Omega'$ such that $\varphi(x) = \varphi'(x') = p$. Then $G_x \approx G'_{x'}$ (a group isomorphism).*

We may assume without loss of generality that $U \subseteq U'$. Let then λ be an injection of $\{\Omega, G, \varphi\}$ into $\{\Omega', G', \varphi'\}$ so that $\lambda(x) = x'$ (e.g. start with just any injection μ, set $\tilde{q}' = \mu(x)$ and (because of $\varphi'(\tilde{q}') = \varphi'(\mu(x)) = \varphi(x) = \varphi'(x')$) choose $\sigma' \in G'$ with $\tilde{q}' = \sigma(x')$. Then $x' = \sigma^{-1}(\mu(x))$ and $\lambda = \sigma^{-1} \circ \mu$ is a new injection, with the desired property). Let $\eta : G \to G'$ be the group monomorphism associated with λ. Then the restriction of η to G_x corestricts to $G'_{x'}$. Indeed, if $\sigma \in G_x$ then

$$\eta(\sigma)(x') = \eta(\sigma)(\lambda(x)) = \lambda(\sigma(x)) = \lambda(x) = x'$$

i.e. $\eta(\sigma) \in G'_{x'}$.

To end the proof of Theorem 9.16 we shall show that $\eta : G_x \to G'_{x'}$ is a group epimorphism. Let $\sigma' \in G'_{x'}$. Then $\sigma'(\lambda(x)) = \lambda(x)$ hence $\sigma'(\lambda(\Omega)) \cap \lambda(\Omega) \neq \emptyset$. It follows (by Theorem 9.11) that $\sigma' \in \eta(G)$. Let $\sigma \in G$ with $\eta(\sigma) = \sigma'$. Finally, it suffices to show that $\sigma \in G_x$. Indeed

$$\lambda(x) = \sigma'(\lambda(x)) = \eta(\sigma)(\lambda(x)) = \lambda(\sigma(x))$$

i.e. $x = \sigma(x)$. Q.e.d.

DEFINITION 9.17. Let $(B, \mathcal{A})$ be an orbifold. Then $p \in B$ is a *singular point* of B if for some $U \in \mathcal{H}$ containing p, and some l.u.s. $\{\Omega, G, \varphi\} \in \mathcal{A}$ of support U, and some $x \in \Omega$ with $\varphi(x) = p$, the isotropy group G_x is nontrivial. □

By Theorem 9.16, the notion of singular point of an orbifold is unambigously defined (the definition does not depend upon the various choises involved).

DEFINITION 9.18. The *singular locus* Σ of B consists of all singular points of B. □

EXAMPLE 9.19. If $D \subset \mathbb{R}^2$ is the unit disc and G the group of rotations (of center the origin $O \in \mathbb{R}^2$) of angles 0, $\pi/2$, π and $3\pi/2$, then the singular locus of the cone $B = D/G$ consists solely of the point $\{O\}$, the orbit of O mod G. □

EXAMPLE 9.20. It may be shown that the quotient space M/Γ of a C^∞ manifold M by a properly discontinuous group Γ of C^∞ diffeomorphisms of M (not necessarily finite) may be endowed with a natural structure of a C^∞ orbifold. For instance $S^n/\mathbb{Z}_p$ is a compact orbifold. In particular, the singular locus of $S^2/\mathbb{Z}_3$ consists of exactly two points. □

Any finite direct product of orbifolds is again an orbifold.

9.5. Vector bundles over orbifolds

Let $(B, \mathcal{A})$ be a n-dimensional orbifold of class C^∞. Let E be a connected paracompact Hausdorff space and $\pi : E \to B$ a continuous surjective map.

DEFINITION 9.21. (E, π, B) is a *vector bundle of standard fibre* K^m, where $K \in \{\mathbb{R}, \mathbb{C}\}$, if the following requirements are fulfilled

1) for any $\{\Omega, G, \varphi\} \in \mathcal{A}$ there is a unique continuous map $\varphi_* : \Omega \times K^m \to E$ such that the following diagram is commutative

$$\begin{array}{ccc} \Omega \times K^m & \xrightarrow{\varphi_*} & E \\ \pi_\Omega \downarrow & & \downarrow \pi \\ \Omega & \xrightarrow{\varphi} & B \end{array}$$

where $\pi_\Omega(x, \xi) = x$, for any $(x, \xi) \in \Omega \times K^m$. Moreover

2) for any injection λ of $\{\Omega, G, \varphi\}$ into $\{\Omega', G', \varphi'\}$ there is a unique C^∞ map $g_\lambda : \Omega \to GL(m, K)$ so that $g_e(x) = x$ for any $x \in \Omega$ and

i) $\{\Omega \times K^m, G_*, \varphi_*\}$ is a l.u.s. of dimension $d(K)m + n$ of E over some open subset of E, where

$$G_* = \{\sigma_* : \sigma \in G\}\ ,\ \sigma_*(x, \xi) = (\sigma(x), g_\sigma(x)\xi)$$

for any $(x, \xi) \in \Omega \times K^m$, and $d(K) = \dim_{\mathbb{R}} K$.

ii) The family of l.u.s.'s $\{\Omega \times K^m, G_*, \varphi_*\}$, obtained as $\{\Omega, G, \varphi\}$ ranges over $\mathcal{A}$, is a defining family for E (thus organizing E as a $(d(K)m+n)$-dimensional orbifold of class C^∞).

iii) The map $\lambda_* : \Omega \times K^m \to \Omega' \times K^m$ given by:

$$\lambda_*(x, \xi) = (\lambda(x, g_\lambda(x)\xi)$$

is an injection of $\{\Omega \times K^m, G_*, \varphi_*\}$ into $\{\Omega' \times K^m, G'_*, \varphi'_*\}$.
Finally

3) for any pair of injections

$$\{\Omega, G, \varphi\} \xrightarrow{\lambda} \{\Omega', G', \varphi'\} \xrightarrow{\mu} \{\Omega'', G'', \varphi''\}$$

one requests that

$$g_\mu(\lambda(x))g_\lambda(x) = g_{\mu\circ\lambda}(x)$$

for any $x \in \Omega$. □

Let B be an orbifold and Σ its singular locus.

DEFINITION 9.22. A point $p \in B \setminus \Sigma$ is a *regular point* of B. □

The notion of vector bundle over B bears a close analogy to that of a vector bundle over an ordinary C^∞ manifold. However, only fibres over regular points of B are isomorphic to K^m. Precisely, we may state

THEOREM 9.23. *Let (E, π, B) be a vector bundle of standard fibre K^m over a C^∞ orbifold B. If $p \in B$ is a regular point then $\pi^{-1}(p) \approx K^m$ (a bijection).*

Proof. Let $p \in B$ be an arbitrary point (eventually singular) and $U \in \mathcal{H}$ such that $p \in U$. Let $\{\Omega, G, \varphi\} \in \mathcal{A}$ be a l.u.s. of support U and $x \in \Omega$ such that $\varphi(x) = p$. Let $\{\Omega_*.G_*, \varphi_*\}$ be a l.u.s. of E corresponding to $\{\Omega, G, \varphi\}$, according to the definition of the notion of a vector bundle over B, where $\Omega_* = \Omega \times K^m$. Then

$$\pi(\varphi_*(x, \xi)) = \varphi(\pi_\Omega(x, \xi)) = \varphi(x) = p$$

hence $\varphi_*(x, \xi) \in \pi^{-1}(p)$ for any $\xi \in K^m$. There is a natural action of G_x on K^m given by

$$G_x \times K^m \to K^m, \quad (\sigma, \xi) \mapsto g_\sigma(x)\xi.$$

At this point, we may define a map

$$K^m/G_x \to \pi^{-1}(p), \quad [\xi] \mapsto \varphi_*(x, \xi), \tag{9.17}$$

where $[\xi]$ is the orbit of $\xi \mod G_x$ i.e. $[\xi] = \{g_\sigma(x)\xi : \sigma \in G_x\}$. We shall show that (9.17) is a bijection. First, the definition does not depend upon the choice of representatives. Indeed, if $[\xi] = [\zeta]$ then $\zeta = g_\sigma(x)\xi$ for some $\sigma \in G_x$ and

$$\varphi_*(x, \zeta) = \varphi_*(x, g_\sigma(x)\xi) = \varphi_*(\sigma(x), g_\sigma(x)\xi) =$$

$$= \varphi_*(\sigma_*(x, \xi)) = \varphi_*(x, \xi).$$

Let us show that (9.17) is an injective map. If $\varphi_*(x, \xi) = \varphi_*(x, \zeta)$, as $\{\Omega_*, G_*, \varphi_*\}$ is a l.u.s. there is $\sigma \in G$ such that

$$(x, \xi) = \sigma_*(x, \xi) = (\sigma(x), g_\sigma(x)\xi)$$

hence $\sigma(x) = x$, i.e. $\sigma \in G_x$, and $g_\sigma(x)\xi = \zeta$ i.e. $\xi \equiv \zeta \mod G_x$.

Finally let us check that (9.17) is surjective. To this end let $f \in \pi^{-1}(p) \subset E$. Also let U_* be the support of the l.u.s. $\{\Omega_*, G_*, \varphi_*\}$ of E (corresponding to $\{\Omega, G, \varphi\}$ as above). As φ_* induces a homeomorphism $\Omega_*/G_* \approx U_*$ there is $\tilde{f} \in \Omega_*$ such that $\varphi_*(\tilde{f}) = f$. As $\Omega_* = \Omega \times K^m$ the element $\tilde{f}$ is of the form $\tilde{f} = (\tilde{g}, \xi)$. Then

$$\varphi(x) = p = \pi(f) = \pi(\varphi_*(\tilde{f})) = \varphi(\pi_\Omega(\tilde{f})) = \varphi(\tilde{q})$$

hence there is $\sigma \in G$ so that $\tilde{q} = \sigma(x)$ i.e. $\tilde{f} = (\sigma(x, \xi)$. At this point let us set $\tilde{f}_* = (\sigma^{-1})_* \tilde{f} \in \Omega_*$. Then

$$\varphi_*(\tilde{f}_*) = \varphi_*((\sigma^{-1})_* \tilde{f}) = \varphi_*(\tilde{f}) = f.$$

Also

$$\tilde{f}_* = (\sigma^{-1})_*(\sigma(x), \xi) = (x, g_{\sigma^{-1}}(\sigma(x))\xi).$$

Hence $\tilde{f}_*$ is an element of the form $\tilde{f}_* = (x, \zeta)$ (with $\zeta = g_{\sigma^{-1}}(\sigma(x))\xi$) and $\varphi_*(x, \zeta) = f$, so we are done.

The discussion of the partition of unity on a C^∞ orbifold is relegated to Appendix B.

9.6. Transition functions

Let (E, π, B) be a vector bundle of standard fibre K^m over the C^∞ orbifold $(B, \mathcal{A})$.

DEFINITION 9.24. The C^∞ functions $g_\lambda : \Omega \to \mathrm{GL}(m, K)$ corresponding to injections $\lambda : \{\Omega, G, \varphi\} \to \{\Omega', G', \varphi'\}$ are the *transition functions* of (E, π, B). □

The aim of the present notion is to show how a vector bundle may be recovered from the transition functions.

Let $(B, \mathcal{A})$ be a C^∞ orbifold, of dimension n. Assume that for any injection $\lambda : \{\Omega, G, \varphi\} \to \{\Omega', G', \varphi'\}$ one is given a C^∞ function $g_\lambda : \Omega \to \mathrm{GL}(m, K)$ such that

$$g_{\mu\circ\lambda}(x) = g_\mu(\lambda(x))g_\lambda(x)$$

for any $x \in \Omega$ and any injections

$$\{\Omega, G, \varphi\} \xrightarrow{\lambda} \{\Omega', G', \varphi'\} \xrightarrow{\mu} \{\Omega'', G'', \varphi''\}.$$

Let then G_* consist of all σ_* given by

$$\sigma_*(x, \xi) = (\sigma(x, g_\sigma(x)), \quad \sigma \in G.$$

Define $\hat{E}$ as the disjoint union of all quotients

$$(\Omega \times K^m)/G_*$$

as $\{\Omega, G, \varphi\}$ ranges over $\mathcal{A}$, i.e.

$$\hat{E} = \bigcup_{\{\Omega, G, \varphi\} \in \mathcal{A}} (\Omega \times K^m)/G_* .$$

Each $(\Omega \times K^m)/G_*$ carries the quotient topology, hence $\hat{E}$ is a topological space in a natural manner (as a disjoint union of topological spaces). We define an equivalence relation $\sim$ on $\hat{E}$ as follows. Let $\hat{p}, \hat{q} \in \hat{E}$. Then we may represent $\hat{p}$ as an orbit mod G_* i.e. $\hat{p} = \mathrm{orb}_{G_*}(x, \xi)$ for some $(x, \xi) \in \Omega \times K^m$ and some l.u.s. $\{\Omega, G, \varphi\} \in \mathcal{A}$. Then $\hat{p} \sim \hat{q}$ if there is an injection $\lambda : \{\Omega, G, \varphi\} \to \{\Omega', G', \varphi'\}$ such that $\hat{q} = \mathrm{orb}_{G'_*}(\lambda(x), g_\lambda(x)\xi)$. We need to check that $\hat{p} \sim \hat{q}$ is well defined i.e. the definition does not depend upon the choice of representatives of $\hat{p}$. Indeed if $(\sigma(x, g_\sigma(x)\xi) \in \hat{p}$ is another representative then

$$\begin{aligned} \mathrm{orb}_{G'_*}(\lambda(\sigma(x)), g_\lambda(\sigma(x))g_\sigma(x)\xi) &= \mathrm{orb}_{G'_*}(\eta(\sigma)\lambda(x), g_{\lambda\circ\sigma}(x)\xi) = \\ = \mathrm{orb}_{G'_*}(\eta(\sigma)\lambda(x), g_{\eta(\sigma)}(\lambda(x))g_\lambda(x)\xi) &= \mathrm{orb}_{G'_*}\eta(\sigma)_*(\lambda(x), g_\lambda(x)\xi) = \\ &= \mathrm{orb}_{G'_*}(\lambda(x), g_\lambda(x)\xi) \end{aligned}$$

where $\eta : G \to G'$ is the group monomorphism associated with λ. Clearly $\sim$ is reflexive and transitive. The only issue which needs a bit of care is the symmetry property. Note that

LEMMA 9.25. *For any injection $\lambda : \{\Omega, G, \varphi\} \to \{\Omega', G', \varphi'\}$ the synthetic object $\{\lambda(\Omega), \eta(G), \psi\}$, where $\psi = \varphi'|_{\lambda(\Omega)}$, is a l.u.s. of support $U = \varphi(\Omega)$.*

Indeed $\eta(G)$ acts on $\lambda(\Omega)$ as a group of C^∞ diffeomorphisms and ψ is $\eta(G)$-invariant, by the G'-invariance of φ'. Moreover $\lambda : \Omega \to \Omega'$ is equivariant hence it induces a homeomorphism $\lambda_G : \Omega/G \approx \lambda(\Omega)/\eta(G)$. Let $\psi_G : \lambda(\Omega)/\eta(G) \to U'$ be the map induced by ψ. Then ψ_G corestricts to U and $\psi_G \circ \lambda_G = \varphi_G$ (where $\varphi_G : \Omega/G \approx U$ is the homeomorphism induced by φ) hence $\psi_G : \lambda(\Omega)/\eta(G) \to U'$

is a homeomorphism. Then $\hat{p} \sim \hat{q}$ yields $\hat{q} \sim \hat{p}$, as we may think of $(\lambda(x), g_\lambda(x)\xi)$ as a representative of $\hat{q}$ with respect to the l.u.s. $\{\lambda(\Omega, \eta(G), \psi\}$ and rewrite $\hat{p}$ as

$$\hat{p} = \mathrm{orb}_{G_*}(\mu(\lambda(x)), g_\mu(\lambda(x)) g_\lambda(x)\xi)$$

where μ is the injection $\mu : \lambda(\Omega) \to \Omega$, $\mu = \lambda^{-1}$.

At this point we set $E = \hat{E}/\sim$. Then E is a topological space with the quotient topology. Let us define $\pi : E \to B$ by setting

$$\pi\left([\mathrm{orb}_{G_*}(x, \xi)]\right) = \varphi(x)$$

where the square brackets indicate classes mod $\sim$ i.e.

$$E = \{[\hat{p}] : \hat{p} \in \hat{E}\}.$$

The definition of $\pi\left([\mathrm{orb}_{G_*}(x, \xi)]\right)$ doesn't depend upon the choice of representatives. Indeed, if $\hat{p} = \mathrm{orb}_{G_*}(x, \xi)$ and $\hat{q} \in [\hat{p}]$ then

$$\hat{q} = \mathrm{orb}_{G'_*}(\lambda(x), g_\lambda(x)\xi)$$

for some injection λ of $\{\Omega, G, \varphi\}$ into $\{\Omega', G', \varphi'\}$. Finally $\varphi'(\lambda(x)) = \varphi(x)$. Q.e.d.

We wish to show that (E, π, B) is a vector bundle of standard fibre K^m (admitting g_λ as transition functions). To this end let $\varphi_* : \Omega \times K^m \to E$ be the (continuous) map defined by

$$\varphi_*(x, \xi) = [\mathrm{orb}_{G_*}(x, \xi)].$$

Then $\pi \circ \varphi_* = \varphi \circ \pi_\Omega$. Also φ_* is G_*-invariant and the induced map $(\varphi_*)_{G_*} : (\Omega \times K^m)/G_* \to E$ is injective. Finally, the reader may check that $\lambda_*(x, \xi) = (\lambda(x), g_\lambda(x)\xi)$ is an injection of $\{\Omega \times K^m, G_*, \varphi_*\}$ into $\{\Omega' \times K^m, G'_*, \varphi'_*\}$ i.e. $\varphi'_* \circ \lambda_* = \varphi_*$, and we are done.

As an application, we may build the tangent bundle $T(B)$ over an orbifold B. Let $(B, \mathcal{A})$ be a n-dimensional orbifold. For any injection $\lambda : \{\Omega, G, \varphi\} \to \{\Omega', G', \varphi'\}$ let $g_\lambda : \Omega \to \mathrm{GL}(n, \mathbb{R})$ be given by

$$g_\lambda(x) = \left[\frac{\partial(u'^i \circ \lambda)}{\partial u^j}(x)\right]$$

for any $x \in \Omega$. Here $(u^1, \cdots, u^n)$ (respectively $(u'^1, \cdots, u'^n)$) are the Cartesian coordinates on Ω (respectively on Ω'). Clearly g_λ satisfies the cocycle condition $g_{\mu\circ\lambda}(x) = g_\mu(\lambda(x)) g_\lambda(x)$. The corresponding bundle (of standard fibre $\mathbb{R}^n$) is denoted by $(T(B), \pi, B)$.

DEFINITION 9.26. $(T(B), \pi, B)$ is the *tangent bundle* over the orbifold B. □

9.7. Compact Hausdorff foliations

Let M be a C^∞ manifold.

DEFINITION 9.27. A foliation $\mathcal{F}$ of M is a *compact Hausdorff foliation* if 1) each leaf of $\mathcal{F}$ is compact, and 2) the leaf space $M/\mathcal{F}$ is Hausdorff (with the quotient topology). □

We may state the following

THEOREM 9.28. (J. Girbau & M. Nicolau, [**120**])
Let M be a real $(p+q)$-dimensional C^∞ manifold and $\mathcal{F}$ a codimension q compact Hausdorff foliation of M. Then the leaf space $B = M/\mathcal{F}$ admits a natural structure of a q-dimensional orbifold.

We shall not prove Theorem 9.28 (and send the reader instead to [**120**], p. 80-85) yet shall show how a defining family $\mathcal{A}$ of B may be built. See also R. Edwards et al., [**99**], where techniques similar to those of D.B.A. Epstein, [**100**], are employed.

If $r > 0$ and $x \in \mathbb{R}^q$ then $B(x,r)$ denotes the ball of center x and radius r in $\mathbb{R}^q$. Let us consider the following data 1) a finite subgroup $G \subset \mathrm{O}(q)$ of the orthogonal group, 2) a compact manifold L, and 3) a C^k action $\alpha : L \times G \to L$ of G on L, $k \geq 1$. We write $\alpha(p,g) = pg$ for any $p \in L$ and $g \in G$. There is a naturally induced action of G on the product manifold $L \times B(0,1)$ i.e.

$$G \times (L \times B(0,1)) \to L \times B(0,1), \quad g \cdot (p,x) = (pg^{-1}, gx)$$

for any $g \in G, p \in L$ and $x \in B(0,1)$. Let

$$L \times_G B(0,1) = (L \times B(0,1))/G$$

be the quotient space. It is a differentiable manifold, in a natural way. Indeed, if Ω is (the geometric zone of) a local chart of L so that $(\Omega g) \cap \Omega = \emptyset$ for any $g \in G \setminus \{e\}$, then the natural map

$$\Omega \times B(0,1) \to \Omega \times_G B(0,1)$$

is a homeomorphism. Let $\mathcal{F}_0$ be the foliation of $L \times B(0,1)$ whose leaves are $\{L \times \{x\}\}_{x \in B(0,1)}$. Note that $R_g(L \times \{x\}) = L \times \{gx\}$ hence $\mathcal{F}_0$ is G-invariant i.e.

$$(d_{(p,x)} R_g) T(\mathcal{F}_0)_{(p,x)} = T(\mathcal{F}_0)_{g \cdot (p,x)} \, .$$

Consequently $\mathcal{F}_0$ pushes forward to a foliation $\hat{\mathcal{F}}_0$ on $L \times_G B(0,1)$. Precisely let $\sigma : L \times B(0,1) \to L \times_G B(0,1)$ be the natural map. We write $\sigma(p,x) = px$. Then we set

$$T(\hat{\mathcal{F}}_0)_w = (d_{(p,x)}\sigma) T(\mathcal{F}_0)_{(p,x)}$$

for any $w \in L \times_G B(0,1)$ and some (p,x) such that $w = px$. The key ingredient in the proof of Theorem 9.28 is the following result

THEOREM 9.29. (D.B.A. Epstein, [**100**])
Let M be a C^∞ manifold of real dimension $p+q$ and $\mathcal{F}$ a codimension q compact Hausdorff foliation of M. Then 1) *there is a leaf L of $\mathcal{F}$ and an open dense subset $U \subseteq M$ such that all leaves of the foliation $\mathcal{F}_U$ induced by $\mathcal{F}$ on U are diffeomorphic to $L \cap U$ and* 2) *given a leaf L_0 of $\mathcal{F}$ there exist* i) *a finite group $G \subset \mathrm{O}(q)$,* ii) *a free C^r action $\alpha : L \times G \to L$, $r \geq 1$,* iii) *an open neighborhood V of L_0 in M, and* iv) *a C^r diffeomorphism $\Phi : L \times_G B(0,1) \to V$ such that Φ is a foliated map of $(L \times_G B(0,1), \hat{\mathcal{F}}_0)$ onto $(V, \mathcal{F}_V)$.*

One builds a defining family $\mathcal{A}$ of B as follows. Let $\pi : M \to B$ be the natural projection. Let L_0 be a leaf of $\mathcal{F}$. By Epstein's theorem one may choose a finite subgroup $G \subset \mathrm{O}(q)$, a free C^r action α of G on L, an open neighborhood V of L_0 in M, and a foliation preserving C^r diffeomorphism Φ of $L \times_G B(0,1)$ onto V. Let

$$\mathbf{a} = \{(V, G, \alpha, \Phi) : L_0 \in B\}$$

be the family of all synthetic objects (V, G, α, Φ) corresponding to all choices of L_0, in the sense of Epstein's theorem. There is a natural injection $L \times_G B(0,1) \to L \times_G B(0,1)$. Let $\mathcal{H}$ be the family of open subsets of B given by

$$\mathcal{H} = \{\pi\Phi(L \times_G B(0,\ 1/2)) : (V, G, \alpha, \Phi) \in \mathbf{a}\}.$$

For each $p \in L$ let us consider the admissible map

$$\sigma_p : B(0,1) \to L \times_G B(0,1), \quad \sigma_p(x) = px.$$

Next let us set $\varphi = \pi\Phi\sigma_p$. *Then* $\{B(0,\ 1/2), G, \varphi\}$ *is a l.u.s. of dimension* q *of* B *of support* $U = \pi\Phi(L \times_G B(0, \frac{1}{2}))$. All we need to check is that $\varphi \circ g = \varphi$ for any $g \in G$. We look at the commutative diagram

$$\begin{array}{ccccc} & B(0,\frac{1}{2}) & \xrightarrow{\sigma_p} & L \times_G B(0,\frac{1}{2}) & \\ & \varphi \downarrow & & \downarrow \Phi & \\ B \supset & U & \xleftarrow{\pi} & V & \subseteq M \end{array}$$

Note that $p(gx) = (pg)x$. Then

$$(\varphi \circ g)(x) = \pi\Phi((pg)x), \quad \varphi(x) = \pi\Phi(px),$$

and

$$(pg)x = \sigma(pg, x) \in \sigma(R_g(L) \times \{x\}) = \sigma(L \times \{x\}),$$
$$px = \sigma(p, x) \in \sigma(L \times \{x\}).$$

Therefore $(pg)x$ and px lie on the same leaf of $\hat{\mathcal{F}}_0$ hence (as Φ is a foliated map) the leaves $\pi\Phi(px)$ and $\pi\Phi((pg)x)$ must coincide. Finally let $\mathcal{A}$ be given by

$$\mathcal{A} = \{\{B(0,\ 1/2),\ G,\ \varphi\} :$$
$$U = \pi\ \Phi\ (L \times_G B(0,\ 1/2)) \in \mathcal{H},\ \ \varphi = \pi\ \Phi\ \sigma_p,\ \ p \in L\}.$$

Then $\mathcal{A}$ is a defining family for B (cf. [**120**], p. 82-85) thus making B into a q-dimensional orbifold.

See also R. Wolak, [**253**], proving that 1) given a transversely complete G-foliation $\mathcal{F}$ of finite type on a compact manifold M, the leaves of $\mathcal{F}$ are compact if and only if $M/\mathcal{F}$ is an orbifold. Also 2) if $\mathcal{F}$ is a ∇-G-foliation (cf. [**252**]) with all leaves compact then the holonomy of any leaf of $\mathcal{F}$ can be linearized if and only if $M/\mathcal{F}$ is an orbifold (cf. *op. cit.*).

CHAPTER 10

Pseudo-differential operators on orbifolds

In this chapter we make use of the theory of pseudo-differential operators, cf. e.g. J.J. Kohn & L. Nirenberg, [**158**]. The reader may also find an excellent practical account of the theory of pseudo-differential operators in $\mathbb{R}^n$ in the monograph by P.B. Gilkey, [**119**]. One purpose of our book being to explain how certain ideas and techniques in classical analysis apply to differential geometry, we briefly review the relevant issues e.g. infinitely smoothing symbols, symbols of compositions of pseudo-differential operators, inverting elliptic differential operators modulo smoothing operators, invariance of pseudo-differential operators under coordinate transformations (see our Appendix C). These notions are used in studying pseudo-differential operators on orbifolds. A more general treatment (rather than on $\mathbb{R}^n$ alone) is that of M.J. Pflaum, [**200**], building a theory of elliptic pseudo-differential operators on Riemannian manifolds. See also T. Tate, [**237**]-[**238**], where a pseudo-differential calculus on the Poincaré disk is developed. Both sources are suitable for applications to orbifolds in the manner indicated by J. Girbau & M. Nicolau, [**120**].

10.1. The Girbau-Nicolau condition

Let $(B, \mathcal{A})$ be a n-dimensional C^∞ orbifold. Let $\mathcal{E}(B)$ be the space of $\mathbb{C}$-valued C^∞ functions on B and $\mathcal{D}(B)$ the space of all $f \in \mathcal{E}(B)$ with compact support.

DEFINITION 10.1. Let $P : \mathcal{D}(B) \to \mathcal{E}(B)$ be a $\mathbb{C}$-linear map. Consider the following data i) a l.u.s. $\{\Omega, G, \varphi\} \in \mathcal{A}$ of support U, ii) a compact set $\tilde{K} \subset \Omega$, and iii) a C^∞ function $F : \Omega \to \mathbb{R}$ of compact support. We say that P satisfies the *Girbau-Nicolau condition* with respect to the data (i)-(iii) if there is a symbol $p(x, \xi)$ of order k on Ω such that

$$F(x)(Pf)_\Omega(x) = \int e^{i(x-y)\cdot\xi} p(x, \xi) f_\Omega(y) dy d\xi$$

for any $f \in \mathcal{D}(B)$ of support contained in $K = \varphi(\tilde{K})$ and any $x \in \Omega$. □

The central concept of this section is captured in the following

DEFINITION 10.2. A $\mathbb{C}$-linear map $P : \mathcal{D}(B) \to \mathcal{E}(B)$ is a *pseudo-differential operator of order* k on B if for any point $x_0 \in B$ there is an open neighborhood $U \in \mathcal{H}$ of x_0 such that P satisfies the Girbau-Nicolau condition with respect to any choice of l.u.s. $\{\Omega, G, \varphi\} \in \mathcal{A}$ of support U, any compact subset $\tilde{K} \subset \Omega$, and any function $F \in C_0^\infty(\Omega, \mathbb{R})$. □

Let P be a pseudo-differential operator on B and let us consider the family $\mathcal{H}_P$ consisting of all $U \in \mathcal{H}$ such that P satisfies the Girbau-Nicolau condition for all choices of l.u.s. $\{\Omega, G, \varphi\} \in \mathcal{A}$ of support U, of a compact subset $\tilde{K} \subset \Omega$, and of a function $F \in C_0^\infty(\Omega, \mathbb{R})$ as in Definition 10.2. Given $U \in \mathcal{H}_P$ we also say loosely that P satisfies the Girbau-Nicolau condition relative to U.

THEOREM 10.3. *Let P be a pseudo-differential operator on B and $U \in \mathcal{H}_P$. If $U' \in \mathcal{H}$ and $U' \subseteq U$ then $U' \in \mathcal{H}_P$.*

Proof. Let $\{\Omega, G, \varphi\}$ and $\{\Omega', G', \varphi'\}$ be l.u.s.'s of supports U and U', respectively. To check whether $U' \in \mathcal{H}_P$ we need to consider a compact subset $\tilde{K}' \subset \Omega'$ and a C^∞ function $F' : \Omega' \to \mathbb{R}$ of compact support. As $U' \subseteq U$ there is an injection λ of $\{\Omega', G', \varphi'\}$ into $\{\Omega, G, \varphi\}$ and, correspondingly, a group monomorphism $\eta : G' \to G$. Consider the compact set $\tilde{K} = \lambda(\tilde{K}') \subset \Omega$. Next let $F : \Omega \to \mathbb{R}$ be given by

$$F = \begin{cases} F' \circ \lambda^{-1} & \text{on } \lambda(\Omega'), \\ 0 & \text{elsewhere.} \end{cases}$$

Then $F \in C_0^\infty(\Omega, \mathbb{R})$. As P is assumed to satisfy the Girbau-Nicolau condition relative to U there is a symbol $p(x, \xi)$ on Ω such that

$$F(x)(Pf)_\Omega(x) = \int e^{i(x-y)\cdot\xi} p(x,\xi) f_\Omega(y) dy d\xi \tag{10.1}$$

for any $f \in \mathcal{D}(B)$ of support in $K = \varphi(\tilde{K}) = \varphi'(\tilde{K}')$. Note that given a C^∞ function $\Psi : \Omega \to \mathbb{R}$ of compact support contained in $\lambda(\Omega')$ and such that $\Psi = 1$ on the support of F, one has

$$F(x)(Pf)_\Omega(x) = \int e^{i(x-y)\cdot\xi} \Psi(x) p(x,\xi) f_\Omega(y) dy d\xi$$

hence one may assume without loss of generality that the x-support of $p(x, \xi)$ is contained in $\lambda(\Omega')$. Let $f \in \mathcal{D}(B)$ of support in K. Consider the function $\tilde{f} : \Omega \to \mathbb{C}$ given by

$$\tilde{f} = \begin{cases} f_{\Omega'} \circ \lambda^{-1} & \text{on } \lambda(\Omega'), \\ 0 & \text{elsewhere.} \end{cases}$$

Then $\tilde{f} \in C^\infty(\Omega)$. Let $\sigma_1, \cdots, \sigma_m$ be representatives of the classes in $G/\eta(G')$ i.e.

$$G/\eta(G') = \{[\sigma_1], \cdots, [\sigma_m]\}$$

such that $\sigma_1 = e$ (and $i \neq j \Longrightarrow [\sigma_i] \neq [\sigma_j]$). Then

$$f_\Omega = \sum_{i=1}^m \tilde{f} \circ \sigma_i^{-1}. \tag{10.2}$$

To check (10.2) we make use of $f_\Omega = f \circ \varphi$ (by definition) and $f_\Omega \circ \lambda = f_{\Omega'}$. Let $\tilde{x} \in \Omega$. We distinguish two cases, as I) $\tilde{x} \in \lambda(\Omega')$ or II) $\tilde{x} \in \Omega - \lambda(\Omega')$. Let us look at case I. We have

$$i \neq 1 \Longrightarrow \sigma_i^{-1}(\tilde{x}) \notin \lambda(\Omega') \Longrightarrow \tilde{f}(\sigma_i^{-1}(\tilde{x})) = 0$$

and

$$\tilde{f}(\tilde{x}) = \left(f_{\Omega'} \circ \lambda^{-1}\right)(\tilde{x}) = f_\Omega(\tilde{x}).$$

As to case II, as $\tilde{K} = \lambda(\tilde{K}') \subset \lambda(\Omega')$ it follows that $\tilde{x} \notin \tilde{K}$. Then $\varphi(\tilde{x}) \notin K$ hence $f(\varphi(\tilde{x})) = 0$, i.e. $f_\Omega(\tilde{x}) = 0$. At this point we distinguish two subcases as II.1) $\sigma_i^{-1}(\tilde{x}) \notin \lambda(\Omega')$ for any $1 \leq i \leq m$ (and then the sum (over i) itself vanishes) or II.2)

$\sigma_i^{-1}(\tilde{x}) \in \lambda(\Omega')$ for some $i \in \{2, \cdots, m\}$. If this is the case then $\sigma_i^{-1}(\tilde{x}) = \lambda(\tilde{x}')$ for some $\tilde{x}' \in \Omega'$. We have

$$\tilde{f}(\sigma_i^{-1}(\tilde{x})) = \tilde{f}(\lambda(\tilde{x}')) = f_{\Omega'}(\tilde{x}') =$$
$$= f_\Omega(\lambda(\tilde{x}')) = f_\Omega(\sigma_i^{-1}(\tilde{x})) = f\varphi\sigma_i^{-1}(\tilde{x}) = f(\varphi(\tilde{x})) = 0$$

and (10.2) is completely proved. Let us substitute from (10.2) into (10.1) such that to obtain

$$F(x)(Pf)_\Omega(x) = \sum_{j=1}^{m} I_j(x) \tag{10.3}$$

where

$$I_j(x) = \int e^{i(x-y)\cdot\xi} p(x,\xi)\tilde{f}(\sigma_j^{-1}(y))dyd\xi.$$

Let $j \in \{2, \cdots, m\}$ be fixed. Consider a function $\varphi_j \in C_0^\infty(\Omega)$ such that $\text{supp}(\varphi_j) \subset \sigma_j(\lambda(\Omega'))$ and $\varphi_j = 1$ on $\sigma_j(\tilde{K})$. Let $y \notin \sigma_j(\tilde{K})$. Then $\sigma_j^{-1}(y) \notin \tilde{K}$ and we distinguish two cases, as I) $\sigma_j^{-1}(y) \notin \lambda(\Omega')$ (and then $\tilde{f}(\sigma_j^{-1}(y)) = 0$) or II) $\sigma_j^{-1}(y) \in \lambda(\Omega')$. If this is the case then $\sigma_j^{-1}(y) = \lambda(\tilde{x}_j')$ for some $\tilde{x}_j' \in \Omega'$, hence

$$\tilde{f}(\sigma_j^{-1}(y)) = f_{\Omega'}(\tilde{x}_j') = f_\Omega(\lambda(\tilde{x}_j')) = f_\Omega(\sigma_j^{-1}(y)) = 0$$

We have proved that

$$y \notin \sigma_j(\tilde{K}) \Longrightarrow \tilde{f}(\sigma_j^{-1}(y)) = 0$$

i.e. $\text{supp}(\tilde{f} \circ \sigma_j^{-1}) \subset \sigma_j(\tilde{K})$.

LEMMA 10.4. *The operator Q_j given by*

$$(Q_j g)(x) = \int e^{i(x-y)\cdot\xi} \varphi_j(y)p(x,\xi)g(y)dyd\xi$$

for any $g \in C_0^\infty(\Omega)$ with $\text{supp}(g) \subset \sigma_j(\tilde{K})$, *is a pseudo-differential operator of order $-\infty$.*

Proof. Let $r(x,\xi,y)$ be C^∞ in (x,ξ,y), with x-support contained in an open set $U \subset \mathbb{R}^n$ of compact closure. Assume that

$$\left|D_x^\alpha D_\xi^\beta D_y^\gamma r\right| \leq C_{\alpha,\beta,\gamma}(1+|\xi|)^{m-|\beta|} \tag{10.4}$$

for any multi-indices α, β, γ. For any $f \in C_0^\infty(U)$ let us set

$$(Rf)(x) = \int e^{i(x-y)\cdot\xi} r(x,\xi,y)f(y)dyd\xi.$$

Then (cf. Lemma 1.2.2 in [**119**], p. 15, or our Appendix C) $R \in \Psi_m(U)$ and the symbol $\sigma(R)$ of R satisfies

$$\sigma(R)(x,\xi) \sim \left(\sum_\alpha \frac{i^{|\alpha|}}{\alpha!} D_\xi^\alpha D_y^\alpha r(x,\xi,y)\right)_{y=x}.$$

This yields Lemma 10.4. Indeed let us set $r(x,\xi,y) = \varphi_j(y)p(x,\xi)$. As $p \in S^m$ and $D^\gamma\varphi_j \in C_0^\infty(\Omega)$ the symbol $r(x,\xi,y)$ satisfies an estimate of the form (10.4) such that Q_j is a pseudo-differential operator (*a priori* of order m). Moreover

$$\sigma(Q_j)(x,\xi) \sim \left\{\sum_\alpha \frac{i^{|\alpha|}}{\alpha!} \left(D_\xi^\alpha p(x,\xi)\right)\left(D_y^\alpha \varphi_j(y)\right)\right\}_{y=x} = 0$$

because p has x-support in $\lambda(\Omega')$, φ_j has support in $\sigma_j(\lambda(\Omega'))$ and, for $j \geq 2$, $\sigma_j(\lambda(\Omega')) \cap \lambda(\Omega') = \emptyset$. Thus Q_j is infinitely smoothing. Q.e.d.

Let $k'_j \in C_0^\infty(\Omega \times \Omega)$ such that

$$(Q_j g)(x) = \int k'_j(x, y) g(y) dy$$

for any $g \in C_0^\infty(\Omega)$ of support in $\sigma_j(\tilde{K})$ (as $Q_j \in \Psi_{-\infty}$ such $k'_j(x, y)$ exists, cf. Lemma 1.2.5 in [**119**], p. 19, or our Appendix C). Clearly one may choose $k'_j(x, y)$ with x-support in $\lambda(\Omega)$ and y-support in $\sigma_j(\lambda(\Omega'))$. Then, with a change of variable $z = \sigma_j^{-1}(y)$, we have

$$I_j(x) = \left(Q_j(\tilde{f} \circ \sigma_j^{-1})\right)(x) = \int k'_j(x, y)(\tilde{f} \circ \sigma_j^{-1})(y) dy =$$

$$= \int k'_j(x, \sigma_j(z)) \tilde{f}(z) J_{\sigma_j}(z) dz.$$

Let us set $k_j(x, y) = J_{\sigma_j}(y) k'_j(x, \sigma_j(y))$. We have obtained

$$I_j(x) = \int k_j(x, y) \tilde{f}(y) dy$$

where $k_j(x, y)$ has both x-support and y-support in $\lambda(\Omega')$. Let us set $k = \sum_{j=1}^m k_j$. Then (10.3) becomes

$$F(x)(Pf)_\Omega(x) = \int e^{i(x-y)\cdot\xi} p(x, \xi) \tilde{f}(y) dy d\xi + \int k(x, y) \tilde{f}(y) dy. \tag{10.5}$$

Again by the preparation in our Appendix C (or by the Lemma 1.2.4 in [**119**], p. 18) the right hand member of (10.5) defines a pseudo-differential operator Q acting on $C_0^\infty(\lambda(\Omega'))$ and (10.5) reads

$$F(x)(Pf)_\Omega(x) = (Q\tilde{f})(x) \tag{10.6}$$

Let us consider the diffeomorphism $\lambda : \Omega' \to \lambda(\Omega')$ and the compact set $\tilde{K}' = \lambda^{-1}(\tilde{K}) \subset \Omega'$. This is precisely the context of the theorem of invariance of pseudo-differential operators under a coordinate transformation. That is, if $\Omega \in C^\infty(\Omega')$ has support in $\tilde{K}'$, let $u = \Omega \circ \lambda^{-1}$. Then $\tilde{Q}$ given by

$$(\tilde{Q}\Omega)(\tilde{x}') = (Qu)(\lambda(\tilde{x}') \tag{10.7}$$

is a pseudo-differential operator. Let us use (10.7) for $\Omega = \tilde{f} \circ \lambda$. Then (as $F(\lambda(\tilde{x}')) = F'(\tilde{x}')$, $\tilde{f} \circ \lambda = f_{\Omega'}$ and $(Pf)_\Omega \circ \lambda = (Pf)_{\Omega'}$) the identity (10.6) becomes

$$F'(\tilde{x}')(Pf)_{\Omega'}(\tilde{x}') = \tilde{Q}(f_{\Omega'})(\tilde{x}')$$

i.e. $U' \in \mathcal{H}_P$. Q.e.d.

COROLLARY 10.5. *Let P_1, P_2 be pseudo-differential operators on the orbifold B. Then $P_1 + P_2$ is a pseudo-differential operator on B.*

Proof. For any $x_0 \in B$ there are $U_1, U_2 \subset B$ so that $x_0 \in U_1 \cap U_2$ and $U_i \in \mathcal{H}_{P_i}, i = 1, 2$. As $U_i \in \mathcal{H}$ there is $U \subseteq U_1 \cap U_2$ such that $U \in \mathcal{H}$. Then (by Theorem 10.3) $U \in \mathcal{H}_{P_1} \cap \mathcal{H}_{P_2}$ hence (by linearity) $U \in \mathcal{H}_{P_1+P_2}$.

10.2. Composition of pseudo-differential operators

Let B be a C^∞ orbifold and P, Q two pseudo-differential operators on B. Given $f \in \mathcal{D}(B)$ we define an operator $PfQ : \mathcal{D}(B) \to \mathcal{E}(B)$ by setting

$$(PfQ)u = P(fQu)$$

for any $u \in \mathcal{D}(B)$.

THEOREM 10.6. *PfQ is a pseudo-differential operator on B.*

Proof. Let $\{U_\alpha\}$ be a locally finite open cover of B so that $U_\alpha \in \mathcal{H}_P \cap \mathcal{H}_Q$. Given $f \in \mathcal{D}(B)$ the set of indices α such that $\text{supp}(f) \cap U_\alpha \neq \emptyset$ is finite. Let $U_1, \cdots, U_k$ be the open sets of the cover which are intersected by $\text{supp}(f)$. Given a C^∞ partition of unity $\{g_\alpha\}$ subordinate to $\{U_\alpha\}$, let $g_1, \cdots, g_k$ be the functions of the partition corresponding to $U_1, \cdots, U_k$. We have

$$(PfQ)u = P(fQu) = P((\sum_\alpha g_\alpha f)Qu) =$$

$$= \sum_{i=1}^{k} P((g_i f)Qu) = \sum_{i=1}^{k} (Pg_i fQ)u.$$

It suffices to prove that $Pg_i fQ$ is a pseudo-differential operator on B. To this end we show that the operator $Pg_i fQ$ satisfies the Girbau-Nicolau condition relative to each open set U_α of the given cover. We distinguish two cases as I) $U_\alpha \cap U_i = \emptyset$ (and then U_α satisfies the Girbau-Nicolau condition, because $Pg_i fQ$ vanishes for this U_α) or II) $U_\alpha \cap U_i \neq \emptyset$. If this is the case we distinguish two subcases as II.1) $\alpha \notin \{1, \cdots, k\}$ (and then U_α satisfies the Girbau-Nicolau conditions, as again $Pg_i fQ$ vanishes for this U_α), or II.2) $\alpha = j$ for some $j \in \{1, \cdots, k\}$. If this is the case let us consider the following data 1) a l.u.s. $\{\Omega_j, G_j, \varphi_j\} \in \mathcal{A}$ of support U_j, 2) a compact subset $\tilde{K}_j \subset \Omega_j$, and 3) a C^∞ function $F_j : \Omega_j \to \mathbb{R}$ of compact support. We must show that $F_j(Pg_i fQu)_{\Omega_j}$ with $u \in \mathcal{D}(B)$ of support in $\tilde{K}_j$ may be expressed by means of a symbol. As P is a pseudo-differential operator and $U_j \in \mathcal{H}_P$ there is a symbol $p(x, \xi)$ on Ω_j such that

$$F_j(x)(P(g_i fQu))_{\Omega_j}(x) = \int e^{i(x-y)\cdot\xi} p(x, \xi)(g_i fQu)_{\Omega_j}(y) dy d\xi. \tag{10.8}$$

On the other hand

$$(g_i fQu)_{\Omega_j} = (g_i fQu) \circ \varphi_j = [(g_i f) \circ \varphi_j] [(Qu) \circ \varphi_j] = (g_i f)_{\Omega_j} (Qu)_{\Omega_j} .$$

As $U_j \in \mathcal{H}_Q$ there is a symbol $q(x, \xi)$ on Ω_j such that

$$(g_i f)_{\Omega_j}(x)(Qu)_{\Omega_j}(x) = \int e^{i(x-y)\cdot\xi} q(x, \xi) u_{\Omega_j}(y) dy d\xi. \tag{10.9}$$

At this point we may substitute from (10.9) into (10.8) and use Lemma 1.2.2 in [**119**], p. 15, to end the proof of Theorem 10.6.

10.3. Elliptic operators on orbifolds

Let B be an orbifold.

DEFINITION 10.7. A linear map $D : \mathcal{E}(B) \to \mathcal{E}(B)$ is a *differential operator of order* k if for any $\{\Omega, G, \varphi\} \in \mathcal{A}$ there is a differential operator $D_\Omega = \sum_{|\alpha|\leq k} a_\alpha(x) D^\alpha$ of order k on Ω with $a_\alpha \in C^\infty(\Omega)$ such that $(Du)_\Omega = D_\Omega u_\Omega$ for any $u \in \mathcal{E}(B)$. □

DEFINITION 10.8. A differential operator D on B is *elliptic* if each D_Ω is elliptic on Ω. □

THEOREM 10.9. *Let $D : \mathcal{E}(B) \to \mathcal{E}(B)$ be a differential operator and $P : \mathcal{D}(B) \to \mathcal{E}(B)$ a pseudo-differential operator. Then both $D \circ P$ and $P \circ D$ are (well defined) pseudo-differential operators on B.*

Proof. If B is compact then Theorem 10.9 is a corollary of Theorem 10.6 (with $f = 1$ on B). Let us look at the noncompact case. Let $\{U_\alpha\}$ be a locally finite open cover of B with $U_\alpha \in \mathcal{H}_P$. We shall show that the operator $D \circ P$ satisfies the Girbau-Nicolau condition relative to each U_α. Given 1) a l.u.s. $\{\Omega_\alpha, G_\alpha, \varphi_\alpha\}$ of support U_α, 2) a compact set $\tilde{K}_\alpha \subset \Omega_\alpha$, and 3) a C^∞ function $F_\alpha : \Omega_\alpha \to \mathbb{R}$ of compact support, let $\phi_\alpha \in C_0^\infty(\Omega)$ such that $\phi_\alpha = 1$ on the support of F_α. For any $u \in \mathcal{D}(B)$ of support in $K_\alpha = \varphi_\alpha(\tilde{K}_\alpha)$ we have

$$F_\alpha (DPu)_{\Omega_\alpha} = F_\alpha D_{\Omega_\alpha} (Pu)_{\Omega_\alpha} = F_\alpha D_{\Omega_\alpha} (\phi_\alpha (Pu)_{\Omega_\alpha}).$$

As $U_\alpha \in \mathcal{H}_P$, the operator $\phi_\alpha (Pu)_{\Omega_\alpha}$ comes from a symbol on Ω_α hence so does $F_\alpha D_{\Omega_\alpha}(\phi_\alpha (Pu)_{\Omega_\alpha})$ (cf. Lemma 1.2.3 in [**119**], p. 15, or our Appendix C). Q.e.d.

Let us show now that the operator $P \circ D$ satisfies the Girbau-Nicolau condition relative to U_α. Consider the data $\{\Omega_\alpha, G_\alpha, \varphi_\alpha\}$, $\tilde{K}_\alpha \subset \Omega_\alpha$ and F_α, as above. For any $u \in \mathcal{D}(B)$ with support in $K_\alpha = \varphi_\alpha(\tilde{K}_\alpha)$ one has

$$F_\alpha(x)(PDu)_{\Omega_\alpha}(x) = \int e^{i(x-y)\cdot\xi} p(x,\xi)(Du)_{\Omega_\alpha}(y)\,dy\,d\xi$$

(because of $U_\alpha \in \mathcal{H}_P$) hence (cf. Lemma 1.2.3 in [**119**], p. 15, or our Appendix C) $F_\alpha (PDu)_{\Omega_\alpha}$ comes from a symbol on Ω_α. Q.e.d.

THEOREM 10.10. *Let $D : \mathcal{E}(B) \to \mathcal{E}(B)$ be an elliptic differential operator and $U \in \mathcal{H}$. Let $\{\Omega, G, \varphi\} \in \mathcal{A}$ be a l.u.s. of support U and $\tilde{K} \subset \Omega$ a G-invariant compact subset. Then there is a pseudo-differential operator $Q_\Omega : C_0^\infty(\Omega) \to C_0^\infty(\Omega)$ so that* 1) *$Q_\Omega \circ D_\Omega \sim I$ and $D_\Omega \circ Q_\Omega \sim I$ on the G-invariant C^∞ functions of support in $\tilde{K}$, and* 2) *Q_Ω maps G-invariant functions in G-invariant functions.*

Proof. As D is elliptic the differential operator D_Ω is elliptic in the ordinary sense hence there is a pseudo-differential operator $Q : C_0^\infty(\Omega) \to C_0^\infty(\Omega)$ such that $Q \circ D_\Omega \sim I$ and $D_\Omega \circ Q \sim I$ on C^∞ functions of support in $\tilde{K}$. Next we modify Q to get an operator satisfying the requirement 2) in Theorem 10.10. Precisely as $D_\Omega \circ Q \sim I$ and $Q \circ D_\Omega \sim I$, there are $k_1, k_2 \in C^\infty(\mathbb{R}^n \times \mathbb{R}^n)$ with compact x-supports and y-supports contained in Ω such that

$$(D_\Omega Q h)(x) = h(x) + \int k_1(x,y) h(y)\,dy,$$

$$(Q D_\Omega h)(x) = h(x) + \int k_2(x,y) h(y)\,dy,$$

for any $h \in C_0^\infty(\Omega)$ of support in $\tilde{K}$. Let $\sigma \in G$ and consider the operator $Q_\sigma : C_0^\infty(\Omega) \to C_0^\infty(\Omega)$ given by

$$(Q_\sigma u)(x) = Q(u \circ \sigma^{-1})(\sigma(x)).$$

As Q_σ comes from Q by a coordinate transformation, it is a pseudo-differential operator on Ω. Next we define Q_Ω by

$$Q_\Omega = \frac{1}{n(G)} \sum_{\sigma \in G} Q_\sigma \,.$$

Clearly Q_Ω maps G-invariant functions in G-invariant functions, for given a G-invariant function $u \in C_0^\infty(\Omega)$ and $\sigma, \tau \in G$ one has

$$[(Q_\sigma u) \circ \tau](x) = (Q_\sigma u)(\tau(x)) = Q(u \circ \sigma^{-1})(\sigma \tau x) =$$
$$= Q(u \circ (\sigma\tau)^{-1})(\sigma\tau x) = (Q_{\sigma\tau} u)(x)$$

i.e. $(Q_\sigma u) \circ \tau = Q_{\sigma\tau} u$ and then

$$(Q_\Omega u) \circ \tau = \frac{1}{n(G)} \sum_{\sigma \in G} Q_{\sigma\tau} u = \frac{1}{n(G)} \sum_{\sigma \in G} Q_\sigma u = Q_\Omega u \,.$$

Yet one has to check that Q_Ω inverts D_Ω. First let us note that for any G-invariant $h \in C_0^\infty(\Omega)$ it follows that $D_\Omega h \in C_0^\infty(\Omega)$ and $D_\Omega h$ is G-invariant. Indeed as there is $u \in \mathcal{D}(B)$ with $h = u_\Omega$ one has

$$(D_\Omega h) \circ \sigma = (D_\Omega u_\Omega) \circ \sigma = (Du)_\Omega \circ \sigma =$$
$$= (Du)_\Omega = D_\Omega u_\Omega = D_\Omega h$$

Let $h \in C_0^\infty(\Omega)$ be a G-invariant function of support in $\tilde{K}$. Then

$$(Q_\sigma D_\Omega h)(x) = Q((D_\Omega h) \circ \sigma^{-1})(\sigma(x)) = Q(D_\Omega h)(\sigma x) =$$
$$= h(\sigma(x)) + \int k_2(\sigma(x), y) h(y) dy = h(x) + \int k_2(\sigma(x), y) h(y) dy \,.$$

Then

$$(Q_\Omega D_\Omega h)(x) = \frac{1}{n(G)} \sum_{\sigma \in G} (Q_\sigma D_\Omega h)(x) =$$
$$= \frac{1}{n(G)} \sum_{\sigma \in G} \left\{ h(x) + \int k_2(\sigma(x), y) h(y) dy \right\} = h(x) + \int K_2(x, y) h(y) dy$$

where

$$K_2(x, y) = \frac{1}{n(G)} \sum_{\sigma \in G} k_2(\sigma(x), y)$$

hence $Q_\Omega \circ D_\Omega \sim I$ over the G-invariant C^∞ functions with support in $\tilde{K}$. Moreover, for any G-invariant $h \in C_0^\infty(\Omega)$ of support in $\tilde{K}$

$$D_\Omega Q_\Omega h = \frac{1}{n(G)} \sum_{\sigma \in G} D_\Omega Q_\sigma h$$

and (because of $Q_\sigma h = (Qh) \circ \sigma$ and $D_\Omega(g \circ \sigma) = (D_\Omega g) \circ \sigma$ for any $g \in C_0^\infty(\Omega)$, cf. [**120**], p. 73)

$$(D_\Omega Q_\Omega h)(x) = \frac{1}{n(G)} \sum_{\sigma \in G} D_\Omega((Qh) \circ \sigma)(x) =$$
$$= \frac{1}{n(G)} \sum_{\sigma \in G} [(D_\Omega Q h) \circ \sigma](x) = h(x) + \int K_1(x, y) h(y) dy$$

where

$$K_1(x, y) = \frac{1}{n(G)} \sum_{\sigma \in G} k_1(\sigma(x), y)$$

hence $D_\Omega \circ Q_\Omega \sim I$ on these h. Q.e.d.

DEFINITION 10.11. Two pseudo-differential operators P and P' on an orbifold B are *equivalent* (and one writes $P \sim P'$) if $P - P'$ is a pseudo-differential operator of order $-\infty$, i.e. of order k for any $k \in \mathbb{R}$, on B. □

THEOREM 10.12. *Let B be a C^∞ orbifold such that $\mathcal{H}$ is a basis of open sets for the topology of B. Then for any elliptic differential operator $D : \mathcal{E}(B) \to \mathcal{E}(B)$ there is a pseudo-differential operator $Q : \mathcal{D}(B) \to \mathcal{E}(B)$ such that $Q \circ D \sim I$ and $D \circ Q \sim I$.*

By Theorem 10.9 both $Q \circ D$ and $D \circ Q$ are well defined pseudo-differential operators on B.

DEFINITION 10.13. The pseudo-differential operator Q on B furnished by Theorem 10.12 is the *parametrix* of the elliptic differential operator D on B. □

To prove Theorem 10.12 let $\{U_i\}$ be a locally finite open cover of B with $U_i \in \mathcal{H}$. Next, let $\{\psi_i\}$ be a C^∞ partition of unity subordinate to $\{U_i\}$ and choose functions $g_i \in \mathcal{D}(B)$ such that each g_i has support in U_i and $g_i = 1$ on the support of ψ_i. Let $\{\Omega_i, G_i, \varphi_i\}$ be a l.u.s. of B of support U_i. By Theorem 10.10 we may invert the elliptic operator D_{Ω_i} over the G_i-invariant C^∞ functions of support in $\tilde{K}_i = \varphi_i^{-1}(\operatorname{supp}(g_i))$ i.e. there is a pseudo-differential operator Q_{Ω_i} on Ω_i such that 1) $Q_{\Omega_i} \circ D_{\Omega_i} \sim I$ and $D_{\Omega_i} \circ Q_{\Omega_i} \sim I$ over the G_i-invariant C^∞ functions of support in $\tilde{K}_i$, and 2) Q_{Ω_i} maps G_i-invariant functions in G_i-invariant functions. Let us think of U_i as an orbifold with the orbifold structure induced from B. Consider the operator $Q_i : \mathcal{D}(U_i) \to \mathcal{D}(U_i)$ defined as follows. Let $U \subseteq U_i$ be an open subset so that $U \in \mathcal{H}$. Let $\{\Omega, G, \varphi\} \in \mathcal{A}$ of support U (such that $\{\Omega, G, \varphi\}$ is also a l.u.s. for U_i) and λ an injection of $\{\Omega, G, \varphi\}$ into $\{\Omega_i, G_i, \varphi_i\}$. Then for any $f \in \mathcal{D}(U_i)$ we set

$$(Q_i f)_\Omega = (Q_{\Omega_i} f_{\Omega_i}) \circ \lambda.$$

LEMMA 10.14. *The operator $(Q_i f)_\Omega$ depends neither on the choice of l.u.s. $\{\Omega_i, G_i, \varphi_i\}$ of support U_i nor on the choice of injection λ of $\{\Omega, G, \varphi\}$ into $\{\Omega_i, G_i, \varphi_i\}$.*

Proof. Let $\{\Omega_i', G_i', \varphi_i'\} \in \mathcal{A}$ be a l.u.s. of support U_i. Let μ be an injection of $\{\Omega, G, \varphi\}$ into $\{\Omega_i', G_i', \varphi_i'\}$. Let λ_i be an injection of $\{\Omega_i, G_i, \varphi_i\}$ into $\{\Omega_i', G_i', \varphi_i'\}$. As $\varphi_i(U_i) = \varphi_i'(U_i') = U_i$ then (by Theorem 9.12) $\lambda_i : \Omega_i \approx \Omega_i'$ (a C^∞ diffeomorphism). Also if $\eta_i : G_i \to G_i'$ is the group monomorphism corresponding to λ_i then (again by Theorem 9.12) $\eta_i : G_i \approx G_i'$ (a group isomorphism). Let $Q_{\Omega_i'}$ be the operator defined by

$$(Q_{\Omega_i'} u')(x') = (Q_{\Omega_i}(u' \circ \lambda_i))(\lambda_i^{-1} x')$$

for any $u' \in C_0^\infty(\Omega_i')$ of support in $\tilde{K}_i' = \lambda_i(\tilde{K}_i)$ and any $x' \in \Omega_i'$. Then $Q_{\Omega_i'}$ inverts $D_{\Omega_i'}$ over the G_i'-invariant C^∞ functions of support in $\tilde{K}_i'$. That is because given an elliptic differential operator $D : \mathcal{E}(B) \to \mathcal{E}(B)$ and Q_Ω inverting D_Ω, a coordinate transformation leaves invariant this situation. Precisely let $\{\Omega, G, \varphi\}$ and $\{\Omega', G', \varphi'\}$ be two l.u.s.'s of (the same) support U and $\tilde{K} \subset \Omega$ a G-invariant compact set. Let λ be an injection of $\{\Omega, G, \varphi\}$ into $\{\Omega', G', \varphi'\}$ (inducing a diffeomorphism $\Omega \approx \Omega'$ (by Theorem 9.12) as recalled above). Let $Q_\Omega : C_0^\infty(\Omega) \to C_0^\infty(\Omega)$ be a pseudo-differential operator inverting D_Ω over the

G-invariant C^∞ functions of support in $\tilde{K}$, which maps G-invariant functions in G-invariant functions. Set $\tilde{K}' = \lambda(\tilde{K}) \subset \Omega'$. Then $\tilde{K}'$ is a G'-invariant (as $\sigma'(\tilde{K}') = \sigma'\lambda\tilde{K} = \eta(\sigma)\lambda\tilde{K} = \lambda\sigma\tilde{K} = \lambda(\tilde{K}) = \tilde{K}')$ compact set. Let $Q_{\Omega'}$ be obtained from Q_Ω by a change of variables $x' = \lambda(x)$ i.e.

$$(Q_{\Omega'}u')(x') = Q_\Omega(u' \circ \lambda)(\lambda^{-1}x')$$

for any $u' \in C_0^\infty(\Omega')$ of support in $\tilde{K}'$. Then $Q_{\Omega'}$ is a pseudo-differential operator. Let us show that $Q_{\Omega'}$ inverts $D_{\Omega'}$ over the G'-invariant functions of support in $\tilde{K}'$. We have

$$(Q_{\Omega'}D_{\Omega'}h')(x') = Q_\Omega((D_{\Omega'}h') \circ \lambda)(\lambda^{-1}x')$$

Also there is $f \in \mathcal{D}(B)$ of support in U such that $f_{\Omega'} = h'$ (cf. Theorem B.10) hence

$$(D_{\Omega'}h') \circ \lambda = (D_{\Omega'}f_{\Omega'}) \circ \lambda = (Df)_{\Omega'} \circ \lambda = (Df)_\Omega = D_\Omega f_\Omega$$

Let us set $h = h' \circ \lambda$, i.e. $h = f_\Omega$. We have shown that $(D_{\Omega'}h') \circ \lambda = D_\Omega h$. Finally

$$(Q_{\Omega'}D_{\Omega'}h')(x') = Q_\Omega(D_\Omega(h' \circ \lambda))(\lambda^{-1}x') =$$
$$= (h' \circ \lambda)(\lambda^{-1}x') + P_{-\infty}(h' \circ \lambda)(\lambda^{-1}x') =$$
$$= h'(x') + \int k(\lambda^{-1}x', y)(h' \circ \lambda)(y)dy$$

for some pseudo-differential operator $P_{-\infty}$ of order $-\infty$ (hence represented as $P_{-\infty} = P(k)$, cf. our Appendix C). By a change of variables $z = \lambda(y)$ we obtain

$$(Q_{\Omega'}D_{\Omega'}h')(x') = h'(x') + \int k'(x', y')h'(y')dy'$$

where

$$k'(x', y') = k(\lambda^{-1}x', \lambda^{-1}y')J_{\lambda^{-1}}(y').$$

The proof of $D_{\Omega'} \circ Q_{\Omega'} \sim I$ is left as an exercise to the reader. Let us go back now to the proof of Lemma 10.14. We have

$$\left[(Q_{\Omega_i'}f_{\Omega_i'}) \circ \mu\right](x) = (Q_{\Omega_i}f_{\Omega_i'} \circ \lambda)(\lambda_i^{-1}\mu x) = (Q_{\Omega_i}f_{\Omega_i})(\lambda_i^{-1}\mu x) =$$
$$= (Q_{\Omega_i}f_{\Omega_i})(\sigma_i\lambda x) = (Q_{\Omega_i}f_{\Omega_i})(\lambda x)$$

where σ_i is the unique element of G_i with the property $\lambda_i^{-1} \circ \mu = \sigma_i \circ \lambda$ and by using $(Q_\Omega u) \circ \tau = Q_\Omega u$ (from the construction of Q_Ω). Note that $\mathcal{D}(U_i)$ injects canonically into $\mathcal{D}(B)$. Indeed if $f \in \mathcal{D}(U_i)$ let $f_{\Omega_i} = f \circ \varphi_i$. As $f_{\Omega_i} \in C_0^\infty(\Omega_i)$ and f_{Ω_i} is G_i-invariant there is (by Theorem B.10) a unique $F \in \mathcal{D}(B)$ such that $F_{\Omega_i} = f_{\Omega_i}$ (hence also $F|_{U_i} = f$).

Let us consider the operator $\psi_i Q_i g_i : \mathcal{D}(B) \to \mathcal{D}(B)$ defined as follows. Let $f \in \mathcal{D}(B)$. Then $g_i f \in \mathcal{D}(U_i)$ and we let $\psi_i Q_i g_i$ map f into $\psi_i Q_i(g_i f)$. Next let us consider the operator $Q : \mathcal{D}(B) \to \mathcal{D}(B)$ given by

$$Q = \sum_i \psi_i Q_i g_i$$

(as $f \in \mathcal{D}(B)$ it follows that $\{i : U_i \cap \operatorname{supp}(f) \neq \emptyset\}$ is a finite set, so that Q is well defined).

LEMMA 10.15. *Q is a pseudo-differential operator on B.*

Proof. Let $p \in B$ and set $I_p = \{j : p \in U_j\}$. Then I_p is a finite set (as $\{p\}$ is a compact set and $\{U_j\}$ locally finite). It follows that

$$A = \bigcup_{j \in I \setminus I_p} \{(\bigcap_{i \in I_p} U_i) \setminus \operatorname{supp}(g_j)\}$$

is an open subset of B. Here I denotes the index set of the open cover $\{U_i\}$. Let $j \notin I_p$. Then $p \notin U_j \supset \operatorname{supp}(g_j)$ hence $p \notin \operatorname{supp}(g_j)$, i.e. $p \in A$. By hypothesis $\mathcal{H}$ is a basis of open neighborhoods for the topology of B hence there is $U \in \mathcal{H}$ such that $p \in U \subseteq A$. To prove Lemma 10.15 it suffices to show that the operator Q satisfies the Girbau-Nicolau condition relative to U. Let us consider the following data 1) a l.u.s. $\{\Omega, G, \varphi\} \in \mathcal{A}$ of support U, 2) a compact set $\tilde{K} \subset \Omega$, and 3) a C^∞ function $F : \Omega \to \mathbb{R}$ of compact support. Let $f \in \mathcal{D}(B)$ with support in $K = \varphi(\tilde{K})$. We wish to compute $F(Qf)_\Omega$. Let $i \in I$. Then either $i \notin I_p$, and then $\psi_i Q_i(g_i f) = 0$, or $i \in I_p$ and then $U \subseteq U_i$ so that there is an injection λ_i of $\{\Omega, G, \varphi\}$ into $\{\Omega_i, G_i, \varphi_i\}$. We have

$$(\psi_i Q_i(g_i f))_\Omega = (\psi_i)_\Omega (Q_i(g_i f))_\Omega = (\psi_i)_\Omega \left[Q_{\Omega_i}(g_i f)_{\Omega_i}\right] \circ \lambda_i \, .$$

Yet Q_{Ω_i} is a pseudo-differential operator on Ω_i hence it comes from a symbol p_i on Ω_i i.e.

$$(Q_{\Omega_i}(g_i f)_{\Omega_i})(\lambda_i(x)) = \int e^{\sqrt{-1}(\lambda_i(x)-y)\cdot\xi} p_i(\lambda_i(x), \xi)(g_i f)_{\Omega_i}(y)dyd\xi.$$

Note that

$$\operatorname{supp}(g_i f)_{\Omega_i} \subseteq \lambda_i(\tilde{K}).$$

Indeed, the commutativity of the following diagram

$$\begin{array}{ccccccc} \tilde{K} \subset & \Omega & \stackrel{\lambda_i}{\to} & \Omega_i & \stackrel{f_{\Omega_i}}{\to} & \mathbb{C} \\ & \varphi \downarrow & & \downarrow \varphi_i & & \\ K \subset & U & \subset & U_i & \stackrel{f}{\to} & \mathbb{C} \end{array}$$

yields

$$\operatorname{supp}(g_i f)_{\Omega_i} = \operatorname{supp}\left[(g_i)_{\Omega_i} f_{\Omega_i}\right] \subset \operatorname{supp}(f_{\Omega_i}) \subset \lambda_i(\tilde{K})$$

where the last inclusion is a consequence of

$$x' \notin \lambda_i(\tilde{K}) \Longrightarrow \varphi_i(x') \notin \varphi_i(\lambda_i \tilde{K}) = \varphi(\tilde{K}) = K \Longrightarrow f(x') = 0 \Longrightarrow$$
$$\Longrightarrow \varphi_i(f(x')) = 0 \Longrightarrow f_{\Omega_i}(x') = 0.$$

By an argument similar to that in the proof of Theorem 10.3 one may show that there is a symbol q_i on Ω such that

$$\left[Q_{\Omega_i}(g_i f)_{\Omega_i}\right](\lambda_i(x)) = \int e^{\sqrt{-1}(x-z)\cdot\xi} q_i(x, \xi)(g_i f)_\Omega(z)dzd\xi.$$

By Lemma 1.2.2 in [**119**], p. 15, with $r(x, \xi, z) = q_i(x, \xi)(g_i)_\Omega(z)$ there is a symbol q_i' such that

$$\left[Q_{\Omega_i}(g_i f)_{\Omega_i}\right](\lambda_i(x)) = \int e^{\sqrt{-1}(x-z)\cdot\xi} q_i'(x, \xi) f_\Omega(z)dzd\xi$$

and Lemma 10.15 is proved. Our calculations also proved that $(Q_i)_\Omega$ is a pseudo-differential operator on Ω (in the sense of Appendix C). To show that $Q \circ D \sim I$ and $D \circ Q \sim I$ we shall need the following

LEMMA 10.16. *Let $U \in \mathcal{H}$ such that $U \subseteq U_i \cap U_j$. Let $K_i = \text{supp}(g_i)$ and $K_j = \text{supp}(g_j)$. Let $\{\Omega, G, \varphi\}$ be a l.u.s. of support U. The pseudo-differential operators $(Q_i)_\Omega$ and $(Q_j)_\Omega$ are equivalent over the G-invariant functions $F \in C_0^\infty(\Omega)$ with $\text{supp}(f) \subseteq K_i \cap K_j \cap U$.*

Here $f : U \to \mathbb{C}$ is the unique continuous function such that $F = f \circ \varphi$. Cf. [**120**], p. 76, for a proof of Lemma 10.16. Let us see how Lemma 10.16 may be used to end the proof of Theorem 10.12. For instance let us check that $D \circ Q \sim I$ (the equivalence $Q \circ D \sim I$ may be proved in a similar manner). Let $p \in B$ and $U \in \mathcal{H}$ such that $p \in U \subset U_i$ for any $i \in I_p$. Let $\{\Omega, G, \varphi\} \in \mathcal{A}$ be a l.u.s. of support U. Consider a compact set $\tilde{K} \subset \Omega$ and a C^∞ function $F : \Omega \to \mathbb{R}$ of compact support. One should prove that

$$F \cdot (D \circ Q)_\Omega \sim F \cdot I_\Omega$$

over the functions f_Ω such that $f \in \mathcal{D}(B)$ with support in $\tilde{K}$. Note that

$$F \cdot (D \circ Q)_\Omega f_\Omega = \sum_{i \in I_p} F \cdot D_\Omega(f_i Q_i g_i)_\Omega f_\Omega =$$

$$= \sum_{i \in I_p} F \cdot D_\Omega((f_i)_\Omega (Q_i)_\Omega ((g_i)_\Omega f_\Omega)) =$$

$$= \sum_{j \in I_p} \sum_{i \in I_p} F \cdot D_\Omega((f_i)_\Omega (Q_i)_\Omega (g_i f_j)_\Omega f_\Omega)$$

(because of $\sum_{j \in I_p} (f_j)_\Omega = 1$). Yet $\text{supp}(g_i f_j f) \subseteq K_i \cap K_j$ hence (by Lemma 10.16)

$$F \cdot (D \circ Q)_\Omega f_\Omega \sim \sum_{j \in I_p} \sum_{i \in I_p} F \cdot D_\Omega((f_i)_\Omega (Q_j)_\Omega (g_i f_j f)_\Omega) =$$

$$= \sum_{i,j \in I_p} F \cdot D_\Omega((f_i)_\Omega (Q_j)_\Omega (f_j f)_\Omega) +$$

$$+ \sum_{i,j \in I_p} F \cdot D_\Omega((f_i)_\Omega (Q_j)_\Omega ((g_i)_\Omega - 1)(f_j f)_\Omega).$$

As

$$\text{supp}(f_i) \cap \text{supp}(g_i - 1) = \emptyset$$

one may easily show that $(f_i)_\Omega (Q_j)_\Omega ((g_i)_\Omega - 1)$ is a pseudo-differential operator of order $-\infty$ on U. Then

$$F \cdot (D \circ Q)_\Omega f_\Omega \sim \sum_{i,j \in I_p} F\, D_\Omega((f_i)_\Omega (Q_j)_\Omega (f_j f)_\Omega) =$$

$$= \sum_{j \in I_p} F\, D_\Omega((Q_j)_\Omega (f_j f)_\Omega).$$

Then $D_{\Omega_j} \circ Q_{\Omega_j} \sim I_{\Omega_j}$ over the C^∞ functions f_{Ω_j} with $\text{supp}(f) \subset K_j$ yields $D_\Omega \circ (Q_j)_\Omega \sim I_\Omega$. Finally

$$F\,(D \circ Q)_\Omega f_\Omega \sim \sum_{j \in I_p} F (f_j f)_\Omega = F\, f_\Omega\,.$$

Q.e.d.

The reader may consult J. Girbau & M. Nicolau, [**120**], p. 77-80, for a treatment of pseudo-differential operators on vector bundles over a C^∞ orbifold, of the relevant

Sobolev spaces, as well as for the decomposition theorem for self-adjoint elliptic operators on a vector bundle over an oriented compact Riemannian orbifold.

CHAPTER 11

Cauchy-Riemann Orbifolds

For any CR orbifold (a notion to be defined below) B, of CR dimension n, we build a vector bundle (in the sense of J. Girbau & M. Nicolau, [**120**], and our Section 9.5) $T_{1,0}(B)$ over B such that $T_{1,0}(B)_p \approx \mathbb{C}^n/G_x$ at any singular point $p = \varphi(x) \in B$ (and the portion of $T_{1,0}(B)$ over the regular part of B is an ordinary CR structure), hence study the tangential Cauchy-Riemann equations on orbifolds. As an application, we build a two-sided parametrix for the Kohn-Rossi laplacian $\Box_\Omega$ (on the domain Ω of a local uniformizing system $\{\Omega, G, \varphi\}$ of B) inverting $\Box_\Omega$ over the G-invariant $(0,q)$-forms $(1 \leq q \leq n-1)$ up to (smoothing) operators of type 1 (in the sense of G.B. Folland & E.M. Stein, [**107**]).

As we saw in Chapter 9 an N-dimensional orbifold (or V-manifold, cf. I. Satake, [**214**]) is a Hausdorff space B looking locally like a quotient of (an open set in) the Euclidean space, by the action of some finite group of C^∞ diffeomorphisms (cf. also [**8**]-[**10**], [**63**], [**213**]-[**216**]). That is, each point $p \in B$ admits a neighborhood U which is uniformized by a domain $\Omega \subset \mathbb{R}^N$ and a continuous map $\varphi : \Omega \to U$, in the sense that there is a finite subgroup $G \subset Diff^\infty(\Omega)$ such that φ is G-invariant and factors to a homeomorphism $\Omega/G \approx U$. Such (local) uniformizing systems $\{\Omega, G, \varphi\}$ (shortly l.u.s.'s) play the role of local coordinate charts in manifold theory, and as well as for ordinary manifolds they are required to agree smoothly on overlaps: if $p \in U' \cap V$ and $\{\Omega', G', \varphi'\}$, $\{D, H, \psi\}$ uniformize U', V respectively, then there is a neighborhood $U \subset U' \cap V$ of p uniformized by some $\{\Omega, G, \varphi\}$, and an injection $\lambda : \Omega \to \Omega'$ i.e. a smooth map which is a C^∞ diffeomorphism on some open subset of Ω' and satisfies $\varphi' \circ \lambda = \varphi$. This being the case, various G-structures of current use in differential geometry, such as Riemannian metrics, complex structures, etc., may be prescribed on orbifolds, by merely assigning an ordinary G-structure to Ω, for each l.u.s. $\{\Omega, G, \varphi\}$, and requiring that injections preserve these (local) G-structures (cf. also [**51**], [**66**], [**172**], [**220**]). For instance, if B is a $(2n+k)$-dimensional orbifold, whose V-manifold structure is described by some fixed family of l.u.s.'s $\mathcal{A}$, then

DEFINITION 11.1. A *CR structure* on B is a set

$$\{T_{1,0}(\Omega) : \{\Omega, G, \varphi\} \in \mathcal{A}\} \tag{11.1}$$

where $T_{1,0}(\Omega)$ is a CR structure (of type (n,k)) on Ω and each injection $\lambda : \Omega \to \Omega'$ is a CR map i.e. $(d_x\lambda)T_{1,0}(\Omega)_x \subseteq T_{1,0}(\Omega')_{\lambda(x)}$, $x \in \Omega$. □

A CR structure (11.1) on B is easily seen to be a vector bundle over B, in the sense of W.L. Baily, [**10**], p. 863, i.e. there is a group monomorphism

$$h_\Omega : G \to \mathrm{Hom}(T_{1,0}(\Omega), T_{1,0}(\Omega))$$

for each l.u.s. $\{\Omega, G, \varphi\} \in \mathcal{A}$, and a bundle map

$$\lambda^* : T_{1,0}(\Omega')|_{\lambda(\Omega)} \to T_{1,0}(\Omega)$$

for each injection $\lambda : \Omega \to \Omega'$, such that

1) $h_\Omega(\sigma)T_{1,0}(\Omega)_x \subseteq T_{1,0}(\Omega)_{\sigma^{-1}(x)}$, $x \in \Omega$,

2) $h_\Omega(\sigma) \circ \lambda^* = \lambda^* \circ h_{\Omega'}(\eta(\sigma))$, $\sigma \in G$,

and

3) $(\mu \circ \lambda)^* = \lambda^* \circ \mu^*$,

for any pair of injections $\lambda : \Omega \to \Omega'$ and $\mu : \Omega' \to \Omega''$, where $\eta : G \to G'$ is a natural group monomorphism associated with λ (cf. our Section 11.3). Indeed

$$h_\Omega(\sigma)_x := d_x\sigma^{-1}, \quad \sigma \in G, \quad x \in \Omega,$$

respectively

$$\lambda^*(v') = (d_{\lambda(x)}\mu)v', \quad v' \in T_{1,0}(\Omega')_{\lambda(x)}, \quad x \in \Omega,$$

where $\mu := (\lambda : \Omega \to \lambda(\Omega))^{-1}$, satisfy the requirements (1) to (3). Each $\sigma \in G$ is in particular an injection, hence $G \subset \mathrm{Aut}_{CR}(\Omega)$. One may proceed to define CR functions as follows.

DEFINITION 11.2. A continuous function $f : B \to \mathbb{C}$ is said to be a *CR function* if each $f_\Omega := f \circ \varphi : \Omega \to \mathbb{C}$ is smooth and

$$\overline{\partial}_\Omega f_\Omega = 0 \text{ in } \Omega \tag{11.2}$$

where $\overline{\partial}_\Omega$ is the tangential Cauchy-Riemann operator on $(\Omega, T_{1,0}(\Omega))$. □

Then the equations (11.2) may then be referred to as the tangential Cauchy-Riemann equations on (the CR orbifold) B and it appears that a satisfactory scheme for recovering CR geometry and analysis, on V-manifolds, has been devised.

The weakness of this approach consists in the lack of relationship between the G-structure (here CR structure) so assigned to B and its singular locus. We recall (cf. our Section 9.4) that a point $p \in B$ is singular if it admits a neighborhood U, uniformized by some l.u.s. $\{\Omega, G, \varphi\}$ for which a point $x \in \Omega$ with nontrivial isotropy group (i.e. $G_x := \{\sigma \in G : \sigma(x) = x\} \neq \{1_\Omega\}$) and lying over p (i.e. $\varphi(x) = p$) may be found.

If Σ is the set of all singular points of B (its singular locus) then $B_{\rm reg} := B \setminus \Sigma$ is an ordinary CR manifold. Although Σ has a quite simple local structure (locally, it is a finite union of real algebraic CR submanifolds) there is no obvious relationship between $T_{1,0}(\Omega)$ and $S := \{x \in \Omega : G_x \neq \{1_\Omega\}\}$, and generally speaking, expressions such as the behavior of the CR structure $T_{1,0}(B_{\rm reg})$ (a bundle over $B \setminus \Sigma$), or of a CR function $f \in CR^\infty(B_{\rm reg})$, near Σ, lack a precise meaning.

To ask a more concrete question, given a CR orbifold B, can one construct a 'bundle' $T_{1,0}(B)$ over the *whole* of B such that $T_{1,0}(B)|_{B_{\rm reg}} = T_{1,0}(B_{\rm reg})$ and the fibres $T_{1,0}(B)_p$ reflect the nature of p (i.e. whether p is singular or regular)? In other words, can one write a set of equations on B reducing to the ordinary Cauchy-Riemann equations $\overline{\partial}_{B_{\rm reg}} f = 0$ on the regular part of B, and exhibiting at Σ a feature related to the nature of Σ?

The scope of Chapter 11 is to answer some fundamental questions of this sort, i.e. regarding (the Cauchy-Riemann equations on) CR orbifolds. Precisely, for each CR orbifold B, we build a bundle $T_{1,0}(B) \to B$ in the sense of J. Girbau & M. Nicolau, [**120**], p. 257-259, such that

$$T_{1,0}(B) \approx \mathbb{C}^n/G_x, \quad p = \varphi(x) \in B, \tag{11.3}$$

a bijection (hence when $p \in \Sigma$, $T_{1,0}(B)_p$ is not even a vector space) and $T_{1,0}(B)_p = T_{1,0}(B_{\rm reg})_p$ for any $p \in B \setminus \Sigma$. Moreover, by adapting (from real to complex geometry) an ideea of I. Satake, [**216**], p. 473, who observed that G_x-invariant tangent vectors at $x \in \Omega$ give rise, in our context, to a subset of $T_{1,0}(B)_p$ depending only on $p = \varphi(x)$ and *possessing* a $\mathbb{C}$-linear space structure, we are led to the equations

$$\sum_{\alpha=1}^{n} \overline{\zeta}^{\alpha} L_{\overline{\alpha}}(f)_x = 0, \tag{11.4}$$

with $f \in C^\infty(\Omega)$, $x \in \Omega$ and

$$\zeta = (\zeta^1, \cdots, \zeta^n) \in \bigcap_{\sigma \in G_x} \mathrm{Ker}[g_\sigma(x) - I_n],$$

where $\{L_\alpha\}$ is a frame of $T_{1,0}(\Omega)$, which may be thought of w.l.o.g. as being defined on the whole of Ω, and $g_\sigma(x) \in \mathrm{GL}(n, \mathbb{C})$ is given by

$$(d_x\sigma) L_{\alpha,x} = g_\sigma(x)^\beta_\alpha L_{\beta,\sigma(x)}\,, \quad x \in \Omega.$$

Clearly (11.4) reduces to (11.2) in $\Omega \setminus S$. We shall show that for each singular point $x \in S$ there is a neighborhood D of x in Ω and an algebraic CR submanifold $F_x \subset S \cap D$ such that each smooth solution f of (11.4) is a CR function on F_x.

Any (smooth) function $f : B \to \mathbb{C}$ gives rise to a G-invariant function $f_\Omega := f \circ \varphi$ on Ω. In general, a (geometric) object prescribed on (each) Ω must be preserved by injections, hence by each $\sigma \in G$, hence it is G-invariant. Therefore, another fundamental feature of any attempt to recover known facts from CR geometry (on CR orbifolds) is, locally, to prove G-invariant analogues of the facts of interest. In view of [**10**] (which uses a G-average of a fundamental solution of an elliptic operator to prove a Kodaira-Hodge-de Rham decomposition theorem on V-manifolds) this part of the task is rather well understood. To illustrate this line of thought, given a domain Ω in $\mathbb{R}^{2n+1}$ carrying a G-invariant strictly pseudoconvex CR structure $T_{1,0}(\Omega)$ and a pseudohermitian structure θ so that G consists of pseudohermitian transformations of (Ω, θ), we build a two-sided parametrix inverting the Kohn-Rossi operator $\Box_\Omega$ on the G-invariant forms of degree $0 < q < n-1$, up to operators of order 1, cf. [**107**]. These are smoothing, in the sense that they are bounded operators $S^p_k(\Omega) \to S^p_{k+1}(\Omega)$ of Folland-Stein spaces. The methods resemble closely those in [**10**], p. 870-874, and [**120**], p. 71-74.

Chapter 11 is organized as follows. Section 11.1 presents the *pseudohermitian normal coordinates* on a strictly pseudoconvex CR manifold, a notion (and essential tool) due to D. Jerison & J.M. Lee, [**143**]. In Section 11.2 we discuss the case of complex orbifolds (CR codimension $k = 0$), the local structure of their singular locus, and V-holomorphic functions. Sections 11.3 and 11.4 are devoted to CR orbifolds of CR codimension 1. In Section 11.5 we prove the main result (inverting the Kohn-Rossi operator over the G-invariant forms).

11.1. Parabolic geodesics

Only for Chapter 11 we adopt the following definitions.

DEFINITION 11.3. A *pseudohermitian transformation* is a CR isomorphism between two CR manifolds M and N on which pseudohermitian structures θ and θ_N

have been fixed, such that $f^*\theta_N = a(f)\theta$, for some $a(f) \in \mathbb{R} \setminus \{0\}$. If $a(f) \equiv 1$ then f is *isopseudohermitian.* $\square$

Note that previously one automatically required $a(f) \equiv 1$ for a pseudohermitian transformation f.

DEFINITION 11.4. Let M be a nondegenerate CR manifold, of CR dimension n, and θ a contact form on M. Let

$$\overline{\partial}_M^* : \Omega^{0,s}(M) \to \Omega^{0,s-1}(M)$$

be the formal adjoint of the tangential Cauchy-Riemann operator $\overline{\partial}_M$ with respect to the L^2 inner product $(\alpha, \beta)_s = \int_M g_\theta^*(\alpha, \overline{\beta})\ \theta \wedge (d\theta)^n$ (where g_θ^* is the pointwise inner product induced on $\Omega^{0,s}(M)$ by the Webster metric g_θ and at leat one of the forms $\alpha, \beta \in \Omega^{0,s}(M)$ is of compact support) i.e.

$$(\overline{\partial}_M^* \alpha, \beta)_{s-1} = (\alpha, \overline{\partial}_M \beta)_s\,,$$

for any $\alpha \in \Omega^{0,s}(M)$ and any $\beta \in \Omega^{0,s-1}(M)$ of compact support. The *Kohn-Rossi operator*

$$\square_M : \Omega^{0,s}(M) \to \Omega^{0,s}(M), \quad s \geq 0,$$

of (M, θ) is defined by

$$\square_M \omega = \overline{\partial}_M^* \overline{\partial}_M \omega + \overline{\partial}_M \overline{\partial}_M^* \omega$$

for any $\omega \in \Omega^{0,s}(M)$. $\square$

In particular $\square u = \overline{\partial}_M^* \overline{\partial}_M u$ for any C^∞ function $u : M \to \mathbb{C}$. It is immediate that

PROPOSITION 11.5. *If $f : M \to N$ is an isopseudohermitian transformation then*

$$\square_M^f v = \square_N v, \quad v \in C^\infty(N), \tag{11.5}$$

where $\square_M^f v := \left(\square_M v^{f^{-1}}\right)^f$ *and* $u^f := u \circ f^{-1}$, $u \in C^\infty(M)$.

Let M be a strictly pseudoconvex CR manifold and θ a contact form with L_θ positive definite.

DEFINITION 11.6. A smooth curve $\gamma(t)$ in M satisfying the ODE

$$(\nabla_{d\gamma/dt} \frac{d\gamma}{dt})_{\gamma(t)} = 2cT_{\gamma(t)}, \tag{11.6}$$

for some $c \in \mathbb{R}$ and any value of the parameter t is a *parabolic geodesic* on M. $\square$

Let $x \in M$ and $W \in H(M)_x$. By standard theorems on ODEs there is $\delta > 0$ such that whenever $g_{\theta,x}(W,W)^{1/2} < \delta$ the unique solution $\gamma_{W,c}(t)$ to (11.6) of initial data (x, W) may be uniquely continued to an interval containing $t = 1$ and the map

$$\Psi_x : B(0, \delta) \subset T_x(M) \to M, \quad \Psi_x(W + cT_x) := \gamma_{W,c}(1), \tag{11.7}$$

is a diffeomorphism of a sufficiently small neighborhood of $0 \in T_x(M)$ onto a neighborhood of $x \in M$.

DEFINITION 11.7. The map (11.7) is called the *parabolic exponential map.* $\square$

The terminology is justified by the fact that Ψ_x maps any parabola $t \mapsto tW + t^2 c T_x$ in the tangent space onto $\gamma_{W,c}$.

Let now $\{T_\alpha\}$ be a local orthonormal frame of $T_{1,0}(M)$, defined on a neighborhood U of x in M. It determines an isomorphism $\lambda_x : T_x(M) \to \mathbb{H}_n$ given by

$$\lambda_x(v) = (\theta_x^\alpha(v) e_\alpha, \theta_x(v)),$$

for any $v \in T_x(M)$. Here $\mathbb{H}_n$ is the Heisenberg group and $\{\theta^\alpha\}$ is the frame of $T_{1,0}(M)^*$ determined by

$$\theta^\alpha(T_\beta) = \delta_\beta^\alpha, \quad \theta^\alpha(T_{\overline{\beta}}) = \theta^\alpha(T) = 0.$$

DEFINITION 11.8. The resulting local coordinates

$$(z,t) := \lambda_x \circ \Psi_x^{-1},$$

defined in some neighborhood of x, are the *pseudohermitian normal coordinates* at x determined by $\{T_\alpha\}$. □

By Prop. 2.5 in [**143**], p. 313, these coordinates are also normal coordinates at x in the sense of G.B. Folland & E.M. Stein (cf. [**107**], p. 471-472). We shall need the following

LEMMA 11.9. *Let M be a nondegenerate CR manifold and θ a contact form on M. Let $\sigma : M \to M$ be a CR automorphism such that $\sigma^*\theta = a(\sigma)\theta$ for some $a(\sigma) \in \mathbb{R} \setminus \{0\}$. Let $\gamma_{W,c}(s)$ be the solution to*

$$\nabla_{d\gamma/dt} \frac{d\gamma}{dt} = 2cT \circ \gamma$$

of initial data (η, W) with $\eta \in M$, $W \in H(M)_\eta$. Then $\sigma \circ \gamma_{W,c} = \gamma_{W_\sigma, a(\sigma)c}$ where $W_\sigma := (d_\eta \sigma) W \in H(M)_{\sigma(\eta)}$ i.e. $\sigma \circ \gamma_{W,c}$ is the solution to

$$\nabla_{d\gamma/dt} \frac{d\gamma}{dt} = 2ca(\sigma) T \circ \gamma$$

of initial data $(\sigma(\eta), W_\sigma)$.

Proof. For each $y \in M$ and $X \in \mathcal{X}(M)$ let us consider

$$(\sigma_* X)_y := (d_{\sigma^{-1}(y)} \sigma) X_{\sigma^{-1}(y)}$$

(hence $\sigma_* : \mathcal{X}(M) \approx \mathcal{X}(M)$, an isomorphism) and let us set

$$\nabla_X^\sigma Y := (\sigma_*)^{-1} \nabla_{\sigma_* X} \sigma_* Y.$$

Then $\nabla^\sigma \theta = 0$. Using

$$\sigma^* g_\theta = a(\sigma) g_\theta + [a(\sigma)^2 - a(\sigma)] \theta \otimes \theta$$

one may show that $\nabla^\sigma g_\theta = 0$. Also it is easy to check that $\nabla^\sigma J = 0$. Next $\sigma_* T = a(\sigma) T$ so that

$$T_{\nabla^\sigma}(Z, W) = 0,$$
$$T_{\nabla^\sigma}(Z, \overline{W}) = 2i L_\theta(Z, \overline{W}) T,$$
$$T_{\nabla^\sigma}(T, JX) + J T_{\nabla^\sigma}(T, X) = 0,$$

for any $Z, W \in T_{1,0}(M)$ and $X \in T(M)$. We may conclude that $\nabla^\sigma = \nabla$, the Tanaka-Webster connection of (M, θ). Let us set $\gamma := \gamma_{W,c}$ and $\gamma_\sigma := \sigma \circ \gamma$. Then $\gamma_\sigma(0) = \sigma(\eta)$ and $(d\gamma_\sigma/ds)(0) = W_\sigma$. Finally

$$\nabla_{d\gamma_\sigma/ds} \frac{d\gamma_\sigma}{ds} = \sigma_* \nabla^\sigma_{d\gamma/ds} \frac{d\gamma}{ds} = \sigma_* \nabla_{d\gamma/ds} \frac{d\gamma}{ds} = \sigma_*(2cT \circ \gamma) = 2ca(\sigma) T \circ \gamma_\sigma,$$

hence $\gamma_\sigma = \gamma_{W_\sigma, a(\sigma)c}$, that is a pseudohermitian transformation σ maps the parabolic geodesic $\gamma_{W,c}$ into the parabolic geodesic $\gamma_{W_\sigma, a(\sigma)c}$. Q.e.d.

We have specified the behaviour (11.5) of the Kohn-Rossi laplacian on functions, with respect to isopseudohermitian transformations. In general

PROPOSITION 11.10. *If φ is a $(0,q)$-form and $\sigma : M \to M$ a pseudohermitian transformation of a nondegenerate CR manifold then*

$$\square_M(\sigma^*\varphi) = a(\sigma)\,\sigma^*\,\square_M\varphi. \tag{11.8}$$

Indeed, on one hand $\sigma^*\,\overline{\partial}_M\varphi = \overline{\partial}_M\sigma^*\varphi$, as it easily follows from the axioms defining $\overline{\partial}_M$. On the other hand

$$\overline{\partial}_M^*\psi = (-1)^{q+1}(q+1)\,h^{\lambda\overline{\mu}}(\nabla_\lambda\psi_{\overline{\alpha}_1\cdots\overline{\alpha}_q\overline{\mu}})\theta^{\overline{\alpha}_1}\wedge\cdots\wedge\theta^{\overline{\alpha}_q}$$

for any $(0,q+1)$-form ψ on M, where covariant derivatives are meant with respect to the Tanaka-Webster connection of (M,θ). For instance if φ is a $(0,1)$-form

$$\overline{\partial}_M^*\varphi = -h^{\lambda\overline{\mu}}\nabla_\lambda\varphi_{\overline{\mu}}$$

hence

$$\overline{\partial}_M^*(\sigma^*\varphi) = -h^{\lambda\overline{\mu}}\left\{T_\lambda\left((g_\sigma)_{\overline{\mu}}^{\overline{\nu}}\right)(\varphi_{\overline{\nu}}\circ\sigma) + \right.$$
$$\left. + (g_\sigma)_{\overline{\mu}}^{\overline{\nu}}(g_\sigma)_\lambda^\rho\left[T_\rho(\varphi_{\overline{\nu}})\circ\sigma\right] - \Gamma_{\lambda\overline{\mu}}^{\overline{\nu}}(g_\sigma)_{\overline{\nu}}^{\overline{\rho}}(\varphi_{\overline{\rho}}\circ\sigma)\right\}$$

and the identity

$$\Gamma_{\alpha\overline{\beta}}^{\overline{\mu}}(g_\sigma)_{\overline{\mu}}^{\overline{\nu}} = T_\alpha\left((g_\sigma)_{\overline{\beta}}^{\overline{\nu}}\right) + (g_\sigma)_\alpha^\mu(g_\sigma)_{\overline{\beta}}^{\overline{\rho}}(\Gamma_{\mu\overline{\rho}}^{\overline{\nu}}\circ\sigma)$$

(a consequence of $\nabla = \nabla^\sigma$) lead to

$$\overline{\partial}_M^*(\sigma^*\varphi) = a(\sigma)\left(\overline{\partial}_M^*\varphi\right)\circ\sigma\,.$$

Q.e.d.. Here Γ^A_{BC} denote the Christoffel symbols (of ∇ with respect to $\{T_\alpha\}$) and $\sigma_* T_\alpha = (g_\sigma)_\alpha^\beta T_\beta$.

11.2. Complex orbifolds

The definitions in Section 11.2 parallel those in Sections 9.1 to 9.2 (the words *orbifold*, *diffeomorphism*, etc., there, should be replaced by *complex orbifold*, *biholomorphism*, etc.). However, the reader may profit from the explicit description of the complex orbifold category.

Let X be a Hausdorff space and $U \subseteq X$ an open subset.

DEFINITION 11.11. A *local uniformizing system* (l.u.s.) *of dimension* n *of* X *over* U is a synthetic object $\{\Omega, G, \varphi\}$ consisting of a domain $\Omega \subseteq \mathbb{C}^n$, a finite subgroup $G \subset \mathrm{Aut}(\Omega)$ of biholomorphisms of Ω in itself, and a continuous map $\varphi : \Omega \to U$ such that the induced map $\varphi_G : \Omega/G \to U$ is a homeomorphism. An *injection* of $\{\Omega, G, \varphi\}$ into $\{\Omega', G', \varphi'\}$ is a C^∞ map $\lambda : \Omega \to \Omega'$ such that λ is a biholomorphism of Ω onto some open subset of Ω' and $\varphi'\circ\lambda = \varphi$. The set $U = \varphi(\Omega)$ is the *support* of the l.u.s. $\{\Omega, G, \varphi\}$. □

Given a family $\mathcal{F}$ of l.u.s.'s of dimension n of X, let $\mathcal{H}$ be the family of all supports of all l.u.s.'s in $\mathcal{F}$.

DEFINITION 11.12. $\mathcal{F}$ is a *defining family* for X if 1) for any l.u.s.'s $\{\Omega, G, \varphi\}$, $\{\Omega', G', \varphi'\} \in \mathcal{F}$ of supports U and U', if $U \subseteq U'$ then there is an injection λ of $\{\Omega, G, \varphi\}$ into $\{\Omega', G', \varphi'\}$, and 2) $\mathcal{H}$ is a basis of open sets for the topology of X. Two defining families $\mathcal{F}$, $\mathcal{F}'$ are *directly equivalent* if there is a third defining family $\mathcal{F}''$ such that $\mathcal{F} \cup \mathcal{F}' \subseteq \mathcal{F}''$. Also $\mathcal{F}$, $\mathcal{F}'$ are *equivalent* if there is a set $\{\mathcal{F}_1, \cdots, \mathcal{F}_r\}$ of defining families such that $\mathcal{F}_1 = \mathcal{F}$, $\mathcal{F}_r = \mathcal{F}''$, and $\mathcal{F}_i$, $\mathcal{F}_{i+1}$ are directly equivalent for each $1 \leq i \leq r-1$. A n-dimensional *complex orbifold* is a connected paracompact Hausdorff space X together with an equivalence class of defining families. □

As in ordinary complex manifold theory it is customary to choose a defining family $\mathcal{F}$ in the class and refer to $(X, \mathcal{F})$ as a complex orbifold. Cf. I. Satake, [**215**], p. 261-262 (where complex orbifolds are referred to as complex analytic V-manifolds). Clearly, any complex orbifold of complex dimension n as above is a real $2n$-dimensional V-manifold (in the sense of [**214**], p. 359-360, or [**10**], p. 862-863).

Let $(X, \mathcal{F})$ be an orbifold. Set $S = \{x \in \Omega : G_x \neq \{e\}\}$ (a closed subset of Ω). Then $\Sigma := \bigcup_{\{\Omega, G, \varphi\} \in \mathcal{F}} \varphi(S)$ is the singular locus of X and $X_{\rm reg} := X \setminus \Sigma$ its regular part. $X_{\rm reg}$ is an ordinary C^∞ manifold.

Let $\mathcal{E}(X)$ be the ring of all complex valued smooth functions on X. We shall prove the following

THEOREM 11.13. (J. Masamune et al., [**86**])
For any complex orbifold $(X, \mathcal{F})$, of complex dimension n, there is a vector bundle $(T_{1,0}(X), \pi, X)$ such that

1) *for any $p \in X$, if $p \in U \in \mathcal{H}$ and $\{\Omega, G, \varphi\} \in \mathcal{F}$ is a l.u.s. over U then $\pi^{-1}(x) \approx \mathbb{C}^n / G_x$ (a bijection) for any $x \in \Omega$ with $\varphi(x) = p$.*

2) *$X_{\rm reg}$ is a complex manifold and $T_{1,0}(X)|_{X_{\rm reg}}$ its holomorphic tangent bundle. The singular locus of $T_{1,0}(X)$ (as a $4n$-dimensional V-manifold) is contained in $\pi^{-1}(\Sigma)$.*

3) *For any section Z in $T_{1,0}(X)$ (i.e. any continuous map $Z : X \to T_{1,0}(X)$ so that $Z(p) \in \pi^{-1}(p)$ for any $p \in X$) and any $f \in \mathcal{E}(X)$ there is a (naturally defined) function $Z(f) : X \to \mathbb{C}$; if $Z(\overline{f}) = 0$ for all sections Z then f_Ω is holomorphic in Ω for any l.u.s. $\{\Omega, G, \varphi\} \in \mathcal{F}$, and conversely.*

We organize the proof in several steps, as follows.

Step 1. *The construction of $T_{1,0}(X)$.*

Define $g_\lambda : \Omega \to \mathrm{GL}(n, \mathbb{C})$ by setting

$$g_\lambda(x)\zeta = \zeta^k \frac{\partial(z^j \circ \lambda)}{\partial z^k}(x) e_j,$$

where (z^j) are the natural complex coordinates on $\mathbb{C}^n$, and $\{e_j\}$ its canonical linear basis. Then $G_* = \{\sigma_* : \sigma \in G\}$ acts on $\Omega \times \mathbb{C}^n$ as a (finite) group of biholomorphisms. Let us set

$$\hat{T}_{1,0}(X) := \bigcup_{\{\Omega, G, \varphi\} \in \mathcal{F}} (\Omega \times \mathbb{C}^n)/G_*$$

(a disjoint union). Then $\hat{T}_{1,0}(X)$ is a Hausdorff space in a natural manner. We define an equivalence relation $\sim$ on $\hat{T}_{1,0}(X)$ as follows.

DEFINITION 11.14. Let $\hat{x}$, $\hat{y} \in \hat{T}_{1,0}(X)$. If $\hat{x}$ is the G_*-orbit $\text{orb}_{G_*}(x,\zeta)$ of some $(x,\zeta) \in \Omega \times \mathbb{C}^n$, for some l.u.s. $\{\Omega, G, \varphi\} \in \mathcal{F}$, then we say that $\hat{x} \sim \hat{y}$ if there is an injection $\lambda : \Omega \to \Omega'$ such that

$$\hat{y} = \text{orb}_{G'_*}(\lambda(x)\,,\, g_\lambda(x)\zeta).$$

□

If $(\sigma(x)\,,\, g_\sigma(x)\zeta) \in \hat{x}$ is another representative of $\hat{x}$ then

$$\text{orb}_{G'_*}(\lambda(\sigma(x))\,,\, g_\lambda(\sigma(x))g_\sigma(x)\zeta) = \text{orb}_{G'_*}(\eta(\sigma)\lambda(x)\,,\, g_{\lambda\circ\sigma}(x)\zeta) =$$
$$= \text{orb}_{G'_*}[\eta(\sigma)_*(\lambda(x)\,,\, g_\lambda(x)\zeta)] = \text{orb}_{G'_*}(\lambda(x)\,,\, g_\lambda(x)\zeta),$$

(where $\eta : G \to G'$ is the group monomorphism associated with λ) hence $\hat{x} \sim \hat{y}$ is well defined. Clearly $\sim$ is refexive and transitive. The only issue which needs a bit of care is the symmetry property. Note that for any injection $\lambda : \Omega \to \Omega'$ the synthetic object $\{\lambda(\Omega), \eta(G), \psi\}$, where $\psi = \varphi'|_{\lambda(\Omega)}$, is a l.u.s. of support $U = \varphi(\Omega)$. Indeed $\eta(G)$ acts on $\lambda(\Omega)$ as a group of complex analytic transformations and ψ is $\eta(G)$-invariant. Moreover λ is equivariant hence it induces a homeomorphism

$$\lambda_G : \Omega/G \approx \lambda(\Omega)/\eta(G).$$

The map $\psi_G : \lambda(\Omega)/\eta(G) \to U'$ (induced by ψ) corestricts to U and $\psi_G \circ \lambda_G = \varphi_G$ hence $\psi_G : \lambda(\Omega)/\eta(G) \approx U$ (a homeomorphism). Then $\hat{x} \sim \hat{y}$ yields $\hat{y} \sim \hat{x}$, as we may think of $(\lambda(x), g_\lambda(x)\zeta)$ as a representative of $\hat{y}$ with respect to the l.u.s. $\{\lambda(\Omega), \eta(G), \psi\}$ and rewrite $\hat{x}$ as

$$\hat{x} = \text{orb}_{G_*}(\mu(\lambda(x))\,,\, g_\mu(\lambda(x))g_\lambda(x)\zeta),$$

where μ is the injection $(\lambda : \Omega \to \lambda(\Omega))^{-1}$.

Next $T_{1,0}(X) := \hat{T}_{1,0}(X)/\sim$ carries the quotient topology and

$$\pi : T_{1,0}(X) \to X, \quad \pi([\text{orb}_{G_*}(x,\zeta)]) := \varphi(x),$$

is continuous. Square brackets indicate classes mod $\sim$ that is

$$T_{1,0}(X) = \{[\hat{x}] : \hat{x} \in \hat{T}_{1,0}(X)\}.$$

The definition doesn't depend upon the choice of representatives. Indeed if $\hat{x} = \text{orb}_{G_*}(x,\zeta)$ and $\hat{y} \in [\hat{x}]$ then $\hat{y} = \text{orb}_{G'_*}(\lambda(x)\,,\, g_\lambda(x)\zeta)$ for some injection $\lambda : \Omega \to \Omega'$, and $\varphi'(\lambda(x)) = \varphi(x)$.

We wish to show that $(T_{1,0}(X), \pi, X)$ is a vector bundle of standard fibre $\mathbb{C}^n$. To this end, let $\varphi_* : \Omega \times \mathbb{C}^n \to T_{1,0}(X)$ be the (continuous) map given by

$$\varphi_*(x,\zeta) = [\text{orb}_{G_*}(x,\zeta)].$$

Then $\pi \circ \varphi_* = \varphi \circ \pi_\Omega$. Also φ_* is G_*-invariant and the induced map

$$(\varphi_*)_{G_*} : (\Omega \times \mathbb{C}^n)/G_* \to T_{1,0}(X)$$

is injective. Finally, it is straightforward that

$$\lambda_*(x,\zeta) = (\lambda(x)\,,\, g_\lambda(x)\zeta)$$

is an injection of $\{\Omega \times \mathbb{C}^n, G_*, \varphi_*\}$ into $\{\Omega' \times \mathbb{C}^n, G'_*, \varphi'_*\}$.

Let $p \in X$ be an arbitrary point (eventually singular) and $U \in \mathcal{H}$ such that $p \in U$. Let $\{\Omega, G, \varphi\} \in \mathcal{F}$ be a l.u.s. of support U and $x \in \Omega$ such that $\varphi(x) = p$. Let $\{\Omega_*, G_*, \varphi_*\}$ be a l.u.s. of $T_{1,0}(X)$ corresponding to $\{\Omega, G, \varphi\}$ as above, where $\Omega_* = \Omega \times \mathbb{C}^n$. Then

$$\pi(\varphi_*(x,\zeta)) = \varphi(x) = p$$

hence $\varphi_*(x,\zeta) \in \pi^{-1}(p)$ for any $\zeta \in \mathbb{C}^n$. There is a natural action of G_x on $\mathbb{C}^n$ given by $(\sigma, \zeta) \mapsto g_\sigma(x)\zeta$. We may consider the map

$$\mathbb{C}^n/G_x \to \pi^{-1}(p), \quad [\zeta] \mapsto \varphi_*(x,\zeta), \tag{11.9}$$

where $[\zeta]$ is the G_x-orbit of ζ. If $[\zeta] = [\xi]$ then $\xi = g_\sigma(x)\zeta$ for some $\sigma \in G$ and

$$\varphi_*(x,\xi) = \varphi_*(\sigma(x), g_\sigma(x)\zeta) = \varphi_*(\sigma_*(x,\zeta)) = \varphi_*(x,\zeta),$$

i.e. (11.9) is well defined. To see that (11.9) is injective let $\varphi_*(x,\xi) = \varphi_*(x,\zeta)$. As $\{\Omega_*, G_*, \varphi_*\}$ is a l.u.s., there is $\sigma \in G$ such that $(x,\zeta) = \sigma_*(x,\xi)$ hence $\sigma \in G_x$ and $g_\sigma(x)\xi = \zeta$, i.e. ξ, ζ are G_x-equivalent. To see that (11.9) is surjective let $f \in \pi^{-1}(p)$. As φ_* induces a bijection $\Omega_*/G_* \approx \pi^{-1}(U)$ there is $\tilde{f} = (y,\xi) \in \Omega_*$ such that $\varphi_*(\tilde{f}) = f$. Then

$$\varphi(x) = p = \pi(f) = \pi(\varphi_*(\tilde{f})) = \varphi(\pi_\Omega(\tilde{f})) = \varphi(y),$$

hence there is $\sigma \in G$ such that $y = \sigma(x)$. At this point let us set

$$\tilde{f}_* := (\sigma^{-1})_* \tilde{f} \in \Omega_* .$$

Then $\varphi_*(\tilde{f}_*) = f$ and $\tilde{f}_*$ is an element of the form (x,ζ) with $\zeta = g_{\sigma^{-1}}(\sigma(x))\xi \in [\xi]$, so we are done.

Step 2 *The image $T_{1,0}(X)_p$ of $T_{1,0}(\Omega)_{G_x} := \{v \in T_{1,0}(\Omega)_x : (d_x\sigma)v = v, \ \forall\, \sigma \in G_x\}$ via the map $T_{1,0}(\Omega) \approx \Omega \times \mathbb{C}^n \xrightarrow{\varphi_*} T_{1,0}(X)$ depends only on p i.e. it doesn't depend upon the choice of $\{\Omega, G, \varphi\} \in \mathcal{F}$ and $x \in \Omega$ with $\varphi(x) = p$. Also $T_{1,0}(X)_p$ has a natural $\mathbb{C}$-vector space structure such that*

$$\dim_{\mathbb{C}} T_{1,0}(X)_p = \dim_{\mathbb{C}} \bigcap_{\sigma \in G_x} \operatorname{Ker}[g_\sigma(x) - I_n] \tag{11.10}$$

Here I_n is the $n \times n$ unit matrix. Let $p \in U' \in \mathcal{H}$ and $\{\Omega', G', \varphi'\} \in \mathcal{F}$ over U', and let us consider $x' \in \Omega'$ such that $\varphi'(x') = p$. As $\mathcal{H}$ is a basis of open sets for the topology of X, let $V \subseteq U \cap U'$ with $p \in V \in \mathcal{H}$ and let $\{D, H, \psi\} \in \mathcal{F}$ be a l.u.s. over V. Then there exist injections $\lambda : D \to \Omega$ and $\lambda' : D \to \Omega'$. Let $y \in D$ such that $\psi(y) = p$. We wish to show that

$$\{\varphi_*(x,\zeta) : \zeta \in (\mathbb{C}^n)_{G_x}\}$$

depends only on p, where

$$(\mathbb{C}^n)_{G_x} := \{\zeta \in \mathbb{C}^n : g_\sigma(x)\zeta = \zeta, \ \forall \sigma \in G_x\}.$$

As $\varphi(\lambda(y)) = \varphi(x)$ there is $\sigma \in G$ with $\lambda(y) = \sigma(x)$ hence

$$(\sigma(x)\,,\, g_\lambda(y)\xi) = \sigma_*(x\,,\, g_{\sigma^{-1}\circ\lambda}(y)\xi)$$

and we have

$$\{\psi_*(y,\xi) : \xi \in (\mathbb{C}^n)_{H_y}\} = \{\varphi_*(\lambda(y), g_\lambda(y)\xi) : \xi \in (\mathbb{C}^n)_{H_y}\} =$$
$$= \{\varphi_*(x, g_{\sigma^{-1}\circ\lambda}(y)\xi) : \xi \in (\mathbb{C}^n)_{H_y}\}.$$

At this point, it suffices to show that the map

$$(\mathbb{C}^n)_{H_y} \to (\mathbb{C}^n)_{G_x}, \ \xi \mapsto g_{\sigma^{-1}\circ\lambda}(y)\xi, \tag{11.11}$$

is a well defined bijection. $\sigma^{-1} \circ \lambda : D \to \Omega$ is an injection. Let $\eta_\sigma : H \to G$ be the corresponding group monomorphism. As $\varphi(x) = p = \psi(y)$, $\eta_\sigma : H_y \to G_x$ is an isomorphism (cf. Prop. 1.5 in [**120**], p.257). Given $\tau \in G_x$ let $\rho \in H_y$ so that $\eta_\sigma(\rho) = \tau$. Then

$$g_\tau(x) g_{\sigma^{-1}\circ\lambda}(y)\xi = g_{\tau\circ\sigma^{-1}\circ\lambda}(y)\xi = g_{\eta_\sigma(\rho)\circ\sigma^{-1}\circ\lambda}(y)\xi =$$

$$= g_{(\sigma^{-1}\circ\lambda)\circ\rho}(y)\xi = g_{\sigma^{-1}\circ\lambda}(y) g_\rho(y)\xi = g_{\sigma^{-1}\circ\lambda}(y)\xi,$$

hence (11.11) is well defined. Also, a similar computation shows that

$$g_{\sigma^{-1}\circ\lambda}(y)(\mathbb{C}^n)_{H_y} = (\mathbb{C}^n)_{G_x}$$

and (11.11) is clearly injective. The same proof applies to λ', so we are done.

Note that $T_{1,0}(X)_p$ is a $\mathbb{C}$-linear space [with $\alpha\, \varphi_*(x,\zeta) + \beta\, \varphi_*(x,\xi) := \varphi_*(x, \alpha\zeta + \beta\xi)$ (while the same operation on the image of the whole $\mathbb{C}^n/G_x$ is not well defined)].

To see that X_{reg} is a complex manifold we need to review the differentiable structure of X_{reg} in some detail. Let $\{D, H, \psi\} \in \mathcal{F}$ be a l.u.s. of X over $V \in \mathcal{H}$. Set $\Omega = \psi^{-1}(U)$ where $U := V \cap X_{\text{reg}}$. Then

LEMMA 11.15. $\sigma \in H \Longrightarrow \sigma(\Omega) = \Omega$.

Proof. Let $x \in \Omega$ and $p := \psi(x)$. Then $p \in U \subseteq X \setminus \Sigma$ hence each point of $\psi^{-1}(p)$ has a trivial isotropy group. Yet $\sigma(x) \in \psi^{-1}(p)$ hence $G_{\sigma(x)} = \{e\}$. It follows that $X \setminus \Sigma \ni \psi(\sigma(x)) = \psi(x) = p \in V$ hence $\psi(\sigma(x)) \in U$, i.e. $\sigma(x) \in \Omega$. Q.e.d.

Let us set $G := \{\sigma|_\Omega : \sigma \in H\}$ and $\varphi := \psi|_\Omega$. Then $\{\Omega, G, \varphi\}$ is a l.u.s. of X_{reg} over U. As $\{D, H, \psi\}$ runs over $\mathcal{F}$, the l.u.s.'s $\{\Omega, G, \varphi\}$ form a defining family of X_{reg}, hence X_{reg} is a $2n$-dimensional V-manifold. To see that it actually possesses a C^∞ manifold structure note first that G acts freely on Ω, as a mere consequence of definitions. Let $y \in \Omega$. Then $\sigma(y) \neq y$ for any $\sigma \in G \setminus \{e\}$ (as $G_y = \{e\}$) hence there is an open neighborhood Ω_σ of y in Ω such that $\sigma(\Omega_\sigma) \cap \Omega_\sigma = \emptyset$. Let us set

$$D_y := \bigcap_{\sigma \in G \setminus \{e\}} \Omega_\sigma .$$

As G is finite D_y is open, $y \in D_y \subseteq \Omega$, and $\sigma(D_y) \cap D_y = \emptyset$ for any $\sigma \in G \setminus \{e\}$, hence G acts on Ω as a properly discontinuous group of C^∞ diffeomorphisms. Thus the quotient Ω/G is a real $2n$-dimensional C^∞ manifold and each

$$U \in \mathcal{H}_{\text{reg}} := \{V \cap (X \setminus \Sigma) : V \in \mathcal{H}\}$$

inherits a manifold structure via φ_G.

Once Ω/G is organized as a manifold the projection $\Omega \to \Omega/G$ is a local diffeomorphism and its local inverses form a C^∞ atlas $\mathcal{F}_\Omega$. Then

$$\mathcal{F}_U := \{\chi \circ \varphi_G^{-1} : \chi \in \mathcal{F}_\Omega\}$$

is an atlas on U and

$$\mathcal{F}_{reg} := \bigcup_{U \in \mathcal{H}_{\text{reg}}} \mathcal{F}_U$$

an atlas on X_{reg}. Also $\varphi : \Omega \to U$ is differentiable (and φ_G a diffeomorphism). As Ω and U are locally diffeomorphic there is a unique complex structure on U such that $T_{1,0}(U)_{\varphi(x)} = (d_x\varphi) T_{1,0}(\Omega)_x$, for any $x \in \Omega$. We need to show that

LEMMA 11.16. *Let $p \in X_{\text{reg}}$ and $U, U' \in \mathcal{H}_{\text{reg}}$ such that $p \in U \cap U'$. Then $T_{1,0}(U)_p = T_{1,0}(U')_p$, i.e. the complex structures $\{T_{1,0}(U) : U \in \mathcal{H}_{\text{reg}}\}$ glue up to a globally defined complex structure on X_{reg}.*

Proof. Let $V \in \mathcal{H}_{\rm reg}$ such that $p \in V \subseteq U \cap U'$ and let $\{D, H, \psi\}$ be a l.u.s. of $X_{\rm reg}$ over V. Let $\lambda : D \to \Omega$ and $\lambda' : D \to \Omega'$ be injections and let $y \in D$ such that $\psi(y) = p$. Let us set $x := \lambda(y) \in \Omega$ and $x' := \lambda'(y) \in \Omega'$. Then

$$T_{1,0}(U)_p = (d_y\psi)T_{1,0}(D)_y = T_{1,0}(U')_p,$$

as both λ, λ' are holomorphic maps and $\varphi \circ \lambda = \psi = \varphi' \circ \lambda'$.

So $X_{\rm reg}$ is a complex manifold, in a natural way. Next $\pi^{-1}(X_{\rm reg}) = T_{1,0}(X_{\rm reg})$ because of the isomorphism

$$T_{1,0}(X)_p \to T_{1,0}(X_{\rm reg})_p, \ \varphi_*(x,\zeta) \mapsto (d_x\varphi)\zeta^j \left.\frac{\partial}{\partial z^j}\right|_x, \quad p \in U \in \mathcal{H}_{\rm reg}.$$

If v is a singular point of $T_{1,0}(X)$ with $p := \pi(v)$, there is $U \in \mathcal{H}$ with $p \in U$, and there is a l.u.s. $\{\Omega, G, \varphi\}$ over U such that $(G_*)_{(x,\zeta)} \neq \{e_*\}$, for some $(x,\zeta) \in \Omega \times \mathbb{C}^n$. That is $\sigma_*(x,\zeta) = (x,\zeta)$ for some $\sigma \in G \setminus \{e\}$, hence $\sigma(x) = x$, i.e. $G_x \neq \{e\}$. It follows that $p \in \Sigma$, i.e. the singular locus of $T_{1,0}(X)$ projects on Σ. Statement 2 in Theorem 11.13 is proved.

It remains that we prove 3. Let $Z : X \to T_{1,0}(X)$ be a continuous map such that $\pi \circ Z = 1_X$. Let $f \in \mathcal{E}(X)$ and $p \in X$. Let $U \in \mathcal{H}$ such that $p \in U$ and let $\{\Omega, G, \varphi\} \in \mathcal{F}$ over U. Let $x \in \Omega$ such that $\varphi(x) = p$ and let us set

$$Z(f)_p := \sum_{j=1}^n \zeta^j \frac{\partial f_\Omega}{\partial z^j}(x).$$

Here $[\zeta] \in \mathbb{C}^n/G_x$ corresponds to $Z_p \in \pi^{-1}(p)$ under the bijection

$$\mathbb{C}^n/G_x \approx \pi^{-1}(p).$$

Step 3 *$Z(f)_p$ is well defined.*

If $[\xi] = [\zeta]$ then $\xi = g_\sigma(x)\zeta$ for some $\sigma \in G_x$ and then

$$\xi^j \frac{\partial f_\Omega}{\partial z^j}(x) = g_\sigma(x)^j_k \zeta^k \frac{\partial f_\Omega}{\partial z^j}(x) = \zeta^k \frac{\partial (f_\Omega \circ \sigma)}{\partial z^k}(x).$$

If another open neighborhood $U' \in \mathcal{H}$ of p is used, let $\{\Omega', G', \varphi'\}$ over U' and $x' \in \Omega'$ with $\varphi'(x') = p$. Then, consider $p \in V \subseteq U \cap U'$ and $\{D, H, \psi\}$ over V, and two injections $\lambda : D \to \Omega$, $\lambda' : D \to \Omega'$. Let $y \in D$ with $\psi(y) = p$. Let $[\zeta] \in \mathbb{C}^n/G_x$ and $[\zeta'] \in \mathbb{C}^n/G'_{x'}$ correspond to Z_p. If $[\xi] \in \mathbb{C}^n/H_y$ corresponds to Z_p then

$$\varphi_*(x,\zeta) = Z_p = \psi_*(y,\xi) = [\text{orb}_{H_*}(y,\xi)] =$$

$$= [\text{orb}_{G_*}(\lambda(y)\,,\, g_\lambda(y)\xi)] = \varphi_*(\lambda(y)\,,\, g_\lambda(y)\xi),$$

hence there is $\tau \in G$ such that

$$\tau_*(x,\zeta) = (\lambda(y)\,,\, g_\lambda(y)\xi),$$

i.e. $\tau(x) = \lambda(y)$ and $\zeta = g_{\tau^{-1}}(\tau(x))g_\lambda(y)\xi$. As $f_\Omega \circ \lambda = f_D$

$$\zeta^j \frac{\partial f_\Omega}{\partial z^j}(x) = g_{\tau^{-1}}(\tau(x))^j_k g_\lambda(y)^k_\ell \xi^\ell \frac{\partial f_\Omega}{\partial z^j}(x) = \frac{\partial (f_\Omega \circ \tau^{-1})}{\partial z^k}(\tau(x)) g_\lambda(y)^k_\ell \xi^\ell =$$

(as f_Ω is G-invariant and $\tau(x) = \lambda(y)$)

$$= \frac{\partial (f_\Omega \circ \lambda)}{\partial z^\ell}(y)\xi^\ell = \xi^\ell \frac{\partial f_D}{\partial z^\ell}(y).$$

The same argument holds for λ' hence

$$\zeta'^j \frac{\partial f_{\Omega'}}{\partial z^j}(x') = \zeta^j \frac{\partial f_\Omega}{\partial z^j}(x),$$

and Step 3 is proved.

Let $Z_p \in \pi^{-1}(p)$ correspond to $[e_j] \in \mathbb{C}^n/G_x$, with $\varphi(x) = p$. Then $Z(\overline{f})_p = 0$ yields $(\partial f_\Omega / \partial \overline{z}^j)(x) = 0$, i.e. $f \in \mathcal{O}(\Omega)$. Q.e.d.

Throughout, if Y is a complex manifold, $\mathcal{O}(Y)$ denotes the space of all holomorphic functions on Y. The last statement in Theorem 11.13 shows that the requirement $Z(\overline{f}) = 0$ for all sections Z in $T_{1,0}(X)$ is too restrictive for our purposes. In the sequel, we restrict ourselves to sections Z such that

$$Z_p \in T_{1,0}(X)_p = \{\varphi_*(x, \zeta) : \zeta \in (\mathbb{C}^n)_{G_x}\},$$

as mentioned earlier in this chapter. Locally, we are led to a new notion, termed V-holomorphic function.

DEFINITION 11.17. Let $\Omega \subseteq \mathbb{C}^n$ be a domain and let $G \subset \mathrm{Aut}(\Omega)$ be a finite group of biholomorphisms. A C^1 function $f : \Omega \to \mathbb{C}$ is called *V-holomorphic* if it is G-invariant and

$$\sum_{j=1}^{n} \overline{\zeta}^j \frac{\partial f}{\partial \overline{z}^j}(x) = 0 \tag{11.12}$$

for any $x \in \Omega$ and any $\zeta \in (\mathbb{C}^n)_{G_x}$. $\square$

Let $\mathcal{O}_V(\Omega)$ be the space of all V-holomorphic functions in Ω. Let $\mathcal{O}_G(\Omega)$ consist of all G-invariant functions $f \in \mathcal{O}(\Omega)$. Then

$$\mathcal{O}_G(\Omega) \subseteq \mathcal{O}_V(\Omega) \subseteq \mathcal{O}_G(\Omega \setminus S).$$

Note that the requirement (11.12) is empty at the points of

$$C := \{x \in \Omega : (\mathbb{C}^n)_{G_x} = (0)\} \subseteq S.$$

When $n = 1$, $\mathcal{O}_V(\Omega) \subseteq \mathcal{O}_G(\Omega \setminus C)$.

The following result describes the local structure of S and the behavior of V-holomorphic functions at the points of $S \setminus C$.

THEOREM 11.18. (J. Masamune et al., [**86**])
For any $x \in S$ there is a neighborhood D of x in Ω so that 1) *$D \cap S$ is a finite union of complex submanifolds of Ω of dimension $< n$.* 2) *For any $y \in D$, G_y is a subgroup of G_x.* 3) *If $x \in S \setminus C$ there is a complex submanifold $F_x \subset D$ passing through x so that* a) *for each G-invariant function $f : \Omega \to \mathbb{C}$, f satisfies* (11.12) *at x if and only if the trace of f on F_x is holomorphic at x. Moreover* b) *$F_x \subset \Omega \setminus C$ and if $f \in \mathcal{O}_V(\Omega)$ then $f|_{F_x} \in \mathcal{O}(F_x)$.*

Proof. Let $x \in S$ and let us set

$$w^j := \frac{1}{|G_x|} \sum_{\sigma \in G_x} g_{\sigma^{-1}}(x)^j_k (z^k \circ \sigma)$$

(here for a set A, $|A|$ denotes its cardinality). Then $(\partial w^j / \partial z^k)(x) = \delta^j_k$ hence there is an open neighborhood V of x in Ω such that

$$\Phi := (w^1, \cdots, w^n) : V \to \mathbb{C}^n$$

is a biholomorphism on its image. Let $\sigma \in G \setminus G_x$. Then $\sigma(x) \neq x$ hence there is an open neighborhood Ω_σ of x in V such that $\sigma(\Omega_\sigma) \cap \Omega_\sigma = \emptyset$. Set

$$D_0 := \bigcap_{\sigma \in G \setminus G_x} \Omega_\sigma , \quad D := \bigcap_{\sigma \in G_x} \sigma(D_0).$$

As G is finite D_0, and then D, are open. What we just built is an open neighborhood D of x in V such that i) $\sigma(D) \subseteq D$ for any $\sigma \in G_x$ and ii) $\sigma(D) \cap D = \emptyset$ for any $\sigma \in G \setminus G_x$. The first statement in Theorem 11.18 is a complex analogue of Prop. 1.1 in [**120**], p. 251-252. For each $\tau \in G_x$ let us set

$$F_\tau = \{y \in D : \tau(y) = y\}.$$

Note that $w^j \circ \tau = g_\tau(x)^j_k \circ w^k$. Consequently

$$\Phi(F_\tau) = \Phi(D) \cap \mathrm{Ker}[g_\tau(x) - I_n],$$

hence F_τ is a complex submanifold of D, of complex dimension $< n$. Next $S \cap D = Y_x$, where

$$Y_x := \bigcup_{\tau \in G_x \setminus \{e\}} F_\tau .$$

To prove the third statement note that

LEMMA 11.19. *One has* $\overline{\zeta}^j (\partial/\partial \overline{z}^j)_x \in T_x(F_\tau) \otimes_{\mathbb{R}} \mathbb{C}$ *if and only if* $\zeta \in \mathrm{Ker}[g_\tau(x) - I_n]$.

Indeed if $\rho^j_\sigma(z) := g_\sigma(x)^j_k w^k - w^j$, $\sigma \in G_x$, then

$$\left(\zeta^k \left. \frac{\partial}{\partial \overline{z}^k} \right|_x \right) (\rho^j_\sigma) = \zeta^k \left[g_\sigma(x)^j_\ell - \delta^j_\ell \right] \frac{\partial w^\ell}{\partial z^k}(x) = \zeta^k g_\sigma(x)^j_k - \zeta^j .$$

Let us set

$$F_x := \bigcap_{\tau \in G_x \setminus \{e\}} F_\tau .$$

If $x \in S \setminus C$ then F_x is a complex manifold of dimension $\dim_{\mathbb{C}}(\mathbb{C}^n)_{G_x}$. Let us prove b). To this end, let $y \in F_x$ and $D' \subset V'$ as in the first part of the proof (got by replacing x by y). Then

$$F'_\sigma \supseteq D' \cap F_x \ni y$$

for any $\sigma \in G_y \setminus \{e\}$ hence (by a dimension argument)

$$T_{1,0}(F'_y)_y = T_{1,0}(F_x)_y \approx (\mathbb{C}^n)_{G_x} \neq (0). \tag{11.13}$$

Thus

$$(\mathbb{C}^n)_{G_y} \approx T_{1,0}(F'_y)_y \neq (0),$$

a fact which yields $y \in \Omega \setminus C$, i.e. $F_x \subset \Omega \setminus C$. Finally, let $f \in \mathcal{O}_V(\Omega)$. Then $f|_{F'_y}$ is holomorphic in y hence (by (11.13)) $f|_{F_x}$ is holomorphic in y. Q.e.d.

DEFINITION 11.20. If $(X, \mathcal{F})$ is a complex orbifold, a function $f \in C^1(X)$ (i.e. a continuous function $f : X \to \mathbb{C}$ such that $f_\Omega \in C^1(\Omega)$ for each l.u.s. $\{\Omega, G, \varphi\} \in \mathcal{F}$) is called *V-holomorphic* if each f_Ω is V-holomorphic in Ω. □

In the sequel, we shall study traces of such functions on smooth real hypersurfaces.

11.3. Real hypersurfaces

Let $D \subset \mathbb{C}^n$ be a bounded pseudoconvex domain with real analytic boundary ∂D and $H \subset \mathrm{Aut}(D)$ a finite (hence compact) group of automorphisms of D. By a result of B. Coupet & A. Sukhov, [**73**]

THEOREM 11.21. *There is a domain Ω such that $\overline{D} \subset \Omega$ and each $\tau \in H$ extends holomorphically on Ω as an automorphism of Ω.*

Let $G_{\partial D}$ consist of all $\tilde{\tau}|_{\partial D}$ for $\tau \in H$ and some holomorphic extension $\tilde{\tau} \in \mathrm{Aut}(\Omega)$ of τ. By the identity principle for holomorphic functions

PROPOSITION 11.22. *$G_{\partial D}$ is a well defined finite group of CR automorphisms of ∂D.*

In general, let $\Omega \subseteq \mathbb{C}^n$ be a domain, $G \subset \mathrm{Aut}(D)$ a finite group of biholomorphims, and $M \subset \Omega$ an embedded real hypersurface such that $\sigma(M) = M$ for each $\sigma \in G$. Let us set $G_M := \{\sigma|_M : \sigma \in G\}$ and $S_M := \{x \in M : (G_M)_x \neq \{1_M\}\}$. Then $S_M = M \cap S$. For any $x \in M$ there is a neighborhood U of x in $\mathbb{C}^n$ and a function $\rho \in C^\infty(U)$ such that $M \cap U = \{z \in U : \rho(z) = 0\}$ and $d\rho(z) \neq 0$ for any $z \in M$. The Cauchy-Riemann equations in $\mathbb{C}^n$ induce on M the tangential Cauchy-Riemann equations

$$\overline{L}_\alpha u(z) \equiv \sum_{j=1}^{n} a^j_\alpha(z) \frac{\partial u}{\partial \overline{z}^j} = 0, \quad 1 \leq \alpha \leq n-1, \tag{11.14}$$

$G \subset \mathrm{Aut}(\Omega)$ yields $G_M \subset \mathrm{Aut}_{CR}(M)$ hence

$$(d_x \tau) L_{\alpha,x} = \sum_{\beta=1}^{n-1} \tau^\beta_\alpha(x) L_{\beta,\tau(x)}, \quad x \in V,$$

for each $\tau \in G_M$ and some (unique) system of C^∞ functions $\tau^\beta_\alpha : V \to \mathbb{C}$. For each $\tau \in G_M$ let $g_{M,\tau} : V \to \mathrm{GL}(n-1, \mathbb{C})$ be given by $g_{M,\tau}(x)\zeta = \tau^\alpha_\beta(x)\zeta^\beta e_\alpha$ for any $\zeta \in \mathbb{C}^{n-1}$. Let us set

$$(\mathbb{C}^{n-1})_{(G_M)_x} = \mathrm{Ker}[g_{M,\tau}(x) - I_{n-1}]$$

and $C_M = \{x \in M : (\mathbb{C}^{n-1})_{(G_M)_x} = (0)\} \subseteq S_M$. We need the following

LEMMA 11.23. *The trace $u = f|_M$ of any V-holomorphic function $f \in \mathcal{O}_V(\Omega)$ satisfies*

$$\sum_{\alpha=1}^{n-1} \overline{\xi}^\alpha L_{\overline{\alpha},x} u = 0 \tag{11.15}$$

for any $x \in V$ and any $\xi \in (\mathbb{C}^{n-1})_{(G_M)_x}$. In particular u is a CR function on $M \setminus S_M$ (and if $n = 2$ then u is CR on $M \setminus C_M$).

Proof. Let $\zeta \in (\mathbb{C}^{n-1})_{(G_M)_x}$, $x \in V$, and let us set $\zeta^j = a^j_\alpha(x)\xi^\alpha$. Then

$$a^j_\alpha(x) g_\sigma(x)^k_j = \tau^\beta_\alpha(x) a^k_\beta(x)$$

yields $\zeta \in (\mathbb{C}^n)_{G_x}$ hence

$$0 = \overline{\zeta}^j \frac{\partial f}{\partial \overline{z}^j}(x) = \overline{\xi}^\alpha L_{\overline{\alpha},x} u.$$

Q.e.d.

In view of the result in [**242**], it is an open problem whether the real analytic solutions to (11.15) extend to V-holomorphic functions on a neighborhood of M in Ω (provided $M \in C^\omega$). We leave this as an open problem.

THEOREM 11.24. (J. Masamune et al., [**86**])
For any $x \in S_M$ there is an open neighborhood D of x in Ω such that $S_M \cap D$ is a finite union of CR manifolds of CR dimension $< n-1$. For any $y \in V := M \cap D$, $(G_M)_y$ is a subgroup of $(G_M)_x$. If $x \in S_M \setminus C_M$ there is a CR manifold $F_{M,x}$ such that a C^1 function $u : V \to \mathbb{C}$ satisfies (11.15) for any $\xi \in (\mathbb{C}^{n-1})_{(G_M)_x}$ if and only if the trace of u on $F_{M,x}$ is CR at x.

The proof of Theorem 11.24 is similar to that of Theorem 11.18 so we only emphasize on the main steps. As $x \in S_M \subseteq S$, let D be a neighborhood of x in Ω as in (the proof of) Theorem 11.18. By eventually shrinking D let (u^a) be local coordinates on $V = M \cap D$ and let us set

$$v^a = \frac{1}{|G_x|} \sum_{\tau \in (G_M)_x} h_{\tau^{-1}}(x)^a_b (u^b \circ \tau), \quad 1 \leq a \leq 2n-1,$$

where $h_\tau(x) = [(\partial(u^a \circ \tau)/\partial u^b)(x)]$. Then $(\partial v^a/\partial u^b)(x) = \delta^a_b$ hence $\phi = (v^1, \cdots, v^{2n-1})$ is a C^∞ diffeomorphism of (a perhaps smaller open neighborhood of x in) V onto its image. Given $\tau \in (G_M)_x \setminus \{1_M\}$ let us set

$$F_{M,\tau} = \{y \in V : \tau(y) = y\}.$$

Then

$$\phi(F_{M,\tau}) = \phi(V) \cap \operatorname{Ker}[h_\tau(x) - I_{2n-1}]$$

hence $F_{M,\tau}$ is a manifold (of dimension $\dim_{\mathbb{R}} \operatorname{Ker}[h_\tau(x) - I_{2n-1}] < 2n-1$ if $\tau \neq 1_M$) and

$$S_M \cap V = \bigcup_{\tau \in (G_M)_x \setminus \{1_M\}} F_{M,\tau}\,.$$

Note that $F_{M,\tau} = M \cap F_\sigma$ for any $\sigma \in G_x$ with $\sigma|_M = \tau$. Hence $F_{M,\tau}$ is a CR submanifold of (the complex manifold) F_σ. If $x \in S_M \setminus C_M \subseteq S \setminus C$ then let us set

$$F_{M,x} = \bigcap_{\tau \in (G_M)_x \setminus \{1_M\}} F_{M,\tau}\,.$$

Then $F_{M,x} = M \cap F_x$ hence $F_{M,x}$ is a CR submanifold of F_x. Let $T_{1,0}(F_{M,x})$ be the CR structure induced from (the complex structure of) F_x. The inclusion $F_{M,x} \subset M$ is a CR immersion (i.e. an immersion and a CR map) and $\overline{\zeta}^\alpha L_{\overline{\alpha},x} \in T_{1,0}(F_{M,x})_x$ if and only if $\zeta \in (\mathbb{C}^{n-1})_{(G_M)_x}$. Q.e.d.

11.4. CR orbifolds

Let $(B, \mathcal{A})$ be a $(2n+k)$-dimensional V-manifold, of class C^∞.

DEFINITION 11.25. A *CR structure* on B is a family

$$\{T_{1,0}(B) = \{T_{1,0}(\Omega) : \{\Omega, G, \varphi\} \in \mathcal{A}\}$$

where each $(\Omega, T_{1,0}(\Omega))$ is a CR manifold, of type (n,k), i.e. of CR dimension n and CR codimension k, and each injection $\lambda : \Omega \to \Omega'$ is a CR map. □

In particular $G \subset \operatorname{Aut}_{CR}(\Omega)$ for any l.u.s. $\{\Omega, G, \varphi\} \in \mathcal{A}$.

DEFINITION 11.26. A pair $(B, T_{1,0}(B))$ is a *CR orbifold* (of type (n,k)). □

When $k = 0$, B is a complex orbifold (of complex dimension n). We shall deal mainly with CR orbifolds of CR codimension $k = 1$.

Let $(B, \mathcal{A})$ be a N-dimensional V-manifold.

DEFINITION 11.27. A continuous map $\Psi : B \to M$ into a C^∞ manifold M is an *immersion* if, for any $\{\Omega, G, \varphi\} \in \mathcal{A}$, the map $\Psi_\Omega := \Psi \circ \varphi : \Omega \to M$ is a C^∞ immersion (i.e. rank $[d_x \Psi_\Omega] = N \leq \dim(M)$, $x \in \Omega$). □

Let us give an example of CR orbifold.

EXAMPLE 11.28. Let us assume that $N = 2n + 1$ and let $\Psi : B \to \mathbb{C}^{n+1}$ be an immersion. Let $T_{1,0}(\Omega)$ be the CR structure on Ω given by

$$(d_x \Psi_\Omega) T_{1,0}(\Omega)_x = T_{1,0}(\mathbb{C}^{n+1})_{\Psi(\varphi(x))} \cap [(d_x \Psi_\Omega) T_x(\Omega) \otimes_{\mathbb{R}} \mathbb{C}], \quad x \in \Omega. \tag{11.16}$$

Note that $\Psi_{\Omega'} \circ \lambda = \Psi_\Omega$, for any injection $\lambda : \Omega \to \Omega'$. As a consequence, it is easy to see that λ must be a CR map, hence B together with the family of CR structures (11.16) is a CR orbifold. □

Let $(B, \mathcal{A}, T_{1,0}(B))$ be a CR orbifold, of CR codimension 1.

DEFINITION 11.29. A family $\theta = \{\theta_\Omega : \{\Omega, G, \varphi\} \in \mathcal{A}\}$ is a *pseudohermitian structure* on B if each θ_Ω is a pseudohermitian structure on Ω and $\lambda^* \theta_{\Omega'} = a(\lambda) \theta_\Omega$ for any injection $\lambda : \Omega \to \Omega'$ and some constant $a(\lambda) \in \mathbb{R} \setminus \{0\}$, i.e. the injections are pseudohermitian maps. □

We shall need the following

LEMMA 11.30. *Let $(B, \mathcal{A}, T_{1,0}(B))$ be a CR orbifold and two pseudohermitian structures θ, $\hat{\theta}$ on B. If each injection $\lambda : \Omega \to \Omega'$ is isopseudohermitian, i.e. $a(\lambda) \equiv 1$, there is a unique C^∞ function $u : B \to \mathbb{R} \setminus \{0\}$ such that $\hat{\theta}_\Omega = u_\Omega \theta_\Omega$, for any l.u.s. $\{\Omega, G, \varphi\} \in \mathcal{A}$.*

Proof. Let $u_\Omega : \Omega \to \mathbb{R} \setminus \{0\}$ be a C^∞ function satisfying $\hat{\theta}_\Omega = u_\Omega \theta_\Omega$. Next let us consider an injection $\lambda : \Omega \to \Omega'$. The identities $\lambda^* \theta_{\Omega'} = \theta_\Omega$ and $\lambda^* \hat{\theta}_{\Omega'} = \hat{\theta}_\Omega$ lead to

$$u_{\Omega'} \circ \lambda = u_\Omega \, . \tag{11.17}$$

In particular u_Ω is G-invariant. Define $u : B \to \mathbb{R} \setminus \{0\}$ as follows. Let $p \in B$ and $U \in \mathcal{H}$ so that $p \in U$. Let $\{\Omega, G, \varphi\} \in \mathcal{A}$ be a l.u.s. of support U. Let $x \in \Omega$ such that $\varphi(x) = p$. Finally, let us set $u(p) := u_\Omega(x)$. One needs to check that the definition of $u(p)$ doesn't depend upon the various choices involved. Let $U' \in \mathcal{H}$ such that $p \in U'$. Then there is $V \in \mathcal{H}$ such that $p \in V \subseteq U \cap U'$. Let $\{\Omega', G', \varphi'\}$ over U' and $x' \in \Omega'$ such that $\varphi'(x') = p$. Let $\{D, H, \psi\}$ be a l.u.s. of support V and let us consider two injections $\lambda : D \to \Omega$ and $\lambda' : D \to \Omega'$. Let $y \in D$ such that $\psi(y) = p$. From $\varphi(x) = \psi(y) = \varphi(\lambda(y))$, there is $\sigma \in G$ such that

$$\lambda(y) = \sigma(x). \tag{11.18}$$

Similarly

$$\lambda'(y) = \sigma'(x'), \tag{11.19}$$

for some $\sigma' \in G'$. Finally, using (11.17)-(11.19), one may conduct the following calculation

$$u_{\Omega'}(x') = u_{\Omega'}((\sigma')^{-1} \lambda'(y)) = u_{\Omega'}(\lambda'(y)) =$$

$$= u_D(y) = u_\Omega(\lambda(y)) = u_\Omega(\sigma(x)) = u_\Omega(x).$$

Q.e.d..

DEFINITION 11.31. A *Riemannian orbifold* is a V-manifold B together with a family $g = \{g_\Omega : \{\Omega, G, \varphi\} \in \mathcal{A}\}$, where g_Ω is a Riemannian metric on Ω, such that each injection $\lambda : \Omega \to \Omega'$ is an isometry ($\lambda^* g_{\Omega'} = g_\Omega$). □

Before proceeding, it should be remarked that there is already a large (and growing) literature concerning the geometry of Riemannian orbifolds (cf. e.g. J.E. Borzellino, [**51**], J.E. Borzellino et al., [**52**]-[**54**]). We quote the following refined result

THEOREM 11.32. (E. Stanhope, [**225**]) *Let $\mathcal{S}$ be a collection of isospectral compact oriented n-dimensional Riemannian orbifolds that share the uniform lower bound $\kappa(n-1)$ on the Ricci curvature for some $\kappa \in \mathbb{R}$. Then there are only finitely many possible isotropy types, up to isomorphism, for points of an orbifold in $\mathcal{S}$.*

Theorem 11.32 together with its proof (a sketch is given below) best illustrates the state of the art in Riemannian geometry on compact oriented orbifolds. It may be shown (cf. E. Stanhope, *op. cit.*) that families of isospectral orbifolds which share a uniform lower bound on the Ricci curvature also share an upper bound on their diameters. Let $D > 0$ be an upper bound on the diameters of the orbifolds in $\mathcal{S}$. Again due to the isopspectrality of the orbifolds in $\mathcal{S}$ it follows (by Weyl's asymptotic formula for Riemannian orbifolds, cf. C. Farsi, [**102**], or Theorem 6.4 in [**225**], p. 366) that each orbifold $B \in \mathcal{S}$ has the same dimension $n = \dim(B)$ and the same volume $v = \mathrm{Vol}(B) > 0$. Let $B \in \mathcal{S}$ and $x \in B$. Let $V_\kappa^n(r)$ denote the volume of the ball of radius $r > 0$ in the simply connected n-dimensional space form of (constant) sectional curvature κ. Let $R > r \geq 0$. By the orbifold relative comparison theorem (cf. Proposition 5.1 in [**225**], p. 361)

$$\frac{\mathrm{Vol}(B(x,r))}{V_\kappa^n(r)} \geq \frac{\mathrm{Vol}(B(x,R))}{V_\kappa^n(R)}.$$

Let us replace R by D in the preceding inequality so that

$$\frac{\mathrm{Vol}(B(x,r))}{V_\kappa^n(r)} \geq \frac{\mathrm{Vol}(B)}{V_\kappa^n(D)} = \frac{v}{V_\kappa^n(D)}.$$

Let us take the limit as $r \to 0^+$ and apply again Proposition 5.1 in [**225**], p. 361, so that

$$\frac{1}{|G_x|} = \lim_{r \to 0^+} \frac{\mathrm{Vol}(B(x,r))}{V_\kappa^n(r)} \geq \frac{v}{V_\kappa^n(D)}.$$

Hence for any $x \in B \in \mathcal{S}$ the isotropy group G_x has order $\leq V_\kappa^n(D)/v$, a universal constant. Therefore G_x can have but one of only finitely many isomorphism types.

Let us go back to CR geometry on orbifolds, which needless to say appears to be far less developed than Riemannian geometry on orbifolds. Nevertheless the reader should see J. Song, [**224**] (cf. section 3 there, dealing with the $\overline{\partial}_b$-equation on an orbifold endowed with a CR structure). It should however be observed that the treatment in [**224**] is rather sketchy and the reader is invited to prove the relevant statements (subelliptic estimates leading to the Hodge decomposition theorem for $\overline{\partial}_b$ on an orbifold) by mere analogy to the classical work by G.B. Folland & J.J. Kohn, [**106**].

DEFINITION 11.33. A CR orbifold $(B, \mathcal{A}, T_{1,0}(B))$ is *strictly pseudoconvex* if each $(\Omega, T_{1,0}(\Omega))$ is a strictly pseudoconvex CR manifold. □

Let θ be a pseudohermitian structure on B. Then each θ_Ω is a contact 1-form on Ω. Let g_Ω be the Webster metric of (Ω, θ_Ω) and set $g := \{g_\Omega : \{\Omega, G, \varphi\} \in \mathcal{A}\}$. If each injection λ is isopseudohermitian then λ preserves the Webster metrics, hence (B, g) is a Riemannian orbifold. The following result is similar to Theorem 11.13.

THEOREM 11.34. (J. Masamune et al., [**86**])
For any CR orbifold $(B, \mathcal{A}, T_{1,0}(B))$, of type $(n, 1)$, there is a vector bundle $(E_{1,0}, \pi, B)$ such that for any $p \in B$, if $p \in U \in \mathcal{H}$ and $\{\Omega, G, \varphi\} \in \mathcal{A}$ is a l.u.s. over U then $\pi^{-1}(p) \approx \mathbb{C}^n / G_x$ for any $x \in \Omega$ with $\varphi(x) = p$. $B_{\rm reg}$ is a CR manifold (of type $(n, 1)$) and $E_{1,0}|_{B_{\rm reg}}$ is its CR structure. $T_{1,0}(B_{\rm reg})$ is contained in $(E_{1,0})_{reg}$, the regular part of $E_{1,0}$ as a V-manifold. The image $T_{1,0}(B)_p \subseteq \pi^{-1}(p)$ of $T_{1,0}(\Omega)_{G_x}$ via the map $T_{1,0}(\Omega) \approx \Omega \times \mathbb{C}^n \to E_{1,0}$ depends only on $p = \varphi(x)$. $T_{1,0}(B)_p$ is a $\mathbb{C}$-vector space of dimension $\dim_{\mathbb{C}}(\mathbb{C}^n)_{G_x}$. If Z is a section in $E_{1,0}$ and $f \in \mathcal{E}(B)$ there is a (naturally defined) function $Z(f) : B \to \mathbb{C}$. If $Z(\overline{f}) = 0$ for any Z then $f_\Omega = f \circ \varphi$ is a CR function on Ω, for any $\{\Omega, G, \varphi\} \in \mathcal{A}$, and conversely.

The bundle $E_{1,0}$ is recovered from the transition functions $g_\lambda(x) = [\lambda^\alpha_\beta(x)]$, where $(d_x\lambda)L_{\alpha,x} = \lambda^\beta_\alpha(x) L'_{\beta,\lambda(x)}$, $x \in \Omega$ (we assume w.l.o.g. that a frame $\{L_\alpha\}$ of $T_{1,0}(\Omega)$, defined on the whole of Ω, is prescribed on each Ω). We omit the details.
Let B be a V-manifold.

DEFINITION 11.35. A linear map $D : \mathcal{E}(B) \to \mathcal{E}(B)$ is a *differential operator* (*of order* k) if for any l.u.s. $\{\Omega, G, \varphi\} \in \mathcal{A}$ there is a differential operator D_Ω of order k on Ω such that $(Du)_\Omega = D_\Omega u_\Omega$ for any $u \in \mathcal{E}(B)$. We say D is *elliptic* (respectively *subelliptic* (*of order* ϵ)) if D_Ω is elliptic (respectively subelliptic of order ϵ, (cf. [**105**], p. 373)) for each l.u.s. $\{\Omega, G, \varphi\}$. □

Let $(B, T_{1,0}(B))$ be a nondegenerate CR orbifold, $\theta = \{\theta_\Omega\}$ a fixed pseudohermitian structure on B, and $\Box_\Omega$ the Kohn-Rossi laplacian of (Ω, θ_Ω). If each injection is isopseudohermitian we may build a differential operator $\Box_B : \mathcal{E}(B) \to \mathcal{E}(B)$ by setting

$$(\Box_B u)_\Omega = \Box_\Omega u_\Omega$$

for any $u \in \mathcal{E}(B)$. Then $\Box_B u$ is a well defined element of $\mathcal{E}(B)$ if the functions $f_\Omega = \Box_\Omega u_\Omega$ satisfy $f_{\Omega'} \circ \lambda = f_\Omega$ for any injection $\lambda : \Omega \to \Omega'$. This may be seen as follows. By applying (11.5) we get $\Box^\lambda_\Omega = \Box_{\lambda(\Omega)}$ or

$$(\Box_\Omega(v \circ \lambda)) \circ \lambda^{-1} = \Box_{\lambda(\Omega)} v,$$

for any $v \in C^\infty(\lambda(\Omega))$. In particular let us consider the functions

$$v = u_{\Omega'}|_{\lambda(\Omega)} \in C^\infty(\lambda(\Omega)).$$

Then

$$\Box_\Omega(u_\Omega|_{\lambda(\Omega)}) \circ \lambda) \circ \lambda^{-1} = \Box_{\lambda(\Omega)} \left(u_{\Omega'}|_{\lambda(\Omega)} \right)$$

may be written as

$$\Box_\Omega u_\Omega = (\Box_{\Omega'} u_{\Omega'}) \circ \lambda.$$

Q.e.d.
Let T_Ω be the characteristic direction of (Ω, θ_Ω). We define a differential operator $T : \mathcal{E}(B) \to \mathcal{E}(B)$ by setting $(Tu)_\Omega = T_\Omega u_\Omega$ for any $u \in \mathcal{E}(\Omega)$. Again, the

functions $T_\Omega u_\Omega$ give rise to a well defined element Tu of $\mathcal{E}(B)$ provided that each injection λ is isopseudohermitian. Indeed if this is the case then $(d_x\lambda)T_{\Omega,x} = T_{\Omega',\lambda(x)}$ for any $x \in \Omega$, and one may perform the calculation

$$T_{\Omega',\lambda(x)}(u_{\Omega'}) = [(d_x\lambda)T_{\Omega,x}](u_{\Omega'}) = T_{\Omega,x}(u_{\Omega'} \circ \lambda) = T_{\Omega,x}(u_\Omega).$$

Q.e.d.

Finally, let $(B, T_{1,0}(B))$ be a strictly pseudoconvex CR orbifold and $\theta = \{\theta_\Omega\}$ a pseudohermitian structure on B such that each Levi form L_{θ_Ω} is positive definite and each injection is isopseudohermitian. Consider the second order differential operator $\Delta_B : \mathcal{E}(B) \to \mathcal{E}(B)$ given by $\Delta_B u = \Box_B u - inT(u)$ for any $u \in B$. Then

PROPOSITION 11.36. *Δ_B is a subelliptic operator of order $1/2$ on B.*

The problem of developing a spectral theory for Δ_B on a CR orbifold is open. E. Barletta, [**18**], has given a CR version (for the spectrum of a sublaplacian on a strictly pseudoconvex CR manifold) of the well known A. Lichnerowicz estimate (cf. [**39**]) from below on the first nonzero eigenvalue of the Laplace-Beltrami operator on a Riemannian manifold. It should be observed that the lower bound on the Ricci curvature in [**39**] is replaced (cf. [**18**]) by a similar assumption on the pseudohermitian Ricci tensor associated to a fixed contact form in the spirit of A. Greenleaf, [**127**]. In the light of a result by K. Richardson, [**206**] (extending the Lichnerowicz-Obata theorem to Riemannian foliations) one expects a similar lower bound on the first nonzero eigenvalue of Δ_B.

J. Girbau & M. Nicolau have developed (cf. [**120**]) a pseudo-differential calculus on V-manifolds (inverting a given elliptic differential operator up to infinitely smoothing operators). The same problem for subelliptic operators on V-manifolds, e.g. for Δ_B on a CR orbifold, is not solved (presumably, one needs to adapt the methods in [**187**]). Also see [**107**], p. 493-498, for a parametrix and the regularity of $\Box_M$ for an ordinary strictly pseudoconvex CR manifold M. The problem of building a parametrix for $\Box_B$ on a strictly pseudoconvex CR orbifold B is open. In the next section we solve the local problem.

11.5. A parametrix for $\Box_\Omega$

Let $\Omega \subset \mathbb{R}^{n+1}$ be a domain and $T_{1,0}(\Omega)$ a G-invariant strictly pseudoconvex CR structure on Ω, for some finite group of CR automorphisms $G \subset \mathrm{Aut}_{CR}(\Omega)$. Let θ be a pseudohermitian structure on Ω such that the corresponding Levi form L_θ is positive definite and $\sigma^*\theta = a(\sigma)\theta$, for any $\sigma \in G$ and some $a(\sigma) \in (0, +\infty)$. Let $\{T_\alpha\}$ be an orthonormal (i.e. $L_\theta(T_\alpha, T_{\overline{\beta}}) = \delta_{\alpha\beta}$) frame of $T_{1,0}(\Omega)$ defined everywhere in Ω. Let $(z,t) = \Theta_x : V_x \to \mathbb{H}_n$ be the pseudohermitian normal coordinates at $x \in \Omega$ determined by $\{T_\alpha\}$ and let us set

$$D := \bigcup_{x\in\Omega} \{x\} \times V_x,$$

a neighborhood of the diagonal in $\Omega \times \Omega$. Next, we set $\Theta(x,y) := \Theta_x(y)$ and $\rho(x,y) := |\Theta(x,y)|$, for any $(x,y) \in D$. Here $|(z,t)| = (\|z\|^4 + t^2)^{1/4}$ is the Heisenberg norm of $(z,t) \in \mathbb{H}_n$.

We recall the following (cf. [**108**])

DEFINITION 11.37. A function $K(x,y)$ on $\Omega \times \Omega$ is a *kernel of type* λ ($\lambda > 0$) if for any $m \in \mathbb{Z}$, $m > 0$

$$K(x,y) = \sum_{i=1}^{N} a_i(x) K_i(x,y) b_i(y) + E_m(x,y) \tag{11.20}$$

where 1) $E_m \in C_0^m(\Omega \times \Omega)$, 2) $a_i, b_i \in C_0^\infty(\Omega)$, $1 \le i \le N$, and 3) K_i is C^∞ away from the diagonal and is supported in $\{(x,y) \in D : \rho(x,y) \le 1\}$ and $K_i(x,y) = k_i(\Theta(y,x))$ for $\rho(x,y)$ sufficiently small, where k_i is homogeneous of degree $\lambda_i := \lambda - 2n - 2 + \mu_i$, i.e.

$$k_i(D_r(z,t)) = r^{\lambda_i} k_i(z,t), \quad r > 0, \quad (z,t) \in \mathbb{H}_n\,,$$

for some $\mu_i \ge 0$. □

Here $D_r(z,t) = (rz, r^2 t)$ is the (parabolic) dilation of factor $r > 0$. Next

DEFINITION 11.38.

$$(Af)(x) = \int_\Omega K(x,y) f(y) dy$$

is an *operator of type* λ ($\lambda > 0$) if $K(x,y)$ is a kernel of type λ. Here dy is short for $\omega(y) := (\theta \wedge (d\theta)^n)(y)$. □

Let us set $X_\alpha := T_\alpha + T_{\overline{\alpha}}$ and $Y_\alpha := i(T_{\overline{\alpha}} - T_\alpha)$ and $\{X_j : 1 \le j \le 2n\} := \{X_\alpha, Y_\alpha\}$, where $X_{\alpha+n} = Y_\alpha$. Also, let us set

$$\mathcal{B}_k = \{X_{j_1} \cdots X_{j_\ell} : 1 \le j_s \le 2n, \quad 1 \le s \le \ell, \quad 1 \le \ell \le k\}$$

and let $\mathcal{A}_k$ be the span over $\mathbb{C}$ of $\mathcal{B}_k \cup \{I\}$, where I is the identity.

DEFINITION 11.39. The *Folland-Stein spaces* are $S_k^p(\Omega) = \{f \in L^p(\Omega) : Lf \in L^p(\Omega),\ \forall\, L \in \mathcal{A}_k\}$ where Lf is intended in distributional sense. □

The Folland-Stein spaces are Banach spaces under the norms

$$\|f\|_{p,k} = \|f\|_p + \sum_{L \in \mathcal{B}_k} \|Lf\|_p\,.$$

An important feature of the operators of type $\lambda = m \in \{1, 2, \cdots\}$ is that they are bounded operators from $S_k^p(\Omega)$ to $S_{k+m}^p(\Omega)$ (and in this sense smoothing) for $k \in \{0, 1, 2, \cdots\}$ and $1 < p < \infty$ (cf. Theorem 15.19 in [**107**], p. 491). We shall prove the following result

THEOREM 11.40. (J. Masamune et al., [**86**])
Let W_0 be a G-invariant compact subset of Ω. For each $0 < q < n$ there is an operator $A_{q,\Omega} : \Gamma_0^\infty(\Lambda^{0,q}(\Omega)) \to \Gamma_0^\infty(\Lambda^{0,q}(\Omega))$, of type 2, *such that* 1) $A_{q,\Omega} \circ \Box_\Omega - I$ *and* $\Box_\Omega \circ A_{q,\Omega} - I$ *are operators of type* 1 *on the G-invariant C^∞ forms of support contained in W_0, and* 2) $A_{q,\Omega}$ *maps G-invariant forms in G-invariant forms.*

A $(0,q)$-form φ on Ω may be written locally

$$\varphi = \varphi_{\overline{I}} \theta^{\overline{I}}$$

where $I = (\alpha_1, \cdots, \alpha_n)$ is a multi-index and $\theta^{\overline{I}} = \theta^{\overline{\alpha}_1} \wedge \cdots \wedge \theta^{\overline{\alpha}_n}$. Since

$$(\sigma^* \theta^\alpha)_x = g_\sigma(x)^\alpha_\beta \, \theta^\beta_x, \quad x \in \Omega,$$

if φ is G-invariant (i.e. $\sigma^*\varphi = \varphi$ for any $\sigma \in G$) then

$$\varphi_{\overline{I}}(x) = g_\sigma(x)^{\overline{J}}_{\overline{I}} \varphi_{\overline{J}}(\sigma(x)), \ \ x \in \Omega, \ \sigma \in G,$$

$$g_\sigma(x)^{\overline{J}}_{\overline{I}} := g_\sigma(x)^{\overline{\beta}_1}_{\overline{\alpha}_1} \cdots g_\sigma(x)^{\overline{\beta}_n}_{\overline{\alpha}_n}, \ \ J = (\beta_1, \cdots, \beta_n).$$

By Proposition 16.5 in [**107**], p. 496, for any $1 \leq q \leq n-1$ we may build an operator A_q of type 2 such that $I - \square_\Omega A_q$ and $I - A_q \square_\Omega$ are operators of type 1 on forms $\varphi \in \Gamma_0^\infty(\Lambda^{0,q}(\Omega))$ of support $\subset W_0$. Assuming this is done, let us set

$$A_{q,\sigma}\varphi := \sigma^* A_q(\sigma^{-1})^*\varphi, \quad A_{q,\Omega} := \frac{1}{|G|}\sum_{\sigma \in G} A_{q,\sigma}.$$

From now on, for the sake of simplicity, we drop the index q. If φ is G-invariant then

$$\tau^* A_\sigma \varphi = (\sigma\tau)^* A(\sigma^{-1})^*\varphi = (\sigma\tau)^* A((\sigma\tau)^{-1})^*\varphi,$$

i.e.

$$\tau^*(A_\sigma\varphi) = A_{\sigma\tau}\varphi.$$

Therefore

$$\tau^* A_\Omega \varphi = \frac{1}{|G|}\sum_{\sigma\in G} \tau^* A_\sigma\varphi = \frac{1}{|G|}\sum_{\sigma\in G} A_{\sigma\tau}\varphi = A_\Omega\varphi,$$

i.e. A_Ω maps G-invariant forms in G-invariant forms.

For each $\xi \in \Omega$ let $\delta(\xi) > 0$ be fixed so that $\Psi_\xi : B(0, \delta(\xi)) \subset T_\xi(\Omega) \to \Omega$ is well defined and a diffeomorphism on its image $V_\xi = \Psi_\xi(B(0,\delta(\xi))$. Next, let us fix a number

$$0 < \delta_G(\xi) \leq \min\{\frac{\delta(\sigma(\xi))}{\sqrt{a(\sigma)^2 + a(\sigma)}} : \sigma \in G\} \cup \{\delta(\xi)\}$$

and let us set

$$V_G(\xi) := \Psi_\xi(B(0, \delta_G(\xi))) \subseteq V_\xi \subset \Omega.$$

LEMMA 11.41. $\sigma\left[V_G(\xi)\right] \subseteq V_{\sigma(\xi)}$.

Proof. Let $\eta \in V_G(\xi) \subset V_\xi$ i.e. there is $W + cT_\xi \in B(0, \delta_G(\xi))$ such that $W \in H(\Omega)_\xi$ and $\eta = \Psi_\xi(W + cT_\xi) = \gamma_{W,c}(1)$. Thus (by Lemma 11.9) $\sigma(\eta) = (\sigma \circ \gamma_{W,c})(1) = \gamma_{W_\sigma, a(\sigma)c}(1)$. On the other hand

$$\|W_\sigma + a(\sigma)cT_{\sigma(\xi)}\|^2 = \|W_\sigma\|^2 + a(\sigma)^2c^2 =$$

$$= a(\sigma)\|W\|^2 + a(\sigma)^2c^2 < [a(\sigma) + a(\sigma)^2]\delta_G(\xi)^2 \leq \delta(\sigma(\xi))^2,$$

hence $\gamma_{W_\sigma, a(\sigma)c}(1) \in V_{\sigma(\xi)}$. Q.e.d.

Let us set

$$D_G := \bigcup_{\xi \in \Omega}\{\xi\} \times V_G(\xi).$$

Let us go back to the construction of A. Let us consider

$$A\varphi(\xi) = \left(\int_\Omega K(\xi,\eta)\varphi_{\overline{J}}(\eta)d\eta\right)\theta^{\overline{J}}_\xi,$$

where K is the kernel of type 2

$$K(\xi,\eta) = \psi(\xi,\eta)\Phi_{n-2q}(\Theta(\eta,\xi)).$$

Here $\psi(\xi,\eta)$ is a C_0^∞ function on $\Omega \times \Omega$, supported in

$$\{(\xi,\eta) \in D_G : \rho(\xi,\eta) \leq r\},$$

where

$$r := \min\{a(\sigma)^{1/2} : \sigma \in G\} \cup \{1\},$$

and such that $\psi(\xi,\eta) = \psi(\eta,\xi)$ and $\psi(\xi,\eta) = 1$ in a neighborhood $\mathcal{N}$ of the diagonal Δ of $W_0 \times W_0$ ($\Delta \subset \mathcal{N} \subseteq \{(\xi,\eta) \in D : \rho(\xi,\eta) < r\}$). Also Φ_α is the fundamental solution ($\mathcal{S}_\alpha \Phi_\alpha = \delta$) to

$$\mathcal{S}_\alpha = -\sum_{j=1}^{n} L_j L_{\bar{j}} + i(\alpha - n)\frac{\partial}{\partial t}, \tag{11.21}$$

(the Folland-Stein operators) where

$$L_j := \frac{\partial}{\partial z^j} + i\bar{z}^j \frac{\partial}{\partial t}$$

(the Lewy operators) i.e.

$$\Phi_\alpha = b_\alpha \left(\|z\|^2 - it\right)^{-\frac{n+\alpha}{2}} \left(\|z\|^2 + it\right)^{-\frac{n-\alpha}{2}}, \tag{11.22}$$

for any $\alpha \in \mathbb{C} \setminus \{\pm n, \pm(n+2), \pm(n+4), \cdots\}$, where

$$b_\alpha = \frac{\Gamma\left(\frac{n+\alpha}{2}\right)\Gamma\left(\frac{n-\alpha}{2}\right)}{2^{2-2n}\pi^{n+1}}.$$

Then

$$A_\sigma \varphi(\xi) = \left(\int K(\sigma(\xi),\eta)\left((\sigma^{-1})^*\varphi\right)_{\bar{I}}(\eta)d\eta\right)\theta^{\bar{I}}_{\sigma(\xi)} \circ (d_\xi \sigma). \tag{11.23}$$

By $\sigma^*\omega = a(\sigma)^{2n+1}\omega$ and a change of coordinates $\eta' = \sigma(\eta)$ in (11.23) we get

$$A_\sigma \varphi(\xi) = a(\sigma)^{2n+1}\left(\int g_\sigma(\xi)^{\bar{I}}_{\bar{J}} K(\sigma(\xi),\sigma(\eta)) g_{\sigma^{-1}}(\sigma(\eta))^{\bar{L}}_{\bar{I}} \varphi_{\bar{L}}(\eta)d\eta\right)\theta^{\bar{J}}_{\xi}.$$

LEMMA 11.42. *For any* $(\xi,\eta) \in D_G$

$$\Theta(\sigma(\xi),\sigma(\eta)) = (g_\sigma(\xi)^\beta_\alpha z^\alpha(\eta)e_\beta\,,\, a(\sigma)t(\eta)),$$

where $(z,t) = \Theta_\xi = \lambda_\xi \circ \Psi_\xi^{-1}$ *are the pseudohermitian normal coordinates centered at* ξ.

Proof. As $(\xi,\eta) \in D_G$ we have $\eta \in V_G(\xi)$ hence (by Lemma 11.41) $\sigma(\eta) \in \sigma\left[V_G(\xi)\right] \subseteq V_{\sigma(\xi)}$ and then

$$\Theta(\sigma(\xi),\sigma(\eta)) = \Theta_{\sigma(\xi)}(\sigma(\eta)) = \lambda_{\sigma(\xi)} \circ \Psi^{-1}_{\sigma(\xi)}(\sigma(\eta))$$

makes sense. As $\eta \in V_G(\xi) \subseteq V_\xi$, let us set $W := z^\alpha(\eta)T_{\alpha,\eta} + z^{\overline{\alpha}}(\eta)T_{\overline{\alpha},\eta}$ and $c := t(\eta)$. Then

$$\begin{aligned}\Psi_{\sigma(\xi)}(W_\sigma + c\,a(\sigma)T_{\sigma(\eta)}) = \gamma_{W_\sigma\,,\,c\,a(\sigma)}(1) = \quad &\text{(by Lemma 11.9)} \\ = \sigma(\gamma_{W,c}(1)) = \sigma(\Psi_\xi(W + cT_\eta)) = \sigma(\eta),&\end{aligned}$$

hence

$$\Theta(\sigma(\xi),\sigma(\eta)) = \lambda_{\sigma(\xi)}(W_\sigma + c\,a(\sigma)T_{\sigma(\eta)}).$$

Q.e.d.

For any $\sigma \in G$, $\sigma^* L_\theta = a(\sigma)L_\theta$ hence

$$\sum_\mu g_\sigma(\eta)^\mu_\alpha g_\sigma(\eta)^{\overline{\mu}}_{\overline{\beta}} = a(\sigma)\delta_{\alpha\beta},$$

i.e. $a(\sigma)^{-1/2} g_\sigma(\eta) \in \mathrm{U}(n)$. Consequently $\|g_\sigma(\eta)z\|^2 = a(\sigma)\|z\|^2$ and (by (11.22) and Lemma 11.42)

$$\Phi_{n-2q}(\Theta(\sigma(\eta), \sigma(\xi))) = a(\sigma)^{-n}\Phi_{n-2q}(\Theta(\eta, \xi)),$$

and we obtain

$$a(\sigma)^{-n-1} A_\sigma \varphi(\xi) =$$

$$= \left(\int g_\sigma(\xi)^{\overline{I}}_{\overline{J}} \psi_\sigma(\xi,\eta)\Phi_{n-2q}(\Theta(\eta,\xi)) g_{\sigma^{-1}}(\sigma(\eta))^{\overline{K}}_{\overline{I}} \varphi_{\overline{K}}(\eta) d\eta \right) \theta^{\overline{J}}_\xi ,$$

where $\psi_\sigma(\xi,\eta) := \psi(\sigma(\xi), \sigma(\eta))$. Note that $\psi_\sigma \in C^\infty_0$ and $\psi_\sigma(\xi,\eta) = \psi_\sigma(\eta,\xi)$. Let $\sigma^2 := \sigma \times \sigma$ (direct product). Let us set

$$\mathcal{N}_G := \bigcap_{\sigma \in G} \sigma^2(\mathcal{N}) \subset \mathcal{N}.$$

As W_0 is G-invariant $\Delta = \sigma^2(\Delta) \subset \sigma^2(\mathcal{N})$ for any $\sigma \in G$, hence $\mathcal{N}_G$ is an open neighborhood of Δ. Also $\psi(\xi,\eta) = 1$ on $\mathcal{N}$ yields $\psi_\sigma(\xi,\eta) = 1$ on $\mathcal{N}_G$.

Let $(\xi,\eta) \in D_G$. Then (by Lemma 11.42)

$$|\Theta(\sigma(\xi), \sigma(\eta))| = |(g_\sigma(\xi)z(\eta) \, , \, a(\sigma)t(\eta))| =$$

$$= \left(\|g_\sigma(\xi)z(\eta)\|^4 + a(\sigma)^2 t(\eta)^2\right)^{1/4} = a(\sigma)^{1/2}|(z(\eta), t(\eta))| =$$

$$= a(\sigma)^{1/2}|\Theta(\xi,\eta)|,$$

that is

$$\rho(\sigma(\xi), \sigma(\eta)) = a(\sigma)^{1/2}\rho(\xi,\eta). \tag{11.24}$$

Let Γ and Γ_σ be respectively the supports of ψ and ψ_σ. Then $\sigma^2(\Gamma_\sigma) \subseteq \Gamma \subset \{(\xi,\eta) \in D_G : \rho(\xi,\eta) \leq r\}$. Also (by Lemma 11.41) $\sigma^{-1}(D_G) \subseteq D$. Thus (by (11.24)) $\Gamma_\sigma \subset \{(\xi,\eta) \in D : \rho(\xi,\eta) \leq 1\}$. Then (as in [**107**], p. 494) we may conclude that

$$K_\sigma(\xi,\eta) = \psi_\sigma(\xi,\eta)\Phi_{n-2q}(\Theta(\eta,\xi))$$

is a kernel of type 2. In general, if $K(\xi,\eta)$ is a kernel of type λ then

$$K^{\overline{I}}_{\overline{J}}(\xi,\eta) := g_\sigma(\xi)^{\overline{L}}_{\overline{J}} K(\xi,\eta) g_{\sigma^{-1}}(\sigma(\eta))^{\overline{I}}_{\overline{L}}$$

is another kernel of type λ, as it easily follows from (11.20). We have proved that A_σ, and therefore A_Ω, is an operator of type 2.

Set $a(G) := \frac{1}{|G|}\sum_{\sigma \in G} a(\sigma) > 0$. We wish to check that $a(G)^{-1}A_\Omega$ inverts $\square_\Omega$. Let us set $B := I - \square_\Omega A$. If φ is a G-invariant $(0,q)$-form then (by (11.8))

$$\square_\Omega A_\Omega \varphi(\xi) = \frac{1}{|G|}\sum_{\sigma \in G} \square_\Omega \sigma^* A(\sigma^{-1})^* \varphi(\xi) =$$

$$= \frac{1}{|G|}\sum_{\sigma \in G} a(\sigma)\, \sigma^* \square_\Omega A \varphi(\xi) = \frac{1}{|G|}\sum_{\sigma \in G} a(\sigma)\, \sigma^*(\varphi - B\varphi)(\xi)$$

that is

$$\square_\Omega A_\Omega \varphi(\xi) = a(G)\, \varphi(\xi) - \frac{1}{|G|}\sum_{\sigma \in G} a(\sigma)\, B_\sigma \varphi(\xi),$$

where $B_\sigma := \sigma^* B(\sigma^{-1})^*$. We shall prove that

LEMMA 11.43. *B_σ is an operator of order* 1.

Proof. Let us set

$$A_\epsilon\varphi(\xi) := \left(\int K_\epsilon(\xi,\eta)\varphi_{\overline{J}}(\eta)d\eta\right)\theta_\xi^{\overline{J}},$$

$$K_\epsilon(\xi,\eta) := \psi(\xi,\eta)\Phi^\epsilon_{n-2q}(\Theta(\eta,\xi)),$$

$$\Phi^\epsilon_\alpha := b_\alpha \rho_\epsilon^{-(n+\alpha)/2}\overline{\rho_\epsilon}^{-(n-\alpha)/2}, \quad \rho_\epsilon(z,t) := \|z\|^2+\epsilon^2-it,$$

for any $\epsilon>0$. For the sake of simplicity, we only look at the case $q=1$. For any $(0,1)$-form ψ on Ω, the Kohn-Rossi laplacian is expressed by

$$\Box_\Omega\psi = \{-h^{\lambda\overline{\mu}}\nabla_\lambda\nabla_{\overline{\mu}}\psi_{\overline{\alpha}} - 2i\nabla_0\psi_{\overline{\alpha}} + \psi_{\overline{\gamma}}R^{\overline{\gamma}}{}_{\overline{\alpha}}\}\theta^{\overline{\alpha}},$$

where $R_{\lambda\overline{\mu}}$ is the pseudohermitian Ricci tensor. This may be written

$$(\Box_\Omega\psi)_{\overline{\alpha}} = \mathcal{L}_{n-2}\psi_{\overline{\alpha}} + \sum_{\mu=1}^{n}\{\Gamma^{\overline{\rho}}_{\overline{\mu}\,\overline{\alpha}}T_\mu\psi_{\overline{\rho}} + \frac{1}{2}\Gamma^{\overline{\rho}}_{\mu\overline{\mu}}T_{\overline{\rho}}\psi_{\overline{\alpha}} + \Gamma^{\overline{\rho}}_{\mu\overline{\alpha}}T_{\overline{\mu}}\psi_{\overline{\rho}}\} + F^{\overline{\gamma}}_{\overline{\alpha}}\psi_{\overline{\gamma}},$$

(the reader may compare the above to (16.1) in [**107**], p. 494) for some C^∞ functions $F^{\overline{\gamma}}_{\overline{\alpha}}$ (expressed in terms of the Christoffel symbols and their derivatives, and whose precise form is unimportant). We have (by the proof of Proposition 16.5 in [**107**])

$$\sigma^*B(\sigma^{-1})^*\varphi(\xi) = \varphi(\xi) - \sigma^*\Box_\Omega A(\sigma^{-1})^*\varphi(\xi) =$$
$$= \varphi(\xi) - \sigma^*\lim_{\epsilon\to 0}\Box_\Omega A_\epsilon(\sigma^{-1})^*\varphi(\xi)$$

that is

$$B_\sigma\varphi(\xi) = \varphi(\xi) - \lim_{\epsilon\to 0}\sigma^*\Box_\Omega A_\epsilon(\sigma^{-1})^*\varphi(\xi)$$

hence it suffices to show that if we let $\epsilon\to 0$ then $\sigma^*\Box_\Omega A_\epsilon(\sigma^{-1})^*\varphi$ goes to φ plus an operator of order 1 applied to φ. We have

$$\sigma^*\Box_\Omega A_\epsilon(\sigma^{-1})^*\varphi(\xi) =$$
$$= \left[\Box_\Omega\left(\int K_\epsilon(\cdot,\eta)((\sigma^{-1})^*\varphi)_{\overline{\alpha}}(\eta)d\eta\right)\theta^{\overline{\alpha}}\right]_{\sigma(\xi)}\circ(d_\xi\sigma) =$$
$$= g_\sigma(\xi)^{\overline{\alpha}}_{\overline{\beta}}\left[\mathcal{L}_{n-2}\psi_{\overline{\alpha}} + \sum_\mu\{\Gamma^{\overline{\rho}}_{\overline{\mu}\,\overline{\alpha}}T_\mu\psi_{\overline{\rho}} + \frac{1}{2}\Gamma^{\overline{\rho}}_{\mu\overline{\mu}}T_{\overline{\rho}}\psi_{\overline{\alpha}} + \right.$$
$$\left. + \Gamma^{\overline{\rho}}_{\mu\overline{\alpha}}T_{\overline{\mu}}\psi_{\overline{\rho}}\} + F^{\overline{\gamma}}_{\overline{\alpha}}\psi_{\overline{\gamma}}\right]_{\sigma(\xi)}\theta^{\overline{\beta}}_\xi$$

where

$$\psi_{\overline{\alpha}}(\xi) := \int K_\epsilon(\xi,\eta)((\sigma^{-1})^*\varphi)_{\overline{\alpha}}(\eta)d\eta.$$

and $\mathcal{L}_{n-2} = -\sum_\alpha T_\alpha T_{\overline{\alpha}} - 2iT$. Therefore, using

$$(T_\mu f)(\sigma(\xi)) = g_{\sigma^{-1}}(\sigma(\xi))^\lambda_\mu T_\lambda(f\circ\sigma)$$

we get

$$\sigma^*\Box_\Omega A_\epsilon(\sigma^{-1})^*\varphi(\xi) = \left\{A^0_{\epsilon,\overline{\beta}}\,\varphi(\xi) + \sum_{\mu=1}^{n}\sum_{i=1}^{3}A^i_{\epsilon,\mu\overline{\beta}}\,\varphi(\xi)\right\}\theta^{\overline{\beta}}_\xi +$$

$$+\left(\int g_\sigma(\xi)^{\overline{\alpha}}_{\overline{\beta}}\left[\mathcal{L}^\zeta_{n-2}K_\epsilon(\zeta,\eta)\right]_{\zeta=\sigma(\xi)} g_{\sigma^{-1}}(\eta)^{\overline{\gamma}}_{\overline{\alpha}}\,\varphi_{\overline{\gamma}}(\sigma^{-1}(\eta))d\eta\right)\theta^{\overline{\beta}}_\xi \tag{11.25}$$

where

$$A^0_{\epsilon,\overline{\beta}}\varphi(\xi) = g_\sigma(\xi)^{\overline{\alpha}}_{\overline{\beta}} F^{\overline{\gamma}}_{\overline{\alpha}}(\sigma(\xi)) \int K_\epsilon(\sigma(\xi),\eta) g_{\sigma^{-1}}(\eta)^{\overline{\rho}}_{\overline{\gamma}}\, \varphi_{\overline{\rho}}(\sigma^{-1}(\eta))d\eta,$$

$$A^1_{\epsilon,\mu\overline{\beta}}\varphi(\xi) = g_\sigma(\xi)^{\overline{\alpha}}_{\overline{\beta}}\, \Gamma^{\overline{\rho}}_{\mu\overline{\alpha}}(\sigma(\xi))\, g_{\sigma^{-1}}(\sigma(\xi))^{\lambda}_{\mu} \cdot$$

$$\cdot \int \left[T^\xi_\lambda K_\epsilon(\sigma(\xi),\eta)\right] g_{\sigma^{-1}}(\eta)^{\overline{\gamma}}_{\overline{\rho}}\, \varphi_{\overline{\gamma}}(\sigma^{-1}(\eta))\, d\eta\,,$$

$$A^2_{\epsilon,\mu\overline{\beta}}\varphi(\xi) = \frac{1}{2} g_\sigma(\xi)^{\overline{\alpha}}_{\overline{\beta}}\, \Gamma^{\overline{\rho}}_{\mu\overline{\mu}}(\sigma(\xi))\, g_{\sigma^{-1}}(\sigma(\xi))^{\overline{\lambda}}_{\overline{\rho}} \cdot$$

$$\cdot \int \left[T^\xi_{\overline{\lambda}} K_\epsilon(\sigma(\xi),\eta)\right]\, g_{\sigma^{-1}}(\eta)^{\overline{\gamma}}_{\overline{\alpha}}\, \varphi_{\overline{\gamma}}(\sigma^{-1}(\eta))\, d\eta\,,$$

$$A^3_{\epsilon,\mu\overline{\beta}}\varphi(\xi) = g_\sigma(\xi)^{\overline{\alpha}}_{\overline{\beta}}\, \Gamma^{\overline{\rho}}_{\mu\overline{\alpha}}(\sigma(\xi))\, g_{\sigma^{-1}}(\sigma(\xi))^{\overline{\lambda}}_{\overline{\mu}} \cdot$$

$$\cdot \int \left[T^\xi_{\overline{\lambda}} K_\epsilon(\sigma(\xi),\eta)\right]\, g_{\sigma^{-1}}(\eta)^{\overline{\gamma}}_{\overline{\rho}}\, \varphi_{\overline{\gamma}}(\sigma^{-1}(\eta))\, d\eta\,.$$

Clearly $A^0_{\epsilon,\overline{\beta}}$ gives, in the limit as $\epsilon \to 0$, an operator of type 2 (and hence of type 1). We claim that $A^i_{\epsilon,\mu\overline{\beta}}$ give (as $\epsilon \to 0$) operators of type 1, as well. For instance, let us look at $A^1_{\epsilon,\mu\overline{\beta}}$ (the remaining operators may be treated in a similar manner). Note that

$$\Phi^\epsilon_\alpha(\Theta(\sigma(\eta)\,,\,\sigma(\xi))) = a(\sigma)^{-n}\Phi^{\epsilon/\sqrt{a(\sigma)}}_\alpha(\Theta(\eta,\xi)). \tag{11.26}$$

Indeed (by Lemma 11.42)

$$\rho_\epsilon(g_\sigma(\eta)z(\xi)\,,\,a(\sigma)t(\xi)) = a(\sigma)\|z(\xi)\|^2 + \epsilon^2 - ia(\sigma)t(\xi) =$$
$$= a(\sigma)\,\rho_{\epsilon/\sqrt{a(\sigma)}}(z(\xi),t(\xi)).$$

Consequently

$$K_\epsilon(\sigma(\xi),\sigma(\eta)) = a(\sigma)^{-n}\psi_\sigma(\xi,\eta)\; \Phi^{\epsilon/\sqrt{a(\sigma)}}_{n-2}(\Theta(\eta,\xi))$$

and a change of variables $\eta' = \sigma^{-1}(\eta)$ leads to

$$A^1_{\epsilon,\mu\overline{\beta}}\varphi(\xi) = a(\sigma)^{n+1} g_\sigma(\xi)^{\overline{\alpha}}_{\overline{\beta}}\, \Gamma^{\overline{\rho}}_{\mu\,\overline{\alpha}}(\sigma(\xi)) g_{\sigma^{-1}}(\sigma(\xi))^{\lambda}_{\mu} \cdot$$

$$\cdot \int T^\xi_\lambda \left[\psi_\sigma(\xi,\eta)\Phi^{\epsilon/\sqrt{a(\sigma)}}_{n-2}(\Theta(\eta,\xi))\right]\, g_{\sigma^{-1}}(\sigma(\eta))^{\overline{\gamma}}_{\overline{\rho}}\, \varphi_{\overline{\gamma}}(\eta)d\eta$$

which goes, as $\epsilon \to 0$, to

$$a(\sigma)^{n+1} g_\sigma(\xi)^{\overline{\alpha}}_{\overline{\beta}}\, \Gamma^{\overline{\rho}}_{\mu\,\overline{\alpha}}(\sigma(\xi))\, g_{\sigma^{-1}}(\sigma(\xi))^{\lambda}_{\mu} \cdot$$

$$\cdot\, T^\xi_\lambda \left[\int \psi_\sigma(\xi,\eta)\Psi_{n-2}(\Theta(\eta,\xi))\, g_{\sigma^{-1}}(\sigma(\eta))^{\overline{\gamma}}_{\overline{\rho}}\, \varphi_{\overline{\gamma}}(\eta)\, d\eta\right].$$

As previously shown, $\psi_\sigma(\xi,\eta)\Phi_{n-2}(\Theta(\eta,\xi))$ is a kernel of type 2. Yet, by Proposition 15.14 in [**107**], p. 487, for any operator A of type 2, $T_\lambda A$ is an operator of type 1. Q.e.d.

To deal with the last term in (11.25) we write

$$\mathcal{L}^\zeta_{n-2} K_\epsilon(\zeta,\eta) =$$

$$= \left[\mathcal{L}^\zeta_{n-2}\psi(\zeta,\eta)\right] \Phi^\epsilon_{n-2}(\Theta(\eta,\zeta)) + \psi(\zeta,\eta)\, \mathcal{L}^\zeta_{n-2}\left[\Phi^\epsilon_{n-2}(\Theta(\eta,\zeta))\right] -$$

$$-\frac{1}{2}\sum_{\alpha=1}^{n}\Big\{\left[T^{\zeta}_{\alpha}\psi(\zeta,\eta)\right]\, T^{\zeta}_{\overline{\alpha}}\left[\Phi^{\epsilon}_{n-2}(\Theta(\eta,\zeta))\right]+$$

$$+\left[T^{\zeta}_{\overline{\alpha}}\psi(\zeta,\eta)\right]\, T^{\zeta}_{\alpha}\left[\Phi^{\epsilon}_{n-2}(\Theta(\eta,\zeta))\right]\Big\}. \tag{11.27}$$

The first term on the right hand side of (11.27), when substituted into (11.25), leads (as $\epsilon\to 0$) to an operator of order 1 applied to φ. We need to recall the notion of Heisenberg-type order.

DEFINITION 11.44. A function $f(\xi,y)$ on $\Omega\times\mathbb{H}_n$ is of *order* O^k, $k=1,2,\cdots$, if $f\in C^\infty$ and for any compact set $K\subset\Omega$ there is a constant $C_K>0$ so that $|f(\xi,y)|\le C_K|y|^k$ (Heisenberg norm). $\square$

If $(z,t)=\Theta_\xi^{-1}$ are pseudohermitian normal coordinates at ξ then (cf. Theorem 4.3 in [**142**], p. 177, a refinement of Theorem 14.10 and Corollary 14.9 in [**107**], p. 475)

$$(\Theta_\xi^{-1})_* T_\alpha=\frac{\partial}{\partial z^\alpha}+i\overline{z}^\alpha\frac{\partial}{\partial t}+O^1\mathcal{E}\left(\frac{\partial}{\partial z},\frac{\partial}{\partial\overline{z}}\right)+O^2\mathcal{E}\left(\frac{\partial}{\partial t}\right),$$

where $O^k\mathcal{E}$ denotes an operator involving linear combinations of the indicated derivatives, with O^k coefficients. Similarly, $(\Theta_\xi^{-1})_*\mathcal{L}_{n-2}$ is the operator $\mathcal{S}_{n-2}$ (given by (11.21) with $\alpha=n-2$) plus higher (Heisenberg-type) order terms.

Let $\delta(\xi,\eta)$ be the distribution on $\Omega\times\Omega$ defined by

$$\int\delta(\xi,\eta)f(\xi)g(\eta)d\xi d\eta=\int f(\xi)g(\xi)d\xi.$$

As to the second term in the right hand side of (11.27), when substituted into (11.25), it gives an integral operator applied to φ, which goes to φ for $\epsilon\to 0$, as desired. Indeed

$$\lim_{\epsilon\to 0}\int g_\sigma(\xi)^{\overline{\alpha}}_{\overline{\beta}}\,\psi(\sigma(\xi),\eta)\cdot$$

$$\cdot\mathcal{L}^{\zeta}_{n-2}\left[\Phi^{\epsilon}_{n-2}(\Theta(\eta,\zeta))\right]_{\zeta=\sigma(\xi)}\, g_{\sigma^{-1}}(\eta)^{\overline{\gamma}}_{\overline{\alpha}}\,\varphi_{\overline{\gamma}}(\sigma^{-1}(\eta))\,d\eta$$

is, up to higher order terms [leading to first order operators applied to φ (cf. also [**107**], p. 495)]

$$\int g_\sigma(\xi)^{\overline{\alpha}}_{\overline{\beta}}\,\psi(\sigma(\xi),\eta)\ \left[\mathcal{S}_{n-2}\Phi_{n-2}\right](\Theta(\eta,\sigma(\xi)))\ g_{\sigma^{-1}}(\eta)^{\overline{\gamma}}_{\overline{\alpha}}\,\varphi_{\overline{\gamma}}(\sigma^{-1}(\eta))d\eta=$$

$$\int g_\sigma(\xi)^{\overline{\alpha}}_{\overline{\beta}}\,\psi(\sigma(\xi),\eta)\ \delta(\sigma(\xi),\eta)g_{\sigma^{-1}}(\eta)^{\overline{\gamma}}_{\overline{\alpha}}\,\varphi_{\overline{\gamma}}(\sigma^{-1}(\eta))\,d\eta=$$

$$=g_\sigma(\xi)^{\overline{\alpha}}_{\overline{\beta}}\,\psi(\sigma(\xi),\sigma(\xi))\ g_{\sigma^{-1}}(\sigma(\xi))^{\overline{\gamma}}_{\overline{\alpha}}\,\varphi_{\overline{\gamma}}(\xi)=\delta^{\gamma}_{\beta}\,\psi_\sigma(\xi,\xi)\ \varphi_{\overline{\gamma}}(\xi)=\varphi_{\overline{\beta}}(\xi).$$

Q.e.d. Finally we deal with the third term in the right hand side of (11.27) (the fourth term may be dealt with in a similar way). It may be written (at $\zeta=\sigma(\xi)$) as

$$g_{\sigma^{-1}}(\sigma(\xi))^{\beta}_{\alpha}\, g_{\sigma^{-1}}(\sigma(\xi))^{\overline{\gamma}}_{\overline{\alpha}}\, T^{\xi}_{\beta}[\psi(\sigma(\xi),\eta)]\ T^{\xi}_{\overline{\gamma}}\left[\Phi^{\epsilon}_{n-2}(\Theta(\eta,\sigma(\xi))\right]$$

hence the corresponding integral is (after a change of variable)

$$a(\sigma)^{n+1}\sum_{\rho}\int g_\sigma(\xi)^{\overline{\alpha}}_{\overline{\beta}}\, g_{\sigma^{-1}}(\sigma(\xi))^{\lambda}_{\rho}\, g_{\sigma^{-1}}(\sigma(\xi))^{\overline{\mu}}_{\overline{\rho}}\, T^{\xi}_{\lambda}\left[\psi_\sigma(\xi,\eta)\right]\ \cdot$$

$$\cdot\, T^{\xi}_{\overline{\mu}}\left[\Phi^{\epsilon/\sqrt{a(\sigma)}}_{n-2}(\Theta(\eta,\xi))\right]\ g_{\sigma^{-1}}(\sigma(\eta))^{\overline{\gamma}}_{\overline{\alpha}}\,\varphi_{\overline{\gamma}}(\eta)\,d\eta.$$

Let us set $\psi_{\lambda,\sigma}(\xi,\eta) := T^\xi_\lambda[\psi_\sigma(\xi,\eta)]$ and note that $\psi_{\lambda,\sigma} \in C^\infty_0$ and (as T_λ is a differential operator) $\mathrm{Supp}(\psi_{\lambda,\sigma}) \subset \mathrm{Supp}(\psi_\sigma) \subset \{(\xi,\eta) \in D : \rho(\xi,\eta) \leq 1\}$. The following result completes the proof

LEMMA 11.45.

$$\int \psi_{\lambda,\sigma}(\xi,\eta)\, T^\xi_{\overline{\mu}} \left[\Phi^{\epsilon/\sqrt{a(\sigma)}}_{n-2}(\Theta(\eta,\xi))\right] g_{\sigma^{-1}}(\sigma(\eta))^{\overline{\gamma}}_{\overline{\alpha}}\, \varphi_{\overline{\gamma}}(\eta)\, d\eta \tag{11.28}$$

goes, as $\epsilon \to 0$, to an operator of order 1 *applied to φ.*

Proof. The kernel of the operator (11.28) is

$$T^\xi_{\overline{\mu}}\left[\Phi^{\epsilon/\sqrt{a(\sigma)}}_{n-2}(\Theta(\eta,\xi))\right] = [(d_\xi\Theta_\eta)T_{\overline{\mu},\xi}]\left(\Phi^{\epsilon/\sqrt{a(\sigma)}}_{n-2}\right) =$$

$$= \left[L_{\overline{\mu}} + O^1\mathcal{E}\left(\frac{\partial}{\partial z}, \frac{\partial}{\partial \overline{z}}\right) + O^2\mathcal{E}\left(\frac{\partial}{\partial t}\right)\right]_{\Theta_\eta(\xi)} \left(\Phi^{\epsilon/\sqrt{a(\sigma)}}_{n-2}\right) =$$

$$= -2\left(z^\mu\, \overline{\rho}^{\,-1}_{\epsilon/\sqrt{a(\sigma)}}\, \Phi^{\epsilon/\sqrt{a(\sigma)}}_{n-2}\right)_{\Theta_\eta(\xi)} + \sum_\lambda O^1\,(\overline{z}^\lambda f_\epsilon \Phi^{\epsilon/\sqrt{a(\sigma)}}_{n-2})(\Theta_\eta(\xi)) +$$

$$+ \sum_\lambda O^1\,(z^\lambda f_\epsilon \Phi^{\epsilon/\sqrt{a(\sigma)}}_{n-2})(\Theta_\eta(\xi)) + O^2\,(i\, f_\epsilon \Phi^{\epsilon/\sqrt{a(\sigma)}}_{n-2})(\Theta_\eta(\xi))$$

where

$$f_\epsilon := -\overline{\rho}^{\,-1}_{\epsilon/\sqrt{a(\sigma)}} - (n-1)\rho^{\,-1}_{\epsilon/\sqrt{a(\sigma)}}\ .$$

The Heisenberg group carries the contact form

$$\theta_0 = dt + 2\sum_j (x^j dy^j - y^j dx^j),$$

$z^j = x^j + iy^j$. Let $dV = \theta_0 \wedge (d\theta_0)^n$ be the natural volume form on $\mathbb{H}_n$. Set $h := \Theta_\xi^{-1}$. Note that $\Theta(h(u),\xi) = -\Theta_\xi(h(u)) = -u$. Also

$$(h^*\omega)(u) = (1 + O^1)dV(u)$$

(cf. again Theorem 4.3 in [**142**], p. 177). Then

$$\int_\Omega \psi_{\lambda,\sigma}(\xi,\eta)(z^\mu\, \overline{\rho}^{\,-1}_{\epsilon/\sqrt{a(\sigma)}}\, \Phi^{\epsilon/\sqrt{a(\sigma)}}_{n-2})(\Theta(\eta,\xi)) g_{\sigma^{-1}}(\sigma(\eta))^{\overline{\gamma}}_{\overline{\alpha}}\, \varphi_{\overline{\gamma}}(\eta)\, d\eta =$$

$$= \int_{\mathbb{H}_n} \psi_{\lambda,\sigma}(\xi,h(u))(z^\mu(u)\overline{\rho}_{\epsilon/\sqrt{a(\sigma)}}(u)^{-1}\Phi^{\epsilon/\sqrt{a(\sigma)}}_{n-2}(u)\ \cdot$$

$$\cdot\, g_{\sigma^{-1}}(\sigma(h(u)))\, \varphi_{\overline{\gamma}}(h(u))\,(1+O^1)\, dV(u) =$$

$$= \epsilon^{-2n-2}\int \psi_{\lambda,\sigma}(\xi,h(u))\, z^\mu(u)\frac{\Phi^1_{n-2}(\epsilon^{-1}u)}{\overline{\rho}_1(\epsilon^{-1}u)}\ \cdot$$

$$\cdot\, g_{\sigma^{-1}}(\sigma(h(u)))\, \varphi_{\overline{\gamma}}(h(u))\,(1+O^1)\, dV(u)$$

where $\epsilon^{-1}u$ is short for $\delta_{\epsilon^{-1}}u$. A change of variable $v = \epsilon^{-1}u$ gives (as $dV(u) = \epsilon^{2n+2}dV(v)$)

$$\epsilon\int \psi_{\lambda,\sigma}(\xi,h(\epsilon v))\, z^\mu(v)\, \frac{\Phi^1_{n-2}(v)}{\overline{\rho}_1(v)}\ \cdot$$

$$\cdot\, g_{\sigma^{-1}}(\sigma(h(\epsilon v)))\, \varphi_{\overline{\gamma}}(h(\epsilon v))\,\big(1 + O^1(\epsilon v)\big)\, dV(v).$$

The absolute value of this integral may be estimated by above by

$$\epsilon \sup_{\rho(\xi,\eta)\leq 1} [\psi_{\lambda,\sigma}(\xi,\eta) g_{\sigma^{-1}}(\sigma(\eta))\varphi_{\overline{\gamma}}(\eta)] \cdot$$

$$\cdot \int_{|v|\leq 1} z^{\mu}(v) \left| \frac{\Phi^1_{n-2}(v)}{\overline{\rho}_1(v)} \right| (1+\epsilon|v|)\, dV(v)$$

which goes to zero, as $\epsilon \to 0$. Moreover, in the limit, the O^1 and O^2 terms are

$$\sum_{\lambda} O^1(\overline{z}^{\lambda} f\Phi_{n-2})(\Theta_\eta(\xi)) + \sum_{\lambda} O^1(z^{\lambda} f\Phi_{n-2})(\Theta_\eta(\xi)) + O^2(f\Phi_{n-2})(\Theta_\eta(\xi))$$

where $f(z,t) = -[n\|z\|^2 + (n-2)it]/[\|z\|^4 + t^2]$. Note that $|f(y)| \leq C_n|y|^{-2}$ hence $O^1\overline{z}^{\lambda} f$, $O^1 z^{\lambda} f$ and $O^2 f$ are bounded. Now, for instance, let us look at $k(y) = (O^1\overline{z}^{\lambda} f\Phi_{n-2})(y)$ (the discussion of the remaining terms is similar). First, note that $\overline{z}^{\lambda} f\Phi_{n-2}$ is homogeneous of degree $-2n-1$, with respect to dilations. The Taylor series expansion (about $0 = \Theta_\eta(\eta)$) of the O^1 coefficients is a sum of homogeneous terms of degree at least 1 (with coefficients depending on η) plus a remainder of arbitrarily high order, hence the 'principal part' of $k(y)$ is homogeneous of degree $-2n$. Therefore $k(\Theta(\eta,\xi))$ is a kernel of type 1. Q.e.d.

To end the proof of Theorem 11.40, we shall show that $A_\Omega \Box_\Omega - a(G)I$ is an operator of type 1. First, note that A_σ, and then A_Ω, is symmetric. Indeed, for any two $(0,1)$-forms φ and ψ

$$(A_\sigma \varphi, \psi) =$$

$$= a(\sigma)^{2n+1} \int g_\sigma(\xi)^{\overline{\alpha}}_{\overline{\beta}} K(\sigma(\xi),\sigma(\eta)) g_{\sigma^{-1}}(\sigma(\eta))^{\overline{\gamma}}_{\overline{\alpha}} \varphi_{\overline{\gamma}}(\eta) \psi^{\overline{\beta}}(\xi) d\eta d\xi.$$

As $\Phi_\alpha(-y) = \overline{\Phi_{\overline{\alpha}}(y)}$, it follows that $\overline{K(\sigma(\xi),\sigma(\eta))} = K(\sigma(\eta),\sigma(\xi))$. Hence

$$(A^*_\sigma \psi)_{\overline{\mu}}(\eta) =$$

$$= a(\sigma)^{2n+1} h_{\gamma\overline{\mu}}(\eta) \int g_{\sigma^{-1}}(\sigma(\eta))^{\gamma}_{\alpha} K(\sigma(\eta),\sigma(\xi)) g_\sigma(\xi)^{\alpha}_{\beta} \psi^{\beta}(\xi) d\xi =$$

$$= a(\sigma)^{2n} \int g_\sigma(\eta)^{\overline{\lambda}}_{\overline{\mu}} h_{\alpha\overline{\lambda}}(\eta) K(\sigma(\eta),\sigma(\xi)) g_\sigma(\xi)^{\alpha}_{\beta} \psi^{\beta}(\xi) d\xi =$$

$$= a(\sigma)^{2n+1} \int g_\sigma(\eta)^{\overline{\lambda}}_{\overline{\mu}} h_{\alpha\overline{\lambda}}(\eta) K(\sigma(\eta),\sigma(\xi)) g_{\sigma^{-1}}(\sigma(\xi))^{\overline{\gamma}}_{\overline{\beta}} h^{\alpha\overline{\beta}}(\xi) \psi_{\overline{\gamma}}(\xi) d\xi.$$

Finally (as $h_{\alpha\overline{\beta}} = \delta_{\alpha\beta}$)

$$(A^*_\sigma \psi)_{\overline{\mu}} = (A_\sigma \psi)_{\overline{\mu}}.$$

Q.e.d. Moreover $\Box_\Omega$ is symmetric on compactly supported forms hence

$$A_\Omega \Box_\Omega \psi = a(G)\psi - \frac{1}{|G|} \sum_{\sigma\in G} a(\sigma) B^*_\sigma \psi$$

and the transpose of B_σ (an operator of type 1) is again of type 1.

APPENDIX A

Holomorphic bisectional curvature

The scope of this appendix is to collect the known results on Kähler manifolds of nonnegative holomorphic bisectional curvature, cf. [**122**].

Let V be a Kählerian manifold, of complex dimension ν. Let J denote the complex structure on V. Let g be a Kähler metric on V and R the curvature tensor field of (V, J, g). A *2-plane* at $x \in V$ is a 2-dimensional subspace $\sigma \subset T_x(V)$. A 2-plane σ at x is *J-invariant* if $J_x(\sigma) = \sigma$ and the set of all J-invariant 2-planes is the total space of a holomorphic bundle $\mathbb{C}P^{\nu-1} \to G_2(M) \to M$ (the *Grassmann bundle*). Given two J-invariant 2-planes $\sigma \subset T_x(V)$ and $\sigma' \subset T_x(V)$ the *bisectional curvature* $H(\sigma, \sigma')$ is defined by

$$H(\sigma, \sigma') = g_x(R_x(Y, J_x Y) J_x X \,,\, X)$$

where $X \in \sigma$ and $Y \in \sigma'$ are unit tangent vectors. The definition of $H(\sigma, \sigma')$ doesn't depend upon the choice of unit vectors in σ and σ'. Holomorphic bisectional curvature generalizes holomorphic sectional curvature (for $H(\sigma, \sigma)$ is nothing but the holomorphic sectional curvature of $\sigma \in G_2(M)$, as introduced for instance in [**155**], Vol. II, p. 165) and for a complex space form $V^\nu(c)$, i.e. a Kählerian manifold endowed with a Kähler metric of constant holomorphic sectional curvature c, the holomorphic bisectional curvature isn't constant but rather $H(\sigma, \sigma')$ lies between $c/2$ and c (the value $c/2$ is reached when σ and σ' are orthogonal while the value c is reached when $\sigma = \sigma'$). A generalization of a result by T. Frankel, [**110**] (requiring positive holomorphic sectional curvature) may be stated as

THEOREM A.1. (S.I. Goldberg & S. Kobayashi, [**122**])
Let V be a compact connected complex ν-dimensional Kählerian manifold endowed with a Kähler metric of positive holomorphic bisectional curvature and let M and N be two compact complex submanifolds of V. If $\dim_{\mathbb{C}} M + \dim_{\mathbb{C}} N \geq \nu$ then $M \cap N \neq \emptyset$.

Similarly, the proof of a result by A. Andreotti & T. Frankel (cf. Theorem 3 in [**110**]) may be easily adapted to show that *any compact Kähler surface with positive holomorphic bisectional curvature is biholomorphic to $\mathbb{C}P^2$*. It is also known that

THEOREM A.2. (S.I. Goldberg & S. Kobayashi, [**122**])
i) *A complex ν-dimensional compact connected Kähler manifold with an Einstein metric of positive holomorphic bisectional curvature is globally isometric to $\mathbb{C}P^\nu$ with the Fubini-Study metric.*
ii) *A complex ν-dimensional compact connected Kähler manifold of positive holomorphic bisectional curvature and constant scalar curvature is holomorphically isometric to $\mathbb{C}P^\nu$.*

Theorem A.2 extends previous results of M. Berger, [**38**], and R.L. Bishop & S.I. Goldberg, [**42**]. As well as in the case of a Kähler manifold of positive holomorphic sectional curvature (cf. [**41**])

THEOREM A.3. (S.I. Goldberg & S. Kobayashi, [**122**])
The second Betti number of a compact connected Kähler manifold V of positive holomorphic bisectional curvature is $b_2(V) = 1$.

Let V be an irreducible[1] compact complex ν-dimensional Kählerian manifold of non-negative holomorphic bisectional curvature, that is $H(\sigma, \sigma') \geq 0$. For any $Z \in T(V)$, $Z \neq 0$, we set

$$H_Z(X,Y) := g(R(Z,JZ)X,JY), \quad X,Y \in T(V).$$

Then H_Z is positive semi-definite. Let $\mathcal{N}_Z$ be the corresponding null distribution i.e.

$$\mathcal{N}_Z = \{X \in T(V) : H_Z(X,Y) = 0, \quad \forall\, Y \in T(V)\}.$$

Clearly $\mathcal{N}_Z$ is J-invariant and $(\mathcal{N}_Z)_x$ is determined by Z_x for any $x \in V$. For each $v \in T_x(V) \setminus \{0\}$ let us set $\operatorname{cdim} \mathcal{N}_v := \nu - \dim_{\mathbb{C}} (\mathcal{N}_Z)_x$ where $Z \in T(V)$ is a vector field such that $Z_x = v$.

DEFINITION A.4. The *complex positivity* of V is defined by

$$\ell(x) = \inf\{\operatorname{cdim} \mathcal{N}_v : v \in T_x(V) \setminus \{0\}\}.$$

for any $x \in V$. □

By a result of N. Mok, [**177**], the complex positivity $\ell(x)$ doesn't depend on the point x. The complex positivity ℓ was computed by M. Kim & J. Wolfson, [**153**], for all compact Hermitian symmetric spaces. In particular

V	ℓ	$\mathrm{Sp}(r)/\mathrm{U}(r)$	r
$\mathbb{C}P^\nu$	ν	$\mathrm{SO}(2r)/\mathrm{U}(r)$	$2r-3$
$\mathbf{Gr}_p(\mathbb{C}^{p+q})$	$p+q-1$	$E_6/(\mathrm{Spin}(10) \times T^1)$	11
$\mathbf{Gr}_2(\mathbb{R}^{p+2})$	$p-1$	$E_7/(E_6 \times T^1)$	17

The recalled notions are needed in the discussion of the Lefschetz type results in Section 4.2 of this monograph.

[1] As a Riemannian manifold, cf. e.g. [**155**], Vol. I, p. 179.

APPENDIX B

Partition of unity on orbifolds

Let $(B, \mathcal{A})$ be a C^∞ orbifold.

DEFINITION B.1. A continuous map $f : B \to N$ of B into an ordinary C^∞ manifold N is a C^∞ *map* if, for any l.u.s. $\{\Omega, G, \varphi\} \in \mathcal{A}$, the map $f_\Omega : \Omega \to N$ given by $f_\Omega = f \circ \varphi$ is C^∞ differentiable. □

Let $f : B \to N$ be a C^∞ map. Then $f_{\Omega'} \circ \lambda = f_\Omega$ for any injection λ of $\{\Omega, G, \varphi\}$ into $\{\Omega', G', \varphi'\}$ (as $f_{\Omega'} \circ \lambda = (f \circ \varphi') \circ \lambda = f \circ \varphi = f_\Omega$). In particular, each f_Ω is G-invariant. Here we adopted the following

DEFINITION B.2. Given a l.u.s. $\{\Omega, G, \varphi\} \in \mathcal{A}$, a map $h : \Omega \to N$ is G-*invariant* if $h \circ \sigma = h$ for any $\sigma \in G$. □

With each map $F : \Omega \to N$ one may associate the G-invariant map $F^G : \Omega \to N$ given by

$$F^G = \frac{1}{n(G)} \sum_{\sigma \in G} F \circ \sigma$$

where $n(G)$ is the order of the group G.

DEFINITION B.3. F^G is called the G-*average* of F. □

If F is G-invariant then $F^G = F$. Any G-invariant C^∞ map $F : \Omega \to N$ defines a C^∞ map $f : U \to N$ on the support U of $\{\Omega, G, \varphi\}$, where U is thought of as a C^∞ orbifold with the orbifold structure induced from B (set $f(p) = F(x)$ for just any $x \in \varphi^{-1}(p)$). If $F : \Omega \to N$ is an arbitrary C^∞ map (not necessarily G-invariant) then the G-average of F does the same job (i.e. F^G induces a C^∞ map $f : U \to N$ as before).

DEFINITION B.4. Let $F : \Omega \to N$ be a G-invariant map. Let $\{\Omega', G', \varphi'\}$ be another l.u.s. of B of support $U' \subseteq U$. The *restriction* F' of F to $\{\Omega', G', \varphi'\}$ is built as follows. Let $\lambda : \{\Omega', G', \varphi'\} \to \{\Omega, G, \varphi\}$ be an injection and set $F' = F \circ \lambda$. □

The definition of F' does not depend upon the choice of injection. Indeed, if μ is another injection of $\{\Omega', G', \varphi'\}$ into $\{\Omega, G, \varphi\}$ then there is a unique $\sigma_1 \in G$ such that $\mu = \sigma_1 \circ \lambda$ and one has

$$F \circ \mu = F \circ (\sigma_1 \circ \lambda) = F \circ \lambda$$

by the G-invariance of F. The restriction of F to $\{\Omega', G', \varphi'\}$ is G'-invariant. Indeed

$$F' \circ \sigma' = (F \circ \lambda) \circ \sigma' = F \circ (\eta(\sigma') \circ \lambda) = F \circ \lambda = F'$$

for any $\sigma' \in G'$ (again by the G-invariance of F) where $\eta : G' \to G$ is the group monomorphism corresponding to λ.

THEOREM B.5. *Let $(B, \mathcal{A})$ be an orbifold and $\{\Omega, G, \varphi\}, \{\Omega', G', \varphi'\} \in \mathcal{A}$ two l.u.s.'s of supports $U \subseteq U'$. Let $F : \Omega \to \mathbb{C}$ be a G-invariant C^∞ function. Assume F to have compact support $\tilde{K} \subset \Omega$. Let λ be an injection of $\{\Omega, G, \varphi\}$ into $\{\Omega', G', \varphi'\}$ and $\eta : G \to G'$ the corresponding group monomorphism. Consider $\sigma_1', \cdots, \sigma_k' \in G'$ such that $\sigma_1' = e'$ and*

$$G'/\eta(G) = \{[\sigma_1'], \cdots, [\sigma_k']\}$$

(and $i \neq j \Longrightarrow [\sigma_i'] \neq [\sigma_j']$). Let $F' : \Omega' \to \mathbb{C}$ be given by

$$F'(x') = \begin{cases} 0, & x' \notin V(\Omega, \Omega') \\ F(\lambda^{-1}(\sigma_i'^{-1} x')) & x' \in \sigma_i'(\lambda(\Omega)) \end{cases} \tag{B.1}$$

where

$$V(\Omega, \Omega') = \bigcup_{\sigma' \in G'} \sigma'(\lambda(\Omega)).$$

Then

1) *F' does not depend upon the choice of representatives σ_i' of the elements in $G'/\eta(G)$.*

2) *F' does not depend upon the choice of injection λ.*

3) *F' is C^∞ differentiable, G'-invariant, and its support is contained in*

$$\tilde{K}' = \bigcup_{\sigma' \in G'} \sigma'(\lambda(\tilde{K})).$$

Theorem B.5 allows us to formulate the following

DEFINITION B.6. The function $F' : \Omega \to N$ given by (B.1) is said to be the *extension* of F to $\{\Omega', G', \varphi'\}$. □

To prove 1) of Theorem B.5 let τ_i' be another representative of $[\sigma_i']$, i.e. $\tau_i' = \sigma_i' \circ \eta(\sigma)$ for some $\sigma \in G$. Then

$$\tau_i'(\lambda(\Omega)) = \sigma_i' \circ \eta(\sigma)(\lambda(\Omega)) = \sigma_i' \lambda \sigma(\Omega) = \sigma_i'(\lambda(\Omega)).$$

Moreover, if $x' \in \tau_i'(\lambda(\Omega))$ then

$$\begin{aligned} F(\lambda^{-1}(\tau_i'^{-1}(x'))) &= F(\lambda^{-1}\eta(\sigma)^{-1}\sigma_i'^{-1}(x')) = \\ = F(\eta(\sigma)\lambda)^{-1}\sigma_i'^{-1}(x')) &= F((\lambda\sigma)^{-1}\sigma_i'^{-1}(x')) = \\ = F(\sigma^{-1}\lambda^{-1}\sigma_i'^{-1}(x')) &= F(\lambda^{-1}\sigma_i'^{-1}(x')) \end{aligned}$$

by the G-invariance of F.

The proof of 2)-3) follows from (9.4) and is left as an exercise to the reader.

THEOREM B.7. *Let $(B, \mathcal{A})$ be a C^∞ orbifold and*

$$\{\Omega, G, \varphi\}, \ \ \{\Omega', G', \varphi'\} \in \mathcal{A}$$

two l.u.s.'s of supports $U \cap U' \neq \emptyset$. Assume that $U \setminus U' \neq \emptyset$ and $U' \setminus U \neq \emptyset$. Let $F' : \Omega' \to \mathbb{C}$ be a G'-invariant C^∞ function of compact support. Then there is a unique G-invariant C^∞ function $F : \Omega \to \mathbb{C}$ such that

1) *$F = 0$ on $\varphi^{-1}(U \setminus U')$.*

2) *Given any $\{\Omega_1, G_1, \varphi_1\} \in \mathcal{A}$ of support $U_1 \subset U \cap U'$ and any injection λ of $\{\Omega_1, G_1, \varphi_1\}$ into $\{\Omega, G, \varphi\}$ we have*

$$F'|_{\Omega_1} = F \circ \lambda$$

where $F'|_{\Omega_1}$ *is the restriction of* F' *to* $\{\Omega_1, G_1, \varphi_1\}$.

Theorem B.7 allows us to adopt the following

DEFINITION B.8. The function $F : \Omega \to \mathbb{C}$ (furnished by Theorem B.7) is the *prolongation* of $F' : \Omega' \to \mathbb{C}$ to $\{\Omega, G, \varphi\}$. $\square$

To prove Theorem 9.9 let $\{\Omega_1, G_1, \varphi_1\} \in \mathcal{A}$ be a l.u.s. of support $U_1 \subseteq U \cap U'$ and consider an injection λ of $\{\Omega_1, G_1, \varphi_1\}$ into $\{\Omega.G, \varphi\}$. Let $\eta_1 : G_1 \to G$ be the group monomorphism corresponding to λ. Let $\sigma_1, \cdots, \sigma_k \in G$ represent the elements of $G/\eta_1(G_1)$ (such that $i \neq j \Longrightarrow [\sigma_i] \neq [\sigma_j]$). Let

$$F(\Omega_1, \Omega) : V(\Omega_1, \Omega) \to \mathbb{C}$$

be defined by

$$F(\Omega_1, \Omega)(x) = F_1(\lambda^{-1}(\sigma_i^{-1}(x)))$$

for any $x \in \sigma_i(\lambda(\Omega_1))$. Here $F_1 = F'|_{\Omega_1}$ is the restriction of F' to $\{\Omega_1, G_1, \varphi_1\}$ and

$$V(\Omega_1, \Omega) = \bigcup_{i=1}^{k} \sigma_i(\lambda(\Omega_1)).$$

An argument similar to that in the proof of Theorem B.5 shows that $F(\Omega_1, \Omega)$ depends neither on the choice of representatives σ_i nor on the choice of injection λ.

Let $\{\Omega_2, G_2, \varphi_2\} \in \mathcal{A}$ of support $U_2 \subseteq U_1 \subseteq U \cap U'$. Then we may consider the function $F(\Omega_2, \Omega) \to \mathbb{C}$ defined by analogy to $F(\Omega_1, \Omega)$ above. Note that $V(\Omega_2, \Omega) \subseteq V(\Omega_1, \Omega)$. We shall need the following

LEMMA B.9. $F(\Omega_1, \Omega) = F(\Omega_2, \Omega)$ *on* $V(\Omega_2, \Omega)$.

Proof. Let μ be an injection of $\{\Omega_2, G_2, \varphi_2\}$ into $\{\Omega_1, G_1, \varphi_1\}$ and $\eta_2 : G_2 \to G_1$ the corresponding group monomorphism. Let $\tau_1, \cdots, \tau_r \in G_1$ be representatives of the elements in $G_1/\eta_2(G_2)$. Then

$$\{\sigma_i \circ \eta_1(\tau_j) : 1 \leq i \leq k\,,\, 1 \leq j \leq r\}$$

represent the elements of $G/\eta_1\eta_2(G_2)$. As the definition of $F(\Omega_2, \Omega)$ does not depend upon the choice of injection (of $\{\Omega_2, G_2, \varphi_2\}$ into $\{\Omega, G, \varphi\}$) we may use the injection $\lambda \circ \mu$. We have

$$F(\Omega_2, \Omega)(x) = F_2((\lambda \circ \mu)^{-1}(\sigma_i \circ \eta_1(\tau_j))^{-1}x)$$

for any $x \in \sigma_i\eta_1(\tau_j)\lambda\mu(\Omega_2)$. Here $F_2 = F'|_{\Omega_2}$ is the restriction of F' to $\{\Omega_2, G_2, \varphi_2\}$. Yet F_2 and the restriction of F_1 to $\{\Omega_2, G_2, \varphi_2\}$ actually coincide i.e. $F_2 = F_1 \circ \mu$. Hence

$$F(\Omega_2, \Omega)(x) = F_2(\mu^{-1}\lambda^{-1}\eta_1(\tau_j)^{-1}\sigma_i^{-1}x) =$$
$$= F_1(\lambda^{-1}\eta_1(\tau_j)^{-1}\sigma_i^{-1}x) = F_1((\lambda \circ \tau_j)^{-1}\sigma_i^{-1}x)$$

for any $x \in \sigma_i\eta_1(\tau_j)\lambda\mu(\Omega_2) \subseteq \sigma_i(\lambda\mu(\Omega_2))$. The definition of $F(\Omega_1, \Omega)$ does not depend upon the choice of injection (of $\{\Omega_1, G_1, \varphi_1\}$ into $\{\Omega, G, \varphi\}$) hence we may use the injection $\lambda \circ \tau_j$. We have

$$F(\Omega_1, \Omega)(x) = F_1((\lambda \circ \tau_j)^{-1}\sigma_i^{-1}(x))$$

for any $x \in \sigma_i(\lambda(\Omega))$. Then $F(\Omega_1, \Omega) = F(\Omega_2, \Omega)$ on

$$\sigma_i\eta_1(\tau_j)\lambda\mu(\Omega_2) \subseteq \sigma_i\lambda\mu(\Omega_2) \subseteq \sigma_i\lambda(\Omega_1)$$

and Lemma B.9 is proved.

At this point we may build the function $F : \Omega \to \mathbb{C}$ aimed to in Theorem B.7. We define F to be zero on $\varphi^{-1}(U \setminus U')$. As to the set $\varphi^{-1}(U \cap U')$ we define F as follows. Let $x \in \varphi^{-1}(U \cap U')$ and set $p = \varphi(x)$. By the properties of $\mathcal{H}$ let $U_1 \in \mathcal{H}$ such that $p \in U_1 \subseteq U \cap U'$ and consider a l.u.s. $\{\Omega_1, G_1, \varphi_1\} \in \mathcal{A}$ of support U_1. Finally let us set

$$F(x) = F(\Omega_1, \Omega)(x).$$

The definition of $F(x)$ doesn't depend upon the choice of $\{\Omega_1, G_1, \varphi_1\}$ as above. Indeed let $U_2 \in \mathcal{H}$ be another open set such that $p \in U_2 \subseteq U \cap U'$ and let $\{\Omega_2, G_2, \varphi_2\} \in \mathcal{A}$ be a l.u.s. of support U_2. Then $p \in U_1 \cap U_2$ hence there is $U_3 \in \mathcal{H}$ such that $p \in U_3 \subseteq U_1 \cap U_2$. Let $\{\Omega_3, G_3, \varphi_3\} \in \mathcal{A}$ be a l.u.s. of support U_3. Then

$$F(\Omega_1, \Omega)(x) = F(\Omega_3, \Omega)(x) = F(\Omega_2, \Omega)(x)$$

by Lemma B.9 (applied to the sets $U_3 \subseteq U_1$, respectively $U_3 \subseteq U_2$ (rather than $U_2 \subseteq U_1$)). Next we show that F is G-invariant. Let $\sigma \in G$ and $x \in \Omega$. We distinguish two cases as I) $x \in \Omega \setminus \varphi^{-1}(U \cap U')$ or II) $x \in \varphi^{-1}(U \cap U')$. In the first case $\sigma(x) \in \varphi^{-1}(U \setminus U')$ and $F(\sigma(x)) = 0 = F(x)$. If the second case occurs then we may consider $\{\Omega_1, G_1, \varphi_1\} \in \mathcal{A}$ of support U_1 with $\varphi(x) \in U_1$. Let λ be an injection of $\{\Omega_1, G_1, \varphi_1\}$ into $\{\Omega, G, \varphi\}$. As $x \in \lambda(\Omega_1)$ we have $\sigma(x) \in \sigma(\lambda(\Omega_1))$ hence $F(\sigma(x)) = F_1(\lambda^{-1}(\sigma^{-1}(x))$. Yet $F(x) = F_1((\sigma \circ \lambda)^{-1}(x))$ as $x \in \lambda(\Omega)$. Therefore $F \circ \sigma = F$. To check that $F \in C^\infty$ it suffices to note that i) F is C^∞ on $\varphi^{-1}(U \cap U')$ by construction, ii) $\tilde{K} = \varphi^{-1}(\varphi'(\tilde{K}')) \subseteq \varphi^{-1}(U \cap U')$ and $\tilde{K}$ is closed in Ω, where $\tilde{K}' = \operatorname{supp}(F') \subseteq \Omega$, iii) $F = 0$ on $\Omega \setminus \tilde{K}$, and iv) $\{\Omega \setminus \tilde{K}, \varphi^{-1}(U \cap U')\}$ is an open cover of Ω. The uniqueness of $F : \Omega \to \mathbb{C}$ (with the properties claimed in Theorem B.7) is plain.

THEOREM B.10. *Let $(B, \mathcal{A})$ be a C^∞ orbifold and $\{\Omega_0, G_0, \varphi_0\} \in \mathcal{A}$ a fixed l.u.s. of B. Let $F : \Omega_0 \to \mathbb{C}$ be a G-invariant C^∞ function of compact support. Then there is a unique C^∞ function $f : B \to \mathbb{C}$ of compact support* $\operatorname{supp}(f) \subset U_0 = \varphi_0(\Omega_0)$ *such that $f \circ \varphi_0 = F$.*

Proof. Let $\{\Omega, G, \varphi\} \in \mathcal{A}$ be an arbitrary l.u.s. of B, of support U. We shall define a function $f_\Omega : \Omega \to \mathbb{C}$. To this end we distinguish four cases as I) $U \subseteq U_0$ or II) $U_0 \subseteq U$ or III) $U \setminus U_0 \neq \emptyset$ and $U_0 \setminus U \neq \emptyset$ and $U \cap U_0 \neq \emptyset$ or IV) $U \cap U_0 = \emptyset$. In case I, let f_Ω be the restriction of F to $\{\Omega, G, \varphi\}$. In case II, let f_Ω be the extension of F to $\{\Omega, G, \varphi\}$ (in the sense of Theorem B.5). In case III, let f_Ω be the prolongation of F to $\{\Omega, G, \varphi\}$ (in the sense of Theorem B.7). Finally, in case IV we set $f_\Omega = 0$. Let then $f : B \to \mathbb{C}$ be defined as follows. Let $p \in B$. Let $U \in \mathcal{H}$ so that $p \in U$. Let then $\{\Omega, G, \varphi\} \in \mathcal{A}$ of support U and $x \in \Omega$ so that $\varphi(x) = p$. Then set $f(p) = f_\Omega(x)$. The function f satisfies the required conditions.

THEOREM B.11. *Let B be a C^∞ orbifold and $p \in B$. For any open neighborhood U' of p there is a C^∞ map $f : B \to \mathbb{R}$ of compact support contained in U' such that $0 \leq f \leq 1$ and $f = 1$ on some compact neighborhood K of p.*

Proof. Let $\{\Omega_0, G_0, \varphi_0\}$ be a l.u.s. of B of support U_0 with $p \in U_0$. Let $x \in \Omega_0$ so that $\varphi_0(x) = p$. The set $\{\sigma(x) : \sigma \in G_0\}$ is finite. Let $\{x_1, \cdots, x_r\}$ be its elements, where $x_1 = x$ (and $r \leq n(G)$). Let $\sigma_i \in G_0$ so that $x_i = \sigma_i(x), 1 \leq i \leq r$. Let V'_x be an open neighborhood of x in Ω_0 such that $\varphi_0(V'_x) \subseteq U'$ and the sets

$V'_{x_i} = \sigma_i(V'_x)$ are mutually disjoint. Let us set

$$V_x = \bigcup_{\sigma \in (G_0)_x} \sigma(V'_x).$$

Let K_0 be a compact neighborhood of x contained in V_x. Let $F : \Omega_0 \to \mathbb{R}$ be a C^∞ function with $F(\Omega_0) \subseteq [0, 1]$ such that $F = 1$ on K_0 and $supp(F) \subset V_x$. Consider $f_{\Omega_0} : \Omega_0 \to \mathbb{C}$ given by

$$f_{\Omega_0} = \frac{1}{n(x)} \sum_{\sigma \in G_0} f \circ \sigma$$

where $n(x)$ is the order of $(G_0)_x$. Clearly f_{Ω_0} is a G_0-invariant C^∞ function with $f_{\Omega_0}(\Omega_0) \subseteq [0, 1]$. Let us set

$$K' = \bigcap_{\sigma \in (G_0)_x} \sigma(K_0).$$

Then K' is a compact neighborhood of x. Also the restriction of f_{Ω_0} to V'_x has compact support and $f_{\Omega_0} = 1$ on K'. Let us set $K = \varphi_0(K')$. As φ_0 is continuous and open K is a compact neighborhood of p. At this point one may apply Theorem B.10. The resulting function satisfies the requirements of Theorem B.11.

A standard argument based on Theorem B.11 leads to the following

THEOREM B.12. *Let B be a C^∞ orbifold and $\{U_\alpha\}_{\alpha \in \Gamma}$ a locally finite open cover of B. There exists a countable partition of unity $\{\psi_i\}_{i \in \mathbb{N}}$ subordinate to $\{U_\alpha\}_{\alpha \in \Gamma}$ and such that each ψ_i is a C^∞ function of compact support.*

Note that the same result is stated in [**66**] only for locally finite open covers $\{U_\alpha\}_{\alpha \in \Gamma}$ whose open sets are supports of locally uniformizing systems. Also the preparatory result in Theorem B.11 is taken there (cf. the proof of Proposition 1.2 in [**66**], p. 319) for granted.

COROLLARY B.13. *Let K be a compact subset of a C^∞ orbifold B. Let U be an open subset of B such that $K \subset U$. Then there is a C^∞ function $f : B \to \mathbb{R}$ such that $f = 1$ on K and $f = 0$ on $B \setminus U$.*

Proof. Let $p \in B$. If $p \in K$ let U_p be an open neighborhood of p such that $U_p \subseteq U$. If $p \notin K$ then let U_p be an open neighborhood of p such that $U_p \cap K = \emptyset$. We produced an open cover $\{U_p\}_{p \in B}$ of B. Let $\{U_\alpha\}$ be a locally finite open refinement of $\{U_p\}_{p \in B}$. By Theorem B.12, let $\{\psi_i\}_{i \in \mathbb{N}}$ be a C^∞ partition of unity subordinate to $\{U_\alpha\}$. Let

$$C = \{i \in \mathbb{N} : \text{supp}(\psi_i) \cap K \neq \emptyset\}.$$

Then C is a finite set and $f = \sum_{i \in C} \psi_i$ is the function aimed to in Corollary B.13.

Let X be a Banach space. By replacing $\mathbb{R}^n$ by X in the definition of the notion of an orbifold one may define the notion of a *Banach orbifold* (modelled on X). Eventually, one may request that the injections $\lambda : \Omega' \to \Omega'$ be norm preserving. Partitions of unity on Banach manifolds (modelled on a Banach space satisfying an additional smoothness condition) have been obtained by J.N. Frampton, [**109**] (by firstly constructing partitions of unity on Lindelöf spaces, cf. Theorem 1 in [**109**], p. 8). It is an open problem to construct partitions of unity on (infinitely dimensional) Banach orbifolds.

APPENDIX C

Pseudo-differential operators on $\mathbb{R}^n$

For the convenience of the reader we briefly review the main notions and results about pseudo-differential operators on $\mathbb{R}^n$ (as employed in Chapters 10 and 11). For any multi-index $\alpha = (\alpha_1, \cdots, \alpha_n)$ we set

$$D^\alpha = (-i)^{|\alpha|} \frac{\partial^{|\alpha|}}{\partial x_1^{\alpha_1} \cdots \partial x_n^{\alpha_n}}.$$

DEFINITION C.1. The *Schwartz class* $\mathcal{S}$ is the set of all C^∞ complex valued functions f on $\mathbb{R}^n$ such that for all multi-indices α, β there is a constant $C_{\alpha,\beta} > 0$ such that

$$|x^\alpha D^\beta f| \leq C_{\alpha,\beta}.$$

□

We denote by dx the measure $dx = (2\pi)^{-n/2} d\mu$ where $d\mu$ is the Lebesque measure on $\mathbb{R}^n$. As $C_0^\infty(\mathbb{R}^n) \subset \mathcal{S}$ (and $C_0^\infty(\mathbb{R}^n)$ is dense in $L^2(\mathbb{R}^n)$) $\mathcal{S}$ is dense in $L^2(\mathbb{R}^n)$.

DEFINITION C.2. The *convolution product* of $f, g \in \mathcal{S}$ is given by

$$(f * g)(x) = \int_{\mathbb{R}^n} f(x - y) g(y) dy.$$

□

The convolution product is associative and commutative. However, only approximate identities exist. Precisely, let $f \in \mathcal{S}$ with $\int f(x)dx = 1$ and set $f_u(x) = u^{-n} f(x/u), u > 0$. Then, for any $g \in \mathcal{S}$, $f_u * g$ converges uniformly to g as $u \to 0$.

DEFINITION C.3. The *Fourier transform* of $f \in \mathcal{S}$ is given by

$$(\mathcal{F} f)(\xi) = \hat{f}(\xi) = \int e^{-ix\cdot\xi} f(x) dx$$

for any $\xi \in \mathbb{R}^n$. □

A standard argument based on integration by parts and Lebesgue dominated convergence theorem shows that $\hat{f} \in \mathcal{S}$ hence the Fourier transform is a map $\mathcal{S} \to \mathcal{S}$. In fact this is also bijective and the Fourier inversion formula gives its inverse (expressing f in terms of $\hat{f}$ by $f(x) = \int e^{ix\cdot\xi} \hat{f}(\xi) d\xi$). The Schwartz class $\mathcal{S}$ may be organized as a Fréchet space with the topology defined by the family of seminorms

$$p_{\alpha,\beta}(f) = \sup_{x \in \mathbb{R}^n} |x^\alpha D^\beta f(x)|.$$

Then $C_0^\infty(\mathbb{R}^n)$ is dense in $\mathcal{S}$ with respect to this topology. Also the Fourier transform is a homeomorphism of topological vector spaces. Both convolution and pointwise multiplication define ring structures on $\mathcal{S}$ and the Fourier transform interchanges these ring structures i.e. $\mathcal{F}(f * g) = \hat{f}\hat{g}$. Finally, the Fourier transform is an isometry with respect to the L^2 inner product

$$(f,g) = \int_{\mathbb{R}^n} f(x)\overline{g(x)}dx$$

and, as $\mathcal{S}$ is dense in $L^2(\mathbb{R}^n)$, it extends to a unitary map $L^2(\mathbb{R}^n) \to L^2(\mathbb{R}^n)$ such that $(\hat{f},\hat{g}) = (f,g)$ (the *Plancherel theorem*).

Let $s \in \mathbb{R}$ and $f \in \mathcal{S}$. Set

$$\|f\|_s = \left(\int_{\mathbb{R}^n} (1+|\xi|^2)^s \left|\hat{f}(\xi)\right|^2 d\xi\right)^{1/2}.$$

DEFINITION C.4. The *Sobolev space* $H_s(\mathbb{R}^n)$ is the completion of $\mathcal{S}$ in the norm $\|\cdot\|_s$. $\square$

Then $H_0(\mathbb{R}^n) \approx L^2(\mathbb{R}^n)$ (an isomorphism) by Plancherel theorem.

For any multi-index α, the operator D^α extends to a continuous map

$$D^\alpha : H_s(\mathbb{R}^n) \to H_{s-|\alpha|}(\mathbb{R}^n).$$

Intuitively, the number s counts the L^2 derivatives. Then one may loosely say that when extending D^α to $H_s(\mathbb{R}^n)$, $|\alpha|$ derivatives are lost.

Derivatives are also measured by means of the sup norm. Let $k \in \mathbb{Z}$, $k \geq 0$, and $f \in \mathcal{S}$. Let us set

$$\|f\|_{\infty,k} = \sup_{x\in\mathbb{R}^n} \sum_{|\alpha|\leq k} |D^\alpha f(x)|.$$

The completion of $\mathcal{S}$ in the norm $\|\cdot\|_{\infty,k}$ is contained in $C^k(\mathbb{R}^n)$.

The norms $\|\cdot\|_s$ and $\|\cdot\|_{\infty,k}$ are related as follows. Let $k \in \mathbb{Z}$, $k \geq 0$, and $s > k + n/2$. If $f \in H_s(\mathbb{R}^n)$ then f is of class C^k and

$$\|f\|_{\infty,k} \leq C\|f\|_s$$

for some $C > 0$ (the *Sobolev lemma*). This turns out to be particularly useful in showing that weak solutions (one produces for certain PDEs) are actually smooth.

Let $s > t$. Then the identity map $\mathcal{S} \to \mathcal{S}$ extends to a norm nonincreasing injection $H_s(\mathbb{R}^n) \to H_t(\mathbb{R}^n)$. This injection is compact if one restricts the supports involved. Precisely, let $f_m \in \mathcal{S}$ be a sequence of functions with supports in a fixed compact set K. Let $s > t$. If there is $C > 0$ such that $\|f_m\|_s \leq C$ for all $m \geq 1$ then there is a subsequence of $\{f_m\}$ which converges in $H_t(\mathbb{R}^n)$ (the *Rellich lemma*). The assumption that supports are uniformly bounded may not be dropped.

The space $C_0^\infty(\mathbb{R}^n)$ is dense in $H_s(\mathbb{R}^n)$ for any $s \in \mathbb{R}$. Each $H_s(\mathbb{R}^n)$ is a Hilbert space hence it is isomorphic to its dual. The following invariant characterization of the dual space $H_s(\mathbb{R}^n)^*$ is also available: the L^2 pairing $\mathcal{S} \times \mathcal{S} \to \mathbb{C}$ extends to a map $H_s(\mathbb{R}^n) \times H_{-s}(\mathbb{R}^n) \to \mathbb{C}$ which identifies $H_s(\mathbb{R}^n)^*$ with $H_{-s}(\mathbb{R}^n)$.

DEFINITION C.5. A *linear partial differential operator* P *of order* m ($m \in \mathbb{Z}$, $m \geq 0$) is given by

$$P = \sum_{|\alpha|\leq m} a_\alpha(x) D^\alpha$$

where $a_\alpha(x)$ are smooth. The *symbol* $p(x,\xi)$ of P is given by

$$p(x,\xi) = \sum_{|\alpha|\leq m} a_\alpha(x)\xi^\alpha.$$

The *leading symbol* $p_L(x,\xi)$ of P is given by

$$p_L(x,\xi) = \sum_{|\alpha|=m} a_\alpha(x)\xi^\alpha.$$

□

The symbol $p(x,\xi)$ is a polynomial in ξ of degree m. The leading symbol $p_L(x,\xi)$ is a homogeneous polynomial of degree m in ξ. One of the most useful properties of the Fourier transform is that it interchanges differentiation and multiplication i.e.

$$D^\alpha f(x) = \int e^{ix\cdot\xi}\xi^\alpha \hat{f}(\xi)d\xi, \quad \xi^\alpha \hat{f}(\xi) = \int e^{-ix\cdot\xi} D^\alpha f(x)dx,$$

for any $f \in \mathcal{S}$. Consequently given a linear partial differential operator P of symbol $p(x,\xi)$ one may use the Fourier inversion formula to get

$$(Pf)(x) = \int e^{ix\cdot\xi} p(x,\xi)\hat{f}(\xi)d\xi = \int e^{i(x-y)\cdot\xi}p(x,\xi)f(y)dyd\xi \tag{C.1}$$

for any $f \in \mathcal{S}$. It is noteworthy that the second integral is not absolutely convergent, hence one may not change the order of integration. Operators acting as in (C.1), yet corresponding to a class of symbols $p(x,\xi)$ wider than that of polynomials, are the pseudo-differential operators. Precisely we adopt the following

DEFINITION C.6. A function $p : \mathbb{R}^n \times \mathbb{R}^n \to \mathbb{C}$ is a *symbol of order* $m \in \mathbb{R}$ if 1) $p(x,\xi)$ is C^∞ in (x,ξ) and has compact x-support (i.e. there is a compact set $K \subset \mathbb{R}^n$ such that $p(x,\xi) = 0$ for any $(x,\xi) \in (\mathbb{R}^n - K) \times \mathbb{R}^n$), and 2) for any multi-indices α,β there is a constant $C_{\alpha,\beta} > 0$ such that

$$\left| D_x^\alpha D_\xi^\beta p(x,\xi)\right| \le C_{\alpha,\beta}\left(1+|\xi|\right)^{m-|\beta|}.$$

□

Let S^m denote the space of all symbols of order m. If $m' \le m$ then $S^{m'} \subseteq S^m$.

DEFINITION C.7. A symbol $p(x,\xi)$ is *infinitely smoothing* if

$$p \in S^{-\infty} = \bigcap_{m\in\mathbb{R}} S^m.$$

□

DEFINITION C.8. Two symbols a,b are *equivalent* (and one writes $a \sim b$) if their difference $a - b$ is infinitely smoothing. □

DEFINITION C.9. Given a symbol $p \in S^m$ we define its associated *pseudo-differential operator* P as the linear operator $\mathcal{S} \to \mathcal{S}$ given by

$$(Pf)(x) = \int e^{ix\cdot\xi}p(x,\xi)\hat{f}(\xi)d\xi \tag{C.2}$$

for any $f \in \mathcal{S}$. □

Then

$$\|Pf\|_{s-m} \le C\|f\|_s$$

hence P extends to a continuous map $H_s(\mathbb{R}^n) \to H_{s-m}(\mathbb{R}^n)$ for all $s \in \mathbb{R}$. Therefore, if $p \in S^{-\infty}$ then $P : H_s(\mathbb{R}^n) \to H_t(\mathbb{R}^n)$ for all $s,t \in \mathbb{R}$. Finally, by the Sobolev lemma $P : H_s(\mathbb{R}^n) \to C_0^\infty(\mathbb{R}^n)$ for all s (hence the term *infinitely smoothing* is appropriate).

DEFINITION C.10. Let $(m_j)_{j\geq 1}$ be a sequence of real numbers such that $m_j \to \infty$ as $j \to \infty$, and $p_j \in S^{m_j}$, $j \geq 1$. Given an arbitrary symbol p one writes

$$p \sim \sum_{j=1}^{\infty} p_j \tag{C.3}$$

if for any $m \in \mathbb{R}$ there is $k(m) \in \mathbb{N}$ such that

$$p - \sum_{j=1}^{k} p_j \in S^m$$

for any $k \geq k(m)$. $\square$

The series $\sum_{j=1}^{\infty} p_j$ is not necessarily convergent and (C.3) means merely that the difference between p and the partial sums of the p_j is as smoothing as one wishes. This is the sense in which one generalizes (from differential to pseudo-differential operators) the formula expressing the symbol of a composition of two differential operators. Precisely, let P and Q be two pseudo-differential operators, of symbols $p \in S^m$ and $q \in S^{m'}$, respectively. Then PQ is a pseudo-differential operator of symbol $\sigma(PQ) \in S^{m+m'}$ satisfying

$$\sigma(PQ) \sim \sum_{\alpha} \frac{i^\alpha}{\alpha!}(D_\xi^\alpha p)(D_x^\alpha q).$$

As long as one wishes to deal with operators on compact orbifolds one may restrict the domain and range of ones operators as follows. Let $U \subseteq \mathbb{R}^n$ be an open subset with compact closure. Let $p \in S^m$ have x-support in U. We restrict the domain of the pseudo-differential operator P (associated to p) to $C_0^\infty(U)$ such that $P : C_0^\infty(U) \to C_0^\infty(U)$. Let $\Psi_m(U)$ denote the space of all such operators. If $m \leq m'$ then $\Psi_m(U) \subseteq \Psi_{m'}(U)$. Let us set

$$\Psi_{-\infty}(U) = \bigcap_{m\in\mathbb{R}} \Psi_m(U).$$

Let $K(x,y)$ be a C^∞ function on $\mathbb{R}^n \times \mathbb{R}^n$ with compact x-support in U. If f is a function with compact support in U one sets

$$[P(K)f](x) = \int_{\mathbb{R}^n} K(x,y)f(y)dy.$$

Then $P(K) \in \Psi_{-\infty}(U)$. The converse is also true. Precisely let P be a pseudo-differential operator that comes from a symbol $p(x,\xi)$ of order $-\infty$. Let $C \subset \mathbb{R}^n$ be a compact subset. Then there is a C^∞ function $K(x,y)$ on $\mathbb{R}^n \times \mathbb{R}^n$, with compact support, such that $Pf = P(K)f$ for any $f \in C_0^\infty(\mathbb{R}^n)$ with support contained in C.

DEFINITION C.11. Let P, Q be two pseudo-differential operators. One says P, Q are *equivalent* (and one writes $P \sim Q$) if their symbols are equivalent. $\square$

DEFINITION C.12. Let P be a differential operator and Q a pseudo-differential operator. Let $U \subset \mathbb{R}^n$ an open subset with compact closure. We say that $PQ \sim I$ *over* $C_0^\infty(U)$ (respectively that $QP \sim I$ *over* $C_0^\infty(U)$) if there exist pseudo-differential operators P', I' such that $P'Q \sim I'$ and $P'Qf = PQf$ (respectively $QP' \sim I'$ and $QP'f = QPf$) and $I'f = f$ for any $f \in C_0^\infty(U)$. Here I is the identity. $\square$

DEFINITION C.13. A differential operator P is *elliptic* if its leading symbol p_L satisfies $p_L(x,\xi) = 0$ if and only if $\xi = 0$. □

We shall need the following theorem. Let P be an elliptic differential operator and $U \subseteq \mathbb{R}^n$ an open subset with compact closure. Then there is a pseudo-differential operator Q such that $PQ \sim I$ and $QP \sim I$ over $C_0^\infty(U)$.

Finally, we need to recall the invariance of pseudo-differential operators under coordinate transformations. Let $V, \tilde{V} \subset \mathbb{R}^n$ be two open sets and $f : \tilde{V} \to V$ a C^∞ diffeomorphism. Let P be a pseudo-differential operator, acting on C^∞ functions u of support contained in a compact set $K \subset V$, given by (C.2), where $p(x,\xi)$ has x-support contained in V. Let us set $\tilde{K} = f^{-1}(K)$. Given a C^∞ function Ω on $\tilde{V}$ of support contained in $\tilde{K}$, let us define $x\Omega$ by setting $(x\Omega)(\tilde{x}) = (Pu)(x)$ where $x = f(\tilde{x})$ and $u = \Omega \circ f^{-1}$. Then $x\Omega$ is a pseudo-differential operator.

Bibliography

[1] N. Abe, *On foliations and exotic characteristic classes*, Kodai Math. Sem. Rep., 28(1977), 324-341.

[2] N. Abe, *Exotic characteristic classes of certain* Γ*-foliations*, Kodai Math. J., (2)2(1979), 254-271.

[3] L.V. Ahlfors, *Lectures on quasiconformal mappings*, D. Van Nostrand Co., Inc., Princeton, New Jersey-Toronto-New York-London, 1966.

[4] T. Akahori, *A new approach to the local embedding theorem of CR structures, the local embedding theorem for* $n \geq 4$, A.M.S. Memoirs, No. 366, 1987.

[5] J.A. Alvarez López, *Duality in the spectral sequence of Riemannian foliations*, Amer. J. Math., 111(1989), 905-925; *A finiteness theorem for the spectral sequence of a Riemannian foliation*, Illinois J. Math., (1)33(1989), 79-92.

[6] A. Andreotti & T. Frankel, *The Lefschetz theorem on hyperplane sections*, Ann. Math., 69(1959), 713-717.

[7] C. Atindogbe & K.L. Duggal, *Conformal screen on lightlike hypersurfaces*, Int. J. Pure Appl. Math., (4)11(2004), 421-442.

[8] W.L. Baily, Jr., *On the imbedding of* V*-manifolds in projective space*, American J. Math., 73(1957), 403-430.

[9] W.L. Baily, Jr., *Satake's compactification of* V_n, American J. Math., 80(1958), 348-364.

[10] W.L. Baily, Jr., *The decomposition theorem for* V*-manifolds*, American J. Math., 76(1965), 862-888.

[11] M.S. Baouendi & L.P. Rothschild & F. Treves, *CR structures with group action and extendability of CR functions*, Invent. Math., 82(1985), 359-396.

[12] M.S. Baouendi & L.P. Rothschild, *Normal forms for generic manifolds and holomorphic extension of CR functions*, J. Diff. Geometry, 25(1987), 431-467.

[13] M.S. Baouendi & L.P. Rothschild, *Geometric properties of mappings between hypersurfaces in complex space*, J. Diff. Geometry, 21(1990), 473-499.

[14] M.S. Baouendi & L.P. Rothschild, *A general reflection principle in* $\mathbb{C}^2$, J. Funct. Analysis, 99(1991), 409-442.

[15] E. Barletta, *On the sections of a CR-bundle over a foliation with complex leaves*, Boll. U.M.I., (7)**10**-A (1996), 597-608.

[16] E. Barletta, *On the transverse Beltrami equation*, Commun. in Partial Differential Equations, (9& 10)21(1996), 1469-1485.

[17] E. Barletta, *CR submanifolds of maximal CR dimension in a complex Hopf manifold*, Annals of Global Analysis and Geometry, 22(2002), 98-118.

[18] E. Barletta, *The Lichnerowicz theorem on CR manifolds*, to appear in Tsukuba Mathematics Journal, 2006.

[19] E. Barletta & S. Dragomir, *On* $\mathcal{G}$*-Lie foliations with transverse CR structure*, Rendiconti di Matem., Roma, 16(1996), 169-188.

[20] E. Barletta & S. Dragomir, *Pseudohermitian immersions, pseudo-Einstein structures, and the Lee class of a CR manifold*, Kodai Math. J., 19(1996), 62-86.

[21] E. Barletta & S. Dragomir, *Transversally CR foliations*, Rendiconti di Matematica, Roma, 17(1997), 51-85.

[22] E. Barletta & S. Dragomir, *On* **C***-tangent foliations of maximum modulus sets*, Atti Sem. Mat. Fis. Univ. Modena, XLV(1997), 143-153.

[23] E. Barletta & S. Dragomir, *On transversally holomorphic maps of Kählerian foliations*, Acta Applicandae Mathematicae, 54(1998), 121-134.

[24] E. Barletta & S. Dragomir, *Differential equations on contact Riemannian manifolds*, Annali della Scuola Normale Superiore di Pisa, Cl. Sci., Ser. IV, (1)XXX(2001), 63-96.

[25] E. Barletta & S. Dragomir & H. Urakawa, *Pseudoharmonic maps from a nondegenerate CR manifold into a Riemannian manifold*, Indiana University Mathematics Journal, (2)50(2001), 719-746.

[26] E. Barletta & S. Dragomir & H. Urakawa, *Yang-Mills fields on CR manifolds*, Journal of Mathematical Physics, (8)47(2006), 1-41.

[27] M. Barros & F. Urbano, *Submanifolds of complex Euclidean space which admit a holomorphic distribution*, Quart. J. Math. Oxford Ser., (2)34(1983), no. 134, 141-143.

[28] D.E. Barrett, *Global convexity properties of some families of three-dimensional compact Levi-flat hypersurfaces*, Trans. A.M.S., (1)332(1992), 459-474.

[29] D.E. Barrett, *Behavior of the Bergman projection on the Diederich-Fornaess worm*, Acta Math., 168(1992), 1-10.

[30] E. Bedford & M. Kalka, *Foliations and complex Monge-Ampère equations*, Comm. Pure Appl. Math., 30(1977), 543-571.

[31] E. Bedford & D. Burns, *Holomorphic mapping of annuli in* $\mathbb{C}^n$ *and the related extremal function*, Ann. Sc. Norm. Sup. Pisa, 3(1979), 381-414.

[32] J.K. Beem & P.E. Ehrlich, *Global Lorentzian geometry*, Marcel Dekker, Inc., New York - Basel, 1981.

[33] A. Bejancu, *CR submanifolds of a Kaehler manifold*, I-II, Proc. A.M.S., 69(1978), 134-142, Trans. A.M.S., 250(1979), 335-345.

[34] A. Bejancu, *Geometry of CR submanifolds*, D. Reidel Publ. Co., Dordrecht-Boston-Lancaster-Tokyo, 1986.

[35] A. Bejancu & K.L. Duggal, *Gauge theory on foliated manifolds*, Rendiconti del Seminario Matematico di Messina, Serie II, 1(1991), 31-68.

[36] A. Bejancu & K.L. Duggal, *Spacetime geometry of CR-structures*, Contemporary Mathematics, 170(1994), 51-63.

[37] A. Bejancu & H.R. Farran, *Lightlike foliations of codimension one*, Publ. Math. Debrecen, (3-4)62(2003), 325-336.

[38] M. Berger, *Sur les variétés d'Einstein compactes*, C.R. Acad. Sci. Paris, 260(1965), 1554-1557.

[39] M. Berger & P. Gauduchon & E. Mazet, *Le spectre d'une varieté Riemannienne*, Lecture Notes in Math., 194, Springer-Verlag, Berlin-New York, 1971.

[40] E. Bishop, *Differentiable manifolds in complex Euclidean space*, Duke Math. J., 32(1965), 1-21.

[41] R.L. Bishop & S.I. Goldberg, *On the second cohomology group of a Kähler manifold of positive curvature*, Proc. Amer. Math. Soc., 16(1965), 119-122.

[42] R.L. Bishop & S.I. Goldberg, *On the toplogy of positively curved Kähler manifolds*, II, Tohoku Math. J., 17(1965), 310-318.

[43] D.E. Blair, *Contact manifolds in Riemannian geometry*, vol. 509, Springer-Verlag, Berlin-Heidelberg-New York, 1976.

[44] D.E. Blair & B.Y. Chen, *On CR submanifolds of Hermitian manifolds*, Israel J. Math., 34(1979), 353-363.

[45] D.E. Blair & S. Dragomir, *Pseudohermitian geometry on contact Riemannan manifolds*, Rendiconti di Matematica, Roma, 22(2002), 275-341.

[46] R.A. Blumenthal, *Transversally homogeneous foliations*, Ann. Inst. Fourier, (4)29(1979), 143-158.

[47] H.P. Boas & E.J. Straube, *Equivalence of regularity for the Bergman projection and the* $\overline{\partial}$*-Neumann operator*, Manuscr. Math., 67(1990), 25-33.

[48] H.P. Boas & E.J. Straube, *Sobolev estimates for the* $\overline{\partial}$*-Neumann operator on domains in* $\mathbb{C}^n$ *admitting a defining function that is plurisubharmonic on the boundary*, Math. Z., 206(1991), 81-88.

[49] H.P. Boas & E.J. Straube, *De Rham cohomology of manifolds containing the points of infinite type, and Sobolev estimates for the* $\overline{\partial}$*-Neumann problem*, J. Geometric Analysis, (3)3(1993), 225-235.

[50] A. Boggess, *CR manifolds and the tangential Cauchy-Riemann complex*, Studies in Advanced Math., CRC Press, Inc., Boca Raton-Ann Arbor-Boston-London, 1991.

[51] J.E. Borzellino, *Riemannian geometry of orbifolds*, Ph.D. Thesis, 1992; *Orbifolds of maximal diameter*, Indiana Univ. Math. J., (1)42(1993), 37-53; *Pinching theorems for teardrops and footballs of revolution*, Bull. Austral. Math. Soc., 49(1994), 353-364; *Orbifolds with lower Ricci curvature bounds*, Proc. A.M.S., (10)125(1997), 3011-3018.

[52] J.E. Borzellino & Shun-hui Zhu, *The splitting theorem for orbifolds*, Illinois J. Math., (4)38(1994), 679-691

[53] J.E. Borzellino & B.G. Lorica, *The closed geodesic problem for compact Riemannian 2-orbifolds*, Pacific J. Math., (1)175(1996), 39-46.

[54] J.E. Borzellino & V. Brunsden, *An automorphism of the orbifold diffeomorphism group which is not an inner automorphism*; *Determination of the topological structure of an orbifold by its group of orbifold diffeomorphisms*, preprints, 2000.

[55] L. Boutet de Monvel, *Intégration des équations de Cauchy-Riemann induites formelles*, Sém. Goulaouic-Lions-Schwartz (1974-75), Centre. Math. Ecole Polytech., Paris (1975).

[56] L. Boutet de Monvel & A. Iordan, *Sur les feuilletages* **C**-*tangents des sous-variétés du bord d'une variété complexe*, preprint, Univ. de Paris, 1994.

[57] D. Burns, *Curvatures of Monge-Ampère foliations and parabolic manifolds*, Ann. of Math., 115(1982), 349-373.

[58] C. Camacho & A. Lins Neto, *Geometric theory of foliations*, (translated by Sue E. Goodman), Birkhäuser, Boston-Basel-Stuttgart, 1985.

[59] A. Candel & L. Conlon, *Foliations I-II*, Graduate Studies in Mathematics, Vol. 23, Amer. Math. Soc., 2000.

[60] J. Cao & M-C. Shaw & L. Wang, *Estimates for the* $\overline{\partial}$-*problem and nonexistence of* C^2 *Levi-flat hypersurfaces in* $\mathbb{C}P^n$, Math. Z., 248(2004), 183-221.

[61] D. Catlin, *Subelliptic estimates for the* $\overline{\partial}$-*Neumann problem on pseudoconvex domains*, Annals of Math., 126(1987), 131-191.

[62] M. Carlotti, *Coomologia di de Rham sulle V-varietà Riemanniane complete*, Università di Pisa, preprint, 1990.

[63] M. Carlotti, *V-manifold maps and the inverse mapping theorem*, Rendiconti del Circolo Matem. di Palermo, XLI(1992), 325-341.

[64] D. Cerveau & R.P. Sad, *Fonctions et feuilletages Levi-flat. Étude locale.*, Ann. SC. Norm. Sup. Pisa, (2)3(2004), 427-445.

[65] S.S. Chern & J.K. Moser, *Real hypersurfaces in complex manifolds*, Acta Math., 133(1974), 219-271.

[66] Y-J. Chiang, *Harmonic maps of V-manifolds*, Ann. Golbal Anal. Geom., (3)8(1990), 315-344.

[67] E.M. Chirka, *Introduction to the geometry of CR manifolds*, Russian Math. Surveys, (1)46(1991), 95-197.

[68] D. Ciampa, *On Cauchy-Riemann circle bundles*, Kodai Math. J., (1)28(2005), 146-180.

[69] J.A. Cima & W.R. Derrick, *Some solutions of the Beltrami equation with* $\|\mu\|_\infty = 1$, Proceedings of Symposia in Pure Math., 52(1991), Part 1, 41-44.

[70] L. Conlon, *Transversally parallelizable foliations of codimension* 2, Trans. A.M.S., 194(1974), 79-102.

[71] L.A. Cordero & P.M. Gadea, *Exotic characteristic classes and subfoliations*, Ann. Inst. Fourier, Grenoble (1)26(1976), 225-237.

[72] L.A. Cordero & R. Wolak, *Properties of the basic cohomology of transversally Kähler foliations*, preprint, Santiago de Compostela, 2005.

[73] B. Coupet & A. Sukhov, *On the uniform extendibility of proper holomorphic mappings*, Complex variables, 28(1996), 243-248.

[74] M. Crampin, *Pseudo-Riemannian and related structures on differentiable manifolds*, Ph.D. thesis, King's College, London, 1967.

[75] J.P. D'Angelo, *Finite type conditions for real hypersurfaces*, J. Differential Geometry, 14(1979), 59-66.

[76] J.P. D'Angelo, *Real hypersurfaces, orders of contact, and applications*, Ann. of Math., 115(1982), 615-637.

[77] J.P. D'Angelo, *Defining equations for real analytic real hypersurfaces in* $\mathbb{C}^n$, Trans. Amer. Math. Soc., 295(1986), 71-84.

[78] J-P. Demailly, *Complex analytic and differential geometry*, preprint, Université de Grenoble I, Institut Fourier, UMR 5582 du CNRS, 38402 Saint-Martin d'Hères, France, 2004.

[79] K. Diederich & J.E. Fornaess, *Pseudoconvex domains: an example with nontrivial Nebenhulle*, Math. Ann., 225(1977), 275-292.

[80] M. Djorić & M. Okumura, *CR submanifolds of maximal CR dimension of complex projective space*, Arch. Math., 71(1998), 148-158.

[81] M. Djorić & M. Okumura, *Levi form of CR submanifolds of maximal CR dimension of complex space forms*, Acta Math. Hungar., 102(2004), 297-304.

[82] S. Donnini & G. Gigante, *Classification of left invariant CR structures on* $GL^+(3,\mathbb{R})$, Riv. Mat. Univ. Parma, (4)16(1990), 343-355.

[83] S. Dragomir, *On a conjecture of J.M. Lee*, Hokkaido Math. J., (1)23(1994), 35-49.

[84] S. Dragomir, *Pseudohermitian immersions between strictly pseudoconvex CR manifolds*, American J. Math., (1)117(1995), 169-202.

[85] S. Dragomir & P. Nagy, *Complex Finsler structures on CR-holomorphic vector bundles*, Rendiconti di Matematica, Roma, Serie VII, 19(1999), 427-447.

[86] S. Dragomir & J. Masamune, *Cauchy-Riemann orbifolds*, Tsukuba J. Math., (2)26(2002), 351-386.

[87] S. Dragomir & S. Nishikawa, *Foliated CR manifolds*, J. Math. Soc. Japan, (4)56(2004), 1031-1068.

[88] S. Dragomir & L. Ornea, *Locally conformal Kähler geometry*, Progress in Mathematics, Vol. 155, Birkhäuser, Boston-Basel-Berlin, 1998.

[89] S. Dragomir & G. Tomassini, *Differential Geometry and Analysis on CR manifolds*, Progress in Mathematics, Vol. 246, Birkhäuser, Boston-Basel-Berlin, 2006.

[90] S. Dragomir & H. Urakawa, *On the inhomogeneous Yang-Mills equation* $d_D^* R^D = f$, Interdisciplinary Information Sciences, (1)6(2000), 41-52.

[91] S. Dragomir & K.L. Duggal, *Indefinite locally conformal Kähler manifolds*, Diff. Geom. and its Appl., 25(2007), 8-22.

[92] T. Duchamp & M. Kalka, *Invariants of complex foliations and the Monge-Ampère equation*, Michigan Math. J., 35(1988), 91-115.

[93] T. Duchamp & E.L. Stout, *Maximum modulus sets*, Ann. Inst. Fourier, (3)31(1981), 37-69.

[94] K.L. Duggal, *CR-structures and Lorentzian geometry*, Acta Appl. Math., 7(1986), 211-223.

[95] K.L. Duggal, *Spacetime manifolds and contact strctures*, Internat. J. Math. & Math. Sci., (3)13(1990), 545-554.

[96] K.L. Duggal, *Lorentzian geometry and globally framed manifolds*, Acta Appl. Math., 19(1990), 131-148.

[97] K.L. Duggal & A. Bejancu, *Lightlike submanifolds of semi-Riemannian manifolds and applications*, Kluwer Academic, vol. 364, 1996.

[98] K.L. Duggal & B. Sahin, *Screen Cauchy-Riemann lightlike submanifolds*, Acta Math. Hungar., (1-2)106(2005), 137-165.

[99] R. Edwards & K. Millett & D. Sullivan, *Foliations with all leaves compact*, Topology, (1)16(1977), 13-32.

[100] D.B.A. Epstein, *Foliations with all leaves compact*, Ann. Inst. Fourier, Grenoble, (1)26(1976), 265-282.

[101] J. Faran, *Local invariants of foliations by real hypersurfaces*, Michigan Math. J., 35(1988), 395-404.

[102] C. Farsi, *Orbifold spectral theory*, Rocky Mountain J. Math., (1)31(2001), 215-235.

[103] F. Fedida, *Sur les feuilletages de Lie*, C.R. Acad. Sci. Paris, 272(1971), 999-1001.

[104] C. Fefferman, *Monge-Ampère equations, the Bergman kernel, and geometry of pseudoconvex domains*, Ann. of Math., (2)103(1976), 395-416, 104(1976), 393-394.

[105] G.B. Folland, *A fundamental solution for a subelliptic operator*, Bull. A.M.S., (2)79(1973), 373-376.

[106] G.B. Folland & J.J. Kohn, *The Neumann problem for the Cauchy-Riemann complex*, Annals of Math. Studies, Vol. 75, University of Tokyo Press, Tokyo, 1972.

[107] G.B. Folland & E.M. Stein, *Estimates for the* $\overline{\partial}_b$*-complex and analysis on the Heisenberg group*, Comm. Pure Appl. Math., 27(1974), 429-522.

[108] J.E. Fornaess & E.L. Stout, *Spreading polydiscs on complex manifolds*, American J. Math., (5)99(1977), 933-960.

[109] J.N. Frampton, *Smooth partitions of unity on Banach manifolds*, Ph.D. Thesis, Yale University, University Microfilms, Inc., Ann Arbor, Michigan, 1965.

[110] T. Frankel, *Manifolds with positive curvature*, Pacific J. Math., 11(1961), 165-174.

[111] M. Freeman, *The Levi form and local complex foliations*, Trans. A.M.S., 57(1976), 369-370.

[112] M. Freeman, *Local complex foliations of real submanifolds*, Math. Ann., 209(1974), 1-30.

[113] M. Freeman, *Local biholomorphic straightening of real submanifolds*, Ann. Mathem., 106(1977), 319-352.

[114] N. Gamara & R. Yacoub, *CR Yamabe conjecture - the conformally flat case*, Pacific J. Math., (1)201(2001), 121-175.

[115] G. Gigante, *Sulla coomologia di varietà pseudohermitiane con torsione nulla*, Riv. Mat. Univ. Parma, (4)2(1986), 157-165.

[116] G. Gigante & G. Tomassini, *CR structures on a real Lie algebra*, Rend. Mat. Accad. Lincei, 2(1991), 203-205, Advances in Math., 94(1992), 67-81.

[117] G. Gigante & A. Tomassini, *$G_{m,h}$-bundles over foliations with complex leaves*, Complex analysis and geometry (Trento, 1993), 213-228, Lecture Notes in Pure and Appl. Math., 173, Dekker, New York, 1996.

[118] G. Gigante & G. Tomassini, *Foliations with complex leaves*, Atti Accad. Naz. Lincei Cl. Sci. Fis. Mat. Natur. Rend. Lincei, (9) Mat. Appl. 4 (1993), no. 2, 115-120; *Foliations with complex leaves*, Differential Geom. Appl., (1)5(1995), 33-49.

[119] P.B. Gilkey, *Invariance theory, the heat equation, and the Atiyah-Singer index theorem*, Mathematics Lecture Series, vol. 11, Publish or Perish, Inc., Wilmington, Delaware, 1984.

[120] J. Girbau & M. Nicolau, *Pseudo-differential operators on V-manifolds and foliations*, I-II, Collectanea Mathem., 30(1979), 247-265, ibid., 31(1980), 63-95.

[121] C. Godbillon, *Feuilletages*, Progress in Mathematics, vol. 98, Birkhäuser, Basel-Boston-Berlin, 1991.

[122] S.I. Goldberg & S. Kobayashi, *Holomorphic bisectional curvature*, J. Diff. Geometry, 1(1967), 225-233.

[123] C.R. Graham, *On Sparling's characterization of Fefferman metrics*, American J. Math., 109(1987), 853-874.

[124] C.R. Graham & J.M. Lee, *Smooth solutions of degenerate Laplacians on strictly pseudoconvex domains*, Duke Math. J., (3)57(1988), 697-720.

[125] C.R. Graham, *The Dirichlet problem for the Bergman Laplacian*, I-II, Communications in Partial Differential Equations, 8(1983), 433-476, 563-641.

[126] C.R. Graham, *Compatibility operators for degenerate elliptic equations on the ball and Heisenberg group*, Math. Z., 187(1984), 289-304.

[127] A. Greenleaf, *The first eigenvalue of a sublaplacian on a pseudohermitian manifold*, Comm. Partial Differential Equations, 10(1985), 191-217.

[128] P.C. Greiner & J.J. Kohn & E.M. Stein, *Necessary and sufficient conditions for the solvability of the Lewy equation*, Proc. Nat. Acad. Sci. U.S.A., (9)72(1975), 3287-3289.

[129] W. Greub & S. Halperin & R. Vanstone, *Connections, curvature, and cohomology*, Academic Press, New York-London, 1976.

[130] A. Haefliger, *Structures feuilletées et cohomologie à valeur dans un faisceau de groupoides*, Comment. Math. Helvet., 32(1958), 248-329.

[131] A. Haefliger, *Variétés feuilletées*, Ann. Scuola Norm. Sup. Pisa, 16(1962), 367-397.

[132] G. Hector & U. Hirsch, *Introduction to the theory of foliations, Part A: Fundamentals, Part B: Codimension one*, Vieweg und Sohn, Braunschweig, Wiesbaden, 2nd Ed., 1986.

[133] S. Helgason, *Differential geometry, Lie groups, and symmetric spaces*, Academic Press, Inc., New York-London-Toronto-Sydney-San Francisco, 1978.

[134] R. Hermann, *Convexity and pseudoconvexity for complex manifolds*, J. Mathem. Mech., 13(1964), 667-672.

[135] C.D. Hill & G. Taiani, *Families of analytic discs in $\mathbb{C}^n$ with boundaries on a prescribed CR-submanifold*, An. Scuola Norm. Sup. Pisa, 4-5(1978), 327-380.

[136] C.D. Hill & M. Nacinovich, *The toplogy of Stein CR manifolds and the Lefschetz theorem*, Ann. Inst. Fourier, Grenoble, (2)43(1993), 459-468.

[137] C.D. Hill & M. Nacinovich, *A necessary condition for global Stein immersion of compact CR manifolds*, Riv. Mat. Univ. Parma,

[138] R.K. Hladky, *Boundary regularity for the $\overline{\partial}_b$-Neumann problem*, Parts I-II, preprint, `arXiv:math.CV/0412308 v1 15 Dec 2004`, `arXiv:math.CV/0412309 v1 15 Dec 2004`.

[139] A. Iordan, *On the non-existence of smooth Levi-flat hypersurfaces in $\mathbb{C}P^n$*, Proceedings of the Memorial Conference of Kiyoshi Oka's Centenial Birthday on Complex Analysis in Several Variables, Kyoto, Nara, 2001.

[140] H. Jacobowitz, *An introduction to CR structures*, Mathem. Surveys and Monographs, No. 32, Providence, RI, 1990.

[141] T.D. Jeffres, *Singular set of some Kähler orbifolds*, Trans. A.M.S., (5)349(1997), 1961-1971.

[142] D. Jerison & J.M. Lee, *The Yamabe problem on CR manifolds*, J. Diff. Geometry, 25(1987), 167-197.

[143] D. Jerison & J.M. Lee, *CR normal coordinates and the Yamabe problem*, J. Diff. Geometry, 29(1989), 303-344.

[144] J. Jost & C-J. Xu, *Subelliptic harmonic maps*, Trans. of A.M.S., (11)350(1998), 4633-4649.

[145] A. El Kacimi-Alaoui, *Opérateurs transversalment elliptiques sur un feuilletage Riemannien et applications*, Compositio Mathem., 73(1990), 57-106.

[146] A. El Kacimi-Alaoui & G. Hector, *Décomposition de Hodge basique pour un feuilletage Riemannien*, Ann. Inst. Fourier, (3)36(1986), 207-227.

[147] A. El Kacimi-Alaoui & V. Sergiescu & G. Hector, *La cohomologie basique d'un feuilletage Riemannien est de dimension finie*, Math. Zeit., 188(1985), 593-599.

[148] A. El Kacimi-Alaoui & E.G. Gomez, *Applications harmoniques feuilletées*, Illinois J. Math., (1)40(1996), 115-122.

[149] F. Kamber & P. Tondeur, *Foliated bundles and characteristic classes*, Lecture Notes in Math., Vol. 493, Springer-Verlag, Berlin-New York, 1975.

[150] F. Kamber & P. Tondeur, *G-foliations and their characteristic classes*, Bull. A.M.S., 84(1978), 1086-1124.

[151] F. Kamber & P. Tondeur, *Duality for Riemannian foliations*, Proc. Symp. Pure Math., Vol. 40, 1983, Part 1, 609-618.

[152] F. Kamber & P. Tondeur, *Foliations and metrics*, in *Differential Geometry*, Proceedings, pp. 103-152, Special Year, Maryland 1981-82, Ed. by R. Brooks & A. Gray & B.L. Reinhart, Progress in Mathem., Vol. 32, Birkhäuser, Boston-Basel-Stuttgart, 1983.

[153] M. Kim & J. Wolfson, *Theorems of Barth-Lefschetz type on Kähler manifolds of non-negative bisectional curvature*, Forum Math., (2)15(2003), 261-273.

[154] H. Kitahara, *On a parametrix form in a certain V-submersion*, Lecture Notes in Math., vol. 792, p. 264-298, Springer, 1980.

[155] S. Kobayashi & K. Nomizu, *Foundations of differential geometry*, Interscience Publishers, New York, vol. I, 1963, vol. II, 1969.

[156] S. Kobayashi & H. Wu, *Complex differential geometry*, Birkhäuser, Basel-Boston, 1983.

[157] J.J. Kohn, *Boundaries of complex manifolds*, Proc. Conf. on Complex Analysis, Minneapolis, 1964, Springer-Verlag, New York, 1965, pp. 81-94.

[158] J.J. Kohn & L. Nirenberg, *An algebra of pseudo-differential operators*, Comm. Pure Appl. Math., 18(1965), 269-305.

[159] A. Korányi & H.M. Reimann, *Quasiconformal mappings on the Heisenberg group*, Invent. Math., 80(1985), 309-338.

[160] A. Korányi & H.M. Reimann, *Quasiconformal mappings on CR manifolds*, Lecture Notes in Math., vol. 1422, 59-75, Ed. by V. Villani, Springer-Verlag, 1988.

[161] S.G. Krantz, *Function theory of several complex variables*, Wiley, New York, 1982.

[162] M. Kuranishi, *Strongly pseudoconvex CR structures over small balls*, I-III, Ann. of Math., 115(1982), 451-500, ibidem, 116(1982), 1-64, ibidem, 116(1982), 249-330.

[163] J-H. Kwon & J-S. Pak, *CR-submanifolds of $(n-1)$ CR-dimension in a complex projective space*, Saitama Math. J., 15(1997), 55-65.

[164] H.B. Lawson, Jr., *Foliations*, Bull. A.M.S., (3)80(1974), 369-418.

[165] C.R. Lebrun, *Twistor CR manifolds and three-dimensional conformal geometry*, Trans. A.M.S., (2)284(1984), 601-616.

[166] C.R. Lebrun, *Foliated CR manifolds*, J. Diff. Geometry, 22(1985), 81-96.

[167] J.M. Lee & R. Melrose, *Boundary behaviour of the complex Monge-Ampère equation*, Acta Mathematica, 148(1982), 159-192.

[168] J.M. Lee, *The Fefferman metric and pseudohermitian invariants*, Trans. A.M.S., (1)296(1986), 411-429.
[169] J.M. Lee, *Pseudo-Einstein structures on CR manifolds*, American J. Math., 110(1988), 157-178.
[170] Jeffrey M. Lee & K. Richardson, *Riemannian foliations and eigenvalue comparison*, Ann. Global Analhysis and Geometry, 16(1998), 497-525; *Lichnerowicz and Obata theorems for foliations*, Pacific J. Math., (2)206(2002), 339-357.
[171] T. Levi-Civita, *Sulle funzioni di due o più varibili complesse*, Atti Accad. Naz. Lincei, 14(1905).
[172] Liang-Khoon Koh, *Ricci curvature and ends of Riemannian orbifolds*, Mathematika, 45(1998), 135-144.
[173] A. Lichnerowicz, *Géométrie des groupes de transformations*, Dunod, Paris, 1958.
[174] A. Lins Neto, *A note on projective Levi flats and minimal sets of algebraic foliations*, Ann. Inst. Fourier, 49(1999), 1369-1385.
[175] K. Matsumoto, *Some geometric properties of Levi form of distance to real hypersurfaces in* $\mathbb{C}^2$, Japan J. Math., (1)30(2004), 75-90.
[176] I. Moerdijk & J. Mrčun, *Introduction to foliations and Lie groupoids*, Cambridge Studies in Advanced Mathematics, 91, Cambridge University Press, Cambridge, 2003.
[177] N. Mok, *The uniformization theorem for compact Kähler manifolds of nonnegative holomorphic bisectional curvature*, J. Diff. Geometry, 27(1988), 179-214.
[178] P. Molino, *Étude des feuilletages transversalment complets et applications*, Ann. Scient. Éc. Norm. Sup., (3)10(1977), 289-307.
[179] P. Molino, *Riemannian foliationns*, Progress in Mathematics, vol. 73, Ed. by J. Oesterlé & A. Weinstein, Birkhäuser, Boston-Basel, 1988.
[180] P. Molino, *Connexions et G-structures sur les variétés feuilletées*, Bull. Sci. Math., 92(1968), 59-63.
[181] P. Molino, *Propriétés chomologiques des feuilletages à connexion transverse projectable*, Topology, 12 (1973), 317-325.
[182] P. Molino, *Géométrie globale des feuilletages Riemanniens*, Proc. Kon. Nederl. Akad., (1)85(1982), 45-76.
[183] P. Molino, *Théorie des G-structures: le problème d'equivalence*, Lecture Notes in Math., vol. 588, Springer-Verlag, Berlin-Heidelberg-New York, 1976.
[184] A. Montesinos, *Conformal curvature for the normal bundle of a conformal foliation*, Ann. Inst. Fourier, 32(1982), 261-274.
[185] C.C. Moore & C. Sschochet, *Golbal analysis on foliated spaces*, MSRI Publications, Vol 9, Springer-Verlag, New York-Berlin, 1988.
[186] E. Musso, *Homogeneous pseudo-hermitian Riemannian manifolds of Einstein type*, American J. Math., 113(1990), 219-241.
[187] A. Nagel & E.M. Stein, *Lectures on pseudo-differential operators*, Princeton University Press, Princeton, N.J., 1979.
[188] L. Ni & J. Wolfson, *The Lefschetz theorem for CR submanifolds and the nonexistence of real anlytic Levi flat submanifolds*, Comm. Anal. Geom., (3)11(2003), 553-564; `arXiv:math.DG/0302081 v1, 7 Feb 2003`.
[189] S. Nishikawa & P. Tondeur, *Transversal infinitesimal automorphisms for harmonic Kähler foliations*, Tôhoku Math. J., 40(1988), 599-611.
[190] S. Nishikawa & M. Ramachandran & P. Tondeur, *The heat equation for Riemannian foliations*, Trans. A.M.S., (2)319(1990), 619-630.
[191] S. Nishikawa & P. Tondeur & L. Vanheke, *Spectral geometry for Riemannian foliations*, Annals of Global Analysis and Geometry, 10(1992), 292-305.
[192] B. O'Neill, *The fundamental equations of a submersion*, Michigan Math. J., 13(1966), 459-469.
[193] B. O'Neill, *Semi-Riemannian geometry*, Academic Press, New York-London-Paris-San Diego-San Francisco-Sao Paulo-Sydney-Tokyo-Toronto, 1983.
[194] L. Nirenberg, *A complex Frobenius theorem. Seminars on analytic functions*, Vol. I, Lecture Notes, Princeton, New Jersey, Institute for Advanced Study, 1957.
[195] A. Noell, *Local versus global convexity of pseudoconvex domains*, preprint.
[196] J. Nuguchi & T. Ochiai, *Geometric function theory in several complex variables*, Translations of Mathematical Monographs, Vol. 80, Amer. Math. Soc., Providence, RI, 1984.

[197] T. Ohsawa, *A Stein domain with smooth boundary which has a product structure*, Publ. Res. Inst. Math. Sci. Kyoto Univ., 18(1982), 1185-1186.

[198] E. Park & K. Richardson, *The basic Laplacian of a Riemannian foliation*, Amer. J. Math., 118(1996), 1249-1275.

[199] R. Penrose, *Physical space-time and nonrealizable CR structures*, Bull. A.M.S., N.S., 8(1983), 427-448; *The twistor programme*, Rep. Math. Phys., 12(1977), 65-76.

[200] M.J. Pflaum, *The normal symbol on Riemannian manifolds*, New York J. Math., 4(1998), 97-125.

[201] P. Piccinni, *A Weizenböck formula for the second fundamental form of a Riemannan foliation*, Atti Acc. Lincei Rend., (3-4)LXXVII(1984), 102-110.

[202] S.I. Pinchuk & S.I. Tsyganov, *CR-straightenings of real manifolds in* $\mathbf{C}^n$, Matem. Zametki, (6)50(1991), 116-121, 159; translation in Math. Notes, (5-6)50(1991), 1295-1298.

[203] C. Rea, *Levi flat submanifolds and holomorphic extension of foliations*, Ann. Sc. Norm. Sup. Pisa, (3)26(1972), 665-681.

[204] B. Reinhart, *Foliated manifolds with bundle-like metrics*, Ann. of Math., 69(1959), 119-132.

[205] B. Reinhart, *Differential geometry of foliations*, Springer-Verlag, New York-Berlin, 1983.

[206] K. Richardson, *Lichnerowicz and Obata theorems for foliations*, Pacific J. Math., (2)206(2002), 339-357.

[207] C. Roger, *Cohomologie* (p,q) *des feuilletages et applications*, Transversal structure of foliations (Toulouse, 1982), Astrisque, 116(1984), 195-213.

[208] H. Rossi, *LeBrun's nonrealizability theorem in higher dimensions*, Duke Math. J., (2)52(1985), 457-474.

[209] V.Y. Rovenskii, *Foliations on Riemannian manifolds and submanifolds*, Birkhäuser, Basel, 1998.

[210] M. Rumin, *Un complexe de formes différentielles sur les variétés de contact*, C.R. Acad. Sci. Paris, 310(1990), 401-404.

[211] H. Rummler, *Quelques notions simples en géométrie Riemannienne et leurs applications aux feuilletages compacts*, Comment. Math. Helv., 54(1979), 224-239.

[212] A. Sanini & F. Tricerri, *Connessioni e varietà fogliettate*, Cooperativa Libraria Universitaria Torinese - Editrice, Torino, 1978.

[213] I. Satake, *On Siegel's modular functions*, Proceedings of the International Symposium on Algebraic Number Theory, Tokyo-Nikko, 1955.

[214] I. Satake, *On a generalization of the notion of manifold*, Proc. Nat. Acad. Sci. U.S.A., 42(1956), 359-363.

[215] I. Satake, *On the compactification of the Siegel space*, Journal of the Indian Math. Society, 20(1956), 259-281.

[216] I. Satake, *The Gauss-Bonnet theorem for* V*-manifolds*, J. Math. Soc. Japan, (4)9(1957), 464-492.

[217] G.W. Schwarz, *On the de Rham cohomology of the leaf space of a foliation*, Topology, 13(1974), 185-187.

[218] V. Sergiescu, *Cohomologie basique et dualité des feuilletages Riemanniens*, Ann. Inst. Fourier, (3)35(1985), 137-158.

[219] F. Severi, *Lezioni sulle funzioni analitiche di più variabili complesse*, C.E.D.A.M., Padova, 1958.

[220] T. Shioya, *Eigenvalues and suspension structure of compact Riemannian orbifolds with positive Ricci curvature*, Manuscripta Math., 99(1999), 509-516.

[221] Y-T. Siu, *Nonexistence of smooth Levi-flat hypersurfaces in complex projective spaces of dimension* ≥ 3, Ann. of Math., 151(2000), 1217-1243.

[222] Y-T. Siu, $\overline{\partial}$*-regularity for weakly pseudoconvex domains in compact Hermitian symmetric spaces with respect to invariant metrics*, Ann. of Math., 156(2002), 595-621.

[223] F. Sommer, *Komplex-analytische Blätterung reeller Mannigfaltigkeiten im* $\mathbb{C}^m$, Math. Ann., 136(1958), 111-133; *Komplex-analytische Blätterung reeller Hyperflächen im* $\mathbb{C}^m$, ibid., 137(1959), 392-411.

[224] J. Song, *The Szegö kernel on an orbifold circle bundle*, `arXiv:math.DG/0405071 v1 5 May 2004`.

[225] E. Stanhope, *Specral bounds on orbifold istropy*, Annals of Global Analysis and Geometry, 27(2005), 355-375.

[226] N.K. Stanton, *Infinitesimal CR automorphisms of rigid hypersurfaces in* $\mathbb{C}^2$, J. Geom. Anal., (3)1(1991), 231-267.

[227] N.K. Stanton, *Infinitesimal CR automorphisms of rigid hypersurfaces*, American Journal of Mathematics, 117(1995), 141-167.

[228] N.K. Stanton, *Infinitesimal CR automorphisms of real hypersurfaces*, American Journal of Mathematics, 118(1996), 209-233.

[229] N.K. Stanton, *Infinitesimal CR automorphisms*, Advanced Studies in Pure Mathematics 25, 1997, CR-Geometry and Overdetermined Systems, pp. 355-363.

[230] S. Sternberg, *Lectures on differential geometry*, Prentince Hall, 1962.

[231] W. Stoll, *Variétés strictement paraboliques*, C.R. Acad. Sci. Paris, 285(1977), 757-759.

[232] W. Stoll, *The characterization of strictly parabolic manifolds*, Ann. Sc. Norm. Sup. Pisa, VII(1980), 87-154.

[233] E.J. Straube & M.K. Sucheston, *Levi foliations in pseudoconvex boundaries and vector fields that commute approximately with* $\overline{\partial}$, Trans. Amer. Math. Soc., (1)355(2002), 143-154.

[234] I. Tamura, *Topology of foliations: an introduction*, (translated by Kiki Hudson), Translations of Mathematical Monographs, Vol. 97, Amer. Math. Soc., Providence, R.I., 1992.

[235] N. Tanaka, *A differential geometric study on strongly pseudo-convex manifolds*, Kinokuniya Book Store Co., Ltd., Kyoto, 1975.

[236] N. Tanaka, *On the pseudo-conformal geometry of hypersurfaces of the space of* n *complex variables*, J. Math. Soc. Japan, 14(1962), 397-429.

[237] T. Tate, *Weyl calculus and Wigner transform on the Poincaré disk*, in *Noncommutative Differential Geometry*, 227-243, Y. Maeda et al. (eds.), Kluwer Acad. Publ., 2000.

[238] T. Tate, *Weyl pseudo-differential operators and Wigner transform on the Poincaré disk*, preprint, Keio University, Yokohama, 2000.

[239] W.P. Thurston, *The geometry and topology of 3-manifolds*, Lecture Notes, Princeton University, 1978-79.

[240] W.L. Ting, *On nontrivial characteristic classes of contact foliations*, Proc. A.M.S., 75(1979), 131-138.

[241] D. Tischler, *On fibering certain foliated manifolds over* S^1, Topology, 9(1970), 153-154.

[242] G. Tomassini, *Tracce delle funzioni olomorfe sulle sottovarietà analitiche reali d'una varietà complessa*, Ann. Sc. Norm. Sup., Pisa, 20(1966), 31-43.

[243] P. Tondeur, *Foliations of Riemannian manifolds*, Springer-Verlag, New York-Heidelberg-London-Paris-Tokyo, 1988.

[244] P. Tondeur, *Geometry of foliations*, Monographs in Mathematics, Vol. 90, Birkhäuser Verlag, Basel-Boston-Berlin, 1997.

[245] H. Urakawa, *Yang-Mills connections over compact strongly pseudo-convex CR manifolds*, Math. Z., 216(1994), 541-573.

[246] H. Urakawa, *Variational problems over strongly pseudoconvex CR manifolds*, Differential Geometry, Proceedings of the Sympos. in honour of Prof. Su Buchin on his 90th birthday, 233-242, Shanghai, China, September 17-23, 1991, Ed. by C.H. Gu & H.S. Hu & Y.L. Xin, World Scientific Publ. Co. Pte. Ltd., Singapore-New Jersey-London-Hong Kong, 1993.

[247] H. Urakawa, *Yang-Mills connections and deformation theory over compact strongly pseudoconvex CR manifolds*, To appear in Proceedings of the Third MSJ International Research Inst. on Geometric Complex Analysis, Hayama, Japan, March 19-29, 1995, World Scientific Publ. Co. Pte. Ltd., 1995.

[248] I. Vaisman, *Transversal twistor spaces of foliations*, Ann. Global Analysis and Geometry, 19(2001), 209-234.

[249] L. Wang, *The globally homeomorphic solutions to Beltrami system in* $\mathbf{C}^n$, Math. Z., 216(1994), 371-377.

[250] S.M. Webster, *Pseudohermitian structures on a real hypersurface*, J. Diff. Geometry, 13(1978), 25-41.

[251] A.M. Whittlesey, *Foliation by graphs of CR mappings and a nonlinear Riemann- Hilbert problem for smoothly bounded domains*, Michigan Math. J., (3)50(2002), 613-648.

[252] R. Wolak, *On* ∇*-G-foliations*, Rend. Circ. Mat. Palermo, 1984, Suppl. No. 6, 329-341; *On transverse structures of foliations*, ibidem, 1985, Suppl. No. 9, 227-243; *Some remarks on* ∇*-G-foliations*, Pittman Research Notes, 131(1985), 276-289; *On G-foliations*, Ann.

Polon. Math., 46(1985), 371-377; *The structure tensor of a transverse G-structure on a foliated manifold*, Boll. U.M.I., (7) **4**-A (1990), 1-15.

[253] R. Wolak, *Leaves of foliations with a transverse geometric structure of finite type*, Publ. Mat., (1)33(1989), 153-162.

[254] R. Wolak, *Pierrot's theorem for singular Riemannian foliations*, Publicacions Matemàtiques, 38(1994), 433-439; *Basic cohomology for singular Riemannian foliations*, Monatshefte Math., 128(1999), 159-163.

[255] R. Wolak, *Geometric structures on foliated manifolds*, Santiago de Compostela, 1989.

[256] Pit-Mann Wong, *Geometry of the complex homogeneous Monge-Ampère equation*, Invent. Math., 67(1982), 261-274.

[257] K. Yano & M. Kon, *CR submanifolds of Kaehlerian and Sasakian manifolds*, Progress in Mathematics, Vol. 30, Birkhäuser, Boston-Basel-Stuttgart, 1983.

Index

Titles in This Series

140 **Elisabetta Barletta, Sorin Dragomir, and Krishan L. Duggal,** Foliations in Cauchy-Riemann geometry, 2007

139 **Michael Tsfasman, Serge Vlăduţ, and Dmitry Nogin,** Algebraic geometric codes: Basic notions, 2007

138 **Kehe Zhu,** Operator theory in function spaces, 2007

137 **Mikhail G. Katz,** Systolic geometry and topology, 2007

136 **Jean-Michel Coron,** Control and nonlinearity, 2007

135 **Bennett Chow, Sun-Chin Chu, David Glickenstein, Christine Guenther, James Isenberg, Tom Ivey, Dan Knopf, Peng Lu, Feng Luo, and Lei Ni,** The Ricci flow: Techniques and applications, Part I: Geometric aspects, 2007

134 **Dana P. Williams,** Crossed products of C^*-algebras, 2007

133 **Andrew Knightly and Charles Li,** Traces of Hecke operators, 2006

132 **J. P. May and J. Sigurdsson,** Parametrized homotopy theory, 2006

131 **Jin Feng and Thomas G. Kurtz,** Large deviations for stochastic processes, 2006

130 **Qing Han and Jia-Xing Hong,** Isometric embedding of Riemannian manifolds in Euclidean spaces, 2006

129 **William M. Singer,** Steenrod squares in spectral sequences, 2006

128 **Athanassios S. Fokas, Alexander R. Its, Andrei A. Kapaev, and Victor Yu. Novokshenov,** Painlevé transcendents, 2006

127 **Nikolai Chernov and Roberto Markarian,** Chaotic billiards, 2006

126 **Sen-Zhong Huang,** Gradient inequalities, 2006

125 **Joseph A. Cima, Alec L. Matheson, and William T. Ross,** The Cauchy Transform, 2006

124 **Ido Efrat, Editor,** Valuations, orderings, and Milnor K-Theory, 2006

123 **Barbara Fantechi, Lothar Göttsche, Luc Illusie, Steven L. Kleiman, Nitin Nitsure, and Angelo Vistoli,** Fundamental algebraic geometry: Grothendieck's FGA explained, 2005

122 **Antonio Giambruno and Mikhail Zaicev, Editors,** Polynomial identities and asymptotic methods, 2005

121 **Anton Zettl,** Sturm-Liouville theory, 2005

120 **Barry Simon,** Trace ideals and their applications, 2005

119 **Tian Ma and Shouhong Wang,** Geometric theory of incompressible flows with applications to fluid dynamics, 2005

118 **Alexandru Buium,** Arithmetic differential equations, 2005

117 **Volodymyr Nekrashevych,** Self-similar groups, 2005

116 **Alexander Koldobsky,** Fourier analysis in convex geometry, 2005

115 **Carlos Julio Moreno,** Advanced analytic number theory: L-functions, 2005

114 **Gregory F. Lawler,** Conformally invariant processes in the plane, 2005

113 **William G. Dwyer, Philip S. Hirschhorn, Daniel M. Kan, and Jeffrey H. Smith,** Homotopy limit functors on model categories and homotopical categories, 2004

112 **Michael Aschbacher and Stephen D. Smith,** The classification of quasithin groups II. Main theorems: The classification of simple QTKE-groups, 2004

111 **Michael Aschbacher and Stephen D. Smith,** The classification of quasithin groups I. Structure of strongly quasithin K-groups, 2004

110 **Bennett Chow and Dan Knopf,** The Ricci flow: An introduction, 2004

109 **Goro Shimura,** Arithmetic and analytic theories of quadratic forms and Clifford groups, 2004

108 **Michael Farber,** Topology of closed one-forms, 2004

107 **Jens Carsten Jantzen,** Representations of algebraic groups, 2003
106 **Hiroyuki Yoshida,** Absolute CM-periods, 2003
105 **Charalambos D. Aliprantis and Owen Burkinshaw,** Locally solid Riesz spaces with applications to economics, second edition, 2003
104 **Graham Everest, Alf van der Poorten, Igor Shparlinski, and Thomas Ward,** Recurrence sequences, 2003
103 **Octav Cornea, Gregory Lupton, John Oprea, and Daniel Tanré,** Lusternik-Schnirelmann category, 2003
102 **Linda Rass and John Radcliffe,** Spatial deterministic epidemics, 2003
101 **Eli Glasner,** Ergodic theory via joinings, 2003
100 **Peter Duren and Alexander Schuster,** Bergman spaces, 2004
99 **Philip S. Hirschhorn,** Model categories and their localizations, 2003
98 **Victor Guillemin, Viktor Ginzburg, and Yael Karshon,** Moment maps, cobordisms, and Hamiltonian group actions, 2002
97 **V. A. Vassiliev,** Applied Picard-Lefschetz theory, 2002
96 **Martin Markl, Steve Shnider, and Jim Stasheff,** Operads in algebra, topology and physics, 2002
95 **Seiichi Kamada,** Braid and knot theory in dimension four, 2002
94 **Mara D. Neusel and Larry Smith,** Invariant theory of finite groups, 2002
93 **Nikolai K. Nikolski,** Operators, functions, and systems: An easy reading. Volume 2: Model operators and systems, 2002
92 **Nikolai K. Nikolski,** Operators, functions, and systems: An easy reading. Volume 1: Hardy, Hankel, and Toeplitz, 2002
91 **Richard Montgomery,** A tour of subriemannian geometries, their geodesics and applications, 2002
90 **Christian Gérard and Izabella Łaba,** Multiparticle quantum scattering in constant magnetic fields, 2002
89 **Michel Ledoux,** The concentration of measure phenomenon, 2001
88 **Edward Frenkel and David Ben-Zvi,** Vertex algebras and algebraic curves, second edition, 2004
87 **Bruno Poizat,** Stable groups, 2001
86 **Stanley N. Burris,** Number theoretic density and logical limit laws, 2001
85 **V. A. Kozlov, V. G. Maz′ya, and J. Rossmann,** Spectral problems associated with corner singularities of solutions to elliptic equations, 2001
84 **László Fuchs and Luigi Salce,** Modules over non-Noetherian domains, 2001
83 **Sigurdur Helgason,** Groups and geometric analysis: Integral geometry, invariant differential operators, and spherical functions, 2000
82 **Goro Shimura,** Arithmeticity in the theory of automorphic forms, 2000
81 **Michael E. Taylor,** Tools for PDE: Pseudodifferential operators, paradifferential operators, and layer potentials, 2000
80 **Lindsay N. Childs,** Taming wild extensions: Hopf algebras and local Galois module theory, 2000
79 **Joseph A. Cima and William T. Ross,** The backward shift on the Hardy space, 2000
78 **Boris A. Kupershmidt,** KP or mKP: Noncommutative mathematics of Lagrangian, Hamiltonian, and integrable systems, 2000

For a complete list of titles in this series, visit the
AMS Bookstore at **www.ams.org/bookstore/**.